Analog and Digital

Control System Design:

Transfer-Function, State-Space, and

Algebraic Methods

Analog and Digital

Control System Design:

Transfer-Function, State-Space, and

Algebraic Methods

Chi-Tsong Chen
State University of New York at Stony Brook

Saunders College Publishing
Harcourt Brace Jovanovich College Publishers
Fort Worth   Philadelphia   San Diego   New York   Orlando   Austin
San Antonio   Toronto   Montreal   London   Sydney   Tokyo

Text Typeface: Times Roman
Compositor: Waldman Graphics
Acquisitions Editor: Emily Barrosse
Assistant Editor: Laura Shur
Managing Editor: Carol Field
Senior Project Manager: Sally Kusch
Copy Editor: Andrew Potter
Manager of Art and Design: Carol Bleistine
Art Director: Doris Bruey
Cover Designer: Lawrence R. Didona
Text Artwork: Tech-Graphics
Director of EDP: Tim Frelick
Production Manager: Jay Lichty
Marketing Manager: Monica Wilson

Printed in the United States of America

ANALOG AND DIGITAL CONTROL SYSTEM DESIGN: TRANSFER-FUNCTION, STATE-SPACE, AND ALGEBRAIC METHODS

ISBN 0-03-094070-2

Library of Congress Catalog Card Number:

3456   118   98765432

To
Bih-Jau
and
Janet, Pauline, Stanley

Preface

PURPOSE AND SCOPE

This book was developed as a text for an introductory course in control systems for junior- and senior-level students in engineering curricula. It can also be used as a reference book for graduate students and practicing engineers. Necessary background includes knowledge of general physics and elementary circuit analysis. A course on systems and signals that covers the Laplace transform and linear algebraic equations is helpful but not essential. These two topics are reviewed in the appendices of this book.

As an introductory text, this book covers only single-variable linear time-invariant lumped systems. Both the transfer function approach and the state-variable approach are introduced. Even though most physical systems to be controlled are analog, compensators have increasingly become digital. Therefore, both analog and digital techniques are discussed. The scope of this text is briefly described in the following.

After discussing some examples of control systems and related basic terminology in Chapter 1, we introduce transfer functions and state-variable equations in Chapter 2. The relationships between them, including the concepts of complete characterization and minimal equations, are also discussed. Chapter 3 develops block diagrams for control systems; the loading problem is considered in the development. Open-loop and closed-loop control systems are analyzed and compared in Chapter 4. This leads to a study of stability: Implications of stability are then discussed. Chapter 5 discusses digital computer computation and operational amplifier (op-amp) circuit realization of state-variable equations and transfer functions. Op-amp realizations can be viewed as the modern version of analog computer simulation. This concludes the analysis part of control systems.

Chapter 6 discusses formulation of design problems and performance specifications of control systems as well as physical constraints in design. Feedback configurations are then shown to be less sensitive to noise and disturbance than open-loop configurations. Chapter 7 introduces the root-locus design method; Chapter 8, the frequency domain method; Chapters 9 and 10, optimal and linear algebraic methods, and Chapter 11, the state-space method. Chronologically, the frequency-domain method, developed in the 1940s, was the earliest method used to design control systems, but because of the difficulties it presented in designing systems with unstable poles, the root-locus method was developed in the 1950s. Both of the methods described in Chapters 7 and 8 are based on transfer functions and were not easily extendable to multi-input and multi-output systems; this difficulty was removed after the development of state-variable equations in the 1960s. This led to a better understanding of the structure of systems by introducing the concepts of controllability and observability and to the design methods of pole placement and state estimators, as discussed in Chapter 11. With the impetus of state-space results, the interests in transfer functions were renewed in the 1970s. By considering transfer functions as ratios of two polynomials, the polynomial fraction approach was born. The approach was translated into solving linear algebraic equations in the 1980s and is discussed in Chapter 10. For pedagogical reasons, these design methods are arranged as in this text rather than in chronological order. These methods are basically independent, however, and can be studied in any order.

With the wide availability of microprocessors, control systems are increasingly controlled by digital computers or special digital hardware. Chapter 12 introduces some preliminary background. Design methods are then introduced in Chapter 13. Dead-beat design is also discussed. In the last chapter, we discuss PID controllers. These controllers are widely employed in industries. Their employment in linear and nonlinear systems is briefly discussed.

FEATURES

At present, there are a large number of introductory control texts on the market; therefore, it is pertinent to describe some of the special features of this text.

1. Every effort has been made to present the concepts and methods as simply and precisely as possible. For example, the presentation of the Routh test is different from the conventional cross-product form and the Routh table (Table 4.2) is easier for the reader to grasp. The discussion of quadratic optimal design (Section 9.4.1) has been praised in a book review[1] for its simplicity.

---

[1]V. Kučera, Book review on ''Control System Design: Conventional, Algebraic, and Optimal Method,'' *Automatica*, vol. 25, No. 2, pp. 322–323, 1989.

2. This book stresses concepts rather than computational mechanics, and design rather than analysis. This approach is justified by the wide availability of personal computers and software. For example, in analyzing position control systems, I show why an additional tachometer feedback may be needed. In discussing the root-locus method, the design and the plot of root loci are addressed separately. Therefore, the student can read the design part, use a computer to plot root loci, and then complete the design.

3. As an engineering text, this book emphasizes practical aspects of control systems. The loading problem in developing block diagrams is discussed. So are noise, disturbances, plant perturbations, and the constraints they impose on design. Saturation nonlinearities of plants are considered and incorporated in optimal design methods.

4. Most concepts and results are supported by numerical examples. For example, the advantages of feedback are established by comparing open-loop and closed-loop amplifiers and velocity control systems (Section 6.8). I believe that the reader can grasp ideas more easily from specific examples than from general statements.

5. Pole placement design is carried out using state-variable equations and transfer functions. The two design methods are compared.

6. Two methods, using similarity transformations and solving Lyapunov equations, are introduced to design full- and reduced-dimensional state estimators.

7. Model-matching design is discussed using transfer functions. The design controls both poles and zeros and can be easily carried out by solving sets of linear algebraic equations. It is generally easier to use model matching to design a good control system than to use pole placement (Section 10.3.3).

8. Exact and imperfect pole-zero cancellations are discussed. Pole-zero cancellations do not really eliminate the poles; the poles merely become hidden. Pole-zero cancellations arise in transfer function design as well as state-space design.

9. Robust tracking and disturbance rejection are considered in model matching.

10. Two approaches to designing digital compensators to analog control plants are introduced. The first approach designs first analog compensators and then discretizes them. Six discretization methods are studied and compared. The second approach computes first equivalent digital plants and then designs digital compensators. We discuss the problem of missing dynamics and the reason for introducing new zeros in equivalent digital plants. Once an equivalent digital plant is computed, most of the analog design methods can be used, without much modification, to design digital compensators. It is argued that the first approach is simpler and is less subject to numerical errors than the second approach.

11. More general versions of analog and digital Lyapunov theorems are introduced. The Routh test is then established.

12. The employment of PID controllers in nonlinear industrial processes is discussed. In linear systems, it is argued that general compensators generally yield better design than PID controllers.

In addition to these features, I have introduced a number of new concepts such as total stability, well-posedness, and implementable transfer functions. These concepts are important in control system design. The design methodology in Chapter 10 is a unique feature of this text and is not available elsewhere. The method is simpler than the state-space method in designing pole placement and state estimators for the class of systems studied in this text (see Section 11.7.1). The method is believed to be as simple as, if not simpler than, the root-locus and Bode plot methods. It has been classroom-tested for many years and has been very well received by the students. A number of engineers in industry have responded positively to the tutorial articles I have written on this method. See References [17], [19], and [20] listed at the end of this text. Therefore, I am convinced that the method is an extremely attractive alternative method in designing control systems and hope that it will become an integral part of the teaching of control system design.

THE ISSUE OF USING COMPUTERS

Analysis and design of control systems can now be greatly facilitated by the use of personal computers and computer-aided design packages. At present, a large number of software packages, such as MATLAB, MATRIX$_x$, and CTRL-C, are available commercially. An even larger number of software packages has been developed in universities and by industry (see Section 5.3). By means of these packages, tedious computations can be avoided. For example, the command step in MATLAB will generate step responses of transfer functions and state-variable equations. Therefore, this text does not dwell on computational mechanics; instead it concentrates on concepts and basic procedures in analysis and design.

However, I believe that solving a number of simple problems by hand will enhance students' understanding of a topic, and give them confidence in using a digital computer. I also believe that only after mastering a topic can one intelligently interpret computer printouts. Therefore, the advent of digital computers should not alter our basic learning process. At the same time, we should take full advantage of digital computers and learn to use them proficiently. The use of version 3.5f and the Student Edition of MATLAB[2] is discussed throughout the text. See the listing in the index. This text also advocates computer-aided design. Section 9.7 shows that computer simulation may yield comparable or better design results than analytical methods.

COURSE OUTLINE

This text provides enough material for a one-year course. For example, the first eight chapters can be covered in the first semester and the remainder, in the second se-

---

[2]*The Student Edition of* MATLAB™, The Math Works, Inc., Englewood Cliffs, NJ: Prentice Hall, 1992.

mester. The one-semester course at Stony Brook covers only the continuous-time case. We skip most of the state-variable material and concentrate on transfer functions as follows:

Chapter 1

Sections 2.1–2.6

Chapter 3

Chapter 4

Chapter 5 (omit Sec. 5.5.2)

Chapter 6

Sections 9.1–9.4.2

Sections 10.1–10.4 (omit Theorem 10.2 and Sec. 10.3.2)

Chapter 7

Chapter 8 (omit Sec. 8.4.1, and the transient part of Sec. 8.6)

We also skip the proofs of the Nyquist stability criterion and Theorem 10.1. Because Chapters 7, 8, 9, and 10 are essentially independent, their order can be altered. After Chapter 6, I move to Sections 9.1–9.4.2 and Sections 10.1–10.4. These sections will not take more than six lectures of 50 minutes each. I then go back to Chapters 7 and 8. Certainly, other arrangements are also possible for a one-semester course.

A solutions manual is available upon adoption of the text.

ACKNOWLEDGEMENTS

This text is a result of over 25 years of teaching and research in control systems. The original text was called *Analysis and Synthesis of Linear Control Systems* and was published by Holt, Rinehart and Winston in 1975. The second edition was retitled *Control System Design: Conventional, Algebraic and Optimal Methods* and was published by Pond Woods Press in 1987. The present edition, with a new title, has been expanded to include state-space and discrete-time design methods. In addition, the linear algebraic method in Chapter 10 has been expanded. From these editions, one can see that the design methodology in Chapter 10 was developed over a period of more than ten years.

In developing the three editions of this book, I owe a great deal to many people. Professors S. S. L. Chang, C. A. Desoer, E. I. Jury, J. G. Truxal, and Dr. P. Barry were most influential in the preparation of the first edition of this text. Professors J. L. Lin and M. C. Tsai of National Cheng Kung University and Professor M. Y. Wu of the University of Colorado gave me many constructive suggestions for the second edition. Mr. Fuyuan Yang, a visiting scholar from China, Ms. Gina Liu, and Mr. Mike Lee helped me greatly in preparing several versions of this manuscript. I would like to thank Mr. C. C. Wang for preparing many of the computer printouts and Dr. C. Y. Huang of Grumman Corporate Technology for carrying out the nonlinear simulation in Figure 14.2. I am grateful to Mr. Gary Bono and Shian-Hung

Huang who checked the entire manuscript for accuracy. My special thanks go to Dr. B. H. Seo for his help in refining the material in Chapter 10. Professors John R. D'Alessandro, Widener University, Thordor Runolfsson, Johns Hopkins University, and Charles P. Neuman, Carnegie University, reviewed the manuscript; their many thoughtful comments and suggestions are greatly appreciated. It has been a great pleasure to work with Robert L. Argentieri, Barbara Gingery, Emily Barrosse, and Laura Shur of Saunders College Publishing. Finally, special thanks go to Sally Kusch, who did an excellent job as Project Manager.

Chi-Tsong Chen
Stony Brook
June 1992

Contents

1 *Introduction*

1.1 EMPIRICAL AND ANALYTICAL METHODS

The ultimate goal of engineering—in particular, that of control engineering—is to design and build real physical systems to perform given tasks. For example, an engineer might be asked to design and install a heat exchanger to control the temperature and humidity of a large building. This has been a longstanding problem in engineering, and much relevant data has been collected. From the total volume and geographical location of the building, we can determine the required capacity of the exchanger and then proceed to install the system. If, after installation, the exchanger is found to be insufficiently powerful to control the building's environment, it can be replaced by a more powerful one. This approach, which relies heavily on past experience and repeated experimentation, is called the *empirical method*. Although the empirical method must be carried out by trial and error, it has been used successfully to design many physical systems.

The empirical method, however, is inadequate if there is no past experience to draw from or if experimentation is not feasible because of high cost or risk. For example, the task of sending astronauts to the moon and bringing them back safely could not have been carried out by the empirical method. Similarly, the design of fusion control in nuclear power plants should not be so dealt with. In these cases, the analytical method becomes indispensable. The *analytical method* generally consists of four steps: modeling, setting up mathematical equations, analysis, and design. The first two steps are closely related. If we use simple mathematics, then the model chosen must be correspondingly simple. If we use sophisticated mathematics, then

1

the model can be more complex and realistic. Modeling is the most critical step in analytical design. If a physical system is incorrectly modeled, subsequent study will be useless. Once a model is chosen, the rest of the analytical design is essentially a mathematical problem.

Repeated experimentation is indispensable in the empirical method. It is also important in the analytical method. In the former, experiments must be carried out using physical devices, which might be expensive and dangerous. In the latter, however, experiments can be carried out using models or mathematical equations. Many computer-aided design packages are available. We may use any of them to simulate the equations on a digital computer, to carry out design, and to test the result on the computer. If the result is not satisfactory, we repeat the design. Only after a design is found to be satisfactory, will we implement it using physical devices.

If the model is adequately chosen, the performance of the implemented system should resemble the performance predicted by analytical design or computer simulation. However, because of unavoidable inaccuracy in modeling, discrepancies often exist between the performance of the implemented physical system and that predicted by the analytical method. Therefore, the performance of physical systems can often be improved by fine adjustments or tunings. This is why a physical system often requires lengthy testing after it is implemented, before it is put into actual operation or mass production. In this sense, experience is also important in the analytical method.

In the analytical approach, experimentation is needed to set up models, and experience is needed (due to the inaccuracy of modeling) to improve the performance of actual physical systems. Thus, experience and experimentation are both used in the empirical and analytical approaches. The major difference between these two approaches is that in the latter, we gain, through modeling, understanding and insight into the structure of systems. The analytical approach also provides systematic procedures for designing systems and reduces the likelihood of designing flawed or disastrous systems.

In this text, we study analytical methods in the analysis and design of control systems.

1.2   CONTROL SYSTEMS

This text is concerned with the analysis and design of control systems; therefore, it is pertinent to discuss first what control systems are. Before giving a formal definition, we discuss a number of examples.

1.2.1   Position Control Systems

The satellite dish in the backyard or on the rooftop of a house has become common in recent years. It is an antenna aimed at a satellite that is stationary with respect to the earth and is used to transmit television or other signals. To increase the number of channels for a television, the dish may be designed to aim at different satellites.

A possible arrangement of such a system is shown in Figure 1.1(a). This system can indeed be designed using the empirical method. If it is to be designed using the analytical method, we must first develop a model for the system, as shown in Figure 1.1(b). The model actually consists of a number of blocks.[1] Each block represents a model of a physical device. Using this model, we can then carry out the design. A large number of systems can be similarly modeled. For example, the system that aims the antennas shown in Figure 1.2 at communication satellites and the systems that control various antennas and solar panels shown in Figure 1.3 can be similarly modeled. These types of systems are called *position control systems*.

There are other types of position control systems. Consider the simplified nuclear power plant shown in Figure 1.4. The intensity of the reaction inside the reactor (and, consequently, the amount of heat generated) is controlled by the vertical position of the control rods. The more deeply the control rods are submerged, the more heat the reactor will generate. There are many other control systems in the nuclear power plant. Maintenance of the boiler's water level and maintenance or regulation of the generated voltage at a fixed voltage all call for control systems. Position control is also needed in the numerical control of machine tools. For example, it is possible to program a machine tool so that it will automatically drill a number of holes, as shown in Figure 1.5.

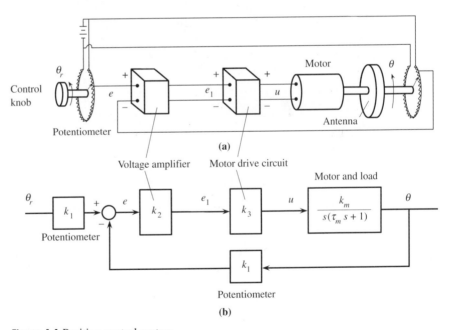

Figure 1.1 Position control system.

Figure 1.2 Radar antenna. (Courtesy of MIT Lincoln Laboratory.)

1.2.2 Velocity Control Systems

Driving tapes in video or audio recorders at a constant speed is important in producing quality output. The problem is complicated because the load of the driver varies from a full reel to an empty reel. A possible control system to achieve this is shown in Figure 1.6(a). This system can indeed be designed using the empirical method. If it is to be designed using the analytical method, we must develop a model, as shown in Figure 1.6(b). Using this model, we can then carry out the design. Velocity control problems also arise in a large number of industrial applications. To "grow" an optical fiber with a uniform diameter from melted glass, the speed of growth must be properly controlled. The speed of the conveyor in a production line must also be precisely controlled. An error in roller speed of just 0.1% in a paper drive system may cause the paper to tear or to pile up on the roller. The velocity of the rotor of the generator in Figure 1.4 is kept constant in order to generate a constant voltage. Velocity control is indeed needed in a wide range of applications.

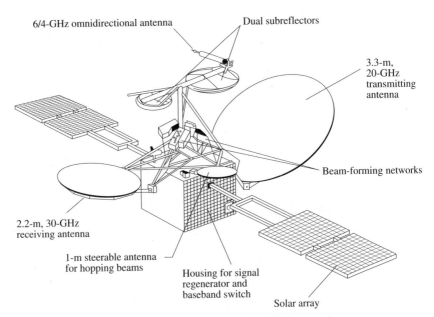

6/4-GHz omnidirectional antenna

Dual subreflectors

3.3-m, 20-GHz transmitting antenna

Beam-forming networks

2.2-m, 30-GHz receiving antenna

1-m steerable antenna for hopping beams

Housing for signal regenerator and baseband switch

Solar array

Figure 1.3 Communications satellite. (Courtesy of *IEEE Spectrum.*)

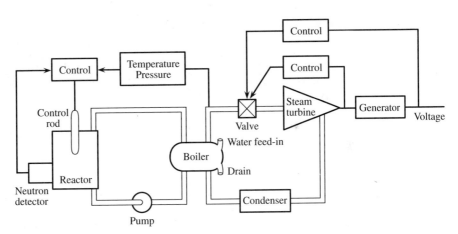

Control

Temperature Pressure

Control

Control

Control rod

Steam turbine

Valve

Generator

Voltage

Water feed-in

Boiler

Neutron detector

Reactor

Drain

Condenser

Pump

Figure 1.4 Nuclear power plant.

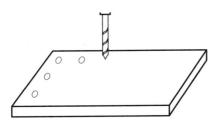

Figure 1.5 Numerical control system.

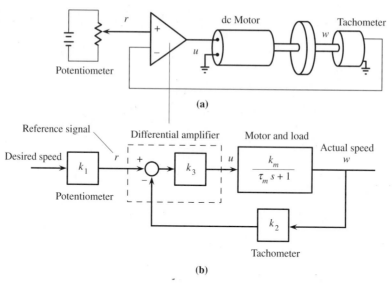

Figure 1.6 Velocity control system.

1.2.3 Temperature Control Systems

We now discuss a different type of example. Consider the temperature control of the enclosed chamber shown in Figure 1.7(a). This problem, which arises in the temperature control of an oven, a refrigerator, an automobile compartment, a house, or the living quarters of a space shuttle, can certainly be approached by using the empirical method. If the analytical method is to be used, we must develop a model as shown in Figure 1.7(b). We can then use the model to carry out analysis and design.

Temperature control is also important in chemical processes. The rate of chemical reactions often depends on the temperature. If the temperature is not properly controlled, the entire product may become useless. In industrial processes, temperature, pressure and flow controls are widely used.

1.2.4 Trajectory Control and Autopilot

The landing of a space shuttle on a runway is a complicated control problem. The desired trajectory is first computed as shown in Figure 1.8. The task is then to bring the space shuttle to follow the desired trajectory as closely as possible. The structure of a shuttle is less stable than that of an aircraft, and its landing speed cannot be controlled. Thus, the landing of a space shuttle is considerably more complicated than that of an aircraft. The landing has been successfully accomplished with the aid of on-board computers and altimeters and of the sensing devices on the ground, as shown in Figure 1.8. In fact, it can even be achieved automatically, without involving astronauts. This is made possible by the use of an autopilot. The autopilot is now widely used on aircrafts and ships to maintain desired altitude and/or heading.

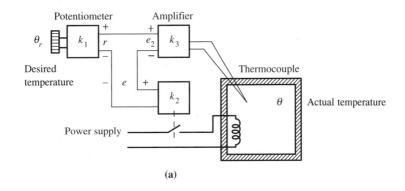

(a)

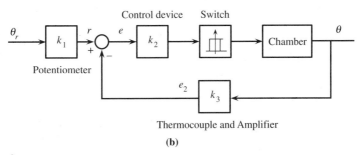

(b)

Figure 1.7 Temperature control system.

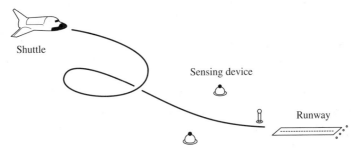

Figure 1.8 Desired landing trajectory of space shuttle.

1.2.5 Miscellaneous Examples

We give two more examples of control systems to conclude this section. Consider the bathroom toilet tank shown in Figure 1.9(a). The mechanism is designed to close the valve automatically whenever the water level reaches a preset height. A schematic diagram of the system is shown in Figure 1.9(b). The float translates the water level into valve position. This is a very simple control problem. Once the mechanism

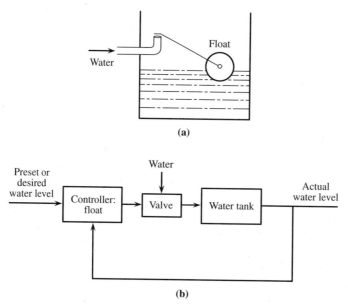

(a)

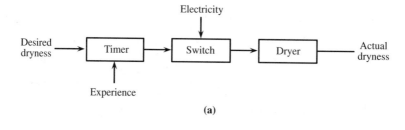

(b)

Figure 1.9 Bathroom toilet tank.

of controlling the valve is understood, the water level can easily be controlled by trial and error.

As a final example, a schematic diagram of the control of clothes dryers is shown in Figure 1.10. Presently there are two types of clothes dryers—manual and automatic. In a manual clothes dryer, depending on the amount of clothes and depending

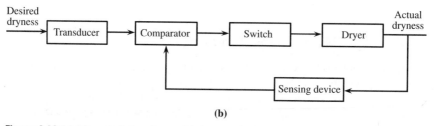

Figure 1.10 (a) Manual clothes dryer. (b) Automatic dryer.

on experience, we set the timer to, say, 40 minutes. At the end of 40 minutes, this dryer will automatically turn off even if the clothes are still damp or wet. Its schematic diagram is shown in Figure 1.10(a). In an automatic dryer, we select a desired degree of dryness, and the dryer will automatically turn off when the clothes reach the desired degree of dryness. If the load is small, it will take less time; if the load is large, it will take more time. The amount of time needed is automatically determined by this type of dryer. Its schematic diagram is shown in Figure 1.10(b). Clearly, the automatic dryer is more convenient to use than a manual one, but it is more expensive. However, in using a manual dryer, we may overset the timer, and electricity may be wasted. Therefore, if we include the energy saved, an automatic dryer may turn out to be more economical.

1.3   PROBLEM FORMULATION AND BASIC TERMINOLOGY

From the examples in the preceding section, we may deduce that a control system is an interconnection of components or devices so that the output of the overall system will follow as closely as possible a desired signal. There are many reasons to design control systems:

1. *Automatic control:*   The temperature of a house can be automatically maintained once we set a desired temperature. This is an automatic control system. Automatic control systems are used widely and are essential in automation in industry and manufacturing.

2. *Remote control:*   The quality of reception of a TV channel can be improved by pointing the antenna toward the emitting station. If the antenna is located at the rooftop, it is impractical to change its direction by hand. If we install an antenna rotator, then we can control the direction *remotely* by turning a knob sitting in front of the TV. This is much more convenient. The Hubble space telescope, which is orbiting over three hundred miles above the earth, is controlled from the earth. This remote control must be done by control systems.

3. *Power amplification:*   The antennas used to receive signals sent by *Voyager 2* have diameters over 70 meters and weights over several tons. Clearly, it is impossible to turn these antennas directly by hand. However, using control systems, we can control them by turning knobs or by typing in command signals on computers. The control systems will then generate sufficient power to turn the antennas. Thus, power amplification is often implicit in many control systems.

In conclusion, control systems are widely used in practice because they can be designed to achieve automatic control, remote control, and power amplification.

We now formulate the control problem in the following. Consider the the position control problem in Figure 1.1, where the objective is to control the direction of the antenna. The first step in the design is to choose a motor to drive the antenna. The motor is called an *actuator*. The combination of the object to be controlled and the actuator is called the *plant*. In a home heating system, the air inside the home is

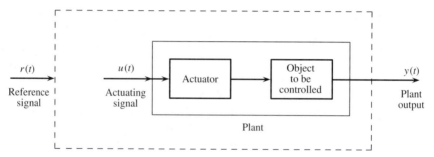

Figure 1.11 Control design problem.

the controlled object and the burner is the actuator. A space shuttle is a controlled object; its actuator consists of a number of thrustors. The input of the plant, denoted by $u(t)$, is called the *control signal* or *actuating signal*; the output of the plant, denoted by $y(t)$, is called the *controlled variable* or *plant output*. The problem is to design an overall system as shown in Figure 1.11 so that the plant output will follow as closely as possible a *desired* or *reference signal*, denoted by $r(t)$. Every example in the preceding section can be so formulated.

There are basically two types of control systems: the open-loop system and the closed-loop or feedback system. In an *open-loop* system, the actuating signal is predetermined by the desired or reference signal; it does not depend on the actual plant output. For example, based on experience, we set the timer of the dryer in Figure 1.10(a). When the time is up, the dryer will stop even if the clothes are still damp or wet. This is an open-loop system. The actuating signal of an open-loop system can be expressed as

$$u(t) = f(r(t))$$

where f is some function. If the actuating signal depends on the reference input and the plant output, or if it can be expressed as

$$u(t) = h(r(t), y(t))$$

where h is some function, then the system is a closed-loop or feedback system. All systems in the preceding section, except the one in Figure 1.10(a), are feedback systems. In every feedback system the plant output must be measured and used to generate the actuating signal. The plant output could be a position, velocity, temperature, or something else. In many applications, it is transformed into a voltage and compared with the reference signal, as shown in Figure 1.12. In these transformations, sensing devices or transducers are needed as shown. The result of the comparison is then used to drive a *compensator* or *controller*. The output of the controller yields an actuating signal. If the controller is designed properly, the actuating signal will drive the plant output to follow the desired signal.

In addition to the engineering problems discussed in the preceding section, a large number of other types of systems can also be considered as control systems.

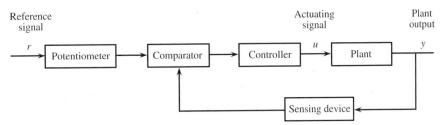

Figure 1.12 Feedback control system.

Our body is in fact a very complex feedback control system. Maintaining our body temperature at 37° Celsius requires perspiration in summer and contraction of blood vessels in winter. Maintaining an automobile in a lane (plant output) is a feedback control system: Our eyes sense the road (reference signal), we are the controller, and the plant is the automobile together with its engine and steering system. An economic system is a control system. Its health is measured by the gross national product (GNP), unemployment rate, average hourly wage, and inflation rate. If the inflation rate is too high or the unemployment rate is not acceptable, economic policy must be modified. This is achieved by changing interest rates, monetary policy, and government spending. The economic system has a large number of interrelated factors whose cause-and-effect relationships are not exactly known. Furthermore, there are many uncertainties, such as consumer spending, labor disputes, or international crises. Therefore, an economic system is a very complex system. We do not intend to solve *every* control problem; we study in this text only a very limited class of control problems.

1.4 SCOPE OF THE TEXT

This text is concerned with the analysis and design of control systems. As it is an introductory text, we study only a special class of control systems. Every system—in particular, every control system—is classified dichotomously as linear or nonlinear, time-invariant or time-varying, lumped or distributed, continuous-time or discrete-time, deterministic or stochastic, and single-variable or multivariable. Roughly speaking, a system is *linear* if it satisfies the additivity and homogeneity properties, *time-invariant* if its characteristics do not change with time, and *lumped* if it has a finite number of state variables or a finite number of initial conditions that can summarize the effect of past input on future output. A system is *continuous-time* if its responses are defined for all time, *discrete-time* if its responses are defined only at discrete instants of time. A system is *deterministic* if its mathematical description does not involve probability. It is called a *single-variable* system if it has only one input and only one output; otherwise it is called a *multivariable* system. For a more detailed discussion of these concepts, see References [15, 18]. *In this text, we study only linear, time-invariant, lumped, deterministic, single-variable systems.* Although this class of control systems is very limited, it is the most important one. Its study is a prerequisite for studying more general systems. Both continuous-time and discrete-time systems are studied.

The class of systems studied in this text can be described by ordinary differential equations with real constant coefficients. This is demonstrated in Chapter 2 by using examples. We then discuss the *zero-input response* and the *zero-state response*. The *transfer function* is developed to describe the zero-state response. Since the transfer function describes only the zero-state response, its use in analysis and design must be justified. This is done by introducing the concept of *complete characterization*. The concepts of *properness, poles,* and *zeros* of transfer functions are also introduced. Finally, we introduce the *state-variable equation* and its discretization. Its relationship with the transfer function is also established.

In Chapter 3, we introduce some control components, their models, and their transfer functions. The *loading problem* is considered in developing the transfer functions. Electrical, mechanical, and electromechanical systems are discussed. We then discuss the manipulation of *block diagrams* and *Mason's formula* to conclude the chapter.

The *quantitative* and *qualitative analyses* of control systems are studied in Chapter 4. Quantitative analysis is concerned with the response of systems due to some specific input, whereas qualitative analysis is concerned with general properties of systems. In quantitative analysis, we also show by examples the need for using feedback and tachometer feedback. The concept of the *time constant* is introduced. In qualitative analysis, we introduce the concept of *stability*, its condition, and a method (the Routh test) of checking it. The problems of *pole-zero cancellation* and *complete characterization* are also discussed.

In Chapter 5, we discuss *digital* and *analog computer simulations*. We show that if the state-variable description of a system is available, then the system can be readily simulated on a digital computer or built using operational amplifier circuits. Because it is simpler and more systematic to simulate transfer functions through state-variable equations, we introduce *the realization problem*—the problem of obtaining state-variable equations from transfer functions. Minimal realizations of vector transfer functions are discussed. The use of MATLAB, a commercially available computer-aided design package, is discussed throughout the chapter.

Chapters 2 through 5 are concerned with modeling and analysis problems; the remaining chapters are concerned with the design problem. In Chapter 6, we discuss the choice of plants. We then discuss physical constraints in the design of control systems. These constraints lead to the concepts of *well-posedness* and *total stability*. The *saturation problem* is also discussed. Finally, we compare the merits of *open-loop* and *closed-loop systems* and then introduce two basic approaches—namely, outward and inward—in the design of control systems. In the *outward approach*, we first choose a configuration and a compensator with open parameters and then adjust the parameters so that the resulting overall system will (we hope) meet the design objective. In the *inward approach*, we first choose an overall system to meet the design objective and then compute the required compensators.

Two methods are available in the outward approach: the *root-locus method* and the *frequency-domain method*. They were developed respectively in the 1950s and

1940s. The root-locus method is introduced in Chapter 7 and the frequency-domain method in Chapter 8. Both methods are trial-and-error methods.

The inward approach consists of two parts: the search for an overall transfer function to meet design specifications and the implementation of that overall transfer function. The first problem is discussed in Chapter 9, where overall systems are chosen to minimize the quadratic performance index and the ITAE (integral of time multiplied by absolute error). It is also shown by examples that good overall transfer functions can also be obtained by computer simulations. The implementation problem is discussed in Chapter 10, where the difference between model matching and pole placement is also discussed. We discuss the implementation in the *unity-feedback* configuration, *two-parameter* configuration, and the *plant input/output feedback* configuration. We also discuss how to increase the degree of compensators to achieve robust tracking and disturbance rejection. The design methods in Chapter 10 are called the *linear algebraic method*, because they are all achieved by solving linear algebraic equations.

The design methods in Chapters 6 through 10 use transfer functions. In Chapter 11, we discuss design methods using state-variable equations. We introduce first the concepts of *controllability* and *observability* and their conditions. Their relationships with pole-zero cancellations are also discussed. We then use a network to illustrate the concept of *equivalent state-variable equations*. Pole placement is then carried out by using equivalent equations. The same procedure is also used to design full-dimensional state estimators. *Reduced-dimensional estimators* are then designed by solving *Lyapunov equations*. The connection of state feedback to the output of state estimators is justified by establishing the *separation property*. Finally, we compare the design of state feedback and state estimator with the linear algebraic method.

Chapters 3 through 11 study continuous-time control systems. The next two chapters study discrete-time counterparts. Chapter 12 first discusses the reasons for using digital compensators to control analog plants and then discusses the interfaces needed to connect analog and digital systems. We introduce the *z-transform, difference equations, state-variable equations, stability*, and the *Jury test*. These are the discrete-time counterparts of the continuous-time case. The relationship between the frequency response of analog and digital transfer functions is also discussed. Chapter 13 discusses two approaches in designing digital compensators. The first approach is to design an analog compensator and then discretize it. Six different discretization methods are introduced. The second approach is to discretize the analog plant into an equivalent digital plant and then design digital compensators. All analog methods, except the frequency-domain method, are directly applicable to design digital compensators without any modification.

If a plant can be modeled as linear, time-invariant, and lumped, then a good control system can be designed by using one of the methods discussed in this text. However, many plants, especially industrial processes, cannot be so modeled. *PID controllers* may be used to control these plants, as discussed in Chapter 14. Various problems in using PID controllers are discussed. The *Laplace transform* and *linear algebraic equations* are reviewed in Appendices A and B: this discussion is not exhaustive, going only to the extent needed in this text.

2 *Mathematical Preliminary*

2.1   PHYSICAL SYSTEMS AND MODELS

This text is concerned with analytical study of control systems. Roughly speaking, it consists of four parts:

1. Modeling
2. Development of mathematical equations
3. Analysis
4. Design

This chapter discusses the first two parts. The distinction between physical systems and models is fundamental in engineering. In fact, the circuits and control systems studied in most texts are models of physical systems. For example, a resistor with a constant resistance is a model; the power limitation of the resistor is often disregarded. An inductor with a constant inductance is also a model; in reality, the inductance may vary with the amount of current flowing through it. An operational amplifier is a fairly complicated device; it can be modeled, however, as shown in Figure 2.1. In mechanical engineering, an automobile suspension system may be modeled as shown in Figure 2.2. In bioengineering, a human arm may be modeled as shown in Figure 2.3(b) or, more realistically, as in Figure 2.3(c). Modeling is an extremely important problem, because the success of a design depends upon whether or not physical systems are adequately modeled.

Depending on the questions asked and depending on operational ranges, a physical system may have different models. For example, an electronic amplifier has

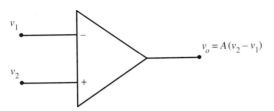

Figure 2.1 Model of operational amplifier.

$$v_o = A(v_2 - v_1)$$

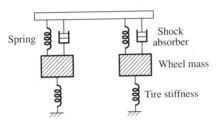

Figure 2.2 Model of automobile suspension system.

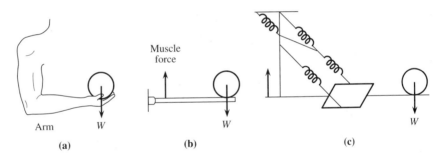

Figure 2.3 Models of arm.

different models at high and low frequencies. A spaceship may be modeled as a particle in the study of trajectory; however, it must be modeled as a rigid body in the study of maneuvering. In order to develop a suitable model for a physical system, we must understand thoroughly the physical system and its operational range. In this text, models of physical systems are also called *systems*. Hence, a *physical system* is a device or a collection of devices existing in the real world; a *system* is a model of a physical system. As shown in Figure 2.4, a system is represented by a unidirectional block with at least one input terminal and one output terminal. We remark that *terminal* does not necessarily mean a physical terminal, such as a wire sticking out of the block, but merely indicates that a signal may be applied or measured from that point. If an excitation or input signal $u(t)$ is applied to the input terminal of a

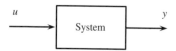

Figure 2.4 System.

system, a unique response or output signal $y(t)$ will be measurable or observable at the output terminal. This *unique* relationship between *excitation* and *response, input* and *output*, or *cause* and *effect* is implicit for every physical system and its model.

A system is called a *single-variable system* if it has only one input terminal and only one output terminal. Otherwise, it is called a *multivariable system*. A multivariable system has two or more input terminals and/or two or more output terminals. We study in this text mainly single-variable systems.

2.2   LINEAR TIME-INVARIANT LUMPED SYSTEMS

The choice of a model for a physical device depends heavily on the mathematics to be used. It is useless to choose a model that closely resembles the physical device but cannot be analyzed using existing mathematical methods. It is also useless to choose a model that can be analyzed easily but does not resemble the physical device. Therefore, the choice of models is not a simple task. It is often accomplished by a compromise between ease of analysis and resemblance to real physical systems.

The systems to be used in this text will be limited to those that can be described by ordinary linear differential equations with constant real coefficients such as

$$3\,\frac{d^2y(t)}{dt^2} + 2\,\frac{dy(t)}{dt} + y(t) = 2\,\frac{du(t)}{dt} - 3u(t)$$

or, more generally,

$$a_n\,\frac{d^ny(t)}{dt^n} + a_{n-1}\,\frac{d^{n-1}y(t)}{dt^{n-1}} + \cdots + a_1\,\frac{dy(t)}{dt} + a_0y(t)$$
$$= b_m\,\frac{d^mu(t)}{dt^m} + b_{m-1}\,\frac{d^{m-1}u(t)}{dt^{m-1}} + \cdots + b_1\,\frac{du(t)}{dt} + b_0u(t)$$

(2.1)

where a_i and b_i are real constants, and $n \geq m$. Such equations are called *nth order linear time-invariant lumped (LTIL) differential equations*. In order to be describable by such an equation, the system must be linear, time-invariant, and lumped. Roughly speaking, a system is linear if it meets the additivity property [that is, the response of $u_1(t) + u_2(t)$ equals the sum of the response of $u_1(t)$ and the response of $u_2(t)$], and the homogeneity property [the response of $\alpha u(t)$ equals α times the response of $u(t)$]. A system is time-invariant if its characteristics—such as mass or moment of inertia for mechanical systems, or resistance, inductance or capacitance for electrical systems—do not change with time. A system is lumped if the effect of any past input $u(t)$, for $t \leq t_0$, on future output $y(t)$, for $t \geq t_0$, can be summarized by a *finite* number of initial conditions at $t = t_0$. For a detailed discussion of these concepts,

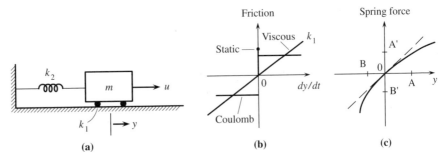

Figure 2.5 Mechanical system.

see References [15, 18]. We now discuss how these equations are developed to describe physical systems.

2.2.1   Mechanical Systems

Consider the system shown in Figure 2.5(a). It consists of a block with mass m connected to a wall by a spring. The input is the applied force $u(t)$, and the output is the displacement $y(t)$ measured from the equilibrium position. Before developing an equation to describe the system, we first discuss the characteristics of the friction and spring. The friction between the block and the floor is very complex. It generally consists of three parts—static, Coulomb, and viscous frictions—as shown in Figure 2.5(b). Note that the coordinates are friction versus velocity. When the mass is stationary or its velocity is zero, we need a certain amount of force to overcome the static friction to start its movement. Once the mass is moving, there is a constant friction, called the *Coulomb friction*, which is independent of velocity. The viscous friction is generally modeled as

$$\text{Viscous friction} = k_1 \times \text{Velocity} \tag{2.2}$$

where k_1 is called the *viscous friction coefficient*. This is a linear equation. Most texts on general physics discuss only static and Coulomb frictions. In this text, however, we consider only viscous friction; static and Coulomb frictions will be disregarded. By so doing, we can model the friction as a linear phenomenon.

In general physics, Hooke's law states that the displacement of a spring is proportional to the applied force, that is

$$\text{Spring force} = k_2 \times \text{Displacement} \tag{2.3}$$

where k_2 is called the *spring constant*. This equation is plotted in Figure 2.5(c) with the dotted line. It implies that no matter how large the applied force is, the displacement equals force$/k_2$. This certainly cannot be true in reality; if the applied force is larger than the elastic limit, the spring will break. In general, the characteristic of a physical spring has the form of the solid line shown in Figure 2.5(c).[1] We see that

---

[1]This is obtained by measurements under the assumption that the mass of the spring is zero and that the spring has no drifting and no hysteresis. See Reference [18].

if the applied force is outside the range [A′, B′], the characteristic is quite different from the dotted line. However, if the applied force lies inside the range [A′, B′], called the *linear operational range*, then the characteristic can very well be represented by (2.3). We shall use (2.3) as a model for the spring.

We now develop an equation to describe the system by using (2.3) and considering only the viscous friction in (2.2). The applied force $u(t)$ must overcome the friction and the spring force, and the remainder is used to accelerate the mass. Thus we have

$$u(t) - k_1 \frac{dy(t)}{dt} - k_2 y(t) = m \frac{d^2 y(t)}{dt^2}$$

or

$$m \frac{d^2 y(t)}{dt^2} + k_1 \frac{dy(t)}{dt} + k_2 y(t) = u(t) \tag{2.4}$$

This is an ordinary linear differential equation with constant coefficients. It is important to remember that this equation is obtained by using the linearized relation in (2.3) and considering only the viscous friction in (2.2). Therefore, it is applicable only for a limited operational range.

Consider now the rotational system shown in Figure 2.6(a). The input is the applied torque $T(t)$ and the output is the angular displacement $\theta(t)$ of the load. The shaft is not rigid and is modeled by a torsional spring. Let J be the moment of inertia of the load and the shaft. The friction between the shaft and bearing may consist of static, Coulomb, and viscous frictions. As in Figure 2.5, we consider only the viscous friction. Let k_1 be the viscous friction coefficient and k_2 the torsional spring constant. Then the torque generated by the friction equals $k_1 d\theta(t)/dt$ and the torque generated by the spring is $k_2 \theta(t)$. The applied torque $T(t)$ must overcome the friction and spring torques; the remainder is used to accelerate the load. Thus we have

$$T(t) - k_1 \frac{d\theta(t)}{dt} - k_2 \theta(t) = J \frac{d^2\theta(t)}{dt^2}$$

or

$$J \frac{d^2\theta(t)}{dt^2} + k_1 \frac{d\theta(t)}{dt} + k_2 \theta(t) = T(t) \tag{2.5a}$$

This differential equation describes the system in Figure 2.6(a).

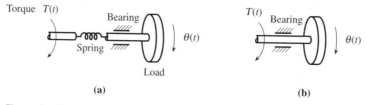

Figure 2.6 Rotational mechanical system.

If we identify the following equivalences:

Translational movement   **Rotational movement**

Linear displacement $y \longleftrightarrow$ Angular displacement θ
Force $u \longleftrightarrow$ Torque T
Mass $m \longleftrightarrow$ Moment of inertia J

then Equation (2.5a) is identical to (2.4). The former describes a rotational movement, the latter, a linear or translational movement.

Exercise 2.2.1

The suspension system of an automobile can be modeled as shown in Figure 2.7. This model is simpler than the one in Figure 2.2, because it neglects the wheel mass and combines the tire stiffness with the spring. The model consists of one spring with spring constant k_2 and one dashpot or shock absorber. The dashpot is a device that provides viscous frictional force. Let k_1 be its viscous friction coefficient and let m be the mass of the car. A vertical force $u(t)$ is applied to the mass when the wheel hits a pothole. Develop a differential equation to describe the system.

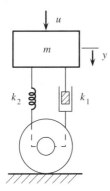

Figure 2.7 Suspension system of automobile.

[**Answer:** Same as (2.4).]

Exercise 2.2.2

Show that the system shown in Figure 2.6(b) where the shaft is assumed to be rigid is described by

$$J \frac{d^2\theta(t)}{dt^2} + k_1 \frac{d\theta(t)}{dt} = T(t) \qquad (2.5b)$$

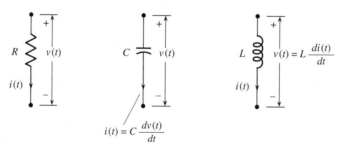

Figure 2.8 Electrical components.

2.2.2 RLC Networks

We discuss in this section circuits that are built by interconnecting resistors, capacitors, inductors, and current and voltage sources, beginning with the three basic elements shown in Figure 2.8. A resistor is generally modeled as $v = Ri$, where R is the resistance, v is the applied voltage, and i is current flowing through it with polarity chosen as shown. In the model $v = Ri$, nothing is said regarding power consumption. In reality, if v is larger than a certain value, the resistor will burn out. Therefore, the model is valid only within the specified power limitation. A capacitor is generally modeled as $Q = Cv$, where C is the capacitance, v is the applied voltage, and Q is the charge stored in the capacitor. The model implies that as the voltage increases to infinity, so does the stored charge. Physically this is not possible. As v increases, the stored charge will saturate and cease to increase, as shown in Figure 2.9. However, for v in a limited range, the model $Q = Cv$ does represent the physical capacitor satisfactorily. An inductor is generally modeled as $\phi = Li$ where L is the inductance, ϕ is the flux, and i is the current. In reality, the flux generated in an inductor will saturate as i increases. Therefore the relationship $\phi = Li$ is again applicable only within a limited range of i. Now if R, L, and C change with time, they are time-varying elements. If R, L, and C are constants, independent of time, then they are time-invariant elements. Using these linear time-invariant models, we can express their voltages and currents as

$$v(t) = Ri(t) \tag{2.6a}$$

$$i(t) = C\,\frac{dv(t)}{dt} \tag{2.6b}$$

$$v(t) = L\,\frac{di(t)}{dt} \tag{2.6c}$$

Now we shall use (2.6) to develop differential equations to describe RLC networks. Consider the network shown in Figure 2.10. The input is a current source $u(t)$ and the output $y(t)$ is the voltage across the capacitor as shown. The current of the capacitor, using (2.6b), is

$$i_c(t) = C\,\frac{dy(t)}{dt} = 2\,\frac{dy(t)}{dt}$$

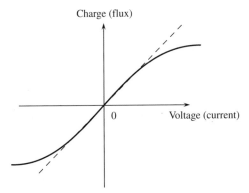

Figure 2.9 Characteristics of capacitor and inductor.

This current also passes through the 1-Ω resistor. Thus the voltage drop across A and B is

$$v_{AB} = i_c(t) \cdot 1 + y(t) = 2 \frac{dy(t)}{dt} + y(t)$$

The current $i_1(t)$ passing through the 0.5-Ω resistor is

$$i_1(t) = \frac{v_{AB}}{0.5} = 4 \frac{dy(t)}{dt} + 2y(t)$$

Thus we have

$$u(t) = i_1(t) + i_c(t) = 4 \frac{dy(t)}{dt} + 2y(t) + 2 \frac{dy(t)}{dt}$$

or

$$6 \frac{dy(t)}{dt} + 2y(t) = u(t) \qquad (2.7)$$

This first-order differential equation describes the network in Figure 2.10.

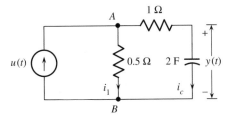

Figure 2.10 RC network.

Exercise 2.2.3

Find differential equations to describe the networks in Figure 2.11. The network in Figure 2.11(b) is called a *phase-lag network*.

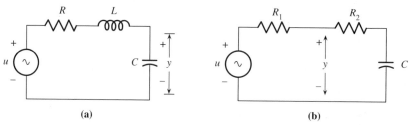

(a)   (b)

Figure 2.11 Networks.

[**Answers:**   (a) $LC\, d^2y(t)/dt^2 + RC\, dy(t)/dt + y(t) = u(t)$.
(b) $C(R_1 + R_2)dy(t)/dt + y(t) = CR_2\, du(t)/dt + u(t)$.]

2.2.3   Industrial Process—Hydraulic Tanks

In chemical plants, it is often necessary to maintain the levels of liquids. A simplified model of such a system is shown in Figure 2.12, in which

$$q_i, q_1, q_2 = \text{rates of the flow of liquid}$$

$$A_1, A_2 = \text{areas of the cross section of tanks}$$

$$h_1, h_2 = \text{liquid levels}$$

$$R_1, R_2 = \text{flow resistance, controlled by valves}$$

It is assumed that q_1 and q_2 are governed by

$$q_1 = \frac{h_1 - h_2}{R_1} \quad \text{and} \quad q_2 = \frac{h_2}{R_2} \tag{2.8}$$

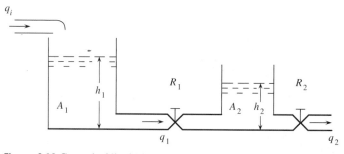

Figure 2.12 Control of liquid levels.

They are proportional to relative liquid levels and inversely proportional to flow resistances. The changes of liquid levels are governed by

$$A_1 dh_1 = (q_i - q_1)dt$$

and

$$A_2 dh_2 = (q_1 - q_2)dt$$

which imply

$$A_1 \frac{dh_1}{dt} = q_i - q_1 \qquad (2.9a)$$

and

$$A_2 \frac{dh_2}{dt} = q_1 - q_2 \qquad (2.9b)$$

These equations are obtained by linearization and approximation. In reality, the flow of liquid is very complex; it may involve turbulent flow, which cannot be described by linear differential equations. To simplify analysis, turbulent flow is disregarded in developing (2.9). Let q_i and q_2 be the input and output of the system. Now we shall develop a differential equation to describe them. The differentiation of (2.8) yields

$$\frac{dh_2}{dt} = R_2 \frac{dq_2}{dt} \quad \text{and} \quad \frac{dh_1}{dt} = R_1 \frac{dq_1}{dt} + \frac{dh_2}{dt} = R_1 \frac{dq_1}{dt} + R_2 \frac{dq_2}{dt}$$

The substitution of these into (2.9) yields

$$q_i - q_1 = A_1 \left(R_1 \frac{dq_1}{dt} + R_2 \frac{dq_2}{dt} \right) \qquad (2.10a)$$

$$q_1 - q_2 = A_2 R_2 \frac{dq_2}{dt} \qquad (2.10b)$$

Now we eliminate q_1 from (2.10a) by using (2.10b) and its derivative:

$$q_i - \left(q_2 + A_2 R_2 \frac{dq_2}{dt} \right) = A_1 R_1 \left(\frac{dq_2}{dt} + A_2 R_2 \frac{d^2 q_2}{dt^2} \right) + A_1 R_2 \frac{dq_2}{dt}$$

which can be simplified as

$$A_1 A_2 R_1 R_2 \frac{d^2 q_2}{dt^2} + (A_1 (R_1 + R_2) + A_2 R_2) \frac{dq_2}{dt} + q_2(t) = q_i(t) \qquad (2.11)$$

This second-order differential equation describes the input q_i and output q_2 of the system in Figure 2.12.

To conclude this section, we mention that a large number of physical systems can be modeled, after simplification and approximation, as linear time-invariant lumped (LTIL) systems over limited operational ranges. These systems can then be described by LTIL differential equations. In this text, we study only this class of systems.

2.3   ZERO-INPUT RESPONSE AND ZERO-STATE RESPONSE

The response of linear, in particular LTIL, systems can always be decomposed into the zero-input response and zero-state response. In this section we shall use a simple example to illustrate this fact and then discuss some general properties of the zero-input response. The Laplace transform in Appendix A is needed for the following discussion.

Consider the differential equation

$$\frac{d^2y(t)}{dt^2} + 3\frac{dy(t)}{dt} + 2y(t) = 3\frac{du(t)}{dt} - u(t) \tag{2.12}$$

Many methods are available to solve this equation. The simplest method is to use the Laplace transform. The application of the Laplace transform to (2.12) yields, using (A.9),

$$s^2Y(s) - sy(0^-) - \dot{y}(0^-) + 3[sY(s) - y(0^-)] + 2Y(s)$$
$$= 3[sU(s) - u(0^-)] - U(s) \tag{2.13}$$

where $\dot{y}(t) := dy(t)/dt$ and capital letters denote the Laplace transforms of the corresponding lowercase letters.[2] Equation (2.13) is an algebraic equation and can be manipulated using addition, subtraction, multiplication, and division. The grouping of $Y(s)$ and $U(s)$ in (2.13) yields

$$(s^2 + 3s + 2)Y(s) = sy(0^-) + \dot{y}(0^-) + 3y(0^-) - 3u(0^-) + (3s - 1)U(s)$$

which implies

$$Y(s) = \underbrace{\frac{(s + 3)y(0^-) + \dot{y}(0^-) - 3u(0^-)}{s^2 + 3s + 2}}_{\text{Zero-Input Response}} + \underbrace{\frac{3s - 1}{s^2 + 3s + 2}U(s)}_{\text{Zero-State Response}} \tag{2.14}$$

This equation reveals that the solution of (2.12) is partly excited by the input $u(t)$, $t \geq 0$, and partly excited by the initial conditions $y(0^-)$, $\dot{y}(0^-)$, and $u(0^-)$. These initial conditions will be called *the initial state*. The initial state is excited by the input applied before $t = 0$. In some sense, the initial state summarizes the effect of the past input $u(t)$, $t < 0$, on the future output $y(t)$, for $t \geq 0$. If different past inputs $u_1(t), u_2(t), \ldots, t \leq 0$, excite the *same* initial state, then their effects on the future output will be identical. Therefore, how the differential equation acquires the initial state at $t = 0$ is immaterial in studying its solution $y(t)$, for $t \geq 0$. We mention that

---

[2]We use A := B to denote that A, by definition, equals B, and A =: B to denote that B, by definition, equals A.

the initial time $t = 0$ is not the absolute time; it is the instant we start to study the system.

Consider again (2.14). The response can be decomposed into two parts. The first part is excited exclusively by the initial state and is called the *zero-input response*. The second part is excited exclusively by the input and is called the *zero-state response*. In the study of LTIL systems, it is convenient to study the zero-input response and the zero-state response separately. We first study the zero-input response and then the zero-state response.

2.3.1 Zero-Input Response—Characteristic Polynomial

Consider the differential equation in (2.12). If $u(t) = 0$, for $t \geq 0$, then (2.12) reduces to

$$\frac{d^2y(t)}{dt^2} + 3\frac{dy(t)}{dt} + 2y(t) = 0$$

This is called the *homogeneous* equation. We now study its response due to a nonzero initial state. The application of the Laplace transform yields, as in (2.13),

$$s^2Y(s) - sy(0^-) - \dot{y}(0^-) + 3[sY(s) - y(0^-)] + 2Y(s) = 0$$

which implies

$$Y(s) = \frac{(s + 3)y(0^-) + \dot{y}(0^-)}{s^2 + 3s + 2} = \frac{(s + 3)y(0^-) + \dot{y}(0^-)}{(s + 1)(s + 2)} \tag{2.15}$$

This can be expanded as

$$Y(s) = \frac{k_1}{s + 1} + \frac{k_2}{s + 2} \tag{2.16}$$

with

$$k_1 = \frac{(s + 3)y(0^-) + \dot{y}(0^-)}{s + 2}\bigg|_{s=-1} = 2y(0^-) + \dot{y}(0^-)$$

and

$$k_2 = \frac{(s + 3)y(0^-) + \dot{y}(0^-)}{s + 1}\bigg|_{s=-2} = -[y(0^-) + \dot{y}(0^-)]$$

Thus the zero-input response is

$$y(t) = k_1e^{-t} + k_2e^{-2t} \tag{2.17}$$

No matter what the initial conditions $y(0^-)$ and $\dot{y}(0^-)$ are, the zero-input response is always a linear combination of the two functions e^{-t} and e^{-2t}. The two functions e^{-t} and e^{-2t} are the inverse Laplace transforms of $1/(s + 1)$ and $1/(s + 2)$. The two roots -1 and -2—or, equivalently, the two roots of the denominator of (2.15) are called the *modes* of the system. The modes govern the form of the zero-input response of the system.

We now extend the preceding discussion to the general case. Consider the *n*th order LTIL differential equation

$$a_n y^{(n)}(t) + a_{n-1} y^{(n-1)}(t) + \cdots + a_1 y^{(1)}(t) + a_0 y(t)$$
$$= b_m u^{(m)}(t) + b_{m-1} u^{(m-1)}(t) + \cdots + b_1 u^{(1)}(t) + b_0 u(t) \qquad (2.18)$$

where

$$y^{(i)}(t) := \frac{d^i}{dt^i} y(t), \quad u^{(i)}(t) := \frac{d^i}{dt^i} u(t)$$

and $\dot{y}(t) := y^{(1)}(t)$, $\ddot{y}(t) := y^{(2)}(t)$. We define

$$D(p) := a_n p^n + a_{n-1} p^{n-1} + \cdots + a_1 p + a_0 \qquad (2.19a)$$

and

$$N(p) := b_m p^m + b_{m-1} p^{m-1} + \cdots + b_1 p + b_0 \qquad (2.19b)$$

where the variable p is the differentiator d/dt defined by

$$py(t) := \frac{d}{dt} y(t) \qquad p^2 y(t) := \frac{d^2}{dt^2} y(t) \qquad p^3 y(t) := \frac{d^3}{dt^3} y(t) \qquad (2.20)$$

and so forth. Using this notation, (2.18) can be written as

$$D(p)y(t) = N(p)u(t) \qquad (2.21)$$

In the study of the zero-input response, we assume $u(t) \equiv 0$. Then (2.21) reduces to

$$D(p)y(t) = 0 \qquad (2.22)$$

This is the homogeneous equation. Its solution is excited exclusively by initial conditions. The application of the Laplace transform to (2.22) yields, as in (2.15),

$$Y(s) = \frac{I(s)}{D(s)}$$

where $D(s)$ is defined in (2.19a) with p replaced by s and $I(s)$ is a polynomial of s depending on initial conditions. We call $D(s)$ the *characteristic polynomial* of (2.21) because it governs the *free, unforced,* or *natural* response of (2.21). The roots of the polynomial $D(s)$ are called the *modes*.[3] For example, if

$$D(s) = (s - 2)(s + 1)^2(s + 2 - j3)(s + 2 + j3)$$

then the modes are 2, -1, -1, $-2 + j3$, and $-2 - j3$. The root 2 and the complex roots $-2 \pm j3$ are simple modes and the root -1 is a repeated mode with multi-

---

[3] In the literature, they are also called the *natural frequencies*. However the ω_n in $D(s) = s^2 + 2\zeta\omega_n s + \omega_n^2$ is also called the natural frequency in this and some other control texts. To avoid possible confusion, we call the roots of $D(s)$ the modes.

plicity 2. Thus for any initial conditions, $Y(s)$ can be expanded as

$$Y(s) = \frac{k_1}{s - 2} + \frac{k_2}{s + 2 - j3} + \frac{k_3}{s + 2 + j3} + \frac{c_1}{s + 1} + \frac{c_2}{(s + 1)^2}$$

and its zero-input response is, using Table A.1,

$$y(t) = k_1 e^{2t} + k_2 e^{-(2 - j3)t} + k_3 e^{-(2 + j3)t} + c_1 e^{-t} + c_2 t e^{-t}$$

This is the general form of the zero-input response and is determined by the modes of the system.

Exercise 2.3.1

Find the zero-input responses of (2.12) due to $y(0^-) = 1$, $\dot{y}(0^-) = -1$, and $y(0^-) = \dot{y}(0^-) = 1$.

[**Answers:** $y(t) = e^{-t}$; $y(t) = 3e^{-t} - 2e^{-2t}$.]

Exercise 2.3.2

Find the modes and the general form of the zero-input responses of

$$D(p)y(t) = N(p)u(t) \tag{2.23}$$

where

$$D(p) = p^2(p - 2)^2(p^2 + 4p + 8) \quad \text{and} \quad N(p) = 3p^2 - 10$$

[**Answers:** $0, 0, 2, 2, -2 + j2$, and $-2 - j2$; $k_1 + k_2 t + k_3 e^{2t} + k_4 t e^{2t} + k_5 e^{-(2 - j2)t} + k_6 e^{-(2 + j2)t}$]

2.4 ZERO-STATE RESPONSE—TRANSFER FUNCTION

Consider the differential equation in (2.12) or

$$\frac{d^2 y(t)}{dt^2} + 3\frac{dy(t)}{dt} + 2y(t) = 3\frac{du}{dt} - u(t) \tag{2.24}$$

The response of (2.24) is partly excited by the initial conditions and partly excited by the input $u(t)$. If all initial conditions equal zero, the response is excited exclusively by the input and is called the *zero-state response*. In the Laplace transform domain, the zero-state response of (2.24) is governed by, setting all initial conditions in (2.14) to zero,

$$Y(s) = \frac{3s - 1}{s^2 + 3s + 2} U(s) =: G(s)U(s) \tag{2.25}$$

where the rational function $G(s) = (3s - 1)/(s^2 + 3s + 2)$ is called the *transfer function*. It is the ratio of the Laplace transforms of the output and input when all initial conditions are zero or

$$G(s) = \left.\frac{Y(s)}{U(s)}\right|_{\text{Initial conditions} = 0} = \left.\frac{\mathcal{L}[\text{Output}]}{\mathcal{L}[\text{Input}]}\right|_{\text{Initial conditions} = 0} \tag{2.26}$$

The transfer function describes only the zero-state responses of LTIL systems.

Example 2.4.1   (Mechanical system)

Consider the mechanical system studied in Figure 2.5(a). As derived in (2.4), it is described by

$$m\frac{d^2y(t)}{dt^2} + k_1\frac{dy(t)}{dt} + k_2 y(t) = u(t)$$

The application of the Laplace transform yields, assuming zero initial conditions,

$$ms^2Y(s) + k_1 sY(s) + k_2 Y(s) = U(s)$$

or

$$(ms^2 + k_1 s + k_2)Y(s) = U(s)$$

Thus the transfer function from u to y of the mechanical system is

$$G(s) = \frac{Y(s)}{U(s)} = \frac{1}{ms^2 + k_1 s + k_2}$$

This example reveals that the transfer function of a system can be readily obtained from its differential-equation description. For example, if a system is described by the differential equation

$$D(p)y(t) = N(p)u(t)$$

where $D(p)$ and $N(p)$ are defined as in (2.19), then the transfer function of the system is

$$G(s) = \frac{N(s)}{D(s)}$$

Exercise 2.4.1

Find the transfer functions from u to y of the networks shown in Figures 2.10 and 2.11.

[**Answers:** $1/(6s + 2), 1/(LCs^2 + RCs + 1), (CR_2 s + 1)/(C(R_1 + R_2)s + 1).$]

Exercise 2.4.2   (Industrial process)

Find the transfer function from q_i to q_2 of the system in Figure 2.12.

[**Answer:** $G(s) = 1/(A_1 A_2 R_1 R_2 s^2 + (A_1(R_1 + R_2) + A_2 R_2)s + 1).]$

RLC Networks

Although the transfer function of an RLC network can be obtained from its differential-equation description, it is generally simpler to compute it by using the concept of the Laplacian impedance or, simply, the impedance. If all initial conditions are zero, the application of the Laplace transforms to (2.6) yields

$$V(s) = RI(s) \qquad (resistor)$$

$$V(s) = \frac{1}{Cs} I(s) \qquad (capacitor)$$

and

$$V(s) = LsI(s) \qquad (inductor)$$

These relationships can be written as $V(s) = Z(s)I(s)$, and $Z(s)$ is called the (Laplacian) *impedance*. Thus the impedances of the resistor, capacitor, and inductor are respectively R, $1/Cs$, and Ls. If we consider $I(s)$ the input and $V(s)$ the output, then the impedance is a special case of the transfer function defined in (2.26). Whenever impedances are used, all initial conditions are implicitly assumed to be zero.

The manipulation involving impedances is purely algebraic, identical to the manipulation of resistances. For example, the resistance of the series connection of two resistances R_1 and R_2 is $R_1 + R_2$; the resistance of the parallel connection of R_1 and R_2 is $R_1 R_2/(R_1 + R_2)$. Similarly, the impedance of the series connection of two impedances $Z_1(s)$ and $Z_2(s)$ is $Z_1(s) + Z_2(s)$; the impedance of the parallel connection of $Z_1(s)$ and $Z_2(s)$ is $Z_1(s)Z_2(s)/(Z_1(s) + Z_2(s))$. The only difference is that now we are dealing with rational functions, rather than real numbers as in the resistive case.

Example 2.4.2

Compute the transfer function from u to i of the network shown in Figure 2.13(a). Its equivalent network using impedances is shown in Figure 2.13(b). The impedance of the parallel connection of $1/2s$ and $3s + 2$ is

$$\frac{\frac{1}{2s}(3s + 2)}{\frac{1}{2s} + (3s + 2)} = \frac{3s + 2}{6s^2 + 4s + 1}$$

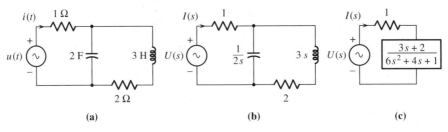

Figure 2.13 Network.

as shown in Figure 2.13(c). Hence the current $I(s)$ shown in Figure 2.13 is given by

$$I(s) = \frac{U(s)}{1 + \dfrac{3s + 2}{6s^2 + 4s + 1}} = \frac{6s^2 + 4s + 1}{6s^2 + 7s + 3} U(s)$$

Thus the transfer function from u to i is

$$G(s) = \frac{I(s)}{U(s)} = \frac{6s^2 + 4s + 1}{6s^2 + 7s + 3}$$

Exercise 2.4.3

Find the transfer functions from u to y of the networks in Figures 2.10 and 2.11 using the concept of impedances.

2.4.1   Proper Transfer Functions

Consider the rational function

$$G(s) = \frac{N(s)}{D(s)}$$

where $N(s)$ and $D(s)$ are two polynomials with real coefficients. We use deg to denote the degree of a polynomial. If

$$\deg N(s) > \deg D(s)$$

$G(s)$ is called an *improper* rational function. For example, the rational functions

$$\frac{s^2 + 1}{s + 1} \qquad s \qquad s^2 + 1 \qquad \text{and} \qquad \frac{s^{10}}{s^9 + s^8 + s^7 - 10}$$

are all improper. If

$$\deg N(s) \le \deg D(s)$$

$G(s)$ is called a *proper* rational function. It is *strictly proper* if deg $N(s) <$ deg $D(s)$; *biproper* if deg $N(s) =$ deg $D(s)$. Thus proper rational functions include both strictly proper and biproper rational functions. If $G(s)$ is biproper, so is $G^{-1}(s) = D(s)/N(s)$. This is the reason for calling it *bi*proper.

Exercise 2.4.4

Classify the following rational functions

$$s^2 + 1 \qquad 2 \qquad \frac{1}{s + 1} \qquad \frac{s^2 - 1}{s + 1} \qquad \frac{s - 1}{s + 1}.$$

How many of them are proper rational functions?

[**Answers:** Improper, biproper, strictly proper, improper, biproper; 3.]

The properness of a rational function $G(s)$ can also be determined from the value of $G(s)$ at $s = \infty$. It is clear that $G(s)$ is improper if $G(\infty) = \pm\infty$, proper if $G(\infty)$ is a finite nonzero or zero constant, biproper if $G(\infty)$ is finite and nonzero, and strictly proper if $G(\infty) = 0$.

The transfer functions we will encounter in this text are mostly proper rational functions. The reason is twofold. First, improper transfer functions are difficult, if not impossible, to build in practice, as will be discussed in Chapter 5. Second, improper transfer functions will amplify high-frequency noise, as will be explained in the following.

Signals are used to carry information. However, they are often corrupted by noise during processing, transmission, or transformation. For example, an angular position can be transformed into an electrical voltage by using the wirewound potentiometer shown in Figure 2.14. The potentiometer consists of a finite number of turns of wiring, hence the contact point moves from turn to turn. Because of brush

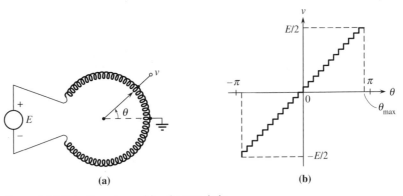

(a)   (b)

Figure 2.14 Potentiometer and its characteristic.

jumps, wire irregularities, or variations of contact resistance, unwanted spurious voltage will be generated. Thus the output voltage $v(t)$ of the potentiometer will not be exactly proportional to the angular displacement $\theta(t)$, but rather will be of the form

$$v(t) = k\theta(t) + n(t) \qquad (2.27)$$

where k is a constant and $n(t)$ is noise. Therefore, in general, every signal is of the form

$$v(t) = i(t) + n(t) \qquad (2.28)$$

where $i(t)$ denotes information and $n(t)$ denotes noise. Clearly in order for $v(t)$ to be useful, we require

$$v(t) \approx i(t)$$

and for any system to be designed, we require

$$\text{Response of the system due to } v(t)$$
$$\approx \text{response of the system due to } i(t) \qquad (2.29)$$

where $\approx$ denotes "roughly equal to." If the response of a system excited by $v(t)$ is drastically different from that excited by $i(t)$, the system is generally useless in practice. Now we show that if the transfer function of a system is improper and if the noise is of high frequency, then the system is useless. Rather than discussing the general case, we study a system with transfer function s and a system with transfer function $1/s$. A system with transfer function s is called a *differentiator* because it performs differentiation in the time domain. A system with transfer function $1/s$ is called an *integrator* because it performs integration in the time domain. The former has an improper transfer function, the latter has a strictly proper transfer function. We shall show that the differentiator will *amplify* high-frequency noise; whereas the integrator will *suppress* high-frequency noise. For convenience of discussion, we assume

$$i(t) = \sin 2t \qquad n(t) = 0.01 \sin 1000t$$

and

$$v(t) = i(t) + n(t) = \sin 2t + 0.01 \sin 1000t \qquad (2.30)$$

The magnitude of the noise is very small, so we have $v(t) \approx i(t)$. If we apply this signal to a differentiator, then the output is

$$\frac{dv(t)}{dt} = 2 \cos 2t + 0.01 \times 1000 \cos 1000t = 2 \cos 2t + 10 \cos 1000t$$

Because the amplitude of the noise term is five times larger than that of the information, we do not have $dv(t)/dt \approx di(t)/dt$ as shown in Figure 2.15. Thus a differentiator—and, more generally, systems with improper transfer functions—cannot be used if a signal contains high-frequency noise.

$\dfrac{di(t)}{dt}$

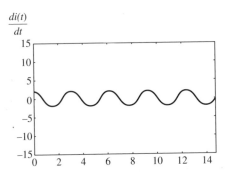

$\dfrac{dv(t)}{dt}$

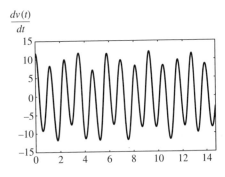

Figure 2.15 Responses of differentiator.

If we apply (2.30) to an integrator, then the output is

$$\int_0^t v(\tau)d\tau = -\frac{1}{2}\cos 2t - \frac{0.01}{1000}\cos 1000t$$

The output due to the noise term is practical zero and

$$\int_0^t v(\tau)d\tau \approx \int_0^t i(\tau)d\tau$$

Thus we conclude that an integrator—and, more generally, systems with strictly proper transfer functions—will suppress high-frequency noise.

In practice, we often encounter high-frequency noise. As was discussed earlier, wirewound potentiometers will generate unwanted high-frequency noise. Thermal noise and shot noise, which are of high-frequency compared to control signals, are always present in electrical systems. Because of splashing and turbulent flow of incoming liquid, the measured liquid level of a tank will consist of high-frequency noise. In order not to amplify high-frequency noise, most systems and devices used in practice have proper transfer functions.

Exercise 2.4.5

Consider

$$v(t) = i(t) + n(t) = \cos 2t + 0.01\cos 0.001t$$

Note that the frequency of the noise $n(t)$ is much smaller than that of the information $i(t)$. Do we have $v(t) \approx i(t)$? Do we have $dv(t)/dt \approx di(t)/dt$? Is it true that a differentiator amplifies any type of noise?

[**Answers:** Yes, yes, no.]

2.4.2 Poles and Zeros

The zero-state response of a system is governed by its transfer function. Before computing the response, we introduce the concepts of poles and zeros. Consider a proper rational transfer function

$$G(s) = \frac{N(s)}{D(s)}$$

where $N(s)$ and $D(s)$ are polynomials with real coefficients and deg $N(s) \leq$ deg $D(s)$.

☐ Definition

A finite real or complex number λ is a *pole* of $G(s)$ if $|G(\lambda)| = \infty$, where $|\cdot|$ denotes the absolute value. It is a *zero* of $G(s)$ if $G(\lambda) = 0$. ∎

Consider the transfer function

$$G(s) = \frac{N(s)}{D(s)} = \frac{2(s^3 + 3s^2 - s - 3)}{(s - 1)(s + 2)(s + 1)^3} \tag{2.31}$$

We have

$$G(-2) = \frac{N(-2)}{D(-2)} = \frac{2[(-2)^3 + 3(-2)^2 - (-2) - 3]}{(-3) \cdot 0 \cdot (-1)} = \frac{6}{0} = \infty$$

Therefore -2 is a pole of $G(s)$ by definition. Clearly -2 is a root of $D(s)$. Does this imply every root of $D(s)$ is a pole of $G(s)$? To answer this, we check $s = 1$, which is also a root of $D(s)$. We compute $G(1)$:

$$G(1) = \frac{N(1)}{D(1)} = \frac{2(1 + 3 - 1 - 3)}{0 \cdot 3 \cdot 8} = \frac{0}{0}$$

It is not defined. However l'Hôpital's rule implies

$$G(1) = \frac{N(s)}{D(s)}\bigg|_{s=1} = \frac{N'(s)}{D'(s)}\bigg|_{s=1}$$

$$= \frac{2(3s^2 + 6s - 1)}{5s^4 + 16s^3 + 12s^2 - 4s - 5}\bigg|_{s=1} = \frac{16}{24} \neq \infty$$

Thus $s = 1$ is not a pole of $G(s)$. Therefore not every root of $D(s)$ is a pole of $G(s)$.

Now we factor $N(s)$ in (2.31) and then cancel the common factors between $N(s)$ and $D(s)$ to yield

$$G(s) = \frac{2(s + 3)(s - 1)(s + 1)}{(s - 1)(s + 2)(s + 1)^3} = \frac{2(s + 3)}{(s + 2)(s + 1)^2} \tag{2.32}$$

We see immediately that $s = 1$ is not a pole of $G(s)$. Clearly $G(s)$ has one zero, -3, and three poles, -2, -1, and -1. The pole -2 is called a *simple* pole and the pole -1 is called a *repeated* pole with multiplicity 2.[4]

From this example, we see that if polynomials $N(s)$ and $D(s)$ have no common factors,[5] then all roots of $N(s)$, and all roots of $D(s)$ are, respectively, the zeros and poles of $G(s) = N(s)/D(s)$. If $N(s)$ and $D(s)$ have no common factor, they are said to be *coprime* and $G(s) = N(s)/D(s)$ is said to be *irreducible*. Unless stated otherwise, every transfer function will be assumed to be irreducible.

We now discuss the computation of the zero-state response. The zero-state response of a system is governed by $Y(s) = G(s)U(s)$. To compute $Y(s)$, we first compute the Laplace transform of $u(t)$. We then multiply $G(s)$ and $U(s)$ to yield $Y(s)$. The inverse Laplace transform of $Y(s)$ yields the zero-state response. This is illustrated by an example.

Example 2.4.3

Find the zero-state response of (2.25) due to $u(t) = 1$, for $t \geq 0$. This is called the unit-step response of (2.25). The Laplace transform of $u(t)$ is $1/s$. Thus we have

$$Y(s) = G(s)U(s) = \frac{3s - 1}{(s + 1)(s + 2)} \cdot \frac{1}{s} \tag{2.33}$$

To compute its inverse Laplace transform, we carry out the partial fraction expansion as

$$Y(s) = \frac{3s - 1}{(s + 1)(s + 2)s} = \frac{k_1}{s + 1} + \frac{k_2}{s + 2} + \frac{k_3}{s}$$

where

$$k_1 = Y(s) \cdot (s + 1)\Big|_{s = -1} = \frac{3s - 1}{(s + 2)s}\Big|_{s = -1} = \frac{-4}{(1)(-1)} = 4$$

$$k_2 = Y(s) \cdot (s + 2)\Big|_{s = -2} = \frac{3s - 1}{(s + 1)s}\Big|_{s = -2} = \frac{-7}{(-1)(-2)} = -3.5$$

and

$$k_3 = Y(s) \cdot s\Big|_{s = 0} = \frac{3s - 1}{(s + 1)(s + 2)}\Big|_{s = 0} = \frac{-1}{2} = -0.5$$

---

[4]If s is very large, (2.32) reduces to $G(s) = 1/s^2$ and $G(\infty) = 0$. Thus ∞ can be considered as a repeated zero with multiplicity 2. Unless stated otherwise, we consider only finite poles and zeros.

[5]Any two polynomials, such as $4s + 2$ and $6s + 2$, have a constant as a common factor. Such a common factor, a polynomial of degree 0, is called a *trivial common factor*. We consider only nontrivial common factors—that is, common factors of degree 1 or higher.

Using Table A.1, the zero-state response is

$$y(t) = \underbrace{4e^{-t} - 3.5e^{-2t}}_{\substack{\text{Due to the} \\ \text{Poles of } G(s)}} \underbrace{- \ 0.5}_{\substack{\text{Due to the} \\ \text{Poles of } U(s)}} \qquad (2.34)$$

for $t \geq 0$. Thus, the use of the Laplace transform to compute the zero-state response is simple and straightforward.

This example reveals an important fact of the zero-state response. We see from (2.34) that the response consists of three terms. Two are the inverse Laplace transforms of $1/(s + 2)$ and $1/(s + 1)$, which are the poles of the system. The remaining term is due to the step input. In fact, for any $u(t)$, the response of (2.33) is generally of the form

$$y(t) = k_1 e^{-t} + k_2 e^{-2t} + (\text{terms due to the poles of } U(s)) \qquad (2.35)$$

(see Problem 2.20). Thus the poles of $G(s)$ determine the basic form of the zero-state response.

Exercise 2.4.6

Find the zero-state response of $1/(s + 1)$ due to e^{-2t}, $t \geq 0$.

[**Answer:** $y(t) = e^{-t} - e^{-2t}$.]

If k_1 in (2.35) is zero, the corresponding pole -1 is not excited. A similar remark applies to k_2. Because both k_1 and k_2 in (2.34) are nonzero, both poles of $G(s)$ in (2.33) are excited by the step input. For other inputs, the two poles may not always be excited. This is illustrated by an example.

Example 2.4.4

Consider the system in (2.33). Find a bounded input $u(t)$ so that the pole -1 will not be excited. If $U(s) = s + 1$, then

$$Y(s) = G(s)U(s) = \frac{3s - 1}{(s + 1)(s + 2)} \cdot (s + 1)$$

$$= \frac{3s - 1}{s + 2} = \frac{3(s + 2) - 7}{s + 2} = 3 - \frac{7}{s + 2}$$

which implies

$$y(t) = 3\delta(t) - 7e^{-2t}$$

This response does not contain e^{-t}, thus the pole -1 is not excited. Therefore if we introduce a zero in $U(s)$ to cancel a pole, then the pole will not be excited by the input $u(t)$.

If $U(s)$ is biproper or improper, as is the case for $U(s) = s + 1$, then its inverse Laplace transform $u(t)$ will contain an impulse and its derivatives and is not bounded. In order for $u(t)$ to be bounded, we choose, rather arbitrarily, $U(s) = (s + 1)/s(s + 3)$, a strictly proper rational function. Its inverse Laplace transform is

$$u(t) = \frac{1}{3} + \frac{2}{3} e^{-3t}$$

for $t \geq 0$ and is bounded. The application of this input to (2.33) yields

$$Y(s) = \frac{3s - 1}{(s + 2)(s + 1)} \cdot \frac{s + 1}{s(s + 3)} = \frac{3s - 1}{(s + 2)(s + 3)s}$$

$$= \frac{7}{2(s + 2)} - \frac{10}{3(s + 3)} - \frac{1}{6s}$$

which implies

$$y(t) = \frac{7}{2} e^{-2t} - \frac{10}{3} e^{-3t} - \frac{1}{6}$$

for $t \geq 0$. The second and third terms are due to the input, the first term is due to the pole -2. The term e^{-1} does not appear in $y(t)$, thus the pole -1 is not excited by the input. Similarly, we can show that the input $(s + 2)/s(s + 1)$ or $(s + 2)/(s + 3)^2$ will not excite the pole -2 and the input $(s + 2)(s + 1)/s(s + 3)^2$ will not excite either pole.

From this example, we see that whether or not a pole will be excited depends on whether $u(t)$ or $U(s)$ has a zero to cancel it. The Laplace transforms of the unit-step function and $\sin \omega_0 t$ are

$$\frac{1}{s} \quad \text{and} \quad \frac{\omega_0}{s^2 + \omega_0^2}$$

They have no zero. Therefore, either input will excite all poles of every LTIL system.

The preceding discussion can be extended to the general case. Consider, for example,

$$Y(s) = G(s)U(s) := \frac{(s + 10)(s + 2)(s - 1)^2}{s^3(s - 2)^2(s + 2 - j2)(s + 2 + j2)} U(s)$$

The transfer function $G(s)$ has poles at 0, 0, 0, 2, 2, and $-2 \pm j2$. The complex poles $-2 \pm j2$ are simple poles, the poles 0 and 2 are repeated poles with multiplicities 3 and 2. If $G(s)$ and $U(s)$ have no pole in common, then the zero-state response of the system due to $U(s)$ is of the form

$$y(t) = k_1 + k_2 t + k_3 t^2 + k_4 e^{2t} + k_5 t e^{2t} + k_6 e^{-(2-j2)t} + k_7 e^{-(2+j2)t}$$
$$+ \text{(Terms due to the poles of } U(s)) \tag{2.36}$$

(see Problem 2.20.) Thus the poles of $G(s)$ determine the basic form of the response.

Do the zeros of $G(s)$ play any role in the zero-state response? Certainly, they do. They affect the values of k_i. Different sets of k_i yield drastically different responses, as is illustrated by the following example.

Example 2.4.5

Consider

$$G_1(s) = \frac{2}{(s + 1)(s + 1 + j)(s + 1 - j)}$$

$$G_2(s) = \frac{0.2(s + 10)}{(s + 1)(s + 1 + j)(s + 1 - j)}$$

$$G_3(s) = \frac{-0.2(s - 10)}{(s + 1)(s + 1 + j)(s + 1 - j)}$$

$$G_4(s) = \frac{10(s^2 + 0.1s + 0.2)}{(s + 1)(s + 1 + j)(s + 1 - j)}$$

The transfer function $G_1(s)$ has no zero, $G_2(s)$ and $G_3(s)$ have one zero, and $G_4(s)$ has a pair of complex conjugate zeros at $-0.05 \pm 0.444j$. They all have the same

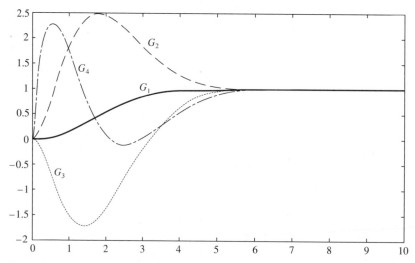

Figure 2.16 Unit-step responses of $G_i(s)$.

set of poles, and their unit-step responses are all of the form

$$y(t) = k_1 e^{-t} + k_2 e^{-(1+j1)t} + k_3 e^{-(1-j1)t} + k_4$$

with k_3 equal to the complex conjugate of k_2. Their responses are shown in Figure 2.16, respectively, with the solid line ($G_1(s)$), dashed line ($G_2(s)$), dotted line ($G_3(s)$), and dash-and-dotted line ($G_4(s)$). They are quite different. In conclusion, even though the poles of $G(s)$ determine the basic form of responses, exact responses are determined by the poles, zeros, and the input. Therefore, the zeros of a transfer function cannot be completely ignored in the analysis and design of control systems.

2.5 BLOCK REPRESENTATION—COMPLETE CHARACTERIZATION

In the analysis and design of control systems, every device is represented by a block as shown in Figure 2.4 or 2.17(a). The block is then represented by its transfer function $G(s)$. If the input is $u(t)$ and the output is $y(t)$, then they are related by

$$Y(s) = G(s)U(s) \tag{2.37}$$

where $Y(s)$ and $U(s)$ are respectively the Laplace transforms of $y(t)$ and $u(t)$. Note that we have mixed the time-domain representation $u(t)$ and $y(t)$ and the Laplace transform representation $G(s)$ in Figure 2.17(a). This convention will be used throughout this text. It is important to know that it is incorrect to write $y(t) = G(s)u(t)$. The correct expression is $Y(s) = G(s)U(s)$.[6]

Equation (2.37) is an algebraic equation. The product of the Laplace transform of the input and the transfer function yields the Laplace transform of the output. The advantage of using this algebraic representation can be seen from the tandem connection of two systems shown in Figure 2.17(b). Suppose the two systems are represented, respectively, by

$$Y_1(s) = G_1(s)U_1(s) \qquad Y_2(s) = G_2(s)U_2(s)$$

In the tandem connection, we have $u_2(t) = y_1(t)$ or $U_2(s) = Y_1(s)$ and

$$Y_2(s) = G_2(s)Y_1(s) = G_2(s)G_1(s)U_1(s) \tag{2.38}$$

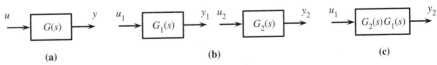

(a)   (b)   (c)

Figure 2.17 (a) A system. (b) Tandem connection of two systems. (c) Reduction of (b).

---

[6]It can also be expressed as $y(t) = \int_0^t g(t - \tau)u(\tau)d\tau$, where $g(t)$ is the inverse Laplace transform of $G(s)$. See Reference [18]. This form is rarely used in the design of control systems.

Thus the tandem connection can be represented by a single block, as shown in Figure 2.17(c), with transfer function $G(s) := G_2(s)G_1(s)$, the product of the transfer functions of the two subsystems. If we use differential equations, then the differential equation description of the tandem connection will be much more complex. Thus the use of transfer functions can greatly simplify the analysis and design of control systems.

The transfer function describes only the zero-state response of a system. Therefore, whenever we use the transfer function in analysis and design, the zero-input response (the response due to nonzero initial conditions) is completely disregarded. However, can we really disregard the zero-input response? This question is studied in this section.

Consider a linear time-invariant lumped (LTIL) system described by the differential equation

$$D(p)y(t) = N(p)u(t) \tag{2.39}$$

where

$$D(p) = a_n p^n + a_{n-1} p^{n-1} + \cdots + a_1 p + a_0$$

$$N(p) = b_m p^m + b_{m-1} p^{m-1} + \cdots + b_1 p + b_0$$

and the variable p is the differentiator defined in (2.19) and (2.20). Then the zero-input response of the system is described by

$$D(p)y(t) = 0 \tag{2.40}$$

and the response is dictated by the roots of $D(s)$, called *the modes of the system* (Section 2.3.1). The zero-state response of the system is described by the transfer function

$$G(s) := \frac{N(s)}{D(s)}$$

and the basic form of its response is governed by the poles of $G(s)$ (Section 2.4.2). The poles of $G(s)$ are defined as the roots of $D(s)$ after canceling the common factors of $N(s)$ and $D(s)$. Thus if $D(s)$ and $N(s)$ have no common factors, then

$$\text{The set of the poles} = \text{The set of the modes} \tag{2.41}$$

In this case, the system is said to be *completely characterized* by its transfer function. If $D(s)$ and $N(s)$ have common factors, say $R(s)$, then the roots of $R(s)$ are nodes of the system but not poles of $G(s)$. In this case the roots of $R(s)$ are called the *missing poles* of the transfer function, and the system is said to be not completely characterized by its transfer function. We use examples to illustrate this concept and discuss its implications.

Example 2.5.1

Consider the system shown in Figure 2.18. The input is a current source. The output y is the voltage across the 2-Ω resistor as shown. The system can be described by the LTIL differential equation

$$\frac{dy(t)}{dt} - 0.75y(t) = \frac{du(t)}{dt} - 0.75u(t) \qquad (2.42a)$$

(Problem 2.21). The equation can also be written, using $p = d/dt$, as

$$(p - 0.75)y(t) = (p - 0.75)u(t) \qquad (2.42b)$$

The mode of the system is the root of $(s - 0.75)$ or 0.75. Therefore its zero-input response is of the form

$$y(t) = ke^{0.75t} \qquad (2.43)$$

where k depends on the initial voltage of the capacitor in Figure 2.18. We see that if the initial voltage is different from zero, then the response will approach infinity as $t \to \infty$.

We will now study its zero-state response. The transfer function of the system is

$$G(s) = \frac{s - 0.75}{s - 0.75} = 1 \qquad (2.44)$$

Because of the common factor, the transfer function reduces to 1. Thus the system has no pole and the zero-state response is $y(t) = u(t)$, for all t. This system is *not* completely characterized by its transfer function because the mode 0.75 does not appear as a pole of $G(s)$. In other words, the transfer function has missing pole 0.75.

If we use the transfer function to study the system in Figure 2.18, we would conclude that the system is acceptable. In reality, the system is not acceptable, because if, for any reason, the voltage of the capacitor becomes nonzero, the response

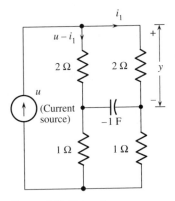

Figure 2.18 Network.

will grow without bound and the system will either become saturated or burn out. Thus the system is of no use in practice.

The existence of a missing pole in Figure 2.18 can easily be explained from the structure of the network. Because of the symmetry of the four resistors, if the initial voltage of the capacitor is zero, its voltage will remain zero no matter what current source is applied. Therefore, the removal of the capacitor will not affect the zero-state response of the system. Thus, the system has a superfluous component as far as the input and output are concerned. These types of systems are not built in practice, except by mistake.

Example 2.5.2

Consider the system described by

$$(p^2 + 2p - 3)y(t) = (p - 2)u(t) \qquad (2.45)$$

The zero-input response of the system is governed by

$$(p^2 + 2p - 3)y(t) = 0$$

Its modes are the roots of $(s^2 + 2s - 3) = (s - 1)(s + 3)$ or 1 and -3. Thus the zero-input response due to any initial conditions is of the form

$$y_{zi}(t) = k_1 e^t + k_2 e^{-3t}$$

where the subscript zi denotes zero-input. The response of mode 1 approaches infinity and the response of mode -3 approaches zero as $t \to \infty$.

The transfer function of the system is

$$G(s) = \frac{s - 2}{s^2 + 2s - 3} = \frac{s - 2}{(s - 1)(s + 3)} \qquad (2.46)$$

Thus the zero-state response of the system due to $u(t)$ will generally be of the form

$$y(t) = \qquad + \bar{k}_2 e^{-3t} + \text{(Terms due to the poles of } U(s))$$

We see that the two modes appear as the poles of $G(s)$. Thus the system has no missing poles and is completely characterized by its transfer function. In this case, the zero-input response due to any initial conditions will appear in the zero-state response, so no essential information is lost in using the transfer function to study the system.

In conclusion, if a system is completely characterized by its transfer function, then the zero-input response will appear essentially as part of the zero-state response. It is therefore permissible to use the transfer function in analysis and design without considering the response due to nonzero initial conditions. If a system is not completely characterized by its transfer function, care must be exercised in using the transfer function to study the system. This point is discussed further in Chapter 4.

Exercise 2.5.1

Which of the following systems are completely characterized by their transfer functions? If not, find the missing poles.

a. $(p^2 + 2p + 1)y(t) = (p + 1)p\, u(t)$

b. $(p^2 - 3p + 2)y(t) = (p - 1)\, u(t)$

c. $(p^2 - 3p + 2)y(t) = u(t)$

[**Answers:** (a) No, -1 (b) No, 1 (c) Yes.]

2.5.1 The Loading Problem

Most control systems are built by interconnecting a number of subsystems as shown in Figures 1.1(b), 1.6(b), and 1.7(b). In analysis and design, every subsystem will be represented by a block and each block will be represented by a transfer function. Thus, most control systems will consist of a number of blocks interconnected with each other. In connecting two or more blocks, the problem of loading may occur. For example, consider a 10-volt voltage source modeled as shown in Figure 2.19(a). It has an internal resistance of 10 Ω; its output voltage is 10 volts when nothing is connected to its terminals. Now consider a device that can be modeled as a 10-Ω resistor. When we connect the device to the voltage source, the output voltage of the source is not 10 volts, but $10 \cdot 10/(10 + 10) = 5$ volts. If we connect a 20-Ω device to the voltage source, the output is $10 \cdot 20/(10 + 20) = 6.7$ volts. We see that the voltages supplied by the same voltage source will be different when it is connected to different devices. In other words, for different loads, the voltages supplied by the same voltage source will be different. In this case, the connection is said to have *a loading problem*. On the other hand, if the voltage source is designed so that its internal resistance is negligible or zero, as shown in Figure 2.19(b), then

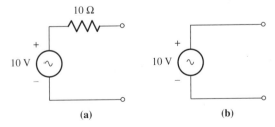

Figure 2.19 Loading problem.

no matter what device is connected to it, the supplied voltage to the device is always 10 volts. In this case, the connection is said to have *no loading problem*.

Roughly speaking, if its transfer function changes after a system is connected to another system, the connection is said to have *a loading effect*. For example, the transfer function from u to y in Figure 2.19(a) is 1 before the system is connected to any device. It becomes $5/10 = 0.5$ when the system is connected a 10-Ω resistor. Thus, the connection has a loading effect. If the tandem connection of two systems has a loading effect, then the transfer function of the tandem connection does not equal the product of the transfer functions of the two subsystems as developed in (2.38). This is illustrated by an example.

Example 2.5.3

Consider the networks shown in Figure 2.20(a). The transfer function from u_1 to y_1 of network M_1 is $G_1(s) = s/(s + 1)$. The transfer function from u_2 to y_2 of network M_2 is $G_2(s) = 2/(2 + 3s)$. Now we connect them together or set $y_1 = u_2$ as shown in Figure 2.20(b) and compute the transfer function from u_1 to y_2. The impedance of the parallel connection of the impedance s and the impedance $(3s + 2)$ is $s(3s + 2)/(s + 3s + 2) = s(3s + 2)/(4s + 2)$. Thus the current I_1 shown in

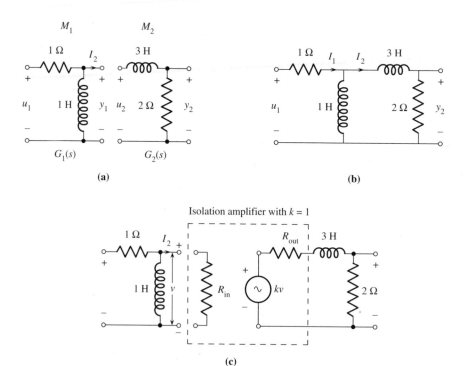

(a)

(b)

(c)

Figure 2.20 Tandem connection of two systems.

Figure 2.20(b) equals

$$I_1(s) = \frac{U_1(s)}{1 + \dfrac{s(3s + 2)}{4s + 2}} \tag{2.47}$$

and the current I_2 equals

$$I_2(s) = \frac{s}{s + 3s + 2} \cdot I_1(s) = \frac{s}{4s + 2 + s(3s + 2)} \cdot U_1(s) \tag{2.48}$$

Thus, the voltage $y_2(t)$ is given by

$$Y_2(s) = I_2(s) \cdot 2 = \frac{2s}{3s^2 + 6s + 2} \cdot U_1(s) \tag{2.49}$$

and the transfer function from u_1 to y_2 of the tandem connection is

$$\frac{Y_2(s)}{U_1(s)} = \frac{2s}{3s^2 + 6s + 2} \tag{2.50}$$

This is different from the product of $G_1(s)$ and $G_2(s)$ or

$$G_1(s)G_2(s) = \frac{s}{s + 1} \cdot \frac{2}{3s + 2} = \frac{2s}{3s^2 + 5s + 2} \tag{2.51}$$

Thus, Equation (2.38) cannot be used for this connection.

The loading of the two networks in Figure 2.20 can be easily explained. The current I_2 in Figure 2.20(a) is zero before the connection; it becomes nonzero after the connection. Thus, the loading occurs. In electrical networks, the loading often can be eliminated by inserting an isolating amplifier, as shown in Figure 2.20(c). The input impedance Z_{in} of an ideal isolating amplifier is infinity and the output impedance Z_{out} is zero. Under these assumptions, I_2 in Figure 2.20(c) remains zero and the transfer function of the connection is $G_2(s)kG_1(s)$ with $k = 1$.

The loading problem must be considered in developing a block diagram for a control system. This problem is considered in detail in the next chapter.

2.6 STATE-VARIABLE EQUATIONS

The transfer function describes only the relationship between the input and output of LTIL systems and is therefore called the *input-output description* or *external description*. In this section we shall develop a different description, called the *state-variable description* or *internal description*. Strictly speaking, this description is the same as the differential equations discussed in Section 2.2. The only difference is that *high-order* differential equations are now written as sets of *first-order* differential equations. In this way, the study can be simplified.

The state-variable description of LTIL systems is of the form

$$\dot{x}_1(t) = a_{11}x_1(t) + a_{12}x_2(t) + a_{13}x_3(t) + b_1u(t)$$
$$\dot{x}_2(t) = a_{21}x_1(t) + a_{22}x_2(t) + a_{23}x_3(t) + b_2u(t) \qquad (2.52a)$$
$$\dot{x}_3(t) = a_{31}x_1(t) + a_{32}x_2(t) + a_{33}x_3(t) + b_3u(t)$$

$$y(t) = c_1x_1(t) + c_2x_2(t) + c_3x_3(t) + du(t) \qquad (2.52b)$$

where u and y are the input and output; x_i, $i = 1, 2, 3$, are called *the state variables*; a_{ij}, b_i, c_i, and d are constants; and $\dot{x}_i(t) := dx_i(t)/dt$. These equations are more often written in matrix form as

$$\dot{\mathbf{x}}(t) = \mathbf{A}\mathbf{x}(t) + \mathbf{b}u(t) \qquad \text{(State equation)} \qquad (2.53a)$$

$$y(t) = \mathbf{c}\mathbf{x}(t) + du(t) \qquad \text{(Output equation)} \qquad (2.53b)$$

with

$$\mathbf{x} = \begin{bmatrix} x_1 \\ x_2 \\ x_3 \end{bmatrix} \qquad \mathbf{A} = \begin{bmatrix} a_{11} & a_{12} & a_{13} \\ a_{21} & a_{22} & a_{23} \\ a_{31} & a_{32} & a_{33} \end{bmatrix} \qquad \mathbf{b} = \begin{bmatrix} b_1 \\ b_2 \\ b_3 \end{bmatrix} \qquad (2.54a)$$

and

$$\mathbf{c} = [c_1 \quad c_2 \quad c_3] \qquad (2.54b)$$

The vector $\mathbf{x}$ is called the *state vector* or simply the *state*. If $\mathbf{x}$ has n state variables or is an $n \times 1$ vector, then $\mathbf{A}$ is an $n \times n$ square matrix, $\mathbf{b}$ is an $n \times 1$ column vector, $\mathbf{c}$ is a $1 \times n$ row vector, and d is a 1×1 scalar. $\mathbf{A}$ is sometimes called the *system matrix* and d the *direct transmission part*. Equation (2.53a) describes the relationship between the input and state, and is called the *state equation*. The state equation in (2.53a) consists of three first-order differential equations and is said to have *dimension 3*. The equation in (2.53b) relates the input, state, and output, and is called the *output equation*. It is an algebraic equation; it does not involve differentiation of $\mathbf{x}$. Thus if $\mathbf{x}(t)$ and $u(t)$ are known, the output $y(t)$ can be obtained simply by multiplication and addition.

Before proceeding, we remark on the notation. Vectors are denoted by boldface lowercase letters; matrices by boldface capital letters. Scalars are denoted by regular-face lowercase letters. In (2.53), $\mathbf{A}$, $\mathbf{b}$, $\mathbf{c}$, and $\mathbf{x}$ are boldface because they are either vectors or matrices; u, y, and d are regular face because they are scalars.

This text studies mainly single-variable systems, that is, systems with single input and single output. For multivariable systems, we have two or more inputs and/or two and more outputs. In this case, $u(t)$ and $y(t)$ will be vectors and the orders of $\mathbf{b}$ and $\mathbf{c}$ and d must be modified accordingly. For example, if a system has three inputs and two outputs, then u will be 3×1; y, 2×1; $\mathbf{b}$, $n \times 3$; $\mathbf{c}$; $2 \times n$; and d, 2×3. Otherwise, the form of state-variable equations remains the same.

The transfer function describes only the zero-state response of systems. Thus, when we use the transfer function, the initial state or initial conditions of the system must be assumed to be zero. In using the state-variable equation, no such assumption is necessary. The equation is applicable even if the initial state is nonzero. The equation describes not only the output but also the state variables. Because the state variables reside within a system and are not necessarily accessible from the input and output terminals, the state-variable equation is also called the *internal description*. Thus the state-variable description is more general than the transfer function description. However, if a system is completely characterized by its transfer function (Section 2.5), the two descriptions are essentially the same. This will be discussed further in a later section.

In developing state-variable equations for LTIL systems, we must first choose state variables. State variables are associated with energy. For example, the potential and kinetic energy of a mass are stored in its position and velocity. Thus, position and velocity can be chosen as state variables for a mass. For RLC networks, capacitors and inductors are energy storage elements because they can store energy in their electric and magnetic fields. Therefore, all capacitor voltages and inductor currents are generally chosen as state variables for RLC networks. Resistors will not store energy; all energy is dissipated as heat. Therefore, resistor voltages or currents are not state-variables. With this brief discussion, we are ready to develop state-variable equations to describe LTIL systems.

Example 2.6.1   (Mechanical system)

Consider the mechanical system shown in Figure 2.5(a). As discussed in (2.4), its differential equation description is

$$m\ddot{y}(t) + k_1\dot{y}(t) + k_2 y(t) = u(t) \tag{2.55}$$

where $u(t)$ is the applied force (input), $y(t)$ is the displacement (output), $\dot{y}(t) := dy(t)/dt$, and $\ddot{y}(t) := d^2y(t)/dt^2$.

The potential energy and kinetic energy of a mass are stored in its position and velocity; therefore the position and velocity will be chosen as state variables. Define

$$x_1(t) := y(t) \tag{2.56a}$$

and

$$x_2(t) := \dot{y}(t) \tag{2.56b}$$

Then we have

$$\dot{x}_1(t) = \dot{y}(t) = x_2(t)$$

This relation follows from the definition of $x_1(t)$ and $x_2(t)$ and is independent of the system. Taking the derivative of $x_2(t)$ yields

$$\dot{x}_2(t) = \ddot{y}(t)$$

which becomes, after the substitution of (2.55) and (2.56),

$$\dot{x}_2(t) = \frac{1}{m}[-k_2 y(t) - k_1 \dot{y}(t) + u(t)]$$

$$= -\frac{k_2}{m} x_1(t) - \frac{k_1}{m} x_2(t) + \frac{1}{m} u(t)$$

These equations can be arranged in matrix form as

$$\begin{bmatrix} \dot{x}_1(t) \\ \dot{x}_2(t) \end{bmatrix} = \begin{bmatrix} 0 & 1 \\ -\dfrac{k_2}{m} & -\dfrac{k_1}{m} \end{bmatrix} \begin{bmatrix} x_1(t) \\ x_2(t) \end{bmatrix} + \begin{bmatrix} 0 \\ \dfrac{1}{m} \end{bmatrix} u(t) \qquad (2.57a)$$

$$y(t) = \begin{bmatrix} 1 & 0 \end{bmatrix} \begin{bmatrix} x_1(t) \\ x_2(t) \end{bmatrix} \qquad (2.57b)$$

This is a state-variable equation of dimension 2. Note that the direct transmission d in (2.53) is zero for this example.

We discuss in the following a procedure for developing state-variable equations for RLC networks that contain a voltage or current source.

Procedure for Developing State-Variable Equations for RLC Networks

Step 1: Assign all capacitor voltages and inductor currents as state variables. Write down capacitor currents and inductor voltages as shown in Figure 2.8. Note that currents flow from high to low potential.

Step 2: Use Kirchhoff's voltage and/or current laws to express every resistor's voltage and current in terms of state variables and, if necessary, the input.

Step 3: Use Kirchhoff's voltage and/or current laws to develop a state equation.

This procedure offers only a general guide. For a more detailed and specific procedure, see Reference [15]. We use an example to illustrate the procedure.

Example 2.6.2 (Electrical system)

Consider the RLC network shown in Figure 2.21. It consists of one resistor, one capacitor, and one inductor. The input $u(t)$ is a voltage source and the voltage across the 3-H inductor is chosen as the output.

Step 1: The capacitor voltage $x_1(t)$ and the inductor currents $x_2(t)$ are chosen as state variables. The capacitor current is $2\dot{x}_1(t)$, and the inductor voltage is $3\dot{x}_2(t)$.

Step 2: The current passing through the 4-Ω resistor clearly equals $x_2(t)$. Thus, the voltage across the resistor is $4x_2(t)$. The polarity of the voltage must be specified, otherwise confusion may occur.

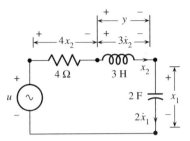

Figure 2.21 Network.

Step 3: From Figure 2.21, we see that the capacitor current $2\dot{x}_1$ equals $x_2(t)$, which implies

$$\dot{x}_1(t) = \frac{1}{2} x_2(t) \qquad (2.58)$$

The voltage across the inductor is, using Kirchhoff's voltage law,

$$3\dot{x}_2(t) = u(t) - x_1(t) - 4x_2(t)$$

or

$$\dot{x}_2(t) = -\frac{1}{3} x_1(t) - \frac{4}{3} x_2(t) + \frac{1}{3} u(t) \qquad (2.59)$$

Equations (2.58) and (2.59) can be arranged in matrix form as

$$\begin{bmatrix} \dot{x}_1(t) \\ \dot{x}_2(t) \end{bmatrix} = \begin{bmatrix} 0 & \frac{1}{2} \\ -\frac{1}{3} & -\frac{4}{3} \end{bmatrix} \begin{bmatrix} x_1(t) \\ x_2(t) \end{bmatrix} + \begin{bmatrix} 0 \\ \frac{1}{3} \end{bmatrix} u(t) \qquad (2.60\text{a})$$

This is the state equation of the network. The output $y(t)$ is

$$y(t) = 3\dot{x}_2$$

which is not in the form of $\mathbf{cx} + du$. However the substitution of (2.59) yields

$$y(t) = -x_1(t) - 4x_2(t) + u(t)$$

$$= [-1 \quad -4] \begin{bmatrix} x_1(t) \\ x_2(t) \end{bmatrix} + u(t) \qquad (2.60\text{b})$$

This is the output equation.

Exercise 2.6.1

Find state-variable descriptions of the networks shown in Figure 2.22 with the state variables and outputs chosen as shown.

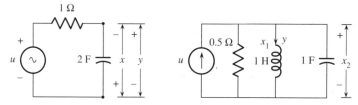

Figure 2.22 Networks.

[**Answers:**   (a) $\dot{x} = -0.5x - 0.5u$, $y = -x$ (b) $\dot{\mathbf{x}} = \begin{bmatrix} 0 & 1 \\ -1 & -2 \end{bmatrix} \mathbf{x} + \begin{bmatrix} 0 \\ 1 \end{bmatrix} u$,

$y = [1 \quad 0]\mathbf{x}.$]

2.7   SOLUTIONS OF STATE EQUATIONS—LAPLACE TRANSFORM METHOD

Consider the n-dimensional state-variable equation

$$\dot{\mathbf{x}}(t) = \mathbf{A}\mathbf{x}(t) + \mathbf{b}u(t) \tag{2.61a}$$

$$y(t) = \mathbf{c}\mathbf{x}(t) + du(t) \tag{2.61b}$$

where $\mathbf{A}$, $\mathbf{b}$, $\mathbf{c}$, and d are respectively $n \times n$, $n \times 1$, $1 \times n$, and 1×1 matrices. In analysis, we are interested in the output $y(t)$ excited by an input $u(t)$ and some initial state $\mathbf{x}(0)$. This problem can be solved directly in the time domain or indirectly by using the Laplace transform. We study the latter in this section. The time-domain method will be studied in the next subsection.

The application of the Laplace transform to (2.61a) yields

$$s\mathbf{X}(s) - \mathbf{x}(0) = \mathbf{A}\mathbf{X}(s) + \mathbf{b}U(s)$$

where $\mathbf{X}(s) = \mathcal{L}[\mathbf{x}(t)]$ and $U(s) = \mathcal{L}[u(t)]$. This is an algebraic equation and can be arranged as

$$(s\mathbf{I} - \mathbf{A})\mathbf{X}(s) = \mathbf{x}(0) + \mathbf{b}U(s) \tag{2.62}$$

where $\mathbf{I}$ is a unit matrix of the same order as $\mathbf{A}$. Note that without introducing $\mathbf{I}$, $(s - \mathbf{A})$ is not defined, for s is a scalar and $\mathbf{A}$ is an $n \times n$ matrix. The premultiplication of $(s\mathbf{I} - \mathbf{A})^{-1}$ to (2.62) yields

$$\mathbf{X}(s) = \underbrace{(s\mathbf{I} - \mathbf{A})^{-1}\mathbf{x}(0)}_{\text{Zero-Input Response}} + \underbrace{(s\mathbf{I} - \mathbf{A})^{-1}\mathbf{b}U(s)}_{\text{Zero-State Response}} \qquad (2.63)$$

This response consists of the zero-input response (the response due to nonzero $\mathbf{x}(0)$) and the zero-state response (the response due to nonzero $u(t)$). If we substitute $\mathbf{X}(s)$ into the Laplace transform of (2.61b), then we will obtain

$$Y(s) = \underbrace{\mathbf{c}(s\mathbf{I} - \mathbf{A})^{-1}\mathbf{x}(0)}_{\text{Zero-Input Response}} + \underbrace{[\mathbf{c}(s\mathbf{I} - \mathbf{A})^{-1}\mathbf{b} + d]U(s)}_{\text{Zero-State Response}} \qquad (2.64)$$

This is the output in the Laplace transform domain. Its inverse Laplace transform yields the time response. Thus, the response of state-variable equations can be easily obtained using the Laplace transform.

Example 2.7.1

Consider the state-variable equation

$$\dot{\mathbf{x}} = \begin{bmatrix} -6 & -3.5 \\ 6 & 4 \end{bmatrix} \mathbf{x} + \begin{bmatrix} -1 \\ 1 \end{bmatrix} u \qquad (2.65a)$$

$$y = [4 \quad 5]x \qquad (2.65b)$$

Find the output due to a unit-step input and the initial state $\mathbf{x}(0) = [-2 \quad 1]'$, where the prime denotes the transpose of a matrix or a vector. First we compute

$$s\mathbf{I} - \mathbf{A} = \begin{bmatrix} s & 0 \\ 0 & s \end{bmatrix} - \begin{bmatrix} -6 & -3.5 \\ 6 & 4 \end{bmatrix} = \begin{bmatrix} s + 6 & 3.5 \\ -6 & s - 4 \end{bmatrix}$$

Its inverse is

$$(s\mathbf{I} - \mathbf{A})^{-1} = \begin{bmatrix} s + 6 & 3.5 \\ -6 & s - 4 \end{bmatrix}^{-1}$$

$$= \frac{1}{(s + 6)(s - 4) + 6 \times 3.5} \begin{bmatrix} s - 4 & -3.5 \\ 6 & s + 6 \end{bmatrix}$$

Thus we have

$$\mathbf{c}(s\mathbf{I} - \mathbf{A})^{-1}\mathbf{x}(0) = [4 \quad 5] \frac{1}{s^2 + 2s - 3} \begin{bmatrix} s - 4 & -3.5 \\ 6 & s + 6 \end{bmatrix} \begin{bmatrix} -2 \\ 1 \end{bmatrix}$$

$$= \frac{1}{s^2 + 2s - 3} [4s + 14 \quad 5s + 16] \begin{bmatrix} -2 \\ 1 \end{bmatrix}$$

$$= \frac{-3s - 12}{s^2 + 2s - 3}$$

and

$$G(s) := \mathbf{c}(s\mathbf{I} - \mathbf{A})^{-1}\mathbf{b}$$

$$= [4 \quad 5] \frac{1}{s^2 + 2s - 3} \begin{bmatrix} s - 4 & -3.5 \\ 6 & s + 6 \end{bmatrix} \begin{bmatrix} -1 \\ 1 \end{bmatrix} \tag{2.66}$$

$$= \frac{s + 2}{s^2 + 2s - 3}$$

Thus, using (2.64) and $U(s) = 1/s$, the output is given by

$$Y(s) = \frac{-3s - 12}{s^2 + 2s - 3} + \frac{s + 2}{s^2 + 2s - 3} \cdot \frac{1}{s} = \frac{-3s^2 - 11s + 2}{(s - 1)(s + 3)s}$$

which can be expanded by partial fraction expansion as

$$Y(s) = \frac{-3}{s - 1} + \frac{2/3}{s + 3} - \frac{2/3}{s}$$

Thus the output is

$$y(t) = -3e^t + \frac{2}{3} e^{-3t} - \frac{2}{3}$$

for $t \geq 0$.

This example shows that the response of state-variable equations can indeed be readily obtained by using the Laplace transform.

2.7.1 Time-Domain Solutions

The solution of state-variable equations can also be obtained directly in the time domain without using the Laplace transform. Following

$$e^{at} = 1 + ta + \frac{t^2}{2!} a^2 + \cdots + \frac{t^n}{n!} a^n + \cdots$$

we define

$$e^{\mathbf{A}t} = \mathbf{I} + t\mathbf{A} + \frac{t^2}{2!} \mathbf{A}^2 + \cdots + \frac{t^n}{n!} \mathbf{A}^n + \cdots \tag{2.67}$$

where $n! = n \cdot (n - 1) \cdots 2 \cdot 1$, $\mathbf{A}^2 = \mathbf{AA}$, $\mathbf{A}^3 = \mathbf{AAA}$, and so forth. If $\mathbf{A}$ is an $n \times n$ matrix, then $e^{\mathbf{A}t}$ is also an $n \times n$ matrix. If $t = 0$, then (2.67) reduces to

$$e^0 = \mathbf{I} \tag{2.68}$$

As in $e^{a(t-\tau)} = e^{at}e^{-a\tau}$, we have

$$e^{\mathbf{A}(t-\tau)} = e^{\mathbf{A}t}e^{-\mathbf{A}\tau} \tag{2.69}$$

which becomes, if $\tau = t$,

$$e^{\mathbf{A}t}e^{-\mathbf{A}t} = e^{\mathbf{A}\cdot 0} = e^0 = \mathbf{I} \qquad (2.70)$$

The differentiation of (2.67) yields

$$\frac{d}{dt} e^{\mathbf{A}t} = \mathbf{0} + \mathbf{A} + \frac{2t}{2!} \mathbf{A}^2 + \frac{3t^2}{3!} \mathbf{A}^3 + \cdots$$

$$= \mathbf{A}(\mathbf{I} + t\mathbf{A} + \frac{t^2}{2!} \mathbf{A}^2 + \cdots) = (\mathbf{I} + t\mathbf{A} + \frac{t^2}{2!} \mathbf{A}^2 + \cdots)\mathbf{A}$$

which implies

$$\frac{d}{dt} e^{\mathbf{A}t} = \mathbf{A}e^{\mathbf{A}t} = e^{\mathbf{A}t}\mathbf{A} \qquad (2.71)$$

Now using (2.67) through (2.71) we can show that the solutions of (2.61) due to $u(t)$ and $\mathbf{x}(0)$ are given by

$$\mathbf{x}(t) = \underbrace{e^{\mathbf{A}t}\mathbf{x}(0)}_{\substack{\text{Zero-Input} \\ \text{Response}}} + \underbrace{e^{\mathbf{A}t} \int_0^t e^{-\mathbf{A}\tau}\mathbf{b}u(\tau)d\tau}_{\text{Zero-State Response}} \qquad (2.72)$$

and

$$y(t) = \overbrace{ce^{\mathbf{A}t}\mathbf{x}(0)} + \overbrace{ce^{\mathbf{A}t} \int_0^t e^{-\mathbf{A}\tau}\mathbf{b}u(\tau)d\tau + du(t)} \qquad (2.73)$$

To show that (2.72) is the solution of (2.61a), we must show that it meets the initial condition and the state equation. Indeed, at $t = 0$, (2.72) reduces to $\mathbf{x}(0) = \mathbf{I}\mathbf{x}(0) + \mathbf{I} \cdot 0 = \mathbf{x}(0)$. Thus it meets the initial condition. The differentiation of (2.72) yields

$$\dot{\mathbf{x}}(t) = \frac{d}{dt} \left[e^{\mathbf{A}t}\mathbf{x}(0) + e^{\mathbf{A}t} \int_0^t e^{-\mathbf{A}\tau}\mathbf{b}u(\tau)d\tau \right]$$

$$= \mathbf{A}e^{\mathbf{A}t}\mathbf{x}(0) + \frac{d}{dt} (e^{\mathbf{A}t}) \int_0^t e^{-\mathbf{A}\tau}\mathbf{b}u(\tau)d\tau) + e^{\mathbf{A}t} \frac{d}{dt} \left(\int_0^t e^{-\mathbf{A}\tau}\mathbf{b}u(\tau)d\tau \right)$$

$$= \mathbf{A} \left[e^{\mathbf{A}t}\mathbf{x}(0) + e^{\mathbf{A}t} \int_0^t e^{-\mathbf{A}\tau}\mathbf{b}u(\tau)d\tau \right] + e^{\mathbf{A}t} \left[e^{-\mathbf{A}\tau}\mathbf{b}u(\tau) \right]_{\tau=t}$$

$$= \mathbf{A}\mathbf{x}(t) + e^{\mathbf{A}t}e^{-\mathbf{A}t}\mathbf{b}u(t) = \mathbf{A}\mathbf{x}(t) + \mathbf{b}u(t)$$

This shows that (2.72) is the solution of (2.61a). The substitution of (2.72) into (2.61b) yields immediately (2.73). If $u = 0$, (2.61a) reduces to

$$\dot{\mathbf{x}}(t) = \mathbf{A}\mathbf{x}(t)$$

This is called the *homogeneous* equation. Its solution due to the initial state $\mathbf{x}(0)$ is, from (2.72),

$$\mathbf{x}(t) = e^{\mathbf{A}t}\mathbf{x}(0)$$

If $u(t) = 0$ or, equivalently, $U(s) = 0$, then (2.63) reduces to

$$\mathbf{X}(s) = (s\mathbf{I} - \mathbf{A})^{-1}\mathbf{x}(0)$$

A comparison of the preceding two equations yields

$$\mathscr{L}(e^{\mathbf{A}t}) = (s\mathbf{I} - \mathbf{A})^{-1} \qquad \text{or} \qquad e^{\mathbf{A}t} = \mathscr{L}^{-1}[(s\mathbf{I} - \mathbf{A})^{-1}] \qquad \text{(2.74)}$$

Thus $e^{\mathbf{A}t}$ equals the inverse Laplace transform of $(s\mathbf{I} - \mathbf{A})^{-1}$.

To use (2.72), we must first compute $e^{\mathbf{A}t}$. The computation of $e^{\mathbf{A}t}$ by using the infinite power series in (2.67) is not simple by hand. An alternative method is to use the Laplace transform in (2.74). Computer computation of (2.67) and (2.72) will be discussed in Section 2.9.

2.7.2 TRANSFER FUNCTION AND CHARACTERISTIC POLYNOMIAL

In this section, we discuss the relationships between state-variable equations and transfer functions. The transfer function of a system is, by definition, the ratio of the Laplace transforms of the output and input under the assumption that all initial conditions are zero. The Laplace transform of the state-variable equation in (2.61) was computed in (2.64). If $\mathbf{x}(0) = 0$, then (2.64) reduces to

$$Y(s) = [\mathbf{c}(s\mathbf{I} - \mathbf{A})^{-1}\mathbf{b} + d]U(s)$$

Thus the transfer function of the state-variable equation in (2.61) is

$$G(s) = \frac{Y(s)}{U(s)} = \mathbf{c}(s\mathbf{I} - \mathbf{A})^{-1}\mathbf{b} + d \qquad \text{(2.75)}$$

This formula was used in (2.66) to compute the transfer function of (2.65). Now we discuss its general property. Let det stand for the determinant and adj for the adjoint of a matrix. See Appendix B. Then (2.75) can be written as

$$G(s) = \mathbf{c}\,\frac{1}{\det(s\mathbf{I} - \mathbf{A})}\,[\text{adj}\,(s\mathbf{I} - \mathbf{A})]\mathbf{b} + d$$

We call det $(s\mathbf{I} - \mathbf{A})$ the *characteristic polynomial* of $\mathbf{A}$. For example, the characteristic polynomial of the $\mathbf{A}$ in (2.65) is

$$\Delta(s) := \det\,(s\mathbf{I} - \mathbf{A}) = \det \begin{bmatrix} s + 6 & 3.5 \\ -6 & s - 4 \end{bmatrix}$$
$$= (s + 6)(s - 4) - (-3.5 \times 6) = s^2 + 2s - 3 \qquad \text{(2.76)}$$

We see that if $\mathbf{A}$ is $n \times n$, then its characteristic polynomial is of degree n and has n roots. These n roots are called the *eigenvalues* of $\mathbf{A}$. The characteristic polynomial and eigenvalues of $\mathbf{A}$ are in fact the same as the characteristic polynomial and modes discussed in Section 2.3.1. They govern the zero-input response of the system. For example, if the characteristic polynomial of **A** is

$$\Delta(s) = s^3(s + 2)(s - 3)$$

then the eigenvalues are 0, 0, 0, -2, and 3, and every entry of $e^{\mathbf{A}t}$ is a linear combination of the time functions 1, t, t^2, e^{-2t}, and e^{3t}. Therefore any zero-input response will be a linear combination of these time functions.

The characteristic polynomial of square matrices often requires tedious computation. However, the characteristic polynomials of the following two matrices

$$\begin{bmatrix} -a_1 & -a_2 & -a_3 & -a_4 \\ 1 & 0 & 0 & 0 \\ 0 & 1 & 0 & 0 \\ 0 & 0 & 1 & 0 \end{bmatrix} \quad \begin{bmatrix} 0 & 1 & 0 & 0 \\ 0 & 0 & 1 & 0 \\ 0 & 0 & 0 & 1 \\ -a_4 & -a_3 & -a_2 & -a_1 \end{bmatrix} \tag{2.77}$$

and their transposes are all

$$\Delta(s) = s^4 + a_1 s^3 + a_2 s^2 + a_3 s + a_4 \tag{2.78}$$

This characteristic polynomial can be read out directly from the entries of the matrices in (2.77). Thus, these matrices are called *companion forms* of the polynomial in (2.78).

Now we use an example to discuss the relationship between the eigenvalues of a state-variable equation and the poles of its transfer function.

Example 2.8.1

Consider the state-variable equation in (2.65), repeated in the following

$$\dot{\mathbf{x}} = \begin{bmatrix} -6 & -3.5 \\ 6 & 4 \end{bmatrix} \mathbf{x} + \begin{bmatrix} -1 \\ 1 \end{bmatrix} u \tag{2.79a}$$

$$y = [4 \quad 5]\mathbf{x} \tag{2.79b}$$

Its characteristic polynomial was computed in (2.76) as

$$\Delta(s) = s^2 + 2s - 3 = (s + 3)(s - 1)$$

Thus its eigenvalues are 1 and -3. The transfer function of (2.79) was computed in (2.66) as

$$G(s) = \frac{s + 2}{s^2 + 2s - 3} = \frac{s + 2}{(s + 3)(s - 1)}$$

Its poles are 1 and -3. The number of the eigenvalues of (2.79) equals the number of the poles of its transfer function.

Example 2.8.2

Consider the state-variable equation

$$\dot{\mathbf{x}} = \begin{bmatrix} -6 & -3.5 \\ 6 & 4 \end{bmatrix} \mathbf{x} + \begin{bmatrix} -1 \\ 1 \end{bmatrix} u \qquad (2.80a)$$

$$y = [-2 \quad -1]\mathbf{x} \qquad (2.80b)$$

Equation (2.80a) is identical to (2.79a); (2.80b), however, is different from (2.79b). The characteristic polynomial of (2.80) is

$$\Delta(s) = s^2 + 2s - 3$$

Its eigenvalues are -3 and 1. The transfer function of (2.80) is, replacing $[4 \quad 5]$ in (2.66) by $[-2 \quad -1]$,

$$G(s) = [-2 \quad -1] \frac{1}{s^2 + 2s - 3} \begin{bmatrix} s - 4 & -3.5 \\ 6 & s + 6 \end{bmatrix} \begin{bmatrix} -1 \\ 1 \end{bmatrix}$$

$$= \frac{1}{s^2 + 2s - 3} [-2 \quad -1] \begin{bmatrix} -s + 0.5 \\ s \end{bmatrix} = \frac{2s - 1 - s}{s^2 + 2s - 3} \qquad (2.81)$$

$$= \frac{s - 1}{(s - 1)(s + 3)}$$

Thus the transfer function of (2.80) is $G(s) = 1/(s + 3)$. It has only one pole at -3, one less than the number of the eigenvalues of (2.80).

A state-variable equation is called *minimal* if the number of its eigenvalues or the dimension of the equation equals the number of the poles of its transfer function. The state-variable equation in (2.79) is such an equation. Every eigenvalue of a minimal state-variable equation will appear as a pole of its transfer function and there is no essential difference between using the state-variable equation or its transfer function to study the system. Thus, the state-variable equation is said to be *completely characterized* by its transfer function. As will be discussed in Chapter 11, every minimal state-variable equation has the properties of controllability and observability. If a state-variable equation is not minimal, such as the one in (2.80), then some eigenvalues will not appear as poles of its transfer function and the transfer function is said to have *missing poles*. In this case, the state-variable equation is not completely characterized by its transfer function and we cannot use the transfer function to carry out analysis and design. This situation is similar to the one in Section 2.5.

2.8   DISCRETIZATION OF STATE EQUATIONS

Consider the state-variable equation

$$\dot{\mathbf{x}}(t) = \mathbf{A}\mathbf{x}(t) + \mathbf{b}u(t) \tag{2.82a}$$

$$y(t) = \mathbf{c}\mathbf{x}(t) + du(t) \tag{2.82b}$$

This equation is defined at every instant of time and is called a *continuous-time equation*. If a digital computer is used to generate the input u, then $u(t)$ is (as will be discussed in Chapter 12) stepwise, as shown in Figure 2.23(a). Let

$$u(t) = u(k) \qquad \text{for } kT \le t < (k + 1)T, k = 1, 2, 3, \ldots \tag{2.83}$$

where T is called the *sampling period*. This type of signal is completely specified by the sequence of numbers $u(k)$, $k = 0, 1, 2, \ldots$, as shown in Figure 2.23(b). The signal in Figure 2.23(a) is called a *continuous-time or analog signal* because it is defined for all time; the signal in Figure 2.23(b) is called a *discrete-time signal* because it is defined only at discrete instants of time. Note that a continuous-time signal may not be a continuous function of time as shown in the figure.

If we apply a stepwise input to (2.82), the output $y(t)$ is generally not stepwise. However, if we are interested in only the output $y(t)$ at $t = kT$, $k = 0, 1, 2, \ldots$, or $y(k) := y(kT)$, then it is possible to develop an equation simpler than (2.82) to describe $y(k)$. We develop such an equation in the following.

The solution of (2.82) was developed in (2.72) as

$$\mathbf{x}(t) = e^{\mathbf{A}t}\mathbf{x}(0) + e^{\mathbf{A}t}\int_0^t e^{-\mathbf{A}\tau}\mathbf{b}u(\tau)d\tau \tag{2.84}$$

This equation holds for any t—in particular, for t at kT. Let $\mathbf{x}(k) := \mathbf{x}(kT)$ and $\mathbf{x}(k + 1) := \mathbf{x}((k + 1)T)$. Then we have

$$\mathbf{x}(k) = e^{\mathbf{A}kT}\mathbf{x}(0) + e^{\mathbf{A}kT}\int_0^{kT} e^{-\mathbf{A}\tau}\mathbf{b}u(\tau)d\tau \tag{2.85}$$

$$\mathbf{x}(k + 1) = e^{\mathbf{A}(k+1)T}\mathbf{x}(0) + e^{\mathbf{A}(k+1)T}\int_0^{(k+1)T} e^{-\mathbf{A}\tau}\mathbf{b}u(\tau)d\tau \tag{2.86}$$

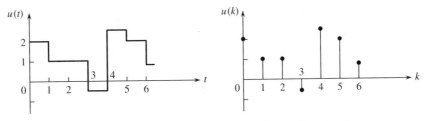

Figure 2.23 (a) Stepwise continuous-time signal. (b) Discrete-time signal.

We rewrite (2.86) as

$$\mathbf{x}(k+1) = e^{\mathbf{A}T}\left[e^{\mathbf{A}kT}\mathbf{x}(0) + e^{\mathbf{A}kT}\int_0^{kT} e^{-\mathbf{A}\tau}\mathbf{b}u(\tau)d\tau\right]$$

$$+ \int_{kT}^{(k+1)T} e^{\mathbf{A}((k+1)T-\tau)}\mathbf{b}u(\tau)d\tau$$

(2.87)

The term inside the brackets equals $\mathbf{x}(k)$. If the input $u(t)$ is stepwise as in (2.83), then $u(\tau)$ in the integrand of the last term of (2.87) equals $u(k)$ and can be moved outside the integration. Define $\alpha = (k+1)T - \tau$. Then $d\alpha = -d\tau$ and (2.87) becomes

$$\mathbf{x}(k+1) = e^{\mathbf{A}T}\mathbf{x}(k) - \left(\int_T^0 e^{\mathbf{A}\alpha}d\alpha\right)\mathbf{b}u(k)$$

(2.88)

$$= e^{\mathbf{A}T}\mathbf{x}(k) + \left(\int_0^T e^{\mathbf{A}\alpha}d\alpha\right)\mathbf{b}u(k)$$

This and (2.82b) can be written as

$$\mathbf{x}(k+1) = \tilde{\mathbf{A}}\mathbf{x}k + \tilde{\mathbf{b}}u(k)$$

(2.89a)

$$y(k) = \tilde{\mathbf{c}}\mathbf{x}(k) + \tilde{d}u(k)$$

(2.89b)

with

$$\tilde{\mathbf{A}} = e^{\mathbf{A}T} \quad \tilde{\mathbf{b}} = \left(\int_0^T e^{\mathbf{A}\tau}d\tau\right)\mathbf{b} \quad \tilde{\mathbf{c}} = \mathbf{c} \quad \tilde{d} = d$$

(2.89c)

This is called a *discrete-time state-variable equation*. It consists of a set of first-order *difference* equations. We use an example to illustrate how to obtain such an equation.

Example 2.9.1

Consider the continuous-time state-variable equation

$$\dot{\mathbf{x}} = \begin{bmatrix} -1 & 1 \\ 0 & -1 \end{bmatrix}\mathbf{x} + \begin{bmatrix} 0 \\ 1 \end{bmatrix}u$$

(2.90a)

$$y = [2 \quad 1]\mathbf{x}$$

(2.90b)

Discretize the equation with sampling period $T = 0.1$.

First we compute

$$\mathcal{L}[e^{\mathbf{A}t}] = (s\mathbf{I} - \mathbf{A})^{-1} = \begin{bmatrix} s + 1 & -1 \\ 0 & s + 1 \end{bmatrix}^{-1}$$

$$= \frac{1}{(s + 1)^2} \begin{bmatrix} s + 1 & 1 \\ 0 & s + 1 \end{bmatrix} = \begin{bmatrix} \dfrac{1}{s + 1} & \dfrac{1}{(s + 1)^2} \\ 0 & \dfrac{1}{s + 1} \end{bmatrix}$$

Its inverse Laplace transform is, using Table A.1,

$$e^{\mathbf{A}t} = \begin{bmatrix} e^{-t} & te^{-t} \\ 0 & e^{-t} \end{bmatrix}$$

Using an integration table, we can compute

$$\int_0^T e^{\mathbf{A}\tau}d\tau = \begin{bmatrix} 1 - e^{-T} & 1 - (1 + T)e^{-T} \\ 0 & 1 - e^{-T} \end{bmatrix}$$

If $T = 0.1$, then

$$e^{0.1\mathbf{A}} = \begin{bmatrix} 0.9048 & 0.0905 \\ 0 & 0.9048 \end{bmatrix} \quad \text{and} \quad \left(\int_0^{0.1} e^{\mathbf{A}\tau}d\tau\right)\mathbf{b} = \begin{bmatrix} 0.0047 \\ 0.0952 \end{bmatrix} \quad (2.91)$$

Thus the discrete-time state-variable equation is

$$\mathbf{x}(k + 1) = \begin{bmatrix} 0.9048 & 0.0905 \\ 0 & 0.9048 \end{bmatrix} \mathbf{x}(k) + \begin{bmatrix} 0.0047 \\ 0.0952 \end{bmatrix} u(k) \quad (2.92a)$$

$$y(k) = [2 \quad 1]\mathbf{x}(k) \quad (2.92b)$$

This completes the discretization. This discretized equation is used on digital computers, as will be discussed in Chapter 5, to compute the response of the continuous-time equation in (2.90).

We have used the Laplace transform to compute $e^{\mathbf{A}T}$ in (2.91). If T is small, the infinite series in (2.67) may converge rapidly. For example, for $e^{\mathbf{A}T}$ in (2.91) with $T = 0.1$, we have

$$\mathbf{I} + T\mathbf{A} + \frac{1}{2}(T\mathbf{A})(T\mathbf{A}) = \begin{bmatrix} 0.9050 & 0.0900 \\ 0 & 0.9050 \end{bmatrix}$$

and

$$\mathbf{I} + T\mathbf{A} + \frac{1}{2}(T\mathbf{A})(T\mathbf{A}) + \frac{1}{6}(T\mathbf{A})(T\mathbf{A})(T\mathbf{A}) = \begin{bmatrix} 0.9048 & 0.0905 \\ 0 & 0.9048 \end{bmatrix}$$

The infinite series converges to the actual values in four terms. Therefore for small T, the infinite series in (2.67) is a convenient way of computing e^{AT}. In addition to the Laplace transform and infinite series, there are many ways (at least 17) of computing e^{AT}. See Reference [47]. In Chapter 5, we will introduce computer software to solve and to discretize state-variable equations.

PROBLEMS

2.1. Consider the pendulum system shown in Figure P2.1, where the applying force u is the input and the angular displacement θ is the output. It is assumed that there is no friction in the hinge and no air resistance of the mass. Is the system linear? Is it time-invariant? Find a differential equation to describe it. Is the differential equation LTIL?

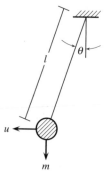

m **Figure P2.1**

2.2. Consider the pendulum system in Figure P2.1. It is assumed that $\sin \theta$ and $\cos \theta$ can be approximated by θ and 1, respectively, if $|\theta| < \pi/4$ radians. Find an LTIL differential equation to describe the system.

2.3. **a.** Consider the system shown in Figure P2.3(a). The two blocks with mass m_1 and m_2 are connected by a rigid shaft. It is assumed that there is no friction between the floor and wheels. Find a differential equation to describe the system.

b. If the shaft is long and flexible, then it must be modeled by a spring with spring constant k_2 and a dashpot. A dashpot is a device that consists of a

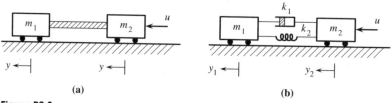

(a)   (b)

Figure P2.3

piston and oil-filled cylinder and is used to provide viscous friction. It is assumed that the force generated by the dashpot equals $k_1(dy_2/dt - dy_1/dt)$. Verify that the system is described by

$$m_2 \frac{d^2y_2(t)}{dt^2} = u(t) - k_1 \left(\frac{dy_2(t)}{dt} - \frac{dy_1(t)}{dt} \right) - k_2(y_2(t) - y_1(t))$$

and

$$m_1 \frac{d^2y_1(t)}{dt^2} = -k_1 \left(\frac{dy_1(t)}{dt} - \frac{dy_2(t)}{dt} \right) - k_2(y_1(t) - y_2(t))$$

2.4. If a robot arm is long, such as the one on a space shuttle, then its employment must be modeled as a flexible system, as shown in Figure P2.4. Define

    T: Applied torque

    k_1: Viscous-friction coefficient

    k_2: Torsional spring constant

    J_i: Moment of inertia

    θ_i: Angular displacement

Find an LTIL differential equation to describe the system.

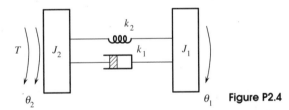

Figure P2.4

2.5. Find LTIL differential equations to describe the networks in Figure P2.5.

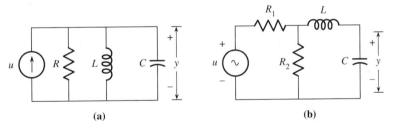

(a)   (b)

Figure P2.5

2.6. Consider the network shown in Figure P2.6(a), in which T is a tunnel diode with characteristics as shown. Show that if v lies between a and b, then the circuit can be modeled as shown in Figure P2.6(b). Show also that if v lies between c and d, then the circuit can be modeled as shown in Figure P2.6(c),

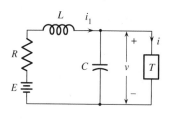
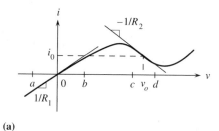

(a)

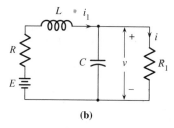
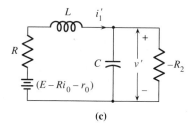

(b)   **(c)**

Figure P2.6

where $i_1' = i_1 - i_o$, $v' = v - v_o$. These linearized models are often called *linear incremental models*, or *small-signal models*.

2.7. Consider the LTIL differential equation

$$\frac{d^2 y(t)}{dt^2} - 2 \frac{dy(t)}{dt} - 3y(t) = 2 \frac{du(t)}{dt} + u(t)$$

a. Find its zero-input response due to $y(0^-) = 1$ and $\dot{y}(0^-) = 2$.

b. What are its characteristic polynomial and modes?

c. Find the set of all initial conditions such that the mode 3 will not be excited.

d. Find the set of all initial conditions such that the mode -1 will not be excited.

e. Plot (c) and (d) on a two-dimensional plane with y and $\dot{y}$ as coordinates. If a set of initial conditions is picked randomly, what is the probability that the two modes will be excited?

2.8. Consider the differential equation in Problem 2.7

a. Find its unit-step response, that is, the zero-state response due to a unit-step input $u(t)$.

b. What are its transfer function, its poles and zeros?

c. Find a bounded input that will not excite pole 3.

d. Find a bounded input that will not excite pole -1.

2.9. Consider the differential equation in Problem 2.7. Find the response due to $y(0^-) = 1$, $\dot{y}(0^-) = 2$, and $u(t) = 2$, for $t \geq 0$.

2.10. Find the transfer function from u to θ of the system in Problem 2.2.

2.11. Find the transfer functions for the networks in Figure P2.5. Compute them from differential equations. Also compute them using the concept of Laplacian impedances.

2.12. Find the transfer functions from u to y, u to y_1, and u to y_2 of the systems in Figure P2.3.

2.13. Consider a system. If its zero-state response due to a unit-step input is measured as $y(t) = 1 - e^{-2t} + \sin t$, where is the transfer function of the system?

2.14. Consider a system. If its zero-state response due to $u(t) = \sin 2t$ is measured as $y(t) = e^{-t} - 2e^{-3t} + \sin 2t + \cos 2t$, what is the transfer function of the system?

2.15. Consider the LTIL system described by

$$\frac{d^2y(t)}{dt^2} + 2\frac{dy(t)}{dt} - 3y(t) = \frac{du(t)}{dt} - u(t)$$

a. Find the zero-input response due to $y(0^-) = 1$ and $\dot{y}(0^-) = 2$. What are the modes of the system?

b. Find the zero-state response due to $u(t) = 1$, for $t \geq 0$.

c. Can you detect all modes of the system from the zero-state response?

d. Is the system completely characterized by its transfer function? Will a serious problem arise if we use the transfer function to study the system?

2.16. Repeat Problem 2.15 for the differential equation

$$\frac{d^2y(t)}{dt^2} + 2\frac{dy(t)}{dt} - 3y(t) = \frac{du(t)}{dt} + 3u(t)$$

2.17. Consider a transfer function $G(s)$ with poles -1 and $-2 \pm j1$, and zero 3. Can you determine $G(s)$ uniquely? If it is also known that $G(2) = -0.1$, can you now determine $G(s)$ uniquely?

2.18. Find the unit-step response of the following systems and plot roughly the responses:

a. $G_1(s) = \dfrac{2}{(s+1)(s+2)}$

b. $G_2(s) = \dfrac{20(s+0.1)}{(s+1)(s+2)}$

c. $G_3(s) = \dfrac{0.2(s+10)}{(s+1)(s+2)}$

d. $G_4(s) = \dfrac{-20(s-0.1)}{(s+1)(s+2)}$

e. $G_5(s) = \dfrac{-0.2(s - 10)}{(s + 1)(s + 2)}$

Which type of zeros, closer to or farther away from the origin, has a larger effect on responses?

2.19. Find the poles and zeros of the following transfer functions:

a. $G(s) = \dfrac{2(s^2 - 9)(s + 1)}{(s + 3)^2(s + 2)(s - 1)^2}$

b. $G(s) = \dfrac{10(s^2 - s + 1)}{s^4 + 2s^3 + s + 2}$

c. $G(s) = \dfrac{2s^2 + 8s + 8}{(s + 1)(s^2 + 2s + 2)}$

2.20. a. Consider a system with transfer function $G(s) = (s - 2)/s(s + 1)$. Show that if $u(t) = e^t, t \geq 0$, then the response of the system can be decomposed as

$$\text{Total response} = \text{Response due to the poles of } G(s)$$
$$+ \text{ Response due to the poles of } U(s)$$

b. Does the decomposition hold if $u(t) = 1$, for $t \geq 0$?

c. Does the decomposition hold if $u(t) = e^{2t}$, for $t \geq 0$?

d. Show that the decomposition is valid if $U(s)$ and $G(s)$ have no common pole and if there are no pole-zero cancellations between $U(s)$ and $G(s)$.

2.21. Show that the network in Figure 2.18 is described by the differential equation in (2.42).

2.22. Consider the simplified model of an aircraft shown in Figure P2.22. It is assumed that the aircraft is dynamically equivalent at the pitched angle θ_0, elevator angle u_0, altitude h_0, and cruising speed v_0. It is assumed that small deviations of θ and u from θ_0 and u_0 generate forces $f_1 = k_1\theta$ and $f_2 = k_2u$, as shown in the figure. Let m be the mass of the aircraft; I, the moment of

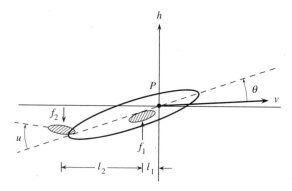

Figure P2.22

inertia about the center of gravity P; $b\dot{\theta}$, the aerodynamic damping and h, the deviation of the altitude from h_0. Show that the transfer function from u to h is, by neglecting the effect of I,

$$G(s) = \frac{k_1 k_2 l_2 - k_2 bs}{ms^2(bs + k_1 l_1)}$$

2.23. Consider a cart with a stick hinged on top of it, as shown in Figure P2.23. This could be a model of a space booster on takeoff. If the angular displacement θ of the stick is small, then the system can be described by

$$\ddot{\theta} = \theta + u$$

$$\ddot{y} = \beta\theta - u$$

where β is a constant, and u and y are expressed in appropriate units. Find the transfer functions from u to θ and from u to y. Is the system completely characterized by the transfer function from u to y? By the one from u to θ? Which transfer function can be used to study the system?

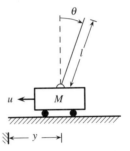

Figure P2.23

2.24. Show that the two tanks shown in Figure P2.24(a) can be represented by the block diagram shown in Figure P2.24(b). Is there any loading problem in the block diagram? The transfer function of the two tanks shown in Figure 2.12 is computed in Exercise 2.4.2. Is it possible to represent the two tanks in Figure 2.12 by two blocks as in Figure P2.24(b) on page 66? Give your reasons.

2.25. Find state-variable equations to describe the systems in Problems 2.1 and 2.2.

2.26. Find state-variable equations to describe the systems in Figure P2.3.

2.27. Find state-variable equations to describe the networks in Figure P2.5.

2.28. The soft landing phase of a lunar module descending on the moon can be modeled as shown in Figure P2.28. It is assumed that the thrust generated is proportional to $\dot{m}$, where m is the mass of the module. The system can be described by $m\ddot{y} = -k\dot{m} - mg$, where g is the gravity constant on the lunar surface. Define the state variables of the system as $x_1 = y$, $x_2 = \dot{y}$, $x_3 = m$, and $u = \dot{m}$. Find a state-variable equation to describe the system. Is it a time-invariant equation?

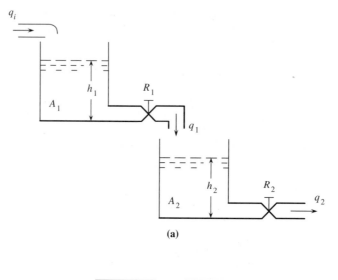

(a)

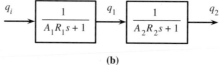

(b)

Figure P2.24

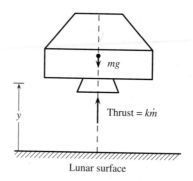

Lunar surface   **Figure P2.28**

2.29. Use the Laplace transform to find the response of

$$\dot{\mathbf{x}} = \begin{bmatrix} 0 & 1 \\ 3 & -2 \end{bmatrix} \mathbf{x} + \begin{bmatrix} -1 \\ 1 \end{bmatrix} u$$

$$y = [4 \quad 5]\mathbf{x}$$

due to $\mathbf{x}(0) = [1 \quad 2]'$ and $u(t) = 1$ for $t \geq 0$.

2.30. Use the infinite series in (2.67) to compute e^{At} with

$$A = \begin{bmatrix} -1 & 0 \\ 0 & -2 \end{bmatrix}$$

Verify your results by using (2.74).

2.31. Compute the transfer functions of the following state-variable equations:

a. $\dot{x} = -x + u \qquad y = 2x$

b. $\dot{x} = \begin{bmatrix} 0 & 0 \\ 1 & -1 \end{bmatrix} x + \begin{bmatrix} 1 \\ 0 \end{bmatrix} u$

$y = [2 \quad -2]x$

c. $\dot{x} = \begin{bmatrix} -3 & 1 \\ -2 & 0 \end{bmatrix} x + \begin{bmatrix} 2 \\ 4 \end{bmatrix} u$

$y = [1 \quad 0]x$

d. $\dot{x} = \begin{bmatrix} -3 & 1 & 0 \\ -2 & 0 & 0 \\ 1 & 0 & 0 \end{bmatrix} x + \begin{bmatrix} 2 \\ 4 \\ 1 \end{bmatrix} u$

$y = [1 \quad 0 \quad 0]x$

Do they have the same transfer functions? Which are minimal equations?

2.32. Find the characteristic polynomials for the matrices

$$\begin{bmatrix} -a_1 & 1 & 0 \\ -a_2 & 0 & 1 \\ -a_3 & 0 & 0 \end{bmatrix} \qquad \begin{bmatrix} 0 & 0 & -a_3 \\ 1 & 0 & -a_2 \\ 0 & 1 & -a_1 \end{bmatrix}$$

2.33. Find the outputs of the equations in Problem 2.31 due to a unit-step input and zero initial state.

2.34. Compute the transfer functions of the state-variable equations in Problems 2.25(b), 2.26(a), and 2.27(a). Are they minimal state-variable equations?

2.35. Compute $y(k)$ for $k = 0, 1, 2, 3, 4$, for the discrete-time state-variable equation

$$x(k + 1) = \begin{bmatrix} -3 & 1 \\ -2 & 0 \end{bmatrix} x(k) + \begin{bmatrix} 0 \\ 4 \end{bmatrix} u(k)$$

$$y(k) = [1 \quad 0]x(k)$$

due to $x(0) = [1 \quad -2]'$ and $u(k) = 1$ for all $k \geq 0$.

2.36. Discretize the continuous-time equation with sampling period 0.1 second

$$\dot{\mathbf{x}} = \begin{bmatrix} -3 & 1 \\ -2 & 0 \end{bmatrix} \mathbf{x} + \begin{bmatrix} 0 \\ 4 \end{bmatrix} u$$

$$y = \begin{bmatrix} 1 & 0 \end{bmatrix} \mathbf{x}$$

2.37. a. The characteristic polynomial of

$$\mathbf{A} = \begin{bmatrix} -6 & -3.5 \\ 6 & 4 \end{bmatrix}$$

is computed in (2.76) as

$$\Delta(s) = s^2 + 2s - 3$$

Verify

$$\Delta(\mathbf{A}) = \mathbf{A}^2 + 2\mathbf{A} - 3\mathbf{I} = \mathbf{0}$$

This is called the *Cayley-Hamilton Theorem*. It states that every square matrix meets its own characteristic polynomial.

b. Show that $\mathbf{A}^2$, $\mathbf{A}^3$, ..., can be expressed as a linear combination of $\mathbf{I}$ and $\mathbf{A}$. In general, for a square matrix $\mathbf{A}$ of order n, $\mathbf{A}^k$ for $k \geq n$ can be expressed as a linear combination of $\mathbf{I}, \mathbf{A}, \mathbf{A}^2, \ldots, \mathbf{A}^{n-1}$.

2.38. Show that if (2.67) is used, $\tilde{\mathbf{b}}$ in (2.89c) can be expressed as

$$\tilde{\mathbf{b}} = \left(\int_0^T e^{\mathbf{A}\tau} d\tau \right) \mathbf{b} = T \left(\mathbf{I} + \frac{T}{2!} \mathbf{A} + \frac{T^2}{3!} \mathbf{A}^2 + \frac{T^3}{4!} \mathbf{A}^3 + \cdots \right) \mathbf{b}$$

This can be used to compute $\tilde{\mathbf{b}}$ on a digital computer if it converges rapidly or if T is small.

3 *Development of Block Diagrams for Control Systems*

3.1 INTRODUCTION

The design of control systems can be divided into two distinct parts. One is concerned with the design of individual components, the other with the design of overall systems by utilizing existing components. The former belongs to the domain of instrumentation engineers; the latter, the domain of control engineers. This is a control text, so we are mainly concerned with utilization of existing components. Consequently, our discussion of control components stresses their functions rather than their structures.

Control components can be mechanical, electrical, hydraulic, pneumatic, or optical devices. Depending on whether signals are modulated or not, electrical devices again are divided into ac (alternating current) or dc (direct current) devices. Even a cursory introduction of these devices can easily take up a whole text, so this will not be attempted. Instead, we select a number of commonly used control components, discuss their functions, and develop their transfer functions. The loading problem will be considered in this development. We then show how these components are connected to form control systems. Block diagrams of these control systems are developed. Finally, we discuss manipulation of block diagrams. Mason's formula is introduced to compute overall transfer functions of block diagrams.

3.2   MOTORS

Motors are indispensable in many control systems. They are used to turn antennas, telescopes, and ship rudders; to close and open valves; to drive tapes in recorders, and rollers in steel mills; and to feed paper in printers. There are many types of motors: dc, ac, stepper, and hydraulic. The magnetic field of a dc motor can be excited by a circuit connected in series with the armature circuit; it can also be excited by a field circuit that is independent of the armature circuit or by a permanent magnet. We discuss in this text only separately excited dc motors. DC motors used in control systems, also called *servomotors*, are characterized by large torque to rotor-inertia ratios, small sizes, and better linear characteristics. Certainly, they are more expensive than ordinary dc motors.

3.2.1   Field-Controlled DC Motor

Most of the dc motors used in control systems can be modeled as shown in Figure 3.1. There are two circuits, called the field circuit and armature circuit. Let i_f and i_a be the field current and armature current, respectively. Then the torque T generated by the motor is given by

$$T(t) = ki_a(t)i_f(t) \tag{3.1}$$

where k is a constant. The generated torque is used to drive a load through a shaft. The shaft is assumed to be rigid. To simplify analysis, we consider only the viscous friction between the shaft and bearing. Let J be the total moment of inertia of the load, the shaft, and the rotor of the motor; θ, the angular displacement of the load; and f, the viscous friction coefficient of the bearing. Then we have, as developed in (2.5),

$$T(t) = J\frac{d^2\theta(t)}{dt^2} + f\frac{d\theta(t)}{dt} \tag{3.2}$$

This describes the relationship between the motor torque and load's angular displacement.

If the armature current i_a is kept constant and the input voltage $u(t)$ is applied

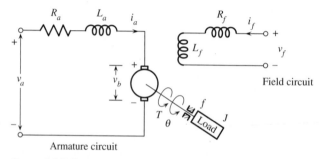

Figure 3.1 DC motor.

to the field circuit, the motor is called a *field-controlled* dc motor. We now develop its block diagram.

In the field-controlled dc motor, the armature current $i_a(t)$ is constant. Therefore, (3.1) can be reduced to

$$T(t) = ki_a(t)i_f(t) =: k_f i_f(t) \tag{3.3}$$

where $k_f = ki_a$. From the field circuit, we have

$$L_f \frac{di_f(t)}{dt} + R_f i_f(t) = v_f(t) = u(t) \tag{3.4}$$

The application of the Laplace transform to (3.3) and (3.4) yields,[1] assuming zero initial conditions,

$$T(s) = k_f I_f(s)$$

$$L_f s I_f(s) + R_f I_f(s) = U(s)$$

which imply

$$I_f(s) = \frac{1}{L_f s + R_f} U(s)$$

and

$$T(s) = \frac{k_f}{L_f s + R_f} U(s)$$

Thus, if the generated torque is considered as the output of the motor, then the transfer function of the field-controlled dc motor is

$$G_m(s) := \frac{T(s)}{U(s)} = \frac{k_f}{L_f s + R_f} \tag{3.5}$$

This transfer function remains the same no matter what load the motor drives. Now we compute the transfer function of the load. The application of the Laplace transform to (3.2) yields, assuming zero initial conditions,

$$T(s) = Js^2 \Theta(s) + fs \Theta(s)$$

which can be written as

$$G_l(s) := \frac{\Theta(s)}{T(s)} = \frac{1}{s(Js + f)}$$

This is the transfer function of the load if we consider the motor torque the input, and the load's angular displacement the output. Thus, the motor and load in

---

[1]Capital letters are used to denote the Laplace transforms of the corresponding lowercase letters. In the case of $T(t)$ and $T(s)$, we use the arguments to differentiate them.

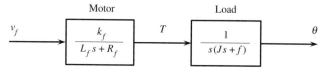

Figure 3.2 Block diagram of field-controlled dc motor.

Figure 3.1 can be modeled as shown in Figure 3.2. It consists of two blocks: One represents the motor; the other, the load. Note that any change of the load will not affect the motor transfer function, so there is no loading problem in the connection.

The transfer function of the field-controlled dc motor driving a load is the product of $G_m(s)$ and $G_l(s)$ or

$$G(s) = G_m(s)G_l(s) = \frac{k_f}{s(L_f s + R_f)(Js + f)} = \frac{k_f/R_f f}{s(\tau_f s + 1)(\tau_m s + 1)}$$

$$=: \frac{k_m}{(\tau_f s + 1)s(\tau_m s + 1)}$$

(3.6)

where $\tau_f := L_f/R_f$, $\tau_m := J/f$ and $k_m := k_f/R_f f$. The constant τ_f depends only on the electric circuit and is therefore called the motor *electrical time constant*. The constant τ_m depends only on the load and is called the motor *mechanical time constant*. The physical meaning of the time constant will be discussed in the next chapter. If the electrical time constant τ_f is much smaller than the mechanical time constant τ_m, as is often the case in practice, then $G(s)$ is often approximated by

$$G(s) = \frac{k_m}{s(\tau_m s + 1)}$$

(3.7)

For a discussion of this type of approximation, see Reference [18]. Note that τ_m depends only on the load.

The field-controlled dc motor is used to drive a load, yet there is no loading problem in Figure 3.2. How could this be possible? Different loads will induce different i_a; even for the same load, i_a will not be constant. Therefore, the loading problem is eliminated by keeping i_a constant. In practice, it is difficult to keep i_a constant, so the field-controlled dc motor is rarely used.

3.2.2   Armature-Controlled DC Motor

Consider the dc motor shown in Figure 3.1. If the field current $i_f(t)$ is kept constant or the field circuit is replaced by a permanent magnetic field, and if the input voltage is applied to the armature circuit, then the motor is called an *armature-controlled dc motor*. We now develop its transfer function. If $i_f(t)$ is constant, (3.1) can be written as

$$T(t) = k_t i_a(t)$$

(3.8)

where $k_t := k i_f(t)$ is a constant. When the motor is driving a load, a back electromotive force (back emf) voltage v_b will develop in the armature circuit to resist the applied voltage. The voltage $v_b(t)$ is linearly proportional to the angular velocity of the motor shaft:

$$v_b(t) = k_b \frac{d\theta(t)}{dt} \qquad (3.9)$$

Thus the armature circuit in Figure 3.1 is described by

$$R_a i_a(t) + L_a \frac{di_a(t)}{dt} + v_b(t) = v_a(t) = u(t) \qquad (3.10)$$

or

$$R_a i_a(t) + L_a \frac{di_a(t)}{dt} + k_b \frac{d\theta(t)}{dt} = u(t) \qquad (3.11)$$

Using (3.2), (3.8), and (3.11), we now develop a transfer function for the motor. The substitution of (3.8) into (3.2) and the application of the Laplace transform to (3.2) and (3.11) yield, assuming zero initial conditions,

$$k_t I_a(s) = Js^2\Theta(s) + fs\Theta(s) \qquad (3.12)$$

$$R_a I_a(s) + L_a s I_a(s) + k_b s\Theta(s) = U(s) \qquad (3.13)$$

The elimination of I_a from these two equations yields

$$G(s) = \frac{\Theta(s)}{U(s)} = \frac{k_t}{s[(Js + f)(R_a + L_a s) + k_t k_b]} \qquad (3.14)$$

This is the transfer function from $v_a = u$ to θ of the armature-controlled dc motor.

Using (3.8), (3.9), (3.10), and (3.12), we can draw a block diagram for the armature-controlled dc motor as shown in Figure 3.3. Although we can draw a block for the motor and a block for the load, the two blocks are not independent as in the case of field-controlled dc motors. A signal from inside the load is fed back to the input of the motor. This essentially takes care of the loading problem. Because of the loading, it is simpler to combine the motor and load into a single block with the transfer function given in (3.14).

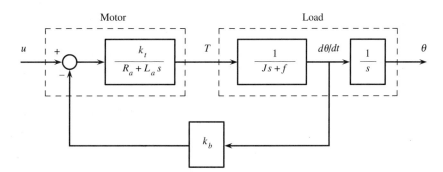

Figure 3.3 Block diagram of armature-controlled dc motor.

In application, the armature inductance L_a is often set to zero.[2] In this case, (3.14) reduces to

$$G(s) = \frac{\Theta(s)}{U(s)} = \frac{k_t}{s(JR_a s + k_t k_b + fR_a)} = \frac{k_m}{s(\tau_m s + 1)} \qquad (3.15)$$

where

$$k_m := \frac{k_t}{k_t k_b + fR_a} =: \text{motor gain constant} \qquad (3.16)$$

and

$$\tau_m := \frac{JR_a}{k_t k_b + fR_a} =: \text{motor time constant} \qquad (3.17)$$

The time constant in (3.17) depends on the load as well as the armature circuit. This is in contrast to (3.7) where τ_m depends only on the load. The electrical and mechanical time constants are well defined in (3.6) and (3.7), but this is not possible in (3.14) and (3.15). We often call (3.14) and (3.15) *motor transfer functions*, although they are actually the transfer functions of the motor *and* load. The transfer function of the same motor will be different if it drives a different load.

Exercise 3.2.1

Consider an armature-controlled dc motor with

$$R_a = 20 \ \Omega \qquad k_t = 1 \ \text{N·m/A} \qquad k_b = 3 \ \text{V·s/rad}$$

If the motor is used to drive a load with moment of inertia 2 N·m·s^2/rad and with viscous friction coefficient 0.1 N·m·s/rad, what is its transfer function? If the moment of inertia of the load and the viscous friction coefficient are doubled, what is its transfer function? Do the time and gain constants equal half of those of the first transfer function?

[**Answers:** $0.2/s(8s + 1), 0.14/s(11.4s + 1)$, no.]

The input and output of the transfer function in (3.15) are the applied electrical voltage and the angular position of the motor shaft as shown in Figure 3.4(a). If motors are used in velocity control systems to drive tapes or conveyers, we are no longer interested in their angular positions θ. Instead we are mainly concerned with

---

[2]In engineering, it is often said that the electrical time constant is much smaller than the mechanical time constant, thus we can set $L_a = 0$. If we write (3.14) as $G(s) = k_m/s(as + 1)(bs + 1)$, then a and b are called time constants. Unlike (3.6), a and b depend on both electrical and mechanical parts of the motor. Therefore, electrical and mechanical time constants in armature-controlled dc motors are no longer as well defined as in field-controlled dc motors.

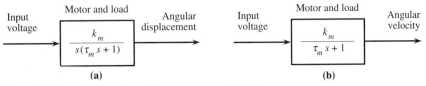

Figure 3.4 Transfer functions of armature-controlled dc motor.

their angular velocities. Let w be the angular velocity of the motor shaft. Then we have $w(t) = d\theta/dt$ and

$$W(s) = s\Theta(s) = s \cdot \frac{k_m}{s(\tau_m s + 1)} \cdot U(s) = \frac{k_m}{\tau_m s + 1} \cdot U(s)$$

The transfer function of the motor from applied voltage u to angular velocity w is

$$\frac{W(s)}{U(s)} = \frac{k_m}{\tau_m s + 1} \tag{3.18}$$

as shown in Figure 3.4(b). Equation (3.18) differs from (3.15) by the absence of one pole at $s = 0$. Thus the transfer function of motors can be (3.15) or (3.18), depending on what is considered as the output. Therefore it is important to specify the input and output in using a transfer function. Different specifications lead to different transfer functions for the same system.

3.2.3 Measurement of Motor Transfer Functions

The transfer function of a motor driving a load is given by (3.15). To use the formula, we must use (3.16) and (3.17) to compute k_m and τ_m which, in turn, require k_t, k_b, R_a, L_a, J, and f. Although the first four constants may be supplied by the manufacturer of the motor, we still need J and f. The moment of inertia J of a load is not easily computable if it is not of regular shape. Neither can f be obtained analytically. Therefore, we may have to obtain J and f by measurements. If measurements are to be carried out, we may as well measure the transfer function directly. We discuss one way of measuring motor transfer functions in the following.

Let the transfer function of a motor driving a load be of the form shown in (3.15). If we apply a voltage and measure its angular velocity, then the transfer function reduces to (3.18). Let the applied voltage be a; the velocity can then be computed as

$$W(s) = \frac{k_m}{\tau_m s + 1} \cdot \frac{a}{s} = \frac{k_m a}{s} - \frac{k_m a \tau_m}{\tau_m s + 1}$$

Its inverse Laplace transform is

$$w(t) = k_m a - k_m a e^{-t/\tau_m} \tag{3.19}$$

for $t \geq 0$. Because τ_m is positive, the term e^{-t/τ_m} approaches zero as t approaches infinity and the response $w(t)$ is as shown in Figure 3.5(a). Thus we have

$$w(\infty) = k_m a \tag{3.20}$$

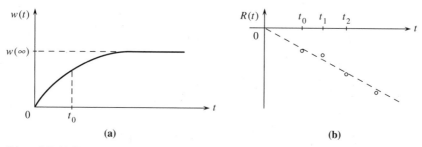

Figure 3.5 (a) Step response of motor. (b) Estimation of τ_m.

This is called the *final speed* or *steady-state speed*. If we apply a voltage of known magnitude a and measure the final speed $w(\infty)$, then the motor gain constant can be easily obtained as $k_m = w(\infty)/a$. To find the motor time constant, we rewrite (3.19) as

$$w(t) = k_m a(1 - e^{-t/\tau_m})$$

which implies, using (3.20),

$$e^{-t/\tau_m} = 1 - \frac{w(t)}{k_m a} = 1 - \frac{w(t)}{w(\infty)}$$

Thus we have

$$-\frac{t}{\tau_m} = \ln\left(1 - \frac{w(t)}{w(\infty)}\right) =: R(t) \tag{3.21}$$

where ln stands for the natural logarithm. Now if we measure the speed at any t, say $t = t_0$, then from (3.21) we have

$$\tau_m = \frac{-t_0}{\ln\left(1 - \dfrac{w(t_0)}{w(\infty)}\right)} \tag{3.22}$$

Thus the motor time constant can be obtained from the final speed and one additional measurement at any t.

Example 3.2.1

Consider an armature-controlled dc motor driving a load. We apply 5 V to the motor. The angular speed of the load at $t_0 = 2$ s is measured as 30 rad/s and the final speed is measured as 70 rad/s. What are the transfer functions from the applied voltage to the speed and from the applied voltage to the displacement?

From (3.20) and (3.22), we have

$$k_m = \frac{w(\infty)}{a} = \frac{70}{5} = 14$$

and

$$\tau_m = \frac{-2}{\ln\left(1 - \frac{30}{70}\right)} = \frac{-2}{\ln 0.57} = \frac{-2}{-0.5596} = 3.57$$

Thus the transfer functions are

$$\frac{W(s)}{U(s)} = \frac{14}{3.57s + 1} \qquad \frac{\Theta(s)}{U(s)} = \frac{14}{s(3.57s + 1)}$$

Exercise 3.2.2

Consider an armature-controlled dc motor driving a load. We apply 10 volts to the motor. The velocity of the load is measured as 20 cycles per second at $t_0 = 5$ s, and its steady-state velocity is measured as 30 cycles per second. What is its transfer function from the applied voltage u to the angular displacement $\theta(t)$? What is its unit-step response? What is $\theta(\infty)$? What does it mean physically?

[Answers: $G(s) = 18.84/s(4.55s + 1) = 4.14/s(s + 0.22)$, $\theta(t) = 85.54e^{-0.22t} - 85.54 + 18.82t$, $\theta(\infty) = \infty$, it means that the load will continue to turn with steady-state speed 18.82 rad/s and the displacement from the original position becomes larger and larger.]

In (3.22), in addition to the final speed, we used only one datum to compute τ_m. A more accurate τ_m can be obtained by using more data. We plot $R(t)$ in (3.21) for a number of t as shown in Figure 3.5(b), and draw from the origin a straight line passing through these points. Then the slope of the straight line equals $1/\tau_m$. This is a more accurate way of obtaining the motor time constant. This method can also check whether or not the simplified transfer functions in (3.18) and (3.15) can be used. If all points in the plot are very close to the straight line, then (3.18) and (3.15) are adequate. If not, more accurate transfer functions such as the one in (3.14) must be used to describe the motor.

The problem of determining τ_m and k_m in (3.18) from measurements is called *parameter estimation*. In this problem, the form of the transfer function is assumed to be known, but its parameters are not known. We then determine the parameters from measurements. This is a special case of the identification problem where neither the form nor the parameters are assumed to be known. If no noise exists in measurements, then the identification problem is not difficult. See Reference [15]. However, in practice, noise often arises in measurements. This makes the problem difficult. For a different method of identification, see Section 8.3.2.

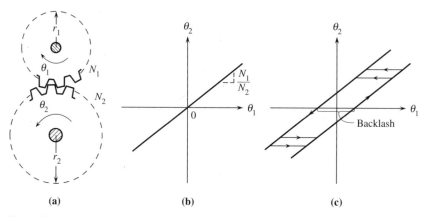

(a)   (b)   (c)

Figure 3.6 Gear train and its characteristics.

3.3   GEARS

Gears are used to convert high speed and small torque into low speed and high torque or the converse. The most popular type of gear is the spur gear shown in Figure 3.6(a). Let r_i be the radius; N_i, the number of teeth; θ_i, the angular displacement, and T_i, the torque for gear i, $i = 1, 2$. Clearly, the number of teeth on a gear is linearly proportional to its radius; hence $N_1/r_1 = N_2/r_2$. The linear distance traveled along the surfaces of both gears are the same, thus $\theta_1 r_1 = \theta_2 r_2$. The linear forces developed at the contact point of both gears are equal, thus $T_1/r_1 = T_2/r_2$. These equalities can be combined to yield

$$\frac{T_1}{T_2} = \frac{N_1}{N_2} = \frac{\theta_2}{\theta_1} \tag{3.23}$$

This is a linear equation as shown in Figure 3.6(b) and is obtained under idealized conditions. In reality, backlash exists between coupled gears. So long as the gears are rotating in one direction, the teeth will remain in contact. However, reversal of the direction of the driving gear will disengage the teeth and the driven gear will remain stationary until reengagement.[3] Therefore, the relationship between θ_1 and θ_2 should be as shown in Figure 3.6(c), rather than in Figure 3.6(b). Keeping the backlash small will increase the friction between the teeth and wear out the teeth faster. On the other hand, an excessive amount of backlash will cause what is called the *chattering, hunting,* or *limited-cycle* problem in control systems. To simplify analysis and design, we use the linear equation in (3.23).

Consider an armature-controlled dc motor driving a load through a gear train as shown in Figure 3.7(a). The numbers of teeth are N_1 and N_2. Let the total moment

---

[3]Here we assume that the mass of the gear is zero, or that the gear has no inertia.

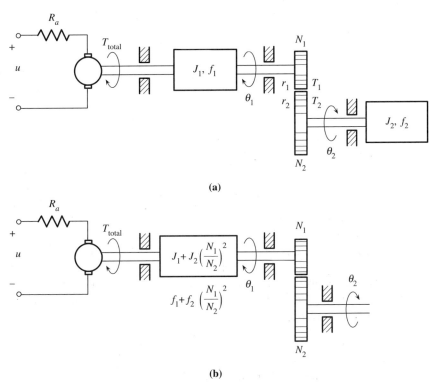

Figure 3.7 Gear train and its equivalence.

of inertia (including rotor, motor shaft, and gear 1) and viscous friction coefficient on the motor shaft be, respectively, J_1 and f_1, and those on the load shaft be J_2 and f_2. The torque generated by the motor must drive J_1, overcome f_1, and generate a torque T_1 at gear 1 to drive the second gear. Thus we have

$$T_{motor} = J_1 \frac{d^2\theta_1(t)}{dt^2} + f_1 \frac{d\theta_1(t)}{dt} + T_1 \tag{3.24}$$

Torque T_1 at gear 1 generates a torque T_2 at gear 2, which in turn drives J_2 and overcomes f_2. Thus we have

$$T_2 = J_2 \frac{d^2\theta_2(t)}{dt^2} + f_2 \frac{d\theta_2(t)}{dt}$$

which, using $\theta_2 = (N_1/N_2)\theta_1$, becomes

$$T_2 = J_2 \frac{N_1}{N_2} \frac{d^2\theta_1(t)}{dt^2} + f_2 \frac{N_1}{N_2} \frac{d\theta_1(t)}{dt} \tag{3.25}$$

The substitution of this T_2 into $T_1 = (N_1/N_2)T_2$ and then into (3.24) yields

$$
\begin{aligned}
T_{\text{motor}} &= J_1 \frac{d^2\theta_1(t)}{dt^2} + f_1 \frac{d\theta_1(t)}{dt} + J_2 \left(\frac{N_1}{N_2}\right)^2 \frac{d^2\theta_1(t)}{dt^2} \\
&\quad + f_2 \left(\frac{N_1}{N_2}\right)^2 \frac{d\theta_1(t)}{dt} \\
&= J_{1\text{eq}} \frac{d^2\theta_1(t)}{dt^2} + f_{1\text{eq}} \frac{d\theta_1(t)}{dt}
\end{aligned}
\tag{3.26}
$$

where

$$
J_{1\text{eq}} := J_1 + J_2 \left(\frac{N_1}{N_2}\right)^2 \qquad f_{1\text{eq}} := f_1 + f_2 \left(\frac{N_1}{N_2}\right)^2
\tag{3.27}
$$

This process transfers the load and friction on the load shaft into the motor shaft as shown in Figure 3.7(b). Using this equivalent diagram, we can now compute the transfer function and develop a block diagram for the motor and load.

Once the load and friction on the load shaft are transferred to the motor shaft, we can simply disregard the gear train and the load shaft. If the armature inductance L_a is assumed to be zero, then the transfer function of the motor and load from u to θ_1 is, as in (3.15),

$$
G(s) = \frac{\Theta_1(s)}{U(s)} = \frac{\bar{k}_m}{s(\bar{\tau}_m s + 1)}
\tag{3.28}
$$

with (3.16) and (3.17) modified as

$$
\bar{k}_m = \frac{k_t}{k_t k_b + f_{1\text{eq}} R_a}
\tag{3.29a}
$$

and

$$
\bar{\tau}_m = \frac{J_{1\text{eq}} R_a}{k_t k_b + f_{1\text{eq}} R_a}
\tag{3.29b}
$$

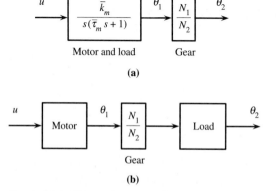

Figure 3.8 (a) Block diagram. (b) Incorrect block diagram.

The transfer function from θ_1 to θ_2 is N_1/N_2. Thus the block diagram of the motor driving a load through a gear train is as shown in Figure 3.8(a). We remark that because of the loading, it is not possible to draw a block diagram as shown in Figure 3.8(b).

3.4 TRANSDUCERS

Transducers are devices that convert signals from one form to another—for example, from a mechanical shaft position, temperature, or pressure to an electrical voltage. They are also called *sensing devices*. There are all types of transducers such as thermocouples, strain gauges, pressure gauges, and others. We discuss in the following only potentiometers and tachometers.

Potentiometers

The potentiometer is a device that can be used to convert a linear or angular displacement into a voltage. Figure 2.14 shows a wire-wound potentiometer with its characteristic. The potentiometer converts the angular displacement $\theta(t)$ into an electric voltage $v(t)$ described by

$$v(t) = k\theta(t) \tag{3.30}$$

where k is a constant and depends on the applied voltage and the type of potentiometer used. The application of the Laplace transform to (3.30) yields

$$V(s) = k\Theta(s) \tag{3.31}$$

Thus the transfer function of the potentiometer is a constant. Figure 3.9 shows three commercially available potentiometers.

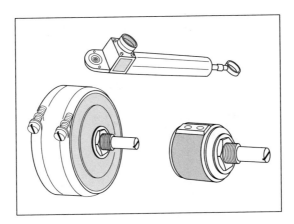

Figure 3.9 Potentiometers.

Tachometers

The tachometer is a device that can convert a velocity into a voltage. It is actually a generator with its rotor connected to the shaft whose velocity is to be measured. Therefore a tachometer is also called a tachogenerator. The output $v(t)$ of the tachometer is proportional to the shaft's angular velocity; that is,

$$v(t) = k \frac{d\theta(t)}{dt} \tag{3.32}$$

where $\theta(t)$ is the angular displacement and k is the sensitivity of the tachometer, in volts per radian per second. The application of the Laplace transform to (3.32) yields

$$G(s) = \frac{V(s)}{\Theta(s)} = ks \tag{3.33}$$

Thus the transfer function from $\theta(t)$ to $v(t)$ of the tachometer is ks.

As discussed in Section 2.4.1, improper transfer functions will amplify high-frequency noise and are not used in practice. The transfer function of tachometers is improper, therefore its employment must be justified. A tachometer is usually attached to a shaft—for example, the shaft of a motor as shown in Figure 3.10(a). Although the transfer function of the tachometer is improper, the transfer function from u to y_1 is

$$\frac{k_m}{s(\tau_m s + 1)} \cdot ks = \frac{kk_m}{\tau_m s + 1}$$

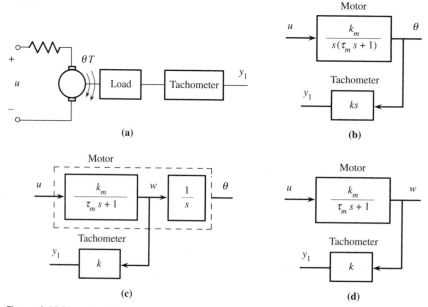

Figure 3.10 Use of tachometer.

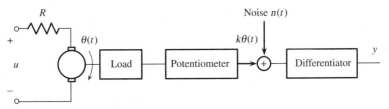

Figure 3.11 Unacceptable way of generating velocity signal.

as shown in Figure 3.10(b). It is strictly proper. Thus electrical noise entered at the armature circuit will not be amplified. The transfer function from motor torque $T(t)$ to y_1 is, from Figure 3.2,

$$\frac{1}{s(Js + f)} \cdot ks = \frac{k}{Js + f}$$

which is again strictly proper. Thus, mechanical noise, such as torque generated by gusts, is smoothed by the moment of inertia of the motor. In conclusion, tachometers will not amplify electrical and mechanical high-frequency noises and are widely used in practice. See also Problem 3.17. Note that the block diagram in Figure 3.10(b) can also be plotted as shown in Figure 3.10(c). This arrangement is useful in computer computation and operational amplifier circuit realization, as will be discussed in Chapter 5. We mention that the arrangement shown in Figure 3.11 cannot be used to measure the motor angular velocity. Although the potentiometer generates signal $k\theta(t)$, it also generates high-frequency noise $n(t)$ due to brush jumps, wire irregularities, and variations of contact resistance. The noise is greatly amplified by the differentiator and overwhelms the desired signal $kd\theta/dt$. Thus the arrangement cannot be used in practice.

The transfer function of a tachometer is ks only if its input is displacement. If its input is velocity $w(t) = d\theta(t)/dt$, then its transfer function is simply k. In velocity control systems, the transfer function of motors is $k_m/(\tau_m s + 1)$ as shown in Figure 3.4(b). In this case, the block diagram of a motor and a tachometer is as shown in Figure 3.10(d). Therefore, it is important to specify what are the input and output of each block.

Error Detectors

Every error detector has two input terminals and one output terminal. The output signal is proportional to the difference of the two input signals. An error detector can be built by connecting two potentiometers as shown in Figure 3.12(a). The two potentiometers may be located far apart. For example, one may be located inside a room and the other, attached to an antenna on the rooftop. The input signals θ_r and θ_o are mechanical positions, either linear or rotational; the output $v(t)$ is a voltage signal. They are related by

$$v(t) = k[\theta_r(t) - \theta_o(t)] \tag{3.34}$$

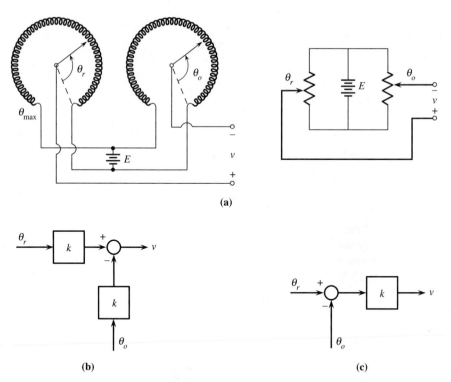

Figure 3.12 Pair of potentiometers and their schematic representations.

or

$$V(s) = k[\Theta_r(s) - \Theta_o(s)]$$

where k is a constant. The pair of potentiometers can be represented schematically as shown in Figure 3.12(b) or (c). The circle where the two signals enter is often called the *summing point*.

3.5 OPERATIONAL AMPLIFIERS (OP-AMPS)

The operational amplifier is one of the most important circuit elements. It is built in integrated-circuit form. It is small, inexpensive, and versatile. It can be used to build buffers, amplifiers, error detectors, and compensating networks, and is therefore widely used in control systems.

The operational amplifier is usually represented as shown in Figure 3.13(a) and is modeled as shown in Figure 3.13(b). It has two input terminals. The one with a "$-$" sign is called the inverting terminal and the one with a "$+$" sign the non-inverting terminal. The output voltage v_o equals $A(v_{i2} - v_{i1})$, and A is called the *open-loop gain*. The resistor R_i in Figure 3.13(b) is called the *input resistance* and

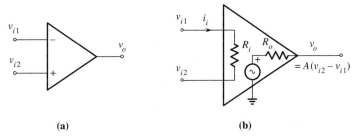

(a)   **(b)**

Figure 3.13 Operational amplifier.

R_o, the *output resistance*. R_i is generally very large, greater than 10^4 Ω, and R_o is very small, less than 50 Ω. The open-loop gain A is very large, usually over 10^5, in low frequencies. Signals in op-amps are limited by supply voltages, commonly ± 15 V. Because of this limitation, if A is very large or infinity, then we have

$$v_{i1} = v_{i2} \tag{3.35}$$

This equation implies that the two input terminals are virtually short-circuited. Because R_i is very large, we have

$$i_i = 0 \tag{3.36}$$

This equation implies that the two input terminals are virtually open-circuited. Thus the two input terminals have the two conflicting properties: open-circuit and short-circuit. Using these two properties, operational amplifier circuits can easily be analyzed.

Consider the operational amplifier circuit shown in Figure 3.14(a). Because of the direct connection of the output terminal and inverting terminal, we have $v_{i1} = v_o$. Thus we have, using the short-circuit property, $v_o = v_{i2}$. It means that the output voltage is identical to the input voltage v_{i2}. One may wonder why we do not connect v_o directly to v_{i2}, rather than through an op-amp. There is an important reason for doing this. The input resistance of op-amps is very large and the output resistance is very small, so op-amps can isolate the circuits before and after them, and thus eliminate the loading problem. Therefore, the circuit is called a *voltage follower*, *buffer*, or *isolating amplifier* and is widely used in practice.

Consider the circuit shown in Figure 3.14(b) where Z_1 and Z_f are two impedances. The open-circuit property $i_i = 0$ implies $i_1 = -i_o$. Thus we have[4]

$$\frac{V_1 - V_{i_1}}{Z_1} = -\frac{V_o - V_{i_1}}{Z_f} \tag{3.37}$$

---

[4]Because impedances are defined in the Laplace transform domain (see Section 2.4), all variables must be in the same domain. We use V_i to denote the Laplace transform of v_i.

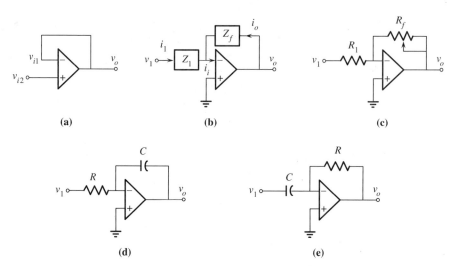

Figure 3.14 Op-amp circuits.

Because the noninverting terminal is grounded and because of the short-circuit property, we have $v_{i_1} = v_{i_2} = 0$. Thus (3.37) becomes

$$\frac{V_o}{V_1} = -\frac{Z_f}{Z_1} \tag{3.38}$$

If $Z_f = R_f$ and $Z_1 = R_1$, then the transfer function from v_1 to v_o is $-R_f/R_1$ and the circuit can be used as an amplifier with fixed gain $-R_f/R_1$. If R_f is replaced by a potentiometer or an adjustable resistor as shown in Figure 3.14(c), then the gain of the amplifier can be easily adjusted.

Exercise 3.5.1

Show that the transfer function of the circuit in Figure 3.14(d) equals $-1/RCs$. Thus, the circuit can act as an integrator. Show that the transfer function of the circuit in Figure 3.14(e) equals $-RCs$. Thus, the circuit can act as a pure differentiator. Integrators and differentiators can be easily built by using operational amplifier circuits. However, differentiators so built may not be stable and cannot be used in practice. See Reference [18]. Integrators so built are stable and are widely used in practice.

Consider the op-amp circuit shown in Figure 3.15. The noninverting terminal is grounded, so $v_{i1} = v_{i2} = 0$. Because of $i_i = 0$, we have

$$i_f = -(i_1 + i_2 + i_3)$$

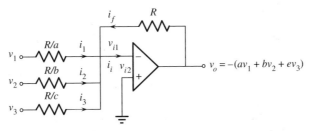

Figure 3.15 Op-amp circuit.

which together with $v_{i1} = 0$ implies

$$\frac{v_o}{R} = -\left(\frac{v_1}{R/a} + \frac{v_2}{R/b} + \frac{v_3}{R/c}\right)$$

Thus we have

$$v_o = -(av_1 + bv_2 + cv_3) \tag{3.39}$$

If $a = 1$ and $b = c = 0$, then (3.39) reduces to $v_o = -v_1$. It is called an *inverting amplifier* as shown in Figure 3.16(a). The op-amp circuit shown in Figure 3.16(b) can serve as an error detector. The output of the first op-amp is

$$e = -(r - v_w)$$

where v_w is the voltage generated by a tachometer. Note the polarity of the tachometer output. The output of the second op-amp, an inverting amplifier, is $u = -e$, thus we have $u = r - v_w$. This is an error detector. In conclusion, op-amp circuits are versatile and widely used. Because of their high input resistances and low output resistances, their connection will not cause the loading problem.

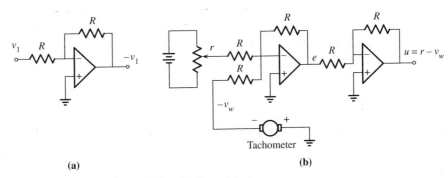

(a) **(b)**

Figure 3.16 (a) Inverting amplifier. (b) Error detector.

3.6   BLOCK DIAGRAMS OF CONTROL SYSTEMS

In this section, we show how block diagrams are developed for control systems.

Example 3.6.1

Consider the control system shown in Figure 3.17(a). The load could be a telescope or an antenna and is driven by an armature-controlled dc motor. The system is designed so that the actual angular position of the load will follow the reference signal. The error e between the reference signal r and the controlled signal y is detected by a pair of potentiometers with sensitivity k_1. The dotted line denotes mechanical coupling, therefore their signals are identical. The error e is amplified by a dc amplifier with gain k_2 and then drives the motor. The block diagram of this system is shown in Figure 3.17(b). The diagram is self-explanatory.

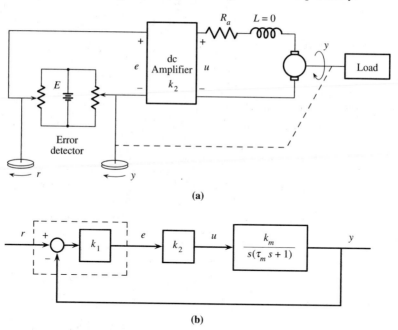

(a)

(b)

Figure 3.17 Position control system.

Example 3.6.2

In a steel or paper mill, the products are moved by rollers, as shown in Figure 3.18(a). In order to maintain a prescribed tension, the roller speeds are kept constant and equal to each other. This can be achieved by using the control system shown in Figure 3.18(b). Each roller is driven by an armature-controlled dc motor. The desired

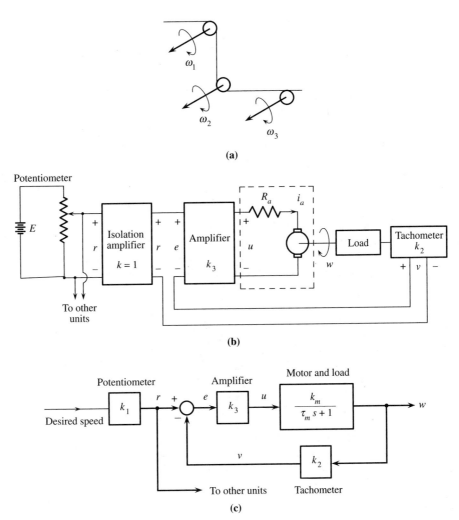

Figure 3.18 Speed control system.

roller speed is transformed by a potentiometer into an electrical voltage. The loading problem associated with the potentiometer is eliminated by using an isolating amplifier. The scale on the potentiometer can be calibrated after the completion of the design. Therefore, there is no need to consider the potentiometer in the design. The roller speed is measured using a tachometer. The block diagram of the system is shown in Figure 3.18(c). For this problem, we are interested in the roller's speed, not its position, so the motor shaft speed is chosen as the output and the transfer function of the motor and load becomes $k_m/(\tau_m s + 1)$. Similarly, the tachometer's transfer function is k_2 rather than $k_2 s$ as shown. We mention that depending on the wiring, the error signal e can be $r - v$, $r + v$, or $v - r$. For the polarities shown in Figure 3.18(b), we have $e = r - v$.

Example 3.6.3

Temperature control is essential in many chemical processes. Consider the temperature control system shown in Figure 3.19(a). The problem is to control the temperature y inside the chamber. The chamber is heated by steam. The flow q of hot steam is proportional to the valve opening x; that is, $q = k_q x$. The valve opening x is controlled by a solenoid and is assumed to be proportional to the solenoid current i; that is, $x = k_s i$. It is assumed that the chamber temperature y and the steam flow q are related by

$$\frac{dy}{dt} = -cy + k_c q \tag{3.40}$$

where c depends on the insulation and the temperature difference between inside and outside the chamber. To simplify analysis and design, c is assumed to be a positive constant. This means that if no steam is pumped into the chamber, the temperature will decrease at the rate of e^{-ct}. The application of the Laplace transform

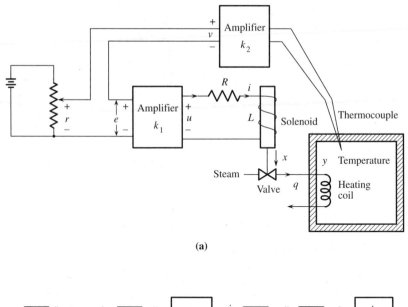

(a)

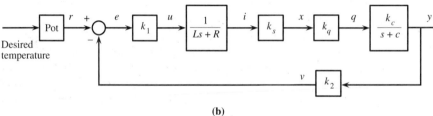

(b)

Figure 3.19 Temperature control system.

to (3.40) yields, assuming zero initial condition,

$$sY(s) = -cY(s) + k_c Q(s)$$

or

$$\frac{Y(s)}{Q(s)} = \frac{k_c}{s + c}$$

Thus the transfer function from q to y is $k_c/(s + c)$.

Figure 3.19(b) shows a block diagram for the heating system. The desired temperature is transformed into an electrical voltage r by the leftmost potentiometer. The scale on the potentiometer can be calibrated after the completion of the design. The chamber temperature is measured using a thermocouple and amplified to yield $v = k_2 y$. From the wiring, we have $e = r - v$ and the wiring is represented by the summer shown. The error signal e is amplified to yield u. The RL circuit is governed by

$$u(t) = Ri(t) + L \frac{di(t)}{dt}$$

The application of the Laplace transform yields, assuming zero initial condition,

$$U(s) = RI(s) + LsI(s)$$

which implies

$$\frac{I(s)}{U(s)} = \frac{1}{R + Ls} \tag{3.41}$$

Thus the transfer function from u to i is $1/(Ls + R)$ as shown. The remainder of the block diagram is self-explanatory.

Exercise 3.6.1

Develop the block diagrams in Figures 1.1, 1.6, and 1.7.

3.6.1 Reaction Wheels and Robotic Arms[5]

In this subsection, we give two more examples of developing block diagrams for control systems.

---

[5]May be skipped without loss of continuity.

Example 3.6.4

In order to point the antenna toward the earth or the solar panels toward the sun, the altitude or orientation of a satellite or space vehicle must be properly controlled. This can be achieved using gas jets or reaction wheels. Because of unlimited supply of electricity through solar panels, reaction wheels are used if the vehicle will travel for long journeys. Three sets of reaction wheels are needed to control the orientation in the three-dimensional space; but they are all identical.

A reaction wheel is actually a flywheel; it may be simply the rotor of a motor. It is assumed to be driven by an armature-controlled dc motor as shown in Figure 3.20(a) and (b). The case of the motor is rigidly attached to the vehicle. Because of the conservation of momentum, if the reaction wheel turns in one direction, the satellite will rotate in the opposite direction. The orientation and its rate of change can be measured using a gyro and a rate gyro. The block diagram of the control system is shown in Figure 3.20(c).

We derive in the following the transfer function $G(s)$ of the space vehicle and

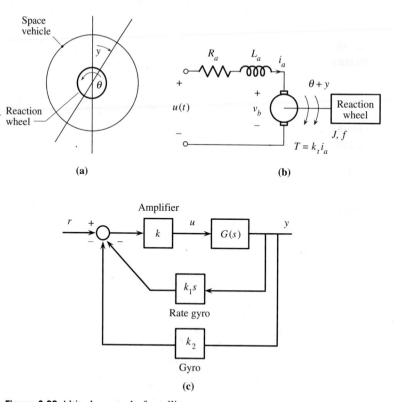

Figure 3.20 Altitude control of satellite.

motor. Let the angular displacements, with respect to the inertial coordinate, of the vehicle and the reaction wheel be respectively y and θ as shown in Figure 3.20(a). They are chosen, for convenience, to be in opposite directions. Clearly, the relative angular displacement of the reaction wheel (or the rotor of the motor) with respect to the vehicle (or the stator or case of the motor) is $y + \theta$. Thus the armature circuit is governed by, as in (3.10),

$$R_a i_a(t) + L_a \frac{di_a(t)}{dt} + v_b(t) = u(t) \tag{3.42}$$

with the back emf voltage $v_b(t)$ given by

$$v_b(t) = k_b \left(\frac{d\theta}{dt} + \frac{dy}{dt} \right) \tag{3.43}$$

Let J and f be the moment of inertia and the viscous friction coefficient on the motor shaft. Because the friction is generated between the motor shaft and the bearing that is attached to the satellite, the friction equals $f(d\theta/dt + dy/dt)$. Thus the torque equation is

$$T(t) = k_t i_a = J \frac{d^2\theta(t)}{dt^2} + f \left(\frac{d\theta}{dt} + \frac{dy}{dt} \right) \tag{3.44}$$

Let the moment of inertia of the vehicle be J_v. Then the conservation of angular momentum implies

$$J_v \frac{dy}{dt} = J \frac{d\theta}{dt} \tag{3.45}$$

Using (3.42) through (3.45), we can show that the transfer function from u to y equals

$$G(s) = \frac{Y(s)}{U(s)} = \frac{k_t J}{s[(L_a s + R_a)(JJ_v s + (J + J_v)f) + k_t k_b(J + J_v)]} \tag{3.46}$$

This completes the block diagram in Figure 3.20.

Example 3.6.5

Consider the industrial robot shown in Figure 3.21(a). The robot has a number of joints. It is assumed that each joint is driven by an armature-controlled dc motor through gears with gear ratio $n = N_1/N_2$. Figure 3.21(b) shows a block diagram of the control of a joint, where the block diagram in Figure 3.3 is used for the motor. The J and f in the diagram are the total moment of inertia and viscous friction coefficient reflected to the motor shaft by using (3.27). The compensator is a

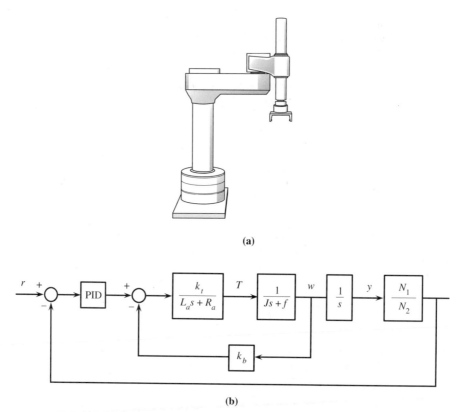

(a)

(b)

Figure 3.21 (a) Industrial robot. (b) Joint control system.

proportional-integral-derivative (PID) compensator, which will be discussed in later chapters.

3.7 MANIPULATION OF BLOCK DIAGRAMS

Once a block diagram for a control system is obtained, the next step in analysis is to simplify the block diagram to a single block or, equivalently, to find the overall transfer function. It is useless to analyze an individual block, because the behavior of the control system is affected only indirectly by individual transfer function. The behavior is dictated completely by its overall transfer function.

Two methods are available to compute overall transfer functions: block diagram manipulation and employment of Mason's formula. We discuss first the former and than the latter. The block diagram manipulation is based on the equivalent diagrams shown in Table 3.1. The first pair is concerned with summers. A summer must have two or more inputs and one and only one output. If a summer has three or more

Table 3.1 Equivalent Block Diagrams

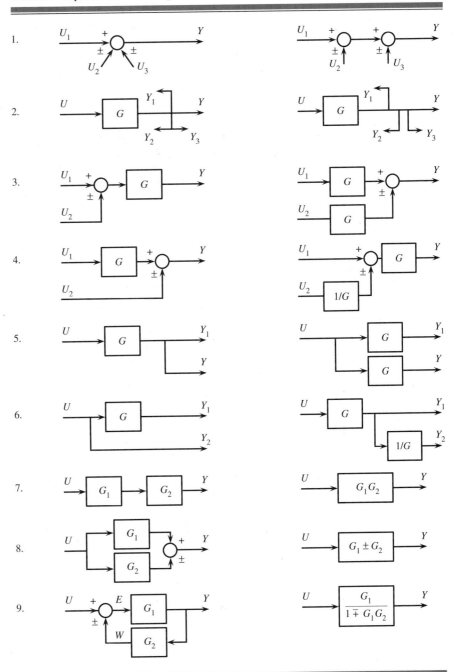

inputs, it can be separated into two summers as shown. A terminal can be branched out, at branching points, into several signals as shown in the second pair of Table 3.1 with all signals equal to each other. A summer or a branching point can be moved around a block as shown from the third to the sixth pair of the table. Their equivalences can be easily verified. For example, for the fourth pair, we have $Y = GU_1 \pm U_2$ for the left-hand-side diagram and $Y = G(U_1 \pm U_2/G) = GU_1 \pm U_2$ for the right-hand-side diagram. They are indeed equivalent.

The last three pairs of Table 3.1 are called, respectively, the *tandem, parallel,* and *feedback connections* of two blocks. The reduction of the tandem and parallel connections to single blocks is very simple. We now reduce the feedback connection to a single block. Note that if W is fed into the summer with a positive sign (that is, $E = U + W$), it is called *positive feedback.* If W is fed into the summer with a negative sign (that is, $E = U - W$), it is called *negative feedback.* We derive only the negative feedback part. Let the input of G_1 be denoted by E and the output of G_2 by W as shown in the left-hand-side diagram of the last pair of Table 3.1. Then we have

$$W = G_2Y \qquad E = U - W = U - G_2Y \qquad Y = G_1E \tag{3.47}$$

The substitution of the second equation into the last yields

$$Y = G_1(U - G_2Y) = G_1U - G_1G_2Y \tag{3.48}$$

which can be written as

$$(1 + G_1G_2)Y = G_1U$$

Thus we have

$$\frac{Y}{U} = \frac{G_1}{1 + G_1G_2} \tag{3.49}$$

This is the transfer function from U to Y, for the negative feedback part, as shown in the right-hand-side diagram of the last pair of Table 3.1. In conclusion, the transfer function of the feedback connection is $G_1/(1 + G_1G_2)$ for negative feedback and $G_1/(1 - G_1G_2)$ for positive feedback. They are important formulas, and should be remembered.

Now we use an example to illustrate the use of Table 3.1 to compute overall transfer functions of block diagrams.

Example 3.7.1

Consider the block diagram shown in Figure 3.22(a). We first use entry 9 in Table 3.1 to simplify the inner positive feedback loop as shown in Figure 3.22(b). Note that the direct feedback in Figure 3.22(a) is the same as feedback with transfer function 1 shown in Figure 3.22(b). Using entries 7 and 9, we have

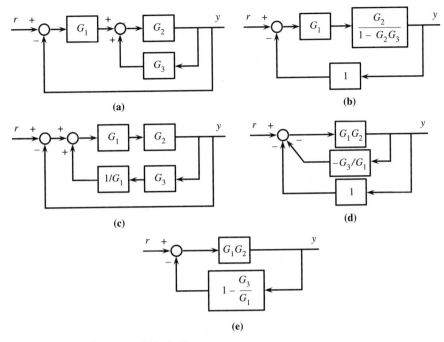

Figure 3.22 Manipulation of block diagram.

$$\frac{Y}{R} = \frac{G_1 \cdot \dfrac{G_2}{1 - G_2 G_3}}{1 + G_1 \cdot \dfrac{G_2}{1 - G_2 G_3} \cdot 1} = \frac{G_1 G_2}{1 - G_2 G_3 + G_1 G_2} \qquad (3.50)$$

This is the overall transfer function from R to Y in Figure 3.22(a).

A block diagram can be manipulated in many ways. For example, we can move the second summer in Figure 3.22(a) to the front of G_1, using entry 4 of Table 3.1, to yield the block diagram in Figure 3.22(c) which can be redrawn, using entry 1 of Table 3.1, as shown in Figure 3.22(d). Note that the positive feedback has been changed to a negative feedback, but we also have introduced a negative sign into the feedback block. The two feedback paths are in parallel and can be combined as shown in Figure 3.22(e). Thus we have

$$\frac{Y}{R} = \frac{G_1 G_2}{1 + G_1 G_2 \cdot \left(1 - \dfrac{G_3}{G_1}\right)} = \frac{G_1 G_2}{1 + G_1 G_2 - G_2 G_3}$$

which is the same as (3.50), as expected.

Exercise 3.7.1

Use block manipulation to find the transfer functions from r to y of the block diagrams in Figure 3.23.

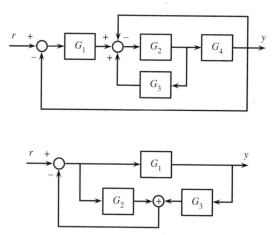

Figure 3.23 Block diagrams.

[**Answers:** $G_1G_2G_4/(1 - G_2G_3 + G_2G_4 + G_1G_2G_4), G_1/(1 + G_2 + G_1G_3).$]

3.7.1 Mason's Formula

Consider the block diagram shown in Figure 3.24. There are two inputs, r and p. The problem is to find the transfer function from r to y, denoted by G_{yr}, and the transfer function from p to y, denoted by G_{yp}. Because the system is linear, when we compute G_{yr}, we may assume $p = 0$ or, equivalently, disregard p. When we compute G_{yp}, we may disregard r. Once they are computed, the output y due to the

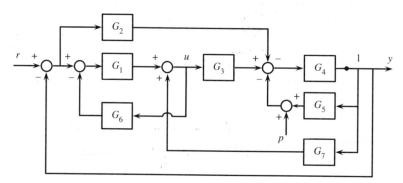

Figure 3.24 Block diagram.

two inputs r and p is given by

$$Y(s) = G_{yr}(s)R(s) + G_{yp}(s)P(s) \qquad (3.51)$$

where Y, R, and P are respectively the Laplace transforms of y, r, and p.

Although G_{yr} and G_{yp} can be obtained by block manipulations, the manipulations are not simple. Furthermore, the manipulation in computing G_{yr} generally cannot be used to compute G_{yp}. Therefore, it requires two separate manipulations. In this subsection, we introduce a formula, called *Mason's formula*, to compute overall transfer functions that does not require any block manipulation.

The employment of Mason's formula requires the concepts of loops and forward paths. Consider the block diagram shown in Figure 3.24. Note that every block and branch is oriented, that is, unidirectional. A loop is any unidirectional path that originates and terminates at the same point and along which no point is encountered more than once. A *loop gain* is the product of all transfer functions along the loop. If a loop contains one or more summers, the sign at each summer must be considered in the loop gain. For example, the loop in Figure 3.25(a) has two summers. One branch enters the summer with a negative sign, one with a positive. The one with a negative sign can be changed to a positive sign by introducing a negative sign into G_2, as is shown in Figure 3.24(b). We remark that a branch with no block is the same as having a block with transfer function 1. Therefore the loop gain of the loop in Figure 3.25(a) or (b) is $G_1(-G_2) = -G_1G_2$. Similarly, the loop in Figure 3.25(c) has loop gain $-G_1(-G_2) = G_1G_2$. Now consider the block diagram in Figure 3.24. It has five loops with loop gains $-G_1G_6$, $-G_4G_5$, $G_3G_4G_7$, $-G_1G_3G_4$, and G_2G_4.

Two loops are said to be *nontouching* if they have no points in common. For example, the four loops in Figure 3.24 with loop gains $-G_4G_5$, $G_3G_4G_7$, $-G_1G_3G_4$, and G_2G_4 all touch each other because they all pass block G_4 or the point denoted by 1 in the diagram. The two loops with loop gains $-G_1G_6$ and $-G_4G_5$ do not touch each other; neither do the two loops with loop gains $-G_1G_6$ and G_2G_4. Now we define

$$
\begin{aligned}
\Delta = 1 \ &- \ (\Sigma \ \text{loop gains}) \\
&+ \ (\Sigma \ \text{products of all possible two nontouching loop gains}) \\
&- \ (\Sigma \ \text{products of all possible three nontouching loop gains}) \\
&+ \ \cdots
\end{aligned}
\qquad (3.52)
$$

Although this formula looks very complicated, it is not necessarily so in its appli-

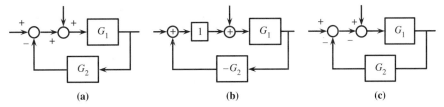

Figure 3.25 Block diagrams with a single loop.

cation. For example, if a block diagram has no loop as in the tandem and parallel connections in Table 3.1, then $\Delta = 1$. If a block diagram has only nontouching loops, then the formula terminates at the first parentheses. If a block diagram does not have three or more mutually nontouching loops, then the formula terminates at the second parentheses. For example, the block diagram in Figure 3.24 does not have three or more mutually nontouching loops, and we have

$$
\begin{aligned}
\Delta = 1 &-(-G_1G_6 - G_4G_5 + G_3G_4G_7 - G_1G_3G_4 + G_2G_4) \\
&+((-G_1G_6)(-G_4G_5) + (-G_1G_6)(G_2G_4))
\end{aligned}
\tag{3.53}
$$

For easy reference, we call Δ the *characteristic function* of the block diagram. It is an inherent property of a block diagram and is independent of input and output terminals.

Now we introduce the concept of forward paths. It is defined for specific input/output pairs. A forward path from input r to output y is any connection of unidirectional branches and blocks from r to y along which no point is encountered more than once. A forward path gain is the product of all transfer functions along the path including signs at summers. For the block diagram in Figure 3.24, the $r-y$ input/output pair has two forward paths, with gains $P_1 = G_1G_3G_4$ and $P_2 = -G_2G_4$. The $p-y$ input/output pair has only one forward path, with gain $-G_4$. A loop touches a forward path if they have at least one point in common.

With these concepts, we are ready to introduce Mason's formula. The formula states that the overall transfer function from input v to output w of a block diagram is given by

$$
G_{wv} = \frac{W(s)}{V(s)} = \frac{\Sigma_i P_i \Delta_i}{\Delta}
\tag{3.54}
$$

where Δ is defined as in (3.52),

$$
P_i = \text{the gain of the } i\text{th forward path from } v \text{ to } w
$$

$$
\Delta_i = \Delta_{\text{set those loop gains to zero if they touch the } i\text{th forward path.}}
$$

and the summation is to be carried out for all forward paths from v to w. If there is no forward path from v to w, then $G_{wv} = 0$. Now we use the formula to compute G_{yr} and G_{py} for the block diagram in Figure 3.24. The Δ for the diagram was computed in (3.53). There are two forward paths from r to y, with $P_1 = G_1G_3G_4$ and $P_2 = -G_2G_4$. Because all loops touch the first forward path, we set all loop gains to zero in (3.53) to yield

$$
\Delta_1 = 1
$$

All loops except the one with loop gain $-G_1G_6$ touch the second forward path, therefore we have

$$
\Delta_2 = 1 - (-G_1G_6) = 1 + G_1G_6
$$

Thus we have

$$G_{yr} = \frac{Y(s)}{R(s)} = \frac{P_1\Delta_1 + P_2\Delta_2}{\Delta}$$ (3.55)

$$= \frac{G_1G_3G_4 \cdot 1 + (-G_2G_4)(1 + G_1G_6)}{1 + G_1G_6 + G_4G_5 - G_3G_4G_7 + G_1G_3G_4 - G_2G_4 + G_1G_6G_4G_5 - G_1G_6G_2G_4}$$

Now we find the transfer function from p to y in Figure 3.24. The Δ for the block diagram was computed in (3.53). There is only one forward path from p to y with gain $P_1 = -G_4$. Because all loops except the loop with loop gain $-G_1G_6$ touch the path, we have

$$\Delta_1 = 1 - (-G_1G_6) = 1 + G_1G_6$$

Thus the transfer function from p to y in Figure 3.24 is, using Mason's formula,

$$G_{yp} = \frac{Y(s)}{P(s)} = \frac{P_1\Delta_1}{\Delta}$$ (3.56)

$$= \frac{(-G_4)(1 + G_1G_6)}{1 + G_1G_6 + G_4G_5 - G_3G_4G_7 + G_1G_3G_4 - G_2G_4 + G_1G_6G_4G_5 - G_1G_6G_2G_4}$$

Exercise 3.7.2

Use Mason's formula to verify entries 7, 8, and 9 in Table 3.1.

Exercise 3.7.3

Find the transfer functions from r to u and from p to u in Figure 3.24.

[**Answers:** $G_{ur} = (G_1(1 + G_4G_5) - G_2G_4G_7)/\Delta$, $G_{up} = (-G_4G_7 + G_4G_1)/\Delta$, where Δ is given in (3.53).]

Exercise 3.7.4

Repeat Exercise 3.7.1 by using Mason's formula. Are the results the same?

3.7.2 Open-Loop and Closed-Loop Transfer Functions

Consider the block diagram shown in Figure 3.26(a). The transfer function from r to y, from p_1 to y, and from p_2 to y are respectively

$$G_{yr} = \frac{G_1G_2}{1 + G_1G_2} \qquad G_{yp_1} = \frac{G_2}{1 + G_1G_2} \qquad G_{yp_2} = \frac{1}{1 + G_1G_2}$$ (3.57)

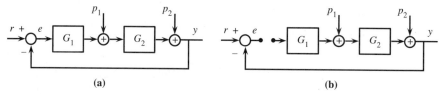

Figure 3.26 Closed-loop and open-loop systems.

These transfer functions are often referred to as *closed-loop transfer functions.* Now we open the loop at e as shown, and the resulting diagram becomes the one in Figure 3.26(b). The transfer functions from r to y, from p_1 to y, and from p_2 to y now become

$$G_{yr} = 0 \qquad G_{yp_1} = G_2 \qquad G_{yp_2} = 1$$

They are called *open-loop transfer functions.* We see that they are quite different from the closed-loop transfer functions in (3.57). This is not surprising, because the block diagrams in Figure 3.26 represent two distinctly different systems. Unless stated otherwise, all transfer functions in this text refer to closed-loop or overall transfer functions.

PROBLEMS

3.1. Consider a generator modeled as shown in Figure P3.1(a). The generator is driven by a diesel engine with a constant speed (not shown). The field circuit is assumed to have resistance R_f and inductance L_f; the generator has internal resistance R_g. The generated voltage $v_g(t)$ is assumed to be linearly proportional to the field current $i_f(t)$—that is, $v_g(t) = k_g i_f(t)$. Strictly speaking, $v_g(t)$ should appear at the generator terminals A and B. However, this would cause a loading problem in developing its block diagram. In order to eliminate this problem,

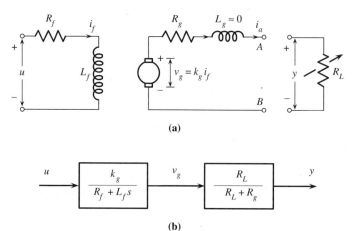

Figure P3.1

we assume that $v_g(t)$ appears as shown in Figure P3.1(a) and combine R_g with the load resistance R_L. Show that the generator can now be represented by the block diagram in Figure P3.1(b) and there is no loading problem in the block diagram. Where does the power of the generator come from? Does the flow of power appear in the block diagram?

3.2. Consider an armature-controlled dc motor driving a load as shown in Figure P3.2. Suppose the motor shaft is not rigid and must be modeled as a torsional spring with constant k_s. What is the transfer function from u to θ? What is the transfer function from u to $w = d\theta/dt$?

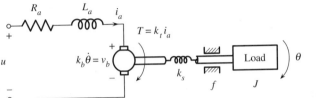

Figure P3.2

3.3. The combination of the generator and motor shown in Figure P3.3 is widely used in industry. It is called the *Ward-Leonard system*. Develop a block diagram for the system. Find the transfer functions from u to v_g and from v_g to θ.

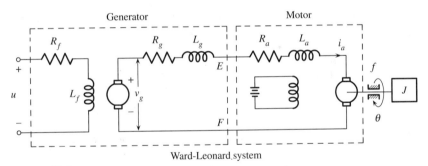

Ward-Leonard system

Figure P3.3

3.4. a. The system in Figure P3.4(a) can be used to control the voltage $v_o(t)$ of a generator that is connected to a load with resistance R_L. A reference signal $r(t)$, which can be adjusted by using a potentiometer, is applied through an amplifier with gain k_1 to the field circuit of the generator. Draw a block diagram for the system. The system is called an *open-loop voltage regulator*.

b. The system in Figure P3.4(b) can also be used to control the voltage $v_o(t)$. It differs from Figure P3.4(a) in that part of the output voltage or $v_1 = k_p v_o$ is subtracted from $r(t)$ to generate an error voltage $e(t)$. The error signal $e(t)$ is applied, through an amplifier with gain k_1, to the field circuit of the generator. Develop a block diagram for the system. The system is called a *closed-loop* or *feedback voltage regulator*.

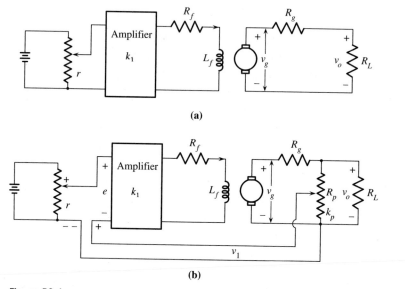

(a)

(b)

Figure P3.4

3.5. Consider the armature-controlled dc motor driving a load through a pair of gears shown in Figure 3.7(a). The following data are given:

$$R_a = 50\ \Omega \qquad k_t = 1\ \text{N·m·A}^{-1} \qquad k_b = 1\ \text{V·rad}^{-1}\text{·s},$$

$$N_1/N_2 = 1/2 \qquad J_2 = 12\ \text{N·m·rad}^{-1}\text{·s}^2 \qquad f_2 = 0.2\ \text{N·m·rad}^{-1}\text{·s},$$

$$J_1 = 0.1\ \text{N·m·rad}^{-1}\text{·s}^2, \qquad f_1 = 0.01\ \text{N·m·rad}^{-1}\text{·s}$$

Draw a block diagram for the system. Also compute their transfer functions.

3.6. Consider the gear train shown in Figure P3.6(a). Show that it is equivalent to the one in Figure P3.6(b) with

$$J_{1\text{eq}} = J_1 + J_2\left(\frac{N_1}{N_2}\right)^2 + J_3\left(\frac{N_1 N_3}{N_2 N_4}\right)^2$$

$$f_{1\text{eq}} = f_1 + f_2\left(\frac{N_1}{N_2}\right)^2 + f_3\left(\frac{N_1 N_3}{N_2 N_4}\right)^2$$

3.7. Find the transfer functions from v_i to v_o of the op-amp circuits shown in Figure P3.7.

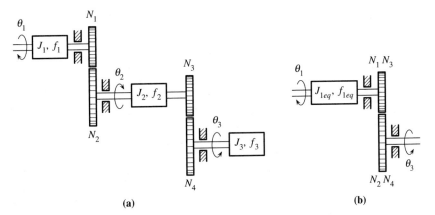

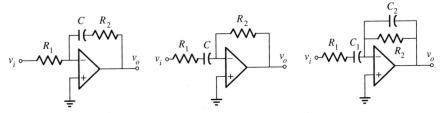

Figure P3.6

Figure P3.7

3.8. Consider the operational amplifier circuit shown in Figure P3.8. Show that the output v_o equals

$$v_o = \frac{R_f}{R}(v_B - v_A)$$

This is called a *differential* amplifier.

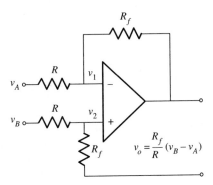

Figure P3.8

3.9. Develop a block diagram for the system shown in Figure P3.9. Compute also the transfer function of each block.

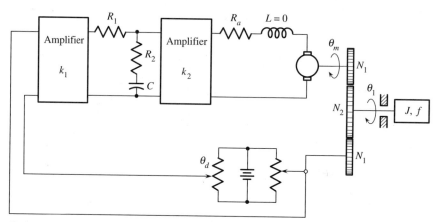

Figure P3.9

3.10. Consider the temperature control system shown in Figure P3.10. It is assumed that the heat q pumped into the chamber is proportional to u and the temperature y inside the chamber is related to q by the differential equation $dy(t)/dt = -0.3y(t) + 0.1q(t)$. Develop a block diagram for the system and compute the transfer function of each block.

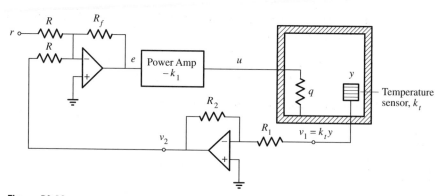

Figure P3.10

3.11. In addition to electromechanical transducers, optical transducers are also used in practice. The transducer shown in Figure P3.11 consists of a pulse generator and a pulse counter. By counting the number of pulses in a fixed interval of time, a signal proportional to the angular speed of the motor shaft can be generated. This signal is in digital form. Using a digital-to-analog converter and neglecting the so-called quantization error, the transfer function from θ to

v shown in Figure P3.10 can be approximated by ks. Draw a block diagram for the system. Indicate also the type of transfer function for each block.

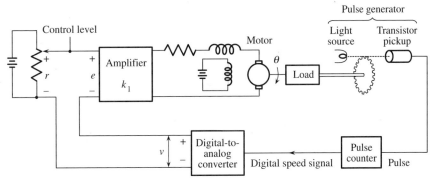

Figure P3.11

3.12. Machine tools can be controlled automatically by the instruction recorded on punched cards or tapes. This kind of control is called the *numerical control* of machine tools. Numerical control can be classified as position control and continuous contour control. A schematic diagram of a position control system is shown in Figure P3.12. A feedback loop is introduced in the D/A (digital-to-analog) converter to obtain a more accurate conversion. Draw a block diagram from θ_d to θ_o for the system. Indicate also the type of transfer function for each block.

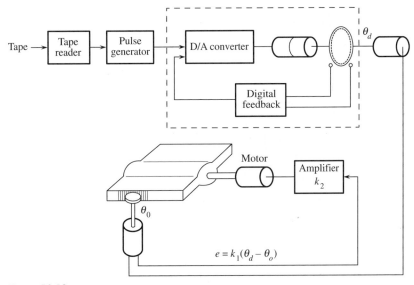

$$e = k_1(\theta_d - \theta_o)$$

Figure P3.12

3.13. In industry, a robot can be designed to replace a human operator to carry out repeated movements. Schematic diagrams for such a robot are shown in Figure P3.13. The desired movement is first applied by an operator to the joystick shown. The joystick activates the hydraulic motor and the mechanical arm. The movement of the arm is recorded on a tape, as shown in Figure P3.13(a). The tape can then be used to control the mechanical arm, as shown in Figure P3.13(b). It is assumed that the signal x is proportional to u, and the transfer function from x to y is $k_m/s(\tau_m s + 1)$. Draw a block diagram for the system in Figure P3.13(b). Indicate also the type of transfer function for each block.

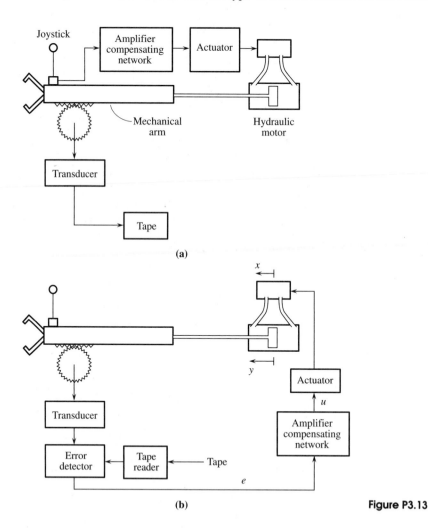

(a)

(b)

Figure P3.13

3.14. Use Table 3.1 to reduce the block diagrams shown in Figure P3.14 to single blocks.

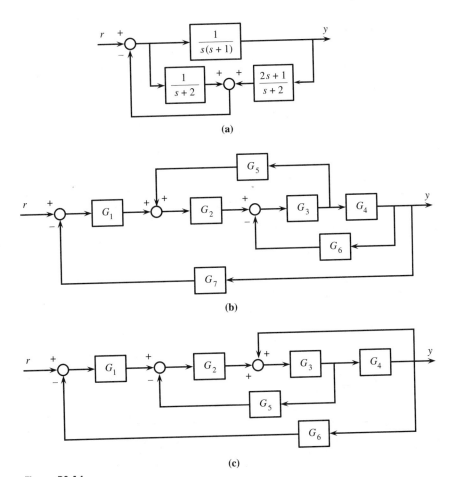

Figure P3.14

3.15. Use Mason's formula to repeat Problem 3.14.

3.16. Use Mason's formula to compute the transfer functions from r_1 to y and r_2 to y of the block diagram shown in Figure P3.16.

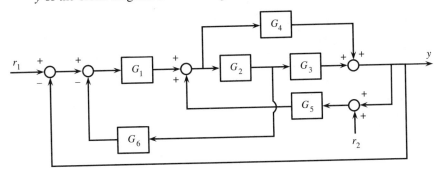

Figure P3.16

3.17. Consider the block diagram shown in Figure P3.17. It is the connection of the block diagram of an armature-controlled dc motor and that of a tachometer. Suppose noises may enter the system as shown. Compute the transfer functions from n_1 to w and from n_2 to w. Are they proper transfer functions?

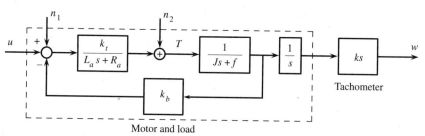

Motor and load

Figure P3.17

4 *Quantitative and Qualitative Analyses of Control Systems*

4.1 INTRODUCTION

Once block diagrams of control systems are developed, the next step is to carry out analysis. There are two types of analysis: quantitative and qualitative. In quantitative analysis, we are interested in the exact response of control systems due to a specific excitation. In qualitative analysis, we are interested in general properties of control systems. We discuss first the former and then the latter.

4.2 FIRST-ORDER SYSTEMS—THE TIME CONSTANT

Consider the armature-controlled dc motor driving a load, such as a video tape, shown in Figure 4.1. The objective is to drive the tape at a constant linear speed. To simplify the discussion, we assume that this can be achieved by maintaining the motor shaft at a constant angular speed.[1] Using a potentiometer, the desired speed is transformed into a reference signal $r(t)$. The reference signal $r(t)$ passes through an amplifier with gain k_1 to generate an actuating signal $u(t) = k_1 r(t)$, which then drives the motor. The transfer function from u to the motor shaft speed w is

---

[1]The radius of the reel changes from a full reel to an empty reel, therefore a constant angular velocity will not generate a constant linear tape speed. For this reason, the angular velocity of motor shafts that drive expensive compact disc (CD) players is not constant. The angular velocity increases gradually as the pick-up reads from the outer rim toward the inner rim so that the linear speed is constant.

111

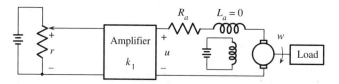

Figure 4.1 Open-loop velocity control system.

$k_m/(\tau_m s + 1)$, as is shown in Figure 3.4(b). Thus the transfer function from r to w is

$$G(s) = \frac{W(s)}{R(s)} = \frac{k_1 k_m}{\tau_m s + 1} \tag{4.1}$$

where k_m and τ_m are the motor gain and time constants (see (3.16) and (3.17)). If we apply a constant voltage, or $r(t) = a$, for $t \geq 0$, then $R(s) = a/s$, and

$$W(s) = \frac{k_1 k_m}{\tau_m s + 1} \cdot \frac{a}{s} = \frac{ak_m k_1}{s} - \frac{ak_m \tau_m k_1}{\tau_m s + 1} = \frac{ak_m k_1}{s} - \frac{ak_m k_1}{s + 1/\tau_m}$$

which implies

$$w(t) = ak_m k_1 - ak_m k_1 e^{-(1/\tau_m)t} \tag{4.2}$$

for $t \geq 0$. This response, shown in Figure 4.2(a), is called the *step response*; it is called the *unit-step response* if $a = 1$. Because e^{-t/τ_m} approaches zero as $t \to \infty$, we have

$$w_s(t) := \lim_{t \to \infty} w(t) = ak_m k_1$$

This is called the *steady-state* or *final speed*. If the desired speed is w_r, by choosing a as $a = w_r/k_1 k_m$, the motor will eventually reach the desired speed.

In controlling the tape, we are interested in not only the final speed but also the speed of response; that is, how fast the tape will reach the final speed. For the first-order transfer function in (4.1), the speed of response is dictated by τ_m, the time constant of the motor. We compute

t	e^{-t/τ_m}
τ_m	$(0.37)^1 = 0.37$
$2\tau_m$	$(0.37)^2 = 0.14$
$3\tau_m$	$(0.37)^3 = 0.05$
$4\tau_m$	$(0.37)^4 = 0.02$
$5\tau_m$	$(0.37)^5 = 0.007$

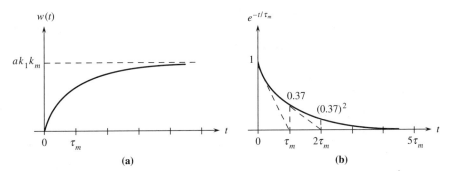

Figure 4.2 Time responses.

and plot e^{-t/τ_m} in Figure 4.2(b). We see that if $t \geq 5\tau_m$, the value of e^{-t/τ_m} is less than 1% of its original value. Therefore, the speed of the motor will reach and stay within 1% of its final speed in 5 time constants. In engineering, the system is often considered to have reached the final speed in 5 time constants.

The system in Figure 4.1 is an open-loop system because the actuating signal $u(t)$ is predetermined by the reference signal $r(t)$ and is independent of the actual motor speed. The motor time constant τ_m of this system depends on the motor and load (see (3.17)). For a given load, once a motor is chosen, the time constant is fixed. If the time constant is very large, for example, $\tau_m = 60$ s, then it will take 300 seconds or 5 minutes for the tape to reach the final speed. This speed of response

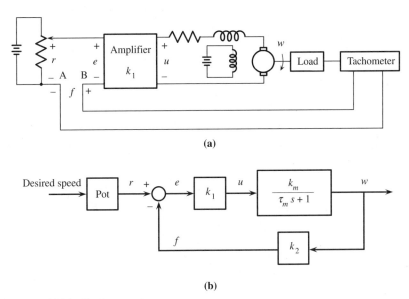

Figure 4.3 Feedback control system.

is much too slow. In this case, the only way to change the time constant is to choose a larger motor. If a motor is properly chosen, a system with good accuracy and fast response can be designed. This type of open-loop system, however, is sensitive to plant perturbation and disturbance, as is discussed in Chapter 6, so it is used only in inexpensive or low-quality speed control systems.

We now discuss a different type of speed control system. Consider the system shown in Figure 4.3(a). A tachometer is connected to the motor shaft and its output is combined with the reference input to generate an error signal. From the wiring shown, we have $e = r - f$. The block diagram is shown in Figure 4.3(b). Because the actuating signal u depends not only on the reference input but also the actual plant output, this is a closed-loop or feedback control system. Note that in developing the transfer function of the motor, if the moment of inertia of the tachometer is negligible compared to that of the load, it may be simply disregarded. Otherwise, it must be included in computing the transfer function of the motor and load.

The transfer function from r to w of the feedback control system in Figure 4.3 is

$$G_o(s) = \frac{W(s)}{R(s)} = \frac{\dfrac{k_1 k_m}{\tau_m s + 1}}{1 + \dfrac{k_1 k_m k_2}{\tau_m s + 1}} = \frac{k_1 k_m}{\tau_m s + k_1 k_2 k_m + 1} \tag{4.3}$$

$$= \frac{\dfrac{k_1 k_m}{k_1 k_2 k_m + 1}}{\left(\dfrac{\tau_m}{k_1 k_2 k_m + 1}\right) s + 1} =: \frac{k_1 k_o}{\tau_o s + 1}$$

where

$$\tau_o := \frac{\tau_m}{k_1 k_2 k_m + 1} \tag{4.4a}$$

$$k_o := \frac{k_m}{k_1 k_2 k_m + 1} \tag{4.4b}$$

This transfer function has the same form as (4.1). If $r(t) = a$, then we have, as in (4.2),

$$w(t) = a k_o k_1 - a k_o k_1 e^{-(1/\tau_o)t} \tag{4.5}$$

and the steady-state speed is $a k_o k_1$. With a properly chosen, the tape can reach a desired speed. Furthermore, it will reach the desired speed in $5 \times \tau_o$ seconds.

The time constant τ_o of the feedback system in Figure 4.3 is $\tau_m/(k_1 k_2 k_m + 1)$. It now can be controlled by adjusting k_1 or k_2. For example, if $\tau_m = 60$ and $k_m = 1$, by choosing $k_1 = 10$ and $k_2 = 4$, we have $\tau_o = 60/(40 + 1) = 1.46$, and the tape will reach the final speed in $5 \times 1.46 = 7.3$ seconds. Thus, unlike the open-loop system in Figure 4.1, the time constant and, more generally, the speed of response of the feedback system in Figure 4.3 can be easily controlled.

4.2.1 Effects of Feedback

We now use the systems in Figures 4.1 and 4.3 to discuss some of the effects of introducing feedback.

1. The time constant of the open-loop system can be changed only by changing the motor. However, if we introduce feedback, the time constant of the resulting system can easily be controlled by merely changing the gain of the amplifier. Thus, a feedback system is more flexible and the choice of a motor is less critical.
2. Although the motor time constant is reduced by a factor of $(k_1 k_2 k_m + 1)$ in the feedback system, as shown in (4.4a) (this is good), the motor gain constant is also reduced by the same factor, as shown in (4.4b) (this is bad). In order to compensate for this loss of gain, the applied reference voltage must be increased by the same factor. This is one of the prices of using feedback.
3. In order to introduce feedback, for example, from mechanical signals to electrical voltages, transducers such as tachometers must be employed. Transducers are expensive. Furthermore, they may introduce noise and consequently inaccuracies into systems. Therefore, feedback systems are more complex and require more components than open-loop systems do.
4. If used properly, feedback may improve the performance of a system. However, if used improperly, it may have a disastrous effect. For example, consider the feedback system in Figure 4.3(a). If the wiring at A and B is reversed, then $u(t) = k_1(r(t) + f(t))$ and the block diagram becomes as shown in Figure 4.4. This is a positive feedback system. Its transfer function from r to w is

$$G_o(s) = \frac{W(s)}{R(s)} = \frac{\dfrac{k_1 k_m}{\tau_m s + 1}}{1 - \dfrac{k_1 k_2 k_m}{\tau_m s + 1}} = \frac{k_1 k_m}{\tau_m s + 1 - k_1 k_2 k_m} \qquad (4.6)$$

If $\tau_m = 1$, $k_1 k_m = 10$ and $k_1 k_2 k_m = 5$, and if $r(t) = a$, for $t \geq 0$, then

$$W(s) = \frac{10}{s - 4} \cdot \frac{a}{s} = \frac{2.5a}{s - 4} - \frac{2.5a}{s}$$

which implies

$$w(t) = 2.5ae^{4t} - 2.5a$$

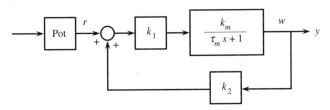

Figure 4.4 Positive feedback system.

We see that if $a \neq 0$, the term $2.5ae^{4t}$ approaches infinity as $t \to \infty$. In other words, the motor shaft speed will increase without bounds and the motor will burn out or disintegrate. For the simple system in Figure 4.3, this phenomenon will not happen for negative feedback. However, for more complex systems, the same phenomenon may happen even for negative feedback. Therefore, care must be exercised in using feedback. This is the stability problem and will be discussed later.

Exercise 4.2.1

Consider (4.6). If $k_1 k_2 k_m = 1$ and $r(t) = 10^{-2}$, for $t \geq 0$, what is the speed $w(t)$ of the motor shaft? What is its final speed?

[**Answers:** $k_1 k_m t/100\tau_m$, infinity.]

4.3   SECOND-ORDER SYSTEMS

Consider the position control of a load such as an antenna. The antenna is to be driven by an armature-controlled dc motor. Open-loop control of such a system is not used in practice because it is difficult to find a reference signal to achieve the control. (See Problem 4.6). A possible feedback design is shown in Figure 3.17. Its transfer function from r to y is

$$G_o(s) = \frac{Y(s)}{R(s)} = \frac{\dfrac{k_1 k_2 k_m}{s(\tau_m s + 1)}}{1 + \dfrac{k_1 k_2 k_m}{s(\tau_m s + 1)}} = \frac{k_1 k_2 k_m}{\tau_m s^2 + s + k_1 k_2 k_m}$$

$$= \frac{k_1 k_2 k_m/\tau_m}{s^2 + \dfrac{1}{\tau_m} s + \dfrac{k_1 k_2 k_m}{\tau_m}} \tag{4.7}$$

If we define $\omega_n^2 := k_1 k_2 k_m/\tau_m$, and $2\zeta\omega_n := 1/\tau_m$, then (4.7) becomes

$$G_o(s) = \frac{Y(s)}{R(s)} = \frac{\omega_n^2}{s^2 + 2\zeta\omega_n s + \omega_n^2} \tag{4.8}$$

This is called a *quadratic transfer function with a constant numerator*. In this section we study its unit-step response.

The transfer function $G_o(s)$ has two poles and no zero. Its poles are

$$-\zeta\omega_n \pm j\omega_n\sqrt{1 - \zeta^2} =: -\sigma \pm j\omega_d \tag{4.9}$$

where $\sigma := \zeta\omega_n$ and $\omega_d := \omega_n\sqrt{1 - \zeta^2}$. They are plotted in Figure 4.5. The constant ζ is called the *damping ratio*; ω_n, the *natural frequency*; σ, the *damping*

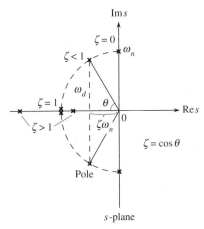

Figure 4.5 Poles of quadratic system.

factor; and ω_d, the *damped* or *actual frequency*. Clearly if the damping ratio ζ is 0, the two poles $\pm j\omega_n$ are pure imaginary. If $0 < \zeta < 1$, the two poles are complex conjugate and so forth as listed in the following:

Damping Ratio	Poles	Remark
$\zeta = 0$	Pure imaginary	Undamped
$0 < \zeta < 1$	Complex conjugate	Underdamped
$\zeta = 1$	Repeated real poles	Critically damped
$\zeta > 1$	Two distinct real poles	Overdamped

In order to see the physical significance of ζ, σ, ω_d, and ω_n, we first compute the unit-step response of (4.8) for $0 \le \zeta \le 1$. If $r(t) = 1$, for $t \ge 0$, then $R(s) = 1/s$ and

$$Y(s) = \frac{\omega_n^2}{s^2 + 2\zeta\omega_n s + \omega_n^2} \cdot \frac{1}{s} = \frac{\omega_n^2}{(s + \sigma + j\omega_d)(s + \sigma - j\omega_d)s}$$

$$= \frac{k_1}{s} + \frac{k_2}{s + \sigma + j\omega_d} + \frac{k_2^*}{s + \sigma - j\omega_d}$$

with

$$k_1 = G_o(s)\big|_{s=0} = 1$$

$$k_2 = \frac{\omega_n^2}{(s + \sigma - j\omega_d)s}\bigg|_{s = -\sigma - j\omega_d} = \frac{\omega_n^2}{(-2j\omega_d)(-\sigma - j\omega_d)} = \frac{\omega_n^2}{(2j\omega_d)(\sigma + j\omega_d)}$$

and

$$k_2^* = \frac{\omega_n^2}{(-2j\omega_d)(\sigma - j\omega_d)}$$

where k_2^* is the complex conjugate of k_2. Thus the unit-step response of (4.8) is

$$y(t) = 1 + k_2 e^{-(\sigma + j\omega_d)t} + k_2^* e^{-(\sigma - j\omega_d)t}$$

which, after some manipulation, becomes

$$y(t) = 1 - \frac{\omega_n}{\omega_d} e^{-\sigma t} \sin(\omega_d t + \theta) \tag{4.10}$$

where

$$\theta = \cos^{-1} \zeta = \tan^{-1} \frac{\sqrt{1 - \zeta^2}}{\zeta} = \sin^{-1} \sqrt{1 - \zeta^2} \tag{4.11}$$

The angle θ is shown in Figure 4.5. We plot $\sin(\omega_d t + \theta)$ and $e^{-\sigma t}$ in Figure 4.6(a) and (b). The point-by-point product of (a) and (b) yields $e^{-\sigma t} \sin(\omega_d t + \theta)$. We see that the frequency of oscillation is determined by ω_d, the imaginary part of the poles in (4.9). Thus, ω_d is called the *actual frequency*; it is the distance of the poles from the real axis. Note that ω_n is called the *natural frequency*; it is the distance of the poles from the origin. The envelope of the oscillation is determined by the damping factor σ, the real part of the poles. Thus, the poles of $G_o(s)$ dictate the response of $e^{-\sigma t} \sin(\omega_d t + \theta)$, and consequently of $y(t)$ in (4.10). The unit-step response $y(t)$ approaches 1 as $t \to \infty$. Thus the final value of $y(t)$ is 1.

We now compute the maximum value or the peak of $y(t)$. The differentiation of $y(t)$ yields

$$\frac{dy(t)}{dt} = \sigma \frac{\omega_n}{\omega_d} e^{-\sigma t} \sin(\omega_d t + \theta) - \omega_n e^{-\sigma t} \cos(\omega_d t + \theta)$$

The peak of $y(t)$ occurs at a solution of $dy(t)/dt = 0$ or, equivalently, a stationary point of $y(t)$. Setting $dy(t)/dt$ to zero yields

$$\frac{\sin(\omega_d t + \theta)}{\cos(\omega_d t + \theta)} = \tan(\omega_d t + \theta) = \frac{\omega_d}{\sigma} = \frac{\omega_n \sqrt{1 - \zeta^2}}{\zeta \omega_n} = \frac{\sqrt{1 - \zeta^2}}{\zeta} \tag{4.12}$$

By comparing (4.11) and (4.12), we conclude that the solutions of (4.12) are

$$\omega_d t = k\pi \qquad k = 0, 1, 2, \ldots$$

Thus the stationary points of $y(t)$ occur at $t = k\pi/\omega_d, k = 0, 1, \ldots$. We plot $y(t)$ in Figure 4.7 for various damping ratios ζ. From the plot, we see that the peak occurs at $k = 1$ or

$$t_p := \frac{\pi}{\omega_d} = \frac{\pi}{\omega_n \sqrt{1 - \zeta^2}}$$

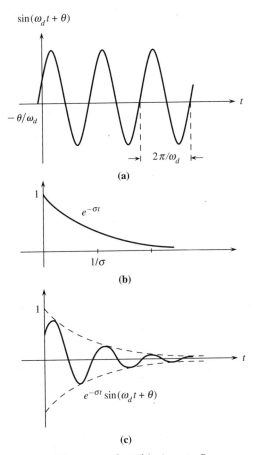

Figure 4.6 Response of $e^{-\sigma t}\sin(\omega_d t + \theta)$.

The substitution of t_p into (4.10) yields

$$y_{\max} := y(t_p) = 1 - \frac{\omega_n}{\omega_d} e^{-\sigma t_p} \sin(\pi + \theta) = 1 + \frac{\omega_n}{\omega_d} e^{-\sigma t_p} \sin\theta$$

which, because $\sin\theta = \omega_d/\omega_n$ (see Figure 4.5), reduces to

$$y_{\max} = 1 + e^{-\sigma t_p} = 1 + e^{-\zeta\omega_n\pi/\omega_n\sqrt{1-\zeta^2}} = 1 + e^{-\zeta\pi/\sqrt{1-\zeta^2}} \quad (4.13)$$

This is the peak of $y(t)$. It depends only on the damping ratio ζ. If $y_{\max}$ is larger than the final value $y(\infty) = 1$, the response is said to have an *overshoot*. From Figure 4.7, we see that the response has an overshoot if $\zeta < 1$. If $\zeta \geq 1$, then there is no overshoot.

We consider again the unit-step response $y(t)$ in (4.10). If the damping ratio ζ is zero, or $\sigma = \zeta\omega_n = 0$, then $e^{-\sigma t}\sin(\omega_d t + \theta)$ reduces to a pure sinusoidal function and $y(t)$ will remain oscillatory for all t. Thus the system in (4.8) is said to

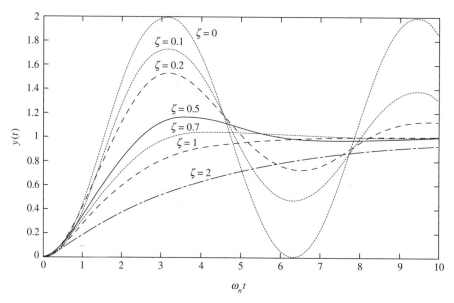

Figure 4.7 Responses of quadratic system with various damping ratios.

be *undamped*. If $0 < \zeta < 1$, the response $y(t)$ contains oscillation whose envelope decreases with time as shown in Figure 4.7. In this case, the system in (4.8) is said to be *underdamped*. If $\zeta > 1$, the two poles of (4.8) are real and distinct and the unit-step response of (4.8) will contain neither oscillation nor overshoot. In this case, the system is said to be *overdamped*. The system is said to be *critically damped* if $\zeta = 1$. In this case, the two poles are real and repeated, and the response is on the verge of having overshoot or oscillation.

The step response of (4.8) is dictated by the natural frequency ω_n and the damping ratio ζ. Because the horizontal coordinate of Figure 4.7 is $\omega_n t$, the larger ω_n, the faster the response. The damping ratio governs the overshoot; the smaller ζ, the larger the overshoot. We see from Figure 4.7 that, if ζ is in the neighborhood of 0.7, then the step response has no appreciable overshoot and has a fairly fast response. Therefore, we often design a system to have a damping ratio of 0.7. Because the response also depends on ω_n, we like to control both ω_n and ζ in the design.

Example 4.3.1

The transfer function of the automobile suspension system shown in Figure 2.7 is

$$G(s) = \frac{1}{ms^2 + k_1 s + k_2} = \frac{1/m}{s^2 + \dfrac{k_1}{m}s + \dfrac{k_2}{m}}$$

If the shock absorber is dead (that is, it does not generate any friction), or $k_1 = 0$, then the damping ratio ζ is zero. In this case, the car will remain oscillatory after hitting a pothole and the car will be difficult to steer. By comparing the transfer function of the suspension system and (4.8), we have

$$\frac{k_2}{m} = \omega_n^2 \qquad \frac{k_1}{m} = 2\zeta\omega_n$$

If we choose $\zeta = 0.7$ and $\omega_n = 2$, then from Figure 4.7, we can see that the automobile will take about 2 seconds to return to the original horizontal position after hitting a pothole and will hardly oscillate. To have these values, k_1 and k_2 must be

$$k_2 = 2^2 m = 4m \qquad k_1 = 2 \cdot 0.7 \cdot 2m = 2.8m$$

Thus, the suspension system of an automobile can be controlled by using suitable k_1 and k_2.

Exercise 4.3.1

Find the damping ratio, damping factor, natural frequency, and actual frequency of the following systems. Also classify them in terms of dampedness.

a. $G(s) = \dfrac{9}{2s^2 + 9}$

b. $G(s) = \dfrac{9}{s^2 + 3s + 9}$

c. $G(s) = \dfrac{9}{s^2 + 12s + 9}$

[**Answers:** (a) 0, 0, $\sqrt{4.5}$, $\sqrt{4.5}$, undamped; (b) 0.5, 1.5, 3, 2.6, underdamped;
(c) $\zeta = 2$, $\omega_n = 3$, the other two not defined, overdamped.]

With the preceding discussion, we are now ready to study the position control system in (4.7). Its block diagram was developed in Figure 3.17(b) and is repeated in Figure 4.8(a). In the block diagram, k_m and τ_m are fixed by the motor and load. The amplifier gain k_2 clearly can be adjusted; so can the sensitivity of the error detector (by changing the power supply E). Although both k_1 and k_2 can be changed, because

$$\omega_n = \sqrt{k_1 k_2 k_m / \tau_m} \qquad \text{and} \qquad \zeta = \frac{1}{2\tau_m \omega_n}$$

only one of ω_n and ζ can be arbitrarily assigned. For example, if k_1 and k_2 are chosen so that $\omega_n = 10$, we may end up with $\zeta = 0.05$. If k_1 and k_2 are chosen so that

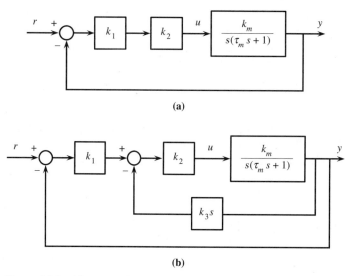

(a)

(b)

Figure 4.8 Position control systems.

$\zeta = 2$, we may end up with $\omega_n = 0.2$. Their step responses are shown in Figure 4.9. The former has too much oscillation; the latter has no oscillation but is much too slow. Thus both responses are not satisfactory. How to choose k_1 and k_2 to yield a satisfactory response is a design problem and will be discussed in later chapters.

Exercise 4.3.2

(a) Consider the position control system in (4.7). Suppose $\tau_m = 4$ and $k_m = 0.25$; find k_1 and k_2 so that $\omega_n = 0.25$. What is ζ? Use Figure 4.7 to sketch roughly its unit-step response. (b) Can you find k_1 and k_2 so that $\omega_n = 0.25$ and $\zeta = 0.7$?

[**Answers:** $k_1 k_2 = 1, \zeta = 0.5$, no.]

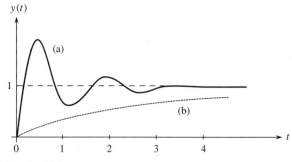

Figure 4.9 Step responses.

Exercise 4.3.3

Suppose the position control system in Figure 4.8(a) cannot achieve the design objective. We then introduce an additional tachometer feedback with sensitivity k_3 as shown in Figure 4.8(b). Show that its transfer function from r to y is

$$G_o(s) = \frac{Y(s)}{R(s)} = \frac{\dfrac{k_1 k_2 k_m}{\tau_m}}{s^2 + \left(\dfrac{1 + k_2 k_3 k_m}{\tau_m}\right) s + \dfrac{k_1 k_2 k_m}{\tau_m}} =: \frac{\omega_n^2}{s^2 + 2\zeta\omega_n s + \omega_n^2}$$

For this system, is it possible to assign ζ and ω_n arbitrarily by adjusting k_1 and k_2? If $\tau_m = 4$, $k_m = 0.25$, and $k_3 = 1$, find k_1 and k_2 so that $\omega_n = 0.25$ and $\zeta = 0.7$.

[**Answers:**   Yes, $k_1 = 1/1.6$, $k_2 = 1.6$.]

4.4   TIME RESPONSES OF POLES

From the preceding two sections, we see that poles of overall transfer functions essentially determine the speed of response of control systems. In this section, we shall discuss further the time response of poles.

Poles can be real or complex, simple or repeated. It is often convenient to plot them on the complex plane or s-plane as shown in Figure 4.10. Their corresponding responses are also plotted. The s-plane can be divided into three parts: the right half plane (RHP), the left half plane (LHP) and the pure imaginary axis or $j\omega$-axis. To avoid possible confusion whether the RHP includes the $j\omega$-axis or not, we shall use the following convention: The *open* RHP is the RHP excluding the $j\omega$-axis and the *closed* RHP is the RHP including the $j\omega$-axis. If a pole lies inside the open LHP, then the pole has a negative real part; its imaginary part can be positive or negative. If a pole lies inside the closed RHP, then the pole has a positive or zero real part.

Poles and zeros are usually plotted on the s-plane using crosses and circles. Note that no zeros are plotted in Figure 4.10. Consider $1/(s + a)^n$ or the pole at $-a$ with multiplicity n. The pole is a simple pole if $n = 1$, a repeated pole if $n > 1$. To simplify the discussion, we assume a to be real. Its time response, using Table A.1, is

$$\frac{1}{n!} t^{n-1} e^{-at} \tag{4.14}$$

If the pole $-a$ is in the open RHP, or $a < 0$, then its response increases exponentially to infinity for $n = 1, 2, \ldots$. If the pole is at the origin, or $a = 0$, and is simple, then its response is a step function. If it is repeated, with multiplicity $n \geq 2$, then its response is $t^{n-1}/n!$, which approaches infinity as $t \to \infty$. If the real pole is in the open LHP, or $a > 0$, and is simple, then its response is e^{-at}, which decreases

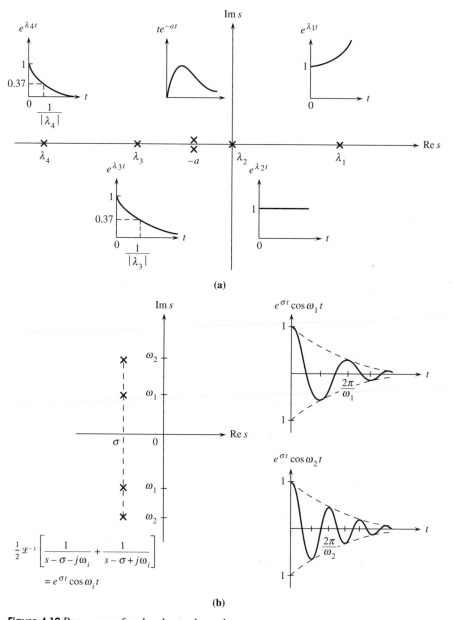

Figure 4.10 Responses of real and complex poles.

exponentially to zero as $t \to \infty$. If it is repeated with multiplicity 2, then its response is

$$te^{-at}$$

the product of t, which goes to ∞, and e^{-at}, which goes to 0 as $t \to \infty$. Therefore, it requires some computation to find its value at $t \to \infty$. We use l'Hôpital's rule to compute

$$\lim_{t \to \infty} te^{-at} = \lim_{t \to \infty} \frac{t}{e^{at}} = \lim_{t \to \infty} \frac{1}{ae^{at}} = 0$$

Thus, as plotted in Figure 4.10(a), the time response te^{-at} approaches zero as $t \to \infty$. Similarly, we can show

$$t^n e^{-at} \to 0 \qquad \text{as } t \to \infty$$

for $a > 0$, and $n = 1, 2, 3, \ldots$. This is due to the fact that the exponential e^{-at}, with $a > 0$, approaches zero with a rate much faster than the rate at which t^n approaches infinity. Thus, we conclude that the time response of any simple or repeated real pole that lies inside the open LHP approaches 0 as $t \to \infty$.

The situation for complex conjugate poles is similar to the case of real poles with the exception that the responses go to 0 or ∞ oscillatorily. Therefore we will not repeat the discussion. Instead we summarize the preceding discussion in the following table:

Table 4.1   Time Responses of Poles as $t \to \infty$

Poles	Simple ($n = 1$)	Repeated ($n \geq 2$)
Open LHP	0	0
Open RHP	$\pm\infty$	$\pm\infty$
Origin (s^n)	A constant	∞
$j\omega$-axis$((s^2 + a^2)^n)$	A sustained oscillation	$\pm\infty$

This table implies the following facts, which will be used later.

1. The time response of a pole, simple or repeated, approaches zero as $t \to \infty$ if and only if the pole lies inside the open LHP or has a negative real part.
2. The time response of a pole approaches a nonzero constant as $t \to \infty$ if and only if the pole is simple and located at $s = 0$.

4.5   STABILITY

In this section, a qualitative property of control systems—namely, stability—will be introduced. The concept of stability is very important because every control system must be stable. If a control system is not stable, it will usually burn out or disintegrate. There are three types of stability, bounded-input bounded-output (BIBO) stability, marginal stability (or stability in the sense of Lyapunov), and asymptotic stability. In this text, we study only BIBO stability. Therefore, the adjective BIBO will be dropped.

A function $u(t)$ defined for $t \geq 0$ is said to be *bounded* if its magnitude does not approach infinity or, equivalently, there exists a constant M such that

$$|u(t)| \leq M < \infty$$

for all $t \geq 0$.

□ Definition   4.1

A system is stable if *every* bounded input excites a bounded output. Otherwise the system is said to be unstable.   ∎

Example 4.5.1

Consider the network shown in Figure 4.11(a). The input u is a current source; the output y is the voltage across the capacitor. Using the equivalent Laplace transform circuit in Figure 4.11(b), we can readily obtain

$$Y(s) = \frac{s \cdot \dfrac{1}{s}}{s + \dfrac{1}{s}} U(s) = \frac{s}{s^2 + 1} U(s) \qquad (4.15)$$

If we apply the bounded input $u(t) = 1$, for $t \geq 0$, then the output is

$$Y(s) = \frac{s}{s^2 + 1} \cdot \frac{1}{s} = \frac{1}{s^2 + 1}$$

which implies

$$y(t) = \sin t$$

It is bounded. If we apply the bounded input $u(t) = \sin at$, for $t \geq 0$, where a is a positive real constant and $a \neq 1$, then the output is

$$Y(s) = \frac{s}{s^2 + 1} \cdot \frac{a}{s^2 + a^2} = \frac{as[(s^2 + a^2) - (s^2 + 1)]}{(a^2 - 1)(s^2 + 1)(s^2 + a^2)}$$

$$= \frac{a}{a^2 - 1} \cdot \frac{s}{s^2 + 1} - \frac{a}{a^2 - 1} \cdot \frac{s}{s^2 + a^2}$$

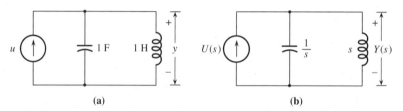

(a)   (b)

Figure 4.11 Network.

which implies

$$y(t) = \frac{a}{a^2 - 1} [\cos t - \cos at]$$

It is bounded for any $a \neq 1$. Thus the outputs due to the bounded inputs $u(t) = 1$ and $\sin at$ with $a \neq 1$ are all bounded. Even so, we cannot conclude the stability of the network because we have not yet checked *every* possible bounded input. In fact, the network is not stable, because the application of $u(t) = \sin t$ yields

$$Y(s) = \frac{s}{s^2 + 1} \cdot \frac{1}{s^2 + 1} = \frac{s}{(s^2 + 1)^2}$$

which, using Table A.1, implies

$$y(t) = \frac{1}{2} t \sin t$$

This output $y(t)$ approaches positive or negative infinity as $t \to \infty$. Thus the bounded input $u(t) = \sin t$ excites an unbounded output, and the network is not stable.

Exercise 4.5.1

Consider a system with transfer function $1/s$. It is called an *integrator*. If we apply the bounded input $\sin at$, will the output be bounded? Can you find a bounded input that excites an unbounded output? Is the system stable?

[**Answers:** Yes, step function, no.]

The instability of a system can be deduced from Definition 4.1 by finding a single bounded input that excites an unbounded output. However, it is difficult to deduce stability from the definition because there are infinitely many bounded inputs to be checked. Fortunately we have the following theorem.

THEOREM 4.1

A system with proper rational transfer function $G(s)$ is stable if and only if every pole of $G(s)$ has a negative real part or, equivalently, lies inside the open left half s-plane. ∎

By open left half s-plane, we mean the left half s-plane excluding the $j\omega$-axis. This theorem implies that a system is unstable if its transfer function has one or more poles with zero or positive real parts. This theorem can be argued intuitively by using Table 4.1. If a transfer function has one or more open right half plane poles, then most bounded inputs will excite these poles and their responses will approach infinity. If the transfer function has a simple pole on the imaginary axis, we may apply a bounded input whose Laplace transform has the same pole. Then its response will approach infinity. Thus a stable system cannot have any pole in the closed right half s-plane. For a proof of the theorem, see Reference [15] or [18].

We remark that the stability of a system depends only on the poles of its transfer function $G(s)$ and does not depend on the zeros of $G(s)$. If all poles of $G(s)$ lie inside the open LHP, the system is stable no matter where the zeros of $G(s)$ are. For convenience, a pole is called a *stable pole* if it lies inside the open LHP or has a negative real part. A pole is called an *unstable pole* if it lies inside the closed RHP or has a zero or positive real part. A zero that lies inside the open LHP (closed RHP) will be called a *minimum-phase* (*nonminimum-phase*) *zero*. The reason for using such names will be given in Chapter 8.

Now we shall employ Theorem 4.1 to study the stability of the network in Figure 4.11. The transfer function, as developed in (4.15), is

$$G(s) = \frac{s}{s^2 + 1}$$

Its poles are $\pm j$; they have zero real part and are unstable poles. Thus, the network is not stable.

Most control systems are built by interconnecting a number of subsystems. In studying the stability of a control system, there is no need to study the stability of its subsystems. All we have to do is to compute the overall transfer function and then apply Theorem 4.1. We remark that a system can be stable with unstable subsystems and vice versa. For example, consider the system in Figure 4.12(a). It consists of two subsystems with transfer functions -2 and $1/(s + 1)$. Both subsystems are stable. However, the transfer function of the overall feedback system is

$$G_o(s) = \frac{\dfrac{-2}{s + 1}}{1 + \dfrac{-2}{s + 1}} = \frac{-2}{s + 1 - 2} = \frac{-2}{s - 1}$$

which is unstable. The overall system shown in Figure 4.12(b) is stable because its transfer function is

$$G_o(s) = \frac{\dfrac{2}{s - 1}}{1 + \dfrac{2}{s - 1}} = \frac{2}{s - 1 + 2} = \frac{2}{s + 1}$$

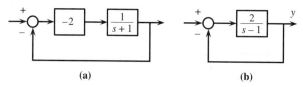

(a)   (b)

Figure 4.12 Stability of overall system and subsystems.

Its subsystem has transfer function $2/(s - 1)$ and is unstable. Thus the stability of a system is independent of the stabilities of its subsystems. Note that a system with transfer function

$$G(s) = \frac{s^2 - 2s - 3}{(s + 2)(s - 3)(s + 10)} \qquad (4.16)$$

is stable, because 3 is not a pole of $G(s)$. Recall that whenever we encounter rational functions, we reduce them to irreducible ones. Only then are the roots of the denominator poles. Thus, the poles of $G(s) = (s - 3)(s + 1)/(s + 2)(s - 3)(s + 10) = (s + 1)/(s + 2)(s + 10)$ are -2 and -10. They are both stable poles, and the transfer function in (4.16) is stable.

Exercise 4.5.2

Are the following systems stable?

a. $\dfrac{s - 1}{s + 1}$

b. $\dfrac{s^2 - 1}{s^2 + 2s + 2}$

c. $\dfrac{s - 1}{s^2 - 1}$

d. The network shown in Figure 4.13(a)

e. The feedback system shown in Figure 4.13(b)

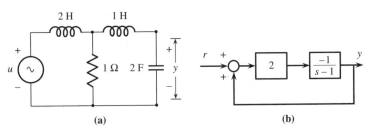

(a)   (b)

Figure 4.13 (a) Network. (b) Feedback system.

[**Answers:** All are stable.]

4.6 THE ROUTH TEST

Consider a system with transfer function $G(s) = N(s)/D(s)$. It is assumed that $N(s)$ and $D(s)$ have no common factor. To determine the stability of $G(s)$ by using Theorem 4.1, we must first compute the poles of $G(s)$ or, equivalently, the roots of $D(s)$. If the degree of $D(s)$ is three or higher, hand computation of the roots is complicated. Therefore, it is desirable to have a method of determining stability *without* solving for the roots. We now introduce such a method, called the *Routh test* or the *Routh-Hurwitz test*.

□ **Definition   4.2**

A polynomial with real coefficients is called a *Hurwitz* polynomial if all its roots have negative real parts.   ■

The Routh test is a method of checking whether or not a polynomial is a Hurwitz polynomial without solving for its roots. The most important application of the test is to check the stability of control systems. Consider the polynomial

$$D(s) = a_n s^n + a_{n-1} s^{n-1} + \cdots + a_1 s + a_0 \qquad a_n > 0 \qquad (4.17)$$

where a_i, $i = 0, 1, \ldots, n$, are real constants. If the leading coefficient a_n is negative, we may simply multiply $D(s)$ by -1 to yield a positive a_n. Note that $D(s)$ and $-D(s)$ have the same set of roots; therefore, $a_n > 0$ does not impose any restriction on $D(s)$.

Necessary Condition for a Polynomial To Be Hurwitz

We discuss first a necessary condition for $D(s)$ to be Hurwitz. If $D(s)$ in (4.17) is Hurwitz, then every coefficient of $D(s)$ must be positive. In other words, if $D(s)$ has a missing term (a zero coefficient) or a negative coefficient, then $D(s)$ is not Hurwitz. We use an example to establish this condition. We assume that $D(s)$ has two real roots and a pair of complex conjugate roots and is factored as

$$\begin{aligned} D(s) &= a_4(s + \alpha_1)(s + \alpha_2)(s + \beta_1 + j\gamma_1)(s + \beta_1 - j\gamma_1) \\ &= a_4(s + \alpha_1)(s + \alpha_2)(s^2 + 2\beta_1 s + \beta_1^2 + \gamma_1^2) \end{aligned} \qquad (4.18)$$

The roots of $D(s)$ are $-\alpha_1$, $-\alpha_2$, and $-\beta_1 \pm j\gamma_1$. If $D(s)$ is Hurwitz, then $\alpha_1 > 0$, $\alpha_2 > 0$, and $\beta_1 > 0$. Note that γ_1 can be positive or negative. Hence, all coefficients in the factors are positive. It is clear that all coefficients will remain positive after multiplying out the factors. This shows that if $D(s)$ is Hurwitz, then its coefficients must be all positive. This condition is also sufficient for polynomials of degree 1 or 2 to be Hurwitz. It is clear that if a_1 and a_0 in $a_1 s + a_0$ are positive, then $a_1 s + a_0$ is Hurwitz. If the three coefficients in

$$D(s) = a_2 s^2 + a_1 s + a_0 \qquad (4.19)$$

are all positive, then it is Hurwitz (see Exercise 4.6.2). In conclusion, for a polynomial of degree 1 or 2 with a positive leading coefficient, the condition that all coefficients are positive is necessary and sufficient for the polynomial to be Hurwitz. However, for a polynomial of degree 3 or higher, the condition is necessary but not sufficient. For example, the polynomial

$$s^3 + 2s^2 + 9s + 68 = (s + 4)(s - 1 + 4j)(s - 1 - 4j)$$

is not Hurwitz although its coefficients are all positive.

Necessary and Sufficient Condition

Now we discuss a necessary and sufficient condition for $D(s)$ to be Hurwitz. For easy presentation, we consider the polynomial of degree 6:

$$D(s) = a_6 s^6 + a_5 s^5 + a_4 s^4 + a_3 s^3 + a_2 s^2 + a_1 s + a_0 \qquad a_6 > 0$$

We form Table 4.2.[2] The first two rows are formed from the coefficients of $D(s)$. The first coefficient (in descending power of s) is put in the $(1, 1)$ position, the second in the $(2, 1)$ position, the third in the $(1, 2)$ position, the fourth in the $(2, 2)$ position and so forth. The coefficients are renamed as shown for easy development of recursive equations. Next we compute $k_5 = b_{61}/b_{51}$; it is the ratio of the first elements of the first two rows. The third row is obtained as follows. We subtract the product of the second row and k_5 from the first row:

$$b_{40} = b_{61} - k_5 b_{51} \qquad b_{41} = b_{62} - k_5 b_{52}$$

$$b_{42} = b_{63} - k_5 b_{53} \qquad b_{43} = b_{64} - k_5 \cdot 0$$

Note that b_{40} is always zero. The result is placed at the right hand side of the second row. We then discard the first element, which is zero, and place the remainder in the third row as shown in Table 4.2. The fourth row is obtained in the same manner from its two previous rows. That is, we compute $k_4 = b_{51}/b_{41}$, the ratio of the first elements of the second and third rows, and then subtract the product of the third row and k_4 from the second row:

$$b_{30} = b_{51} - k_4 b_{41} \qquad b_{31} = b_{52} - k_4 b_{42} \qquad b_{32} = b_{53} - k_4 b_{43}$$

We drop the first element, which is zero, and place the remainder in the fourth row as shown in Table 4.2. We repeat the process until the row corresponding to $s^0 = 1$ is obtained. If the degree of $D(s)$ is n, there should be a total of $(n + 1)$ rows. The table is called the *Routh table*.

We remark on the size of the table. If $n = \deg D(s)$ is even, the first row has one more entry than the second row. If n is odd, the first two rows have the same number of entries. In either case, the number of entries decreases by one at odd powers of s. For example, the number of entries in the rows of s^5, s^3, and s is one less than that of their preceding rows. We also remark that the rightmost entries of the rows corresponding to even powers of s are the same. For example, in Table 4.2, we have $b_{64} = b_{43} = b_{22} = b_{01} = a_0$.

---

[2]The presentation is slightly different from the cross-product method; it requires less computation and is easier to program on a digital computer. See Problem 4.16.

Table 4.2 The Routh Table

	s^6	$b_{61} := a_6$	$b_{62} := a_4$	$b_{63} := a_2$	$b_{64} := a_0$	
$k_5 = \dfrac{b_{61}}{b_{51}}$	s^5	$b_{51} := a_5$	$b_{52} := a_3$	$b_{53} := a_1$		(1st row) $- k_5$(2nd row) $= [b_{40} \quad b_{41} \quad b_{42} \quad b_{43}]$
$k_4 = \dfrac{b_{51}}{b_{41}}$	s^4	b_{41}	b_{42}	b_{43}		(2nd row) $- k_4$(3rd row) $= [b_{30} \quad b_{31} \quad b_{32} \quad b]$
$k_3 = \dfrac{b_{41}}{b_{31}}$	s^3	b_{31}	b_{32}			(3rd row) $- k_3$(4th row) $= [b_{20} \quad b_{21} \quad b_{22}]$
$k_2 = \dfrac{b_{31}}{b_{21}}$	s^2	b_{21}	b_{22}			(4th row) $- k_2$(5th row) $= [b_{10} \quad b_{11}]$
$k_1 = \dfrac{b_{21}}{b_{11}}$	s^1	b_{11}				(5th row) $- k_1$(6th row) $= [b_{00} \quad b_{01}]$
	s^0	b_{01}				

THEOREM 4.2 (The Routh Test)

A polynomial with a positive leading coefficient is a Hurwitz polynomial if and only if every entry in the Routh table is positive or, equivalently, if and only if every entry in the first column of the table (namely, b_{61}, b_{51}, b_{41}, b_{31}, b_{21}, b_{11}, b_{01}) is positive. ∎

It is clear that if all the entries of the table are positive, so are all the entries in the first column. It is rather surprising that the converse is also true. In employing the theorem, either condition can be used. A proof of this theorem is beyond the scope of this text and can be found in Reference [18]. This theorem implies that if a zero or a negative number appears in the table, then the polynomial is not Hurwitz. In this case, it is unnecessary to complete the table.

Example 4.6.1

Consider $2s^4 + s^3 + 5s^2 + 3s + 4$. We form

$$k_3 = \frac{2}{1} \quad \begin{array}{c|ccc} s^4 & 2 & 5 & 4 \\ s^3 & 1 & 3 \\ s^2 & -1 & 4 \end{array} \qquad [0 \quad -1 \quad 4] = \text{(1st row)} - k_3\text{(2nd row)}$$

Clearly we have $k_3 = 2/1$, the ratio of the first entries of the first two rows. The result of subtracting the product of the second row and k_3 from the first row is placed on the right hand side of the s^3-row. We drop the first element, which is zero, and put the rest in the s^2-row. A negative number appears in the table, therefore the polynomial is not Hurwitz.

Example 4.6.2

Consider $2s^5 + s^4 + 7s^3 + 3s^2 + 4s + 2$. We form

$$
k_4 = \frac{2}{1} \quad
\begin{array}{c|ccc}
s^5 & 2 & 7 & 4 \\
s^4 & 1 & 3 & 2 \\
s^3 & 1 & 0 &
\end{array}
\qquad [0 \quad 1 \quad 0] = (\text{1st row}) - k_4(\text{2nd row})
$$

A zero appears in the table, thus the polynomial is not Hurwitz. The reader is advised to complete the table and verify that a negative number appears in the first column.

Example 4.6.3

Consider $2s^5 + s^4 + 7s^3 + 3s^2 + 4s + 1.5$. We form

$$
\begin{array}{c|ccc}
s^5 & 2 & 7 & 4 \\
\end{array}
$$

$$
k_4 = \frac{2}{1} \quad
\begin{array}{c|ccc}
s^4 & 1 & 3 & 1.5 \\
\end{array}
\qquad [0 \quad 1 \quad 1] = (\text{1st row}) - k_4(\text{2nd row})
$$

$$
k_3 = \frac{1}{1} \quad
\begin{array}{c|cc}
s^3 & 1 & 1 \\
\end{array}
\qquad [0 \quad 2 \quad 1.5] = (\text{2nd row}) - k_3(\text{3rd row})
$$

$$
k_2 = \frac{1}{2} \quad
\begin{array}{c|cc}
s^2 & 2 & 1.5 \\
\end{array}
\qquad [0 \quad 0.25] = (\text{3rd row}) - k_2(\text{4th row})
$$

$$
k_1 = \frac{2}{0.25} \quad
\begin{array}{c|c}
s^1 & 0.25 \\
\end{array}
\qquad [0 \quad 1.5] = (\text{4th row}) - k_1(\text{5th row})
$$

$$
\begin{array}{c|c}
s^0 & 1.5 \\
\end{array}
$$

Every entry in the table is positive, therefore the polynomial is Hurwitz.

Exercise 4.6.1

Are the following polynomials Hurwitz?

a. $2s^4 + 2s^3 + 3s + 2$

b. $s^4 + s^3 + s^2 + s + 1$

c. $2s^4 + 2s^3 + s^2 + 3s + 2$

d. $2s^4 + 5s^3 + 5s^2 + 2s + 1$

e. $s^5 + 3s^4 + 10s^3 + 12s^2 + 7s + 3$

[**Answers:** No, no, no, yes, yes.]

Exercise 4.6.2

Show that a polynomial of degree 2 is Hurwitz if and only if the three coefficients of $a_2s^2 + a_1s + a_0$ are of the same sign.

Exercise 4.6.3

Show that the polynomial

$$s^3 + a_2s^2 + a_1s + a_0$$

is Hurwitz if and only if

$$a_2 > 0 \qquad a_0 > 0 \qquad \text{and} \qquad a_1 - \frac{a_2}{a_0} > 0$$

The most important application of the Routh test is to check the stability of control systems. This is illustrated by an example.

Example 4.6.4

Consider the system shown in Figure 4.14. The transfer function from r to y is, using Mason's formula,

$$G_o(s) = \frac{Y(s)}{R(s)}$$

$$= \frac{\dfrac{2s + 1}{s + 2} \cdot \dfrac{1}{s(s^2 + 2s + 2)}}{1 - \left[\dfrac{-2(s - 1)}{(s + 1)(s^2 + 2s + 2)} - \dfrac{2s + 1}{(s + 2)s(s^2 + 2s + 2)}\right]}$$

$$= \frac{(2s + 1)(s + 1)}{(s + 1)(s + 2)s(s^2 + 2s + 2) + 2(s - 1)s(s + 2) + (2s + 1)(s + 1)}$$

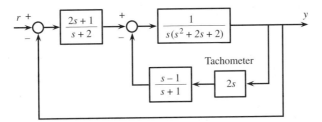

Figure 4.14 Feedback system.

which can be simplified as

$$G_o(s) = \frac{(2s + 1)(s + 1)}{s^5 + 5s^4 + 12s^3 + 14s^2 + 3s + 1}$$

We form the Routh table for the denominator of $G_o(s)$:

	s^5	1	12	3		
$k_4 = \dfrac{1}{5} = 0.2$	s^4	5	14	1	[0	9.2	2.8]
$k_3 = \dfrac{5}{9.2} = 0.54$	s^3	9.2	2.8		[0	12.48	1]
$k_2 = 0.74$	s^2	12.48	1		[0	2.06]
$k_1 = 6.05$	s^1	2.06			[0	1]
	s^0	1			

Because every entry is positive, the denominator of $G_o(s)$ is a Hurwitz polynomial. Thus all poles of $G_o(s)$ lie inside the open left half s-plane and the feedback system is stable.

To conclude this section, we mention that the Routh test can be used to determine the number of roots of $D(s)$ lying in the open right half s-plane. To be more specific, if none of the entries in the first column of the Routh table is zero, then the number of changes of signs in the first column equals the number of open RHP roots of $D(s)$. This property will not be used in this text and will not be discussed further.

4.6.1 Stability Range

In the design of a control system, we sometimes want to find the range of a parameter in which the system is stable. This stability range can be computed by using the Routh test. This is illustrated by examples.

Example 4.6.5

Consider the system shown in Figure 4.15. If

$$G(s) = \frac{8}{(s + 1)(s^2 + 2s + 2)} \qquad (4.20)$$

then the transfer function from r to y is

$$G_o(s) = \frac{8k}{(s + 1)(s^2 + 2s + 2) + 8k} = \frac{8k}{s^3 + 3s^2 + 4s + (8k + 2)}$$

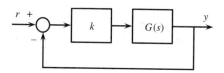

Figure 4.15 Unity-feedback system.

We form the Routh table for its denominator:

$$
\begin{array}{c|cc}
s^3 & 1 & 4 \\
s^2 & 3 & 2 + 8k \\
s^1 & \dfrac{10 - 8k}{3} & \\
s^0 & 2 + 8k &
\end{array}
$$

$k_2 = \dfrac{1}{3}$

$k_1 = \dfrac{9}{10 - 8k}$

$$\left[0 \quad 4 - \frac{2 + 8k}{3} \right] = \left[0 \quad \frac{10 - 8k}{3} \right]$$

$$[0 \quad 2 + 8k]$$

The conditions for $G_o(s)$ to be stable are

$$\frac{10 - 8k}{3} > 0 \qquad \text{and} \qquad 2 + 8k > 0$$

These two inequalities imply

$$1.25 = \frac{10}{8} > k \qquad \text{and} \qquad k > \frac{-2}{8} = -0.25 \qquad (4.21)$$

They are plotted in Figure 4.16(a). From the plot we see that if $1.25 > k > -0.25$, then k meets both inequalities, and the system is stable.

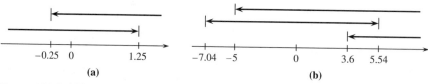

(a)   (b)

Figure 4.16 Stability ranges.

Example 4.6.6

Consider again Figure 4.15. If

$$G(s) = \frac{(s - 1 + j2)(s - 1 - j2)}{(s - 1)(s + 3 + j3)(s + 3 - j3)} = \frac{s^2 - 2s + 5}{s^3 + 5s^2 + 12s - 18} \qquad (4.22)$$

then the overall transfer function is

$$G_o(s) = \frac{kG(s)}{1 + kG(s)} = \frac{k \cdot \dfrac{s^2 - 2s + 5}{s^3 + 5s^2 + 12s - 18}}{1 + k \cdot \dfrac{s^2 - 2s + 5}{s^3 + 5s^2 + 12s - 18}}$$

$$= \frac{k(s^2 - 2s + 5)}{s^3 + (5 + k)s^2 + (12 - 2k)s + 5k - 18}$$

We form the Routh table for its denominator:

$$
\begin{array}{c|ccc}
s^3 & 1 & 12 - 2k & \\
s^2 & 5 + k & 5k - 18 & \left[0 \quad (12 - 2k) - \dfrac{5k - 18}{5 + k}\right] =: [0 \quad x] \\
s^1 & x & & [0 \quad 5k - 18] \\
s^0 & 5k - 18 & &
\end{array}
$$

with multipliers $\dfrac{1}{5 + k}$ for s^2 and $\dfrac{5 + k}{x}$ for s^1.

The x in the table requires some manipulation:

$$x = \frac{(12 - 2k)(5 + k) - (5k - 18)}{5 + k} = \frac{-2k^2 - 3k + 78}{5 + k}$$

$$= \frac{-2(k + 7.04)(k - 5.54)}{5 + k}$$

Thus the conditions for $G_o(s)$ to be stable are

$$5 + k > 0 \qquad 5k - 18 > 0$$

and

$$x = \frac{-2(k + 7.04)(k - 5.54)}{5 + k} > 0$$

These three inequalities imply

$$k > -5 \qquad k > \frac{18}{5} = 3.6 \qquad (4.23a)$$

and

$$(k + 7.04)(k - 5.54) < 0 \qquad (4.23b)$$

Note that, due to the multiplication by -1, the inequality is reversed in (4.23b). In order to meet (4.23b), k must lie between $(-7.04, 5.54)$. The three conditions in (4.23) are plotted in Figure 4.16(b). In order to meet them simultaneously, k must lie inside $(3.6, 5.54)$. Thus, the system is stable if and only if

$$3.6 < k < 5.54$$

Exercise 4.6.4

Find the stability ranges of the systems shown in Figure 4.17.

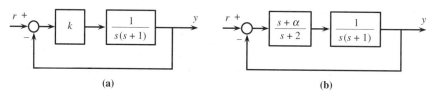

        (a)   (b)

Figure 4.17 Feedback systems.

[**Answers:** (a) $0 < k < \infty$. (b) $0 < \alpha < 9$.]

4.7   STEADY-STATE RESPONSE OF STABLE SYSTEMS—POLYNOMIAL INPUTS

Generally speaking, every control system is designed so that its output $y(t)$ will track a reference signal $r(t)$. For some problems, the reference signal is simply a step function, a polynomial of degree 0. For others, the reference signal may be more complex. For example, the desired altitude of the landing trajectory of a space shuttle may be as shown in Figure 4.18. Such a reference signal can be approximated by

$$r(t) = r_0 + r_1 t + r_2 t^2 + \cdots + r_m t^m$$

a polynomial of t of degree m. Clearly, the larger m, the more complex the reference signal that the system can track. However, the system will also be more complex.

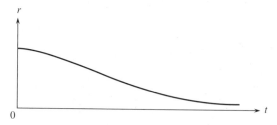

Figure 4.18 Time function.

In practice, control systems are designed to track

$$r(t) = a \qquad \text{(step function)}$$

$$r(t) = at \qquad \text{(ramp function)}$$

or

$$r(t) = at^2 \qquad \text{(acceleration function)}$$

where a is a constant. These are polynomials of degree 0, 1 and 2. The first two are used more often. For example, the temperature setting in a thermostat provides a step reference input, called the *set point* in industry. *To reset the set point* means *to change the amplitude of the step reference input.*

Although it is desirable to design a system so that its output $y(t)$ will track the reference input $r(t)$ immediately—that is, $y(t) = r(t)$, for all $t \geq 0$—it is not possible to achieve this in practice. The best we can hope for is

$$\lim_{t \to \infty} y(t) = r(t)$$

that is, $y(t)$ will track $r(t)$ as t approaches infinity. This is called *asymptotic tracking*, and the response

$$y_s(t) := \lim_{t \to \infty} y(t)$$

is called the *steady-state* response. We will now compute the steady-state response of stable systems due to polynomial inputs.

Consider a system with transfer function

$$G_o(s) = \frac{Y(s)}{R(s)} = \frac{\beta_0 + \beta_1 s + \cdots + \beta_m s^m}{\alpha_0 + \alpha_1 s + \cdots + \alpha_n s^n} \qquad (4.24)$$

with $n \geq m$. It is assumed that $G_o(s)$ is stable, or that all the poles of $G_o(s)$ have negative real parts. If we apply the reference input $r(t) = a$, for $t \geq 0$, then the output is given by

$$Y(s) = G_o(s)R(s) = \frac{\beta_0 + \beta_1 s + \cdots + \beta_m s^m}{\alpha_0 + \alpha_1 s + \cdots + \alpha_n s^n} \cdot \frac{a}{s}$$

$$= \frac{k}{s} + \text{(terms due to the poles of } G_o(s))$$

with k given by, using (A.8b),

$$k = G_o(s) \frac{a}{s} \cdot s \bigg|_{s=0} = G_o(0)a = \frac{\beta_0}{\alpha_0} \cdot a$$

If the system is stable, then the response due to every pole of $G_o(s)$ will approach zero as $t \to \infty$. Thus the steady-state response of the system due to $r(t) = a$ is

$$y_s(t) = \lim_{t \to \infty} y(t) = G_o(0)a = \frac{\beta_0}{\alpha_0} \cdot a \qquad (4.25)$$

This steady-state response depends only on the coefficients of $G_o(s)$ associated with s^0.

Now we consider the ramp reference input. Let $r(t) = at$. Then

$$R(s) = \frac{a}{s^2}$$

and

$$Y(s) = G_o(s) \frac{a}{s^2}$$

$$= \frac{k_1}{s} + \frac{k_2}{s^2} + \text{(terms due to the poles of } G_o(s)\text{)}$$

with, using (A.8c) and (A.8d),

$$k_2 = \left. G_o(s) \frac{a}{s^2} \cdot s^2 \right|_{s=0} = G_o(0)a$$

and

$$k_1 = \left. \frac{d}{ds} G_o(s)a \right|_{s=0}$$

$$= a \left[\frac{(\alpha_0 + \alpha_1 s + \cdots + \alpha_n s^n)(\beta_1 + \cdots + m\beta_m s^{m-1})}{(\alpha_0 + \alpha_1 s + \cdots + \alpha_n s^n)^2} \right.$$

$$\left. - \frac{(\beta_0 + \beta_1 s + \cdots + \beta_m s^m)(\alpha_1 + \cdots + n\alpha_n s^{n-1})}{(\alpha_0 + \alpha_1 s + \cdots + \alpha_n s^n)^2} \right]_{s=0}$$

$$= a \frac{\alpha_0 \beta_1 - \beta_0 \alpha_1}{\alpha_0^2}$$

Thus the steady-state response of the system due to $r(t) = at$ equals

$$y_s(t) = G_o(0)at + G_o'(0)a \tag{4.26a}$$

or

$$y_s(t) = \frac{\beta_0}{\alpha_0} \cdot at + a \cdot \frac{\alpha_0 \beta_1 - \beta_0 \alpha_1}{\alpha_0^2} \tag{4.26b}$$

This steady-state response depends only on the coefficients of $G_o(s)$ associated with s^0 and s.

We discuss now the implications of (4.25) and (4.26). If $G_o(0) = 1$ or $a_0 = \beta_0$ and if $r(t) = a$, $t \geq 0$, then

$$y_s(t) = a = r(t)$$

Thus the output $y(t)$ will track asymptotically any step reference input. If $G_o(0) = 1$ and $G_o'(0) = 0$ or $\alpha_0 = \beta_0$ and $\alpha_1 = \beta_1$, and if $r(t) = at$, $t \geq 0$, then (4.26) reduces to

$$y_s(t) = at$$

that is, $y(t)$ will track asymptotically any ramp reference input. Proceeding forward, if

$$\alpha_0 = \beta_0 \qquad \alpha_1 = \beta_1 \quad \text{and} \quad \alpha_2 = \beta_2 \qquad (4.27)$$

then the output of $G_o(s)$ will track asymptotically any acceleration reference input at^2. Note that in the preceding discussion, the stability of $G_o(s)$ is essential. If $G_o(s)$ is not stable, the output of $G_o(s)$ will not track any $r(t)$.

Exercise 4.7.1

Find the steady-state responses of

a. $G_o(s) = \dfrac{1}{1 - s^2}$ due to $r(t) = a$

b. $G_o(s) = \dfrac{2}{s + 1}$ due to $r(t) = a$

c. $G_o(s) = \dfrac{2 + 3s}{2 + 3s + s^2}$ due to $r(t) = 2 + t$

d. $G_o(s) = \dfrac{2}{s + 1}$ due to $r(t) = 3t$

e. $G_o(s) = \dfrac{68 + 9s + 9s^2}{68 + 9s + 9s^2 + s^3}$ due to $r(t) = a$

[**Answers:** (a) ∞; (b) $2a$; (c) $y_s(t) = 2 + t$; (d) $6t - 6$; (e) ∞.]

4.7.1 Steady-State Response of Stable Systems—Sinusoidal Inputs

Consider a system with proper transfer function $G_o(s) = Y(s)/R(s)$. It is assumed that $G_o(s)$ is stable. Now we shall show that if $r(t) = a \sin \omega_o t$, then the output $y(t)$ will approach a sinusoidal function with the same frequency as $t \to \infty$.

If $r(t) = a \sin \omega_o t$, then, using Table A.1,

$$R(s) = \frac{a\omega_o}{s^2 + \omega_o^2} \qquad (4.28)$$

Hence, we have

$$Y(s) = G_o(s)R(s) = G_o(s) \cdot \frac{a\omega_o}{s^2 + \omega_o^2} = G_o(s) \cdot \frac{a\omega_o}{(s + j\omega_o)(s - j\omega_o)}$$

Because $G_o(s)$ is stable, $s = \pm j\omega_o$ are simple poles of $Y(s)$. Thus $Y(s)$ can be expanded as, using partial fraction expansion,

$$Y(s) = \frac{k_1}{s - j\omega_o} + \frac{k_1^*}{s + j\omega_o} + \text{terms due to the poles of } G_o(s)$$

with

$$k_1 = G_o(s) \cdot \frac{a\omega_o}{s + j\omega_o}\bigg|_{s=j\omega_o} = G_o(j\omega) \cdot \frac{a\omega_o}{2j\omega_o} = \frac{a}{2j} G_o(j\omega_o)$$

and

$$k_1^* = \frac{a}{-2j} G_o(-j\omega_o)$$

Since all the poles of $G_o(s)$ have negative real parts, their time responses will approach zero as $t \to \infty$. Hence, the steady-state response of the system due to $r(t) = a \sin \omega_o t$ is given by

$$y_s(t) = \mathcal{L}^{-1} \left[\frac{aG_o(j\omega_0)}{2j(s - j\omega_o)} - \frac{aG_o(-j\omega_o)}{2j(s + j\omega_o)} \right] \tag{4.29}$$

All coefficients of $G_o(s)$ are implicitly assumed to be real. Even so, the function $G_o(j\omega_o)$ is generally complex. We express it in polar form as

$$G_o(j\omega_o) = A(\omega_o)e^{j\theta(\omega_o)} \tag{4.30}$$

where

$$A(\omega_o) := |G_o(j\omega_o)| = [(\mathrm{Re}\, G_o(j\omega_o))^2 + (\mathrm{Im}\, G_o(j\omega_o))^2]^{1/2}$$

and

$$\theta(\omega_o) := \sphericalangle G_o(j\omega_o) := \tan^{-1} \frac{\mathrm{Im}\, G_o(j\omega_o)}{\mathrm{Re}\, G_o(j\omega_o)}$$

where Im and Re denote, respectively, the imaginary and real parts. $A(\omega_o)$ is called the *amplitude* and $\theta(\omega_o)$, the *phase* of $G_o(s)$. If all coefficients of $G_o(s)$ are real, then $A(\omega_o)$ is an even function of ω_o, and $\theta(\omega_o)$ is an odd function of ω_o; that is, $A(-\omega_o) = A(\omega_o)$ and $\theta(-\omega_o) = -\theta(\omega_o)$. Consequently we have

$$G_o(-j\omega_o) = A(-\omega_o)e^{j\theta(-\omega_o)} = A(\omega_o)e^{-j\theta(\omega_o)} \tag{4.31}$$

The substitution of (4.30) and (4.31) into (4.29) yields

$$\begin{aligned}
y_s(t) &= \frac{aA(\omega_o)e^{j\theta(\omega_o)}}{2j} \cdot e^{j\omega_o t} - \frac{aA(\omega_o)e^{-j\theta(\omega_o)}}{2j} \cdot e^{-j\omega_o t} \\
&= aA(\omega_o) \frac{e^{j[\omega_o t + \theta(\omega_o)]} - e^{-j[\omega_o t + \theta(\omega_o)]}}{2j} \\
&= aA(\omega_o) \sin (\omega_o t + \theta(\omega_o))
\end{aligned} \tag{4.32}$$

This shows that if $r(t) = a \sin \omega_o t$, then the output will approach a sinusoidal function of the same frequency. Its amplitude equals $a|G_o(j\omega_o)|$; its phase differs from the phase of the input by $\tan^{-1}[\mathrm{Im}\, G_o(j\omega_o)/\mathrm{Re}\, G_o(j\omega_o)]$. We stress again that (4.32) holds only if $G_o(s)$ is stable.

Example 4.7.1

Consider $G_o(s) = 3/(s + 0.4)$. It is stable. In order to compute its steady-state response due to $r(t) = \sin 2t$, we compute

$$G_o(j2) = \frac{3}{j2 + 0.4} = \frac{3}{2.04e^{j1.37}} = 1.47e^{-j1.37} \qquad (4.33)$$

Thus the steady-state response is

$$y_s(t) = \lim_{t \to \infty} y(t) = 1.47 \sin (2t - 1.37) \qquad (4.34)$$

Note that the phase -1.37 is in radians, not in degrees. This computation is very simple, but it does not reveal how fast the system will approach the steady state. This problem is discussed in the next subsection.

Exercise 4.7.2

Find the steady-state response of $2/(s + 1)$ due to (a) $\sin 2t$ (b) $1 + \sin 2t$ (c) $2 + 3 \sin 2t - \sin 3t$.

[**Answers:**   (a) $0.89 \sin (2t - 1.1)$. (b) $2 + 0.89 \sin (2t - 1.1)$. (c) $4 + 2.67 \sin (2t - 1.1) - 0.63 \sin (3t - 1.25)$.]

The steady-state response of a stable $G_o(s)$ due to $\sin \omega_o t$ is completely determined by the value of $G_o(s)$ at $s = j\omega_o$. Thus $G(j\omega)$ is called the *frequency response* of the system. Its amplitude $A(\omega)$ is called the *amplitude characteristic*, and its phase $\theta(\omega)$, the *phase characteristic*. For example, if $G_o(s) = 2/(s + 1)$, then $G_o(0) = 2$, $G_o(j1) = 2/(j1 + 1) = 2/(1.4e^{j45°}) = 1.4e^{-j45°}$, $G_o(j10) = 2/(j10 + 1) = 0.2e^{-j84°}$, and so forth. The amplitude and phase characteristics of $G_o(s) = 2/(s + 1)$ can be plotted as shown in Figure 4.19. From the plot, the steady-state response due to $\sin \omega_o t$, for any ω_o, can be read out.

Exercise 4.7.3

Plot the amplitude and phase characteristics of $G_o(s) = 2/(s - 1)$. What is the steady-state response of the system due to $\sin 2t$?

[**Answers:**   Same as Figure 4.19 except the sign of the phase is reversed, infinity. The amplitude and phase characteristics of unstable transfer functions do not have any physical meaning and, strictly speaking, are not defined.]

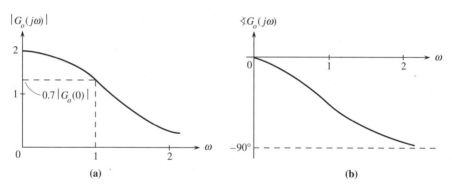

Figure 4.19 Amplitude and phase characteristics.

The frequency response of a stable $G_o(s)$ can be obtained by measurement. We apply $r(t) = \sin \omega_o t$ and measure the steady-state response. From the amplitude and phase of the response, we can obtain $A(\omega_o)$ and $\theta(\omega_o)$. By varying or sweeping ω_o, $G_o(j\omega)$ over a frequency range can be obtained. Special devices called *frequency analyzers*, such as the HP 3562A Dynamic System Analyzer, are available to carry out this measurement. Some devices will also generate a transfer function from the measured frequency response.

We introduce the concept of *bandwidth* to conclude this section. The *bandwidth* of a stable $G_o(s)$ is defined as the frequency range in which[3]

$$|G_o(j\omega)| \geq 0.707| \, G_o(0)| \tag{4.35}$$

For example, the bandwidth of $G_o(s) = 2/(s + 1)$ can be read from Figure 4.19 as 1 radian per second. Thus, the amplitude of $G_o(j\omega)$ at every frequency within the bandwidth is at least 70.7% of that at $\omega = 0$. Because the power is proportional to the square of the amplitude, the power of $G_o(j\omega)$ in the bandwidth is at least $(0.707)^2 = 0.5 = 50\%$ of that at $\omega = 0$. Thus, the bandwidth is also called the *half-power* bandwidth. (It is also called the -3-dB bandwidth as is discussed in Chapter 8.) Note that if $G_o(s)$ is not stable, its bandwidth has no physical meaning and is not defined.

4.7.2 Infinite Time

The steady-state response is defined as the response as $t \to \infty$. Mathematically speaking, we can never reach $t = \infty$ and therefore can never obtain the steady-state response. In engineering, however, this is not the case. For some systems a response may be considered to have reached the steady state in 20 seconds. It is a very short infinity indeed!

---

[3]This definition applies only to $G_o(s)$ with lowpass characteristic as shown in Figure 4.19. More generally, the bandwidth of stable $G_o(s)$ is defined as the frequency range in which the amplitude of $G_o(j\omega)$ is at least 70.7% of the largest amplitude of $G_o(j\omega)$.

Consider $G_o(s) = 3/(s + 0.4)$. If we apply $r(t) = \sin 2t$, then $y_s(t) = 1.47 \sin (2t - 1.37)$ (see [4.34]). One may wonder how fast $y(t)$ will approach $y_s(t)$. In order to find this out, we shall compute the total response of $G_o(s)$ due to $\sin 2t$. The Laplace transform of $r(t) = \sin 2t$ is $2/(s^2 + 4)$. Thus we have

$$Y(s) = \frac{3}{s + 0.4} \cdot \frac{2}{s^2 + 4} = \frac{6}{(s + 0.4)(s + 2j)(s - 2j)} \tag{4.36}$$

$$= \frac{1.44}{s + 0.4} + \frac{1.47e^{-j1.37}}{2j(s - 2j)} - \frac{1.47e^{j1.37}}{2j(s + 2j)}$$

which implies

$$y(t) = \underbrace{1.44e^{-0.4t}}_{\substack{\text{Transient} \\ \text{Response}}} + \underbrace{1.47 \sin (2t - 1.37)}_{\substack{\text{Steady-State} \\ \text{Response}}} \tag{4.37}$$

The second term on the right hand side of (4.37) is the same as (4.34) and is the steady-state response. The first term is called the *transient* response, because it appears right after the application of $r(t)$ and will eventually die out. Clearly the faster the transient response approaches zero, the faster $y(t)$ approaches $y_s(t)$. The transient response in (4.37) is governed by the real pole at -0.4 whose time constant is defined as $1/0.4 = 2.5$. As shown in Figure 4.2(b), the time response of $1/(s + 0.4)$ decreases to less than 1% of its original value in five time constants or $5 \times 2.5 = 12.5$ seconds. Thus the response in (4.37) may be considered to have reached the steady state in five time constants or 12.5 seconds.

Now we shall define the time constant for general proper transfer functions. The time constant can be used to indicate the speed at which a response reaches its steady state. Consider

$$G(s) = \frac{N(s)}{D(s)} = \frac{N(s)}{(s + a_1)(s + a_2)(s + \sigma_1 + j\omega_{d1})(s + \sigma_1 - j\omega_{d1}) \cdots} \tag{4.38}$$

If $G(s)$ is not stable, the response due to its poles will not die out and the time constant is not defined. If $G(s)$ is stable, then $a_i > 0$ and $\sigma_1 > 0$. For each real pole $(s + a_i)$, the time constant is defined as $1/a_i$. For the pair of complex conjugate poles $(s + \sigma_1 \pm j\omega_{d1})$, the time constant is defined as $1/\sigma_1$; this definition is reasonable, because σ_1 governs the envelope of its time response as shown in Figure 4.6. The time constant of $G(s)$ is then defined as the largest time constant of all poles of $G(s)$. Equivalently, it is defined as the inverse of the smallest distance of all poles of $G(s)$ from the imaginary axis. For example, suppose $G(s)$ has poles -1, -3, $-0.1 \pm j2$. The time constants of the poles are 1, $1/3 = 0.33$ and $1/0.1 = 10$. Thus, the time constant of $G(s)$ is 10 seconds. In engineering, the response of $G(s)$ due to a step or sinusoid input will be considered to have reached the steady state in five time constants. Thus the smaller the time constant or, equivalently, the farther away the closest pole from the imaginary axis, the faster the response reaches the steady-state response.

Exercise 4.7.4

Find the time constants of

a. $\dfrac{s + 10}{(s + 2)(s^2 + 8s + 20)}$

b. $\dfrac{s}{(s + 0.2)(s + 2)(s + 3)}$

c. A system with the pole-zero pattern shown in Figure 4.20.
Which system will respond fastest?

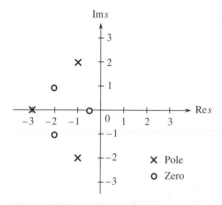

Figure 4.20 Pole-zero pattern.

[**Answers:**   (a) 0.5; (b) 5; (c) 1; (a).]

The time constant of a stable transfer function $G(s)$ as defined is open to argument. It is possible to find a transfer function whose step response will not reach the steady state in five time constants. This is illustrated by an example.

Example 4.7.2

Consider $G(s) = 1/(s + 1)^3$. It has three poles at $s = -1$. The time constant of $G(s)$ is 1 second. The unit-step response of $G(s)$ is

$$Y(s) = \frac{1}{(s + 1)^3} \cdot \frac{1}{s} = \frac{1}{s} + \frac{-1}{(s + 1)^3} + \frac{-1}{(s + 1)^2} + \frac{-1}{(s + 1)}$$

or

$$y(t) = 1 - 0.5t^2 e^{-t} - te^{-t} - e^{-t}$$

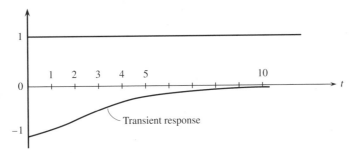

Figure 4.21 Step response.

Its steady-state response is 1 and its transient response is

$$-0.5t^2 e^{-t} - te^{-t} - e^{-t} = -(0.5t^2 + t + 1)e^{-t}$$

These are plotted in Figure 4.21. At five time constants, or $t = 5$, the value of the transient response is -0.126; it is about 13% of the steady-state response. At $t = 9$, the value of the transient response is 0.007 or 0.7% of the steady-state response. For this system, it is more appropriate to claim that the response reaches the steady state in nine time constants.

This example shows that if a transfer function has repeated poles or, more generally, a cluster of poles in a small region close to the imaginary axis, then the rule of five time constants is not applicable. The situation is actually much more complicated. The zeros of a transfer function also affect the transient response. See Example 2.4.5 and Figure 2.16. However, the zeros are not considered in defining the time constant, so it is extremely difficult to state precisely how many time constants it will take for a response to reach the steady state. The rule of five time constants is useful in pointing out that infinity in engineering does not necessarily mean mathematical infinity.

PROBLEMS

4.1. Consider the open-loop voltage regulator in Figure P3.4(a). Its block diagram is repeated in Figure P4.1 with numerical values.

a. If $R_L = 100 \ \Omega$ and if $r(t)$ is a unit-step function, what is the response $v_o(t)$? What is its steady-state response? How many seconds will $v_o(t)$ take to reach and stay within 1% of its steady state?

b. What is the required reference input if the desired output voltage is 20 V?

Figure P4.1

   c. Are the power levels at the reference input and plant output necessarily the same? If they are the same, is the system necessary?

   d. If we use the reference signal computed in (b), and decrease R_L from 100 Ω to 50 Ω, what is the steady-state output voltage?

4.2. Consider the closed-loop voltage regulator shown in Figure P3.4(b). Its block diagram is repeated in Figure P4.2 with numerical values.

   a. If R_L = 100 Ω and if $r(t)$ is a unit-step function, what is the response $v_o(t)$? What is its steady-state response? How many seconds will $v_o(t)$ take to reach the steady state?

   b. What is the required $r(t)$ if the desired output voltage is 20 V?

   c. If we use the $r(t)$ in (b) and decrease R_L from 100 Ω to 50 Ω, what is the steady-state output voltage?

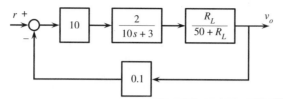

Figure P4.2

4.3. Compare the two systems in Problems 4.1 and 4.2 in terms of (a) the time constants or speeds of response, (b) the magnitudes of the reference signals, and (c) the deviations of the output voltages from 20 V as R_L decreases from 100 Ω to 50 Ω. Which system is better?

4.4. The transfer function of a motor and load can be obtained by measurement. Let the transfer function from the applied voltage to the angular displacement be of the form $k_m/s(\tau_m s + 1)$. If we apply an input of 100 V, the speed (*not* displacement) is measured as 2 rad/s at 1.5 seconds. The speed eventually reaches 3 rad/s. What is the transfer function of the system?

4.5. Maintaining a liquid level at a fixed height is important in many process-control systems. Such a system and its block diagram are shown in Figure P4.5, with

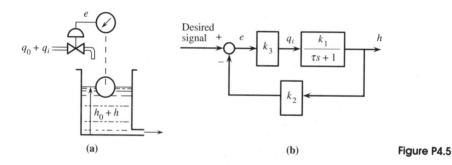

(a)   (b)   **Figure P4.5**

$\tau = 10$, $k_1 = 10$, $k_2 = 0.95$, and $k_3 = 2$. The variables h and q_i in the block diagram are the deviations from the nominal values h_o and q_o; hence, the desired or reference signal for this system is zero. This kind of system is called a *regulating system*. If $h(0) = -1$, what is the response of $h(t)$ for $t > 0$?

4.6. **a.** Consider a motor and load with transfer function $G(s) = Y(s)/U(s) = 1/s(s + 2)$, where the unit of the input is volts and that of the output is radians. Compute the output $y(t)$ due to $u(t) = 1$, for $t \geq 0$. What is $y(t)$ as $t \to \infty$?

b. Show that the response of the system due to

$$u(t) = \begin{cases} a & \text{for } 0 \leq t \leq b \\ 0 & \text{for } t > b \end{cases}$$

a pulse of magnitude a and duration b, is given by

$$y(t) = \frac{ab}{2} + \frac{a}{2} e^{-2t}(1 - e^{2b})$$

for $t > b$. What is the steady-state response?

c. If $a = 1$, what is the duration b necessary to move y 30 degrees?

d. If $b = 1$, what is the amplitude a necessary to move y 30 degrees?

4.7. Consider the position control system shown in Figure 4.8(a). Let the transfer function of the motor and load be $1/s(s + 2)$. The error detector is a pair of potentiometers with sensitivity $k_1 = 3$. The reference input is to be applied by turning a knob.

a. If $k_2 = 1$, compute the response due to a unit-step reference input. Plot the response. Roughly how many seconds will y take to reach and stay within 1% of its final position?

b. If it is required to turn y 30 degrees, how many degrees should you turn the control knob?

c. Find a k_2 so that the damping ratio equals 0.7. Can you find a k_2 so that the damping ratio equals 0.7 and the damping factor equals 3?

4.8. Consider the position control system shown in Figure 4.8(b). Let the transfer function of the motor and load be $1/s(s + 2)$. The error detector is a pair of potentiometers with sensitivity $k_1 = 3$. The reference input is to be applied by turning a knob. A tachometer with sensitivity k_3 is introduced as shown.

a. If $k_2 = 1$ and $k_3 = 1$, compute the response due to a unit-step reference input. Plot the response. Roughly how many seconds will y take to reach and stay within 1% of its final position?

b. If it is required to turn y 30 degrees, how many degrees should you turn the control knob?

c. If $k_3 = 1$, find a k_2 so that the damping ratio equals 0.7. If k_3 is adjustable, can you find a k_2 and a k_3 so that $\zeta = 0.7$ and $\zeta \omega_n = 3$?

d. Compare the system with the one in Problem 4.7 in terms of the speed of response.

4.9. Consider a dc motor. It is assumed that its transfer function from the input to the angular position is $1/s(s + 2)$. Is the motor stable? If the angular velocity of the motor shaft, rather than the displacement, is considered as the output, what is its transfer function? With respect to this input and output, is the system stable?

4.10. A system may consist of a number of subsystems. The stability of a system depends only on the transfer function of the overall system. Study the stability of the three unity-feedback systems shown in Figure P4.10. Is it true that a system is stable if and only if its subsystems are all stable? Is it true that negative feedback will always stabilize a system?

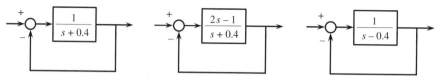

Figure P4.10

4.11. Which of the following are Hurwitz polynomials?

a. $-2s^2 - 3s - 5$

b. $2s^2 + 3s - 5$

c. $s^5 + 3s^4 + s^2 + 2s + 10$

d. $s^4 + 3s^3 + 6s^2 + 5s + 3$

e. $s^6 + 3s^5 + 7s^4 + 8s^3 + 9s^2 + 5s + 3$

4.12. Check the stability of the following systems:

a. $G(s) = \dfrac{s^3 - 1}{s^5 + s^4 + 2s^3 + 2s^2 + 5s + 5}$

b. $G(s) = \dfrac{s^3 - 1}{s^4 + 14s^3 + 71s^2 + 154s + 120}$

c. The system shown in Figure P4.12.

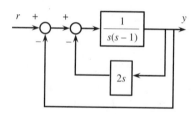

Figure P4.12

4.13. Find the ranges of k in which the systems in Figure P4.13 are stable.

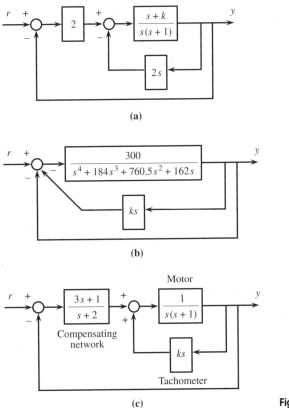

(a)

(b)

(c)   **Figure P4.13**

4.14. Consider a system with transfer function $G(s)$. Show that if we apply a unit-step input, the output approaches a constant if and only if $G(s)$ is stable. This fact can be used to check the stability of a system by measurement.

4.15. In a modern rapid transit system, a train can be controlled manually or automatically. The block diagram in Figure P4.15 shows a possible way to control

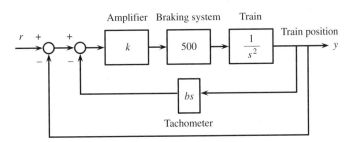

Figure P4.15

the train automatically. If the tachometer is not used in the feedback (that is, if $b = 0$), is it possible for the system to be stable for some k? If $b = 0.2$, find the range of k so that the system is stable.

4.16. Let $D(s) = a_n s^n + a_{n-1} s^{n-1} + \cdots + a_1 s + a_0$, with $a_n > 0$. Let $[j/2]$ be the integer part of $j/2$. In other words, if j is an even integer, then $[j/2] = j/2$. If j is an odd integer, then $[j/2] = (j-1)/2$. Define

$$b_{n,i} = a_{n+2-2i} \qquad i = 1, 2, \ldots, [n/2] + 1$$

$$b_{n-1,i} = a_{n+1-2i} \qquad i = 1, 2, \ldots, [(n-1)/2] + 1$$

For $j = n - 2, n - 3, \ldots, 2, 1$, compute

$$k_{j+1} = \frac{b_{j+2,1}}{b_{j+1,1}}$$

$$b_{j,i} = b_{j+2,i+1} - k_{j+1} b_{j+1,i+1} \qquad i = 1, 2, \ldots, [j/2] + 1$$

Verify that the preceding algorithm computes all entries of the Routh table.

4.17. What are the time constants of the following transfer functions?

a. $\dfrac{s - 2}{s^2 + 2s + 1}$

b. $\dfrac{s - 1}{(s + 1)(s^2 + 2s + 2)}$

c. $\dfrac{s^2 + 2s - 2}{(s^2 + 2s + 4)(s^2 + 2s + 10)}$

d. $\dfrac{s + 10}{s + 1}$

e. $\dfrac{s - 10}{s^2 + 2s + 2}$

Do they all have the same time constant?

4.18. What are the steady-state responses of the system with transfer function $1/(s^2 + 2s + 1)$ due to the following inputs:

a. $u_1(t) = $ a unit-step function.

b. $u_2(t) = $ a ramp function.

c. $u_3(t) = u_1(t) + u_2(t)$.

d. $u_4(t) = 2 \sin 2\pi t$, for $t \geq 0$.

4.19. Consider the system with input $r(t)$, output $y(t)$, and transfer function

$$G(s) = \frac{s + 8}{(s + 2)(s + 4)}$$

Find the steady-state response due to $r(t) = 2 + 3t$. Compute

$$\lim_{t \to \infty} e(t) = \lim_{t \to \infty} (r(t) - y(t))$$

This is called the steady-state error between $r(t)$ and $y(t)$.

4.20. Derive (4.25) by using the final-value theorem (see Appendix A). Can you use the theorem to derive (4.26)?

4.21. Consider a system with transfer function $G(s)$. It is assumed that $G(s)$ has no poles in the closed right half s-plane except a simple pole at the origin. Show that if the input $u(t) = a \sin \omega_o t$ is applied, the steady-state response excluding the dc part is given by Equation (4.32).

4.22. What is the time constant of the system in Problem 4.19? Is it true that the response reaches the steady state in roughly five time constants?

4.23. Consider a system with transfer function

$$G(s) = \frac{s + 1}{(s + 2)(s + 4)}$$

Compute its unit-step response. Is it true that the response reaches the steady state in roughly five time constants?

4.24. What are the bandwidths of the transfer function $1/(s + 3)$ and the transfer functions in Problems 4.19 and 4.23?

5 *Computer Simulation and Realization*

5.1 INTRODUCTION

In recent years computers have become indispensable in the analysis and design of control systems. They can be used to collect data, to carry out complicated computations, and to simulate mathematical equations. They can also be used as control components, and as such are used in space vehicles, industrial processes, autopilots in airplanes, numerical controls in machine tools, and so on. Thus, a study of the use of computers is important in control engineering.

Computers can be divided into two classes: *analog* and *digital*. An interconnection of a digital and an analog computer is called a *hybrid computer*. Signals on analog computers are defined at every instant of time, whereas signals on digital computers are defined only at discrete instants of time. Thus, a digital computer can accept only sequences of numbers, and its outputs again consist only of sequences of numbers. Because digital computers yield more accurate results and are more flexible and versatile than analog computers, the use of general-purpose analog computers has been very limited in recent years. Therefore, general-purpose analog computers are not discussed in this text; instead, we discuss simulations using operational amplifier (op-amp) circuits, which are essentially special-purpose or custom-built analog computers.

We first discuss digital computer computation of state-variable equations. We use the Euler forward algorithm and show its simplicities in programming and computation. We then introduce some commercially available programs. Op-amp circuit implementations of state-variable equations are then discussed. We discuss the reasons for not computing transfer functions directly on digital computers and then introduce the realization problem. After discussing the problem, we show how transfer functions can be simulated on digital computers or built using op-amp circuits through state-variable equations.

5.2   COMPUTER COMPUTATION OF STATE-VARIABLE EQUATIONS

Consider the state-variable equation

$$\dot{\mathbf{x}}(t) = \mathbf{A}x(t) + \mathbf{b}u(t) \tag{5.1}$$

$$y(t) = \mathbf{c}x(t) + du(t) \tag{5.2}$$

where $u(t)$ is the input; $y(t)$ the output, and $\mathbf{x}(t)$ the state. If $\mathbf{x}(t)$ has n components or n state variables, then $\mathbf{A}$ is an $n \times n$ matrix, $\mathbf{b}$ is an $n \times 1$ vector, $\mathbf{c}$ is a $1 \times n$ vector, and d is a 1×1 scalar. Equation (5.2) is an algebraic equation. Once $\mathbf{x}(t)$ and $u(t)$ are available, $y(t)$ can easily be obtained by multiplication and addition. Therefore we discuss only the computation of (5.1) by using digital computers.

Equation (5.1) is a continuous-time equation; it is defined at every instant of time. Because every time interval has infinitely many points and because no digital computer can compute them all, we must discretize the equation before computation. By definition, we have

$$\dot{\mathbf{x}}(t_0) := \frac{\mathbf{x}(t_0)}{dt} := \lim_{\alpha \to 0} \frac{\mathbf{x}(t_0 + \alpha) - \mathbf{x}(t_0)}{\alpha}$$

The substitution of this into (5.1) at $t = t_0$ yields

$$\mathbf{x}(t_0 + \alpha) - \mathbf{x}(t_0) = [\mathbf{A}\mathbf{x}(t_0) + \mathbf{b}u(t_0)]\alpha$$

or

$$\mathbf{x}(t_0 + \alpha) = \mathbf{x}(t_0) + \alpha\mathbf{A}\mathbf{x}(t_0) + \alpha\mathbf{b}u(t_0)$$
$$= \mathbf{I}\mathbf{x}(t_0) + \alpha\mathbf{A}\mathbf{x}(t_0) + \alpha\mathbf{b}u(t_0)$$

where $\mathbf{I}$ is a unit matrix with the same order as $\mathbf{A}$. Note that $\mathbf{x} + \alpha\mathbf{A}\mathbf{x} = (1 + \alpha\mathbf{A})\mathbf{x}$ is not well defined (why?). After introducing the unit matrix, the equation becomes

$$\mathbf{x}(t_0 + \alpha) = (\mathbf{I} + \alpha\mathbf{A})\mathbf{x}(t_0) + \mathbf{b}u(t_0)\alpha \tag{5.3}$$

This is a discrete-time equation, and α is called the *integration step size*. Now, if $\mathbf{x}(t_0)$ and $u(t_0)$ are known, then $\mathbf{x}(t_0 + \alpha)$ can be computed algebraically from (5.3). Using this equation repeatedly or recursively, the solution of (5.1) due to any $\mathbf{x}(0)$ and any $u(t)$, $t \geq 0$, can be computed. For example, from the given $\mathbf{x}(0)$ and $u(0)$, we can compute

$$\mathbf{x}(\alpha) = (\mathbf{I} + \alpha\mathbf{A})\mathbf{x}(0) + \mathbf{b}u(0)\alpha$$

We then use this $\mathbf{x}(\alpha)$ and $u(\alpha)$ to compute

$$\mathbf{x}(2\alpha) = (\mathbf{I} + \alpha\mathbf{A})\mathbf{x}(\alpha) + \mathbf{b}u(\alpha)\alpha$$

Proceeding forward, $\mathbf{x}(k\alpha)$, $k = 0, 1, 2, \ldots$, can be obtained. This procedure can be expressed in a programmatic format as

$$\text{DO } 10 \ k = 0, N$$

$$10 \quad \mathbf{x}((k + 1)\alpha) = (\mathbf{I} + \alpha\mathbf{A})\mathbf{x}(k\alpha) + \mathbf{b}u(k\alpha)\alpha$$

where N is the number of points to be computed. Clearly this can easily be programmed on a personal computer. Equation (5.3) is called the *Euler forward algorithm*. It is the simplest but the least accurate method of computing (5.1).

In using (5.3), we will encounter the problem of choosing the integration step size α. It is clear that the smaller α, the more accurate the result. However, the smaller α, the larger the number of points to be computed for the same time interval. For example, to compute $\mathbf{x}(t)$ from $t = 0$ to $t = 10$, we need to compute 10 points if $\alpha = 1$, and 1000 points if $\alpha = 0.01$. Therefore, the choice of α must be a compromise between accuracy and amount of computation. In actual programming, α may be chosen as follows. We first choose an arbitrary α_0 and compute the response. We then repeat the computation by using $\alpha_1 = \alpha_0/2$. If the result of using α_1 is close to the result of using α_0, we stop, and the result obtained by using α_1 is probably very close to the actual solution. If the result of using α_1 is quite different from the result of using α_0, α_0 is not small enough and cannot be used. Whether or not α_1 is small enough cannot be answered at this point. Next we choose $\alpha_2 = \alpha_1/2$ and repeat the computation. If the result of using α_2 is close to the result of using α_1, we may conclude that α_1 is small enough and stop the computation. Otherwise we continue the process until the results of two consecutive computations are sufficiently close.

Example 5.2.1

Compute the output $y(t)$, from $t = 0$ to $t = 10$ seconds, of

$$\dot{\mathbf{x}}(t) = \begin{bmatrix} 0 & 1 \\ -0.5 & -1.5 \end{bmatrix} \mathbf{x}(t) + \begin{bmatrix} 0 \\ 1 \end{bmatrix} u(t)$$

$$y(t) = [1 \quad -1]\mathbf{x}(t)$$

(5.4)

due to the initial condition $\mathbf{x}(0) = [2 \quad -1]'$ and the input $u(t) = 1$, for $t \geq 0$, where the prime denotes the transpose.

For this equation, (5.3) becomes

$$\begin{bmatrix} x_1(t_0 + \alpha) \\ x_2(t_0 + \alpha) \end{bmatrix} = \left(\begin{bmatrix} 1 & 0 \\ 0 & 1 \end{bmatrix} + \alpha \begin{bmatrix} 0 & 1 \\ -0.5 & -1.5 \end{bmatrix} \right) \begin{bmatrix} x_1(t_0) \\ x_2(t_0) \end{bmatrix} + \begin{bmatrix} 0 \\ 1 \end{bmatrix} \cdot 1 \cdot \alpha$$

$$= \begin{bmatrix} 1 & \alpha \\ -0.5\alpha & 1 - 1.5\alpha \end{bmatrix} \begin{bmatrix} x_1(t_0) \\ x_2(t_0) \end{bmatrix} + \begin{bmatrix} 0 \\ 1 \end{bmatrix} \cdot 1 \cdot \alpha$$

which implies

$$x_1((k + 1)\alpha) = x_1(k\alpha) + \alpha x_2(k\alpha) \tag{5.5a}$$

$$x_2((k + 1)\alpha) = -0.5\alpha x_1(k\alpha) + (1 - 1.5\alpha)x_2(k\alpha) + \alpha \tag{5.5b}$$

The output equation is

$$y(k\alpha) = x_1(k\alpha) - x_2(k\alpha) \tag{5.5c}$$

Arbitrarily, we choose $\alpha_0 = 1$. We compute (5.5) from $k = 0$ to $k = 10$. A FORTRAN program for this computation is as follows:

```
        REAL X1(0:1500), X2(0:1500), Y(0:1500), A
        INTEGER K
        A = 1.0
        X1(0) = 2.0
        X2(0) = -1.0
        DO 10 K = 0, 10
          X1(K + 1) = X1(K) + A*X2(K)
          X2(K + 1) = -0.5*A*X1(K) + (1.0 - 1.5*A)*X2(K) + A
          Y(K) = X1(K) - X2(K)
          PRINT*, 'T =',K, 'Y =', Y(K)
10      CONTINUE
        END
```

where A stands for α. The result is printed in Table 5.1 and plotted in Figure 5.1 using $+$. We then repeat the computation by using $\alpha_1 = \alpha_0/2 = 0.5$ and compute 21 points. The result is plotted in Figure 5.1 using $\circ$, but we print only 11 points in Table 5.1. The two results are quite different, therefore we repeat the process for

Table 5.1 Computation of (5.4) Using Four Different α

	$\alpha = 1.0$	$\alpha = 0.5$	$\alpha = 0.25$	$\alpha = 0.125$	Exact
T = 0.0	Y = 3.0000	Y = 3.0000	Y = 3.0000	Y = 3.0000	Y = 3.0000
T = 1.0	Y = 0.5000	Y = 1.3125	Y = 1.6259	Y = 1.6468	Y = 1.6519
T = 2.0	Y = 1.2500	Y = 1.3008	Y = 1.4243	Y = 1.4350	Y = 1.4377
T = 3.0	Y = 1.6250	Y = 1.5286	Y = 1.5275	Y = 1.5292	Y = 1.5297
T = 4.0	Y = 1.8125	Y = 1.7153	Y = 1.6702	Y = 1.6677	Y = 1.6673
T = 5.0	Y = 1.9063	Y = 1.8350	Y = 1.7851	Y = 1.7814	Y = 1.7807
T = 6.0	Y = 1.9531	Y = 1.9059	Y = 1.8647	Y = 1.8612	Y = 1.8605
T = 7.0	Y = 1.9766	Y = 1.9468	Y = 1.9164	Y = 1.9135	Y = 1.9130
T = 8.0	Y = 1.9883	Y = 1.9700	Y = 1.9488	Y = 1.9467	Y = 1.9463
T = 9.0	Y = 1.9941	Y = 1.9831	Y = 1.9689	Y = 1.9673	Y = 1.9672
T = 10.0	Y = 1.9971	Y = 1.9905	Y = 1.9811	Y = 1.9800	Y = 1.9799

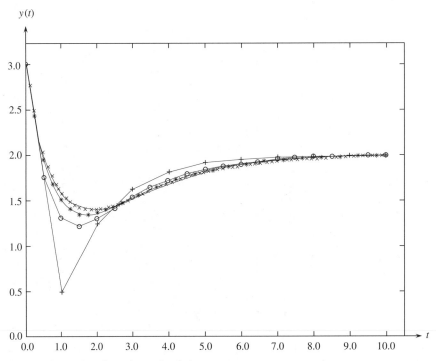

Figure 5.1 Results of computer simulation.

$\alpha = 0.25$ and then for $\alpha = 0.125$. The last two results are very close, therefore we stop the computation.

The exact solution of (5.4) can be computed, using the procedure in Section 2.7, as

$$y(t) = 2 + 4e^{-t} - 3e^{-0.5t}$$

The computed result is very close to the exact solution.

Exercise 5.2.1

Discretize and then compute the solution of $\dot{x}(t) = -0.5x(t) + u(t)$ due to $x(0) = 1$ and $u(t) = 1$, for $t \geq 0$. Compare your result with the exact solution.

The purpose of introducing the preceding algorithm is to show that state-variable equations can easily be programmed on a digital computer. The introduced method is the simplest but yields the least accurate result for the chosen integration step size. There are many other methods. For example, we may discretize a continuous-time equation into a discrete-time state-variable equation as shown in (2.89) and then carry out the computation. This is discussed in the next section.

5.3 EXISTING COMPUTER PROGRAMS

Before the age of personal computers, control systems were mostly simulated on mainframe computers. Many computer programs are available at computing centers. Two of the most widely available are

IBM Scientific Routine Package
LSMA (Library of Statistics and Mathematics Association)

In these packages, there are subroutines for solving differential equations, linear algebraic equations, and roots of polynomials. These subroutines have been thoroughly tested and can be employed with confidence. Most computing centers also have LINPACK and Eispack, which were developed under the sponsorship of the National Science Foundation and are considered to be the best for solving linear algebraic problems. These programs are often used as a basis for developing other computer programs.

Many specialized digital computer programs have been available for simulating continuous-time systems in mainframe computers: CSMP (Continuous System Modeling Program), MIDAS (Modified Integration Digital Analog Simulator), MIMIC (an improved version of MIDAS), and others. These programs were major simulation tools for control systems only a few years ago. Now they are probably completely replaced by the programs to be introduced in the following.

Personal computers are now widely available. A large amount of computer software has been developed in universities and industries. About 90 programs are listed in Reference [30]. We list in the following some commercially available packages:

CTRL-C (Systems Control Technology)

EASY5 (Boeing Computer Services Company)

MATLAB (Math Works Inc.)

$MATRIX_x$ (Integrated Systems Inc.)

Program CC (Systems Technology)

Simnon (SSPA Systems)

For illustration, we discuss only the use of MATLAB.* The author is familiar with version 3.1 and the Student Edition of MATLAB, which is a simplified edition of version 3.5. Where no version is mentioned, the discussion is applicable to either

---

*The author has experience only with PC-MATLAB and has no knowledge of the relative strength or weakness of the programs listed. MATLAB is now available in many universities. The Student Edition of MATLAB™ is now available from Prentice Hall.

version. In MATLAB, a matrix is represented row by row separated by semicolons; entries of each row are separated by spaces or commas. For example, the following matrices

$$\mathbf{A} = \begin{bmatrix} -1.5 & -2 & -0.5 \\ 1 & 0 & 0 \\ 0 & 1 & 0 \end{bmatrix} \quad \mathbf{b} = \begin{bmatrix} 1 \\ 0 \\ 0 \end{bmatrix} \quad \mathbf{c} = [1.5 \quad 0 \quad 0.5] \quad d = 0 \quad (5.6)$$

are represented, respectively, as

a=[−1.5 −2 −0.5;1 0 0;0 1 0]; b=[1;0;0]; c=[1.5 0 0.5]; d=[0];

After each statement, if we type the ENTER key, the statement will be executed, stored, and displayed on the monitor. If we type a semicolon, the statement will be executed and stored but not displayed. Because b has three rows, entries of b are separated by semicolons. The row vector c can also be represented as c=[1.5,0,0.5] and the scalar d as d=0 without brackets.

The state-variable equation

$$\dot{\mathbf{x}} = \mathbf{A}\mathbf{x} + \mathbf{b}u \tag{5.7a}$$

$$y = \mathbf{c}\mathbf{x} + du \tag{5.7b}$$

is represented as (a, b, c, d, iu), where iu denotes the ith input. In our case, we have only one input, thus we have iu = 1.

Suppose we wish to compute the unit-step response of (5.7). In using version 3.1 of MATLAB, we must specify the initial time t_0, the final time t_f, and the time interval α at which the output will be printed or plotted. For example, if we choose $t_0 = 0$, $t_f = 20$, and $\alpha = 1$, then we type

t=0:1:20;

The three numbers are separated by colons. The first number denotes the initial time, the last number denotes the final time, and the middle number denotes the time interval at which the output will appear. Now the following commands

y=step(a,b,c,d,1,t);
plot(t,y)

will produce the solid line shown in Figure 5.2. It plots the output at $t = 0, 1, 2, \ldots, 10$; they are connected by straight lines. The following commands

t=0:0.05:20;
plot(t,step(a,b,c,d,1,t))

will generate the dotted line shown in Figure 5.2, where the output is plotted every 0.05 second. In using version 3.5 or the Student Edition of MATLAB, if we type

step(a,b,c,d,1)

then the response will appear on the screen. There is no need to specify t_0, t_f, and α. They are chosen automatically by the computer.

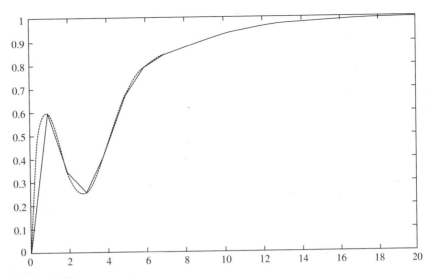

Figure 5.2 Step responses.

We discuss how MATLAB computes the step response of (5.7). It first transforms the continuous-time state-variable equation into a discrete-time equation as in (2.89). This can be achieved by using the command c2d which stands for continuous to discrete. Note that the discretization involves only **A** and **b**; **c** and d remain unchanged. If the sampling period is chosen as 1, then the command

$$[da,db] = c2d(a,b,1)$$

will yield

$$da = \begin{bmatrix} -0.1375 & -0.8502 & -0.1783 \\ 0.3566 & 0.3974 & -0.1371 \\ 0.2742 & 0.7678 & 0.9457 \end{bmatrix} \quad db = \begin{bmatrix} 0.3566 \\ 0.2742 \\ 0.1086 \end{bmatrix}$$

Thus, the discretized equation of (5.6) is

$$\mathbf{x}(k+1) = \begin{bmatrix} -0.1375 & -0.8502 & -0.1783 \\ 0.3566 & 0.3974 & -0.1371 \\ 0.2742 & 0.7678 & 0.9457 \end{bmatrix} \mathbf{x}(k) + \begin{bmatrix} 0.3566 \\ 0.2742 \\ 0.1086 \end{bmatrix} u(k) \quad \text{(5.8a)}$$

$$y(k) = [1.5 \quad 0 \quad 0.5]\mathbf{x}(k) \quad \text{(5.8b)}$$

This equation can be easily computed recursively.

If the input $u(t)$ is stepwise, as shown in Figure 2.19(a), the discretized equation in (5.8) will give the exact response of (5.7) at sampling time. Clearly, a step function is stepwise, so the two responses in Figure 5.2 are the same at the sampling instants even though one sample period is 1 second and the other is 0.05 second. We mention that several methods are listed in MATLAB for computing e^{AT}. One of them is to use the infinite series in (2.67). In using the series, there is no need to compute the eigenvalues of A.

5.4   BASIC BLOCK DIAGRAMS AND OP-AMP CIRCUITS

A basic block diagram is any diagram that is obtained by interconnecting the three types of elements shown in Figure 5.3. The three types of elements are multipliers, adders, and integrators. The gain k of a multiplier can be positive or negative, larger or smaller than 1. An adder or a summer must have two or more inputs and one and only one output. The output is simply the sum of all inputs. If the input of an integrator is $x(t)$, then its output equals $\int_0^t x(\tau)d\tau$. This choice of variable is not as convenient as assigning the output of the integrator as $x(t)$. Then the input of the integrator is $\dot{x}(t) := dx/dt$ as shown in Figure 5.3. These three elements can be easily built using operational amplifier (op-amp) circuits. For example, a multiplier with gain k can be built as shown in Figure 5.4(a) or (b) depending on whether k is positive or negative. The adder can be built as shown in Figure 5.4(c). Figure 5.4(d) shows an implementation of the integrator with $R = 1\,\text{k}\Omega = 1000\,\Omega$, $C = 10^{-3}\,\text{F}$ or $R = 1\,\text{M}\Omega = 10^6\,\Omega$, $C = 1\,\mu\text{F} = 10^{-6}\,\text{F}$. For simplicity, the grounded inverting terminals are not plotted in Figure 5.4(b through f).

Now we show that every state-variable equation can be represented by a basic block diagram. The procedure is simple and straightforward. If an equation has n state variables or, equivalently, has dimension n, we need n integrators. The output of each integrator is assigned as a state variable, say, x_i; then its input is $\dot{x}_i$. If it is assigned as $-x_i$, then its input is $-\dot{x}_i$. Finally, we use multipliers and adders to build up the state-variable equation. This is illustrated by an example. Consider

$$\begin{bmatrix} \dot{x}_1(t) \\ \dot{x}_2(t) \end{bmatrix} = \begin{bmatrix} 2 & -0.3 \\ 1 & -8 \end{bmatrix} \begin{bmatrix} x_1(t) \\ x_2(t) \end{bmatrix} + \begin{bmatrix} -2 \\ 0 \end{bmatrix} u(t) \tag{5.9a}$$

$$y(t) = \begin{bmatrix} -2 & 3 \end{bmatrix} \begin{bmatrix} x_1(t) \\ x_2(t) \end{bmatrix} + 5u(t) \tag{5.9b}$$

It has dimension 2 and needs two integrators. The outputs of the two integrators are assigned as x_1 and x_2 as shown in Figure 5.5. Their inputs are $\dot{x}_1$ and $\dot{x}_2$. The first equation of (5.9a) is $\dot{x}_1 = 2x_1 - 0.3x_2 - 2u$. It is generated in Figure 5.5 using the solid line. The second equation of (5.9a) is $\dot{x}_2 = x_1 - 8x_2$ and is generated using the dashed line. The output equation in (5.9b) is generated using the dashed-and-

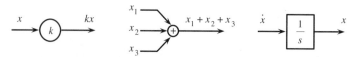

Figure 5.3 Three basic elements.

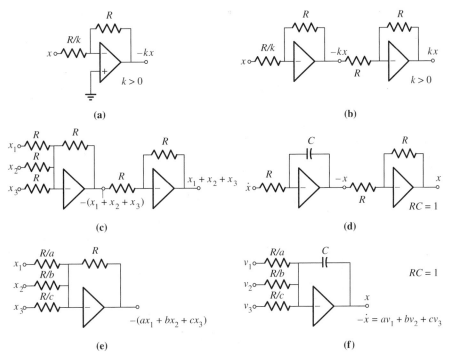

Figure 5.4 Implementation of basic elements.

dotted line. It is indeed simple and straightforward to develop a basic block diagram for any state-variable equation. Conversely, given a basic block diagram, after assigning the output of each integrator as a state variable, a state-variable equation can be easily obtained. See Problem 5.5.

Every basic element can be built using an op-amp circuit as shown in Figure 5.4; therefore every state-variable equation can be so built through its basic block

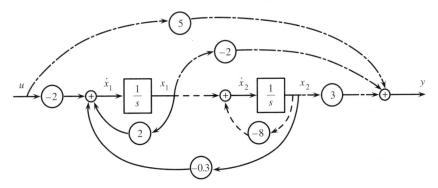

Figure 5.5 Basic block diagram of (5.9).

diagram. For example, Figure 5.6(a) shows an op-amp circuit implementation of (5.9).

The implementation in Figure 5.6(a), however, is not desirable, because it uses unnecessarily large numbers of components. In practice, op-amp circuits are built to perform several functions simultaneously. For example, the circuit in Figure 5.4(e) can act as an adder and multipliers. If its inputs are x_1, x_2, and x_3, then its output is

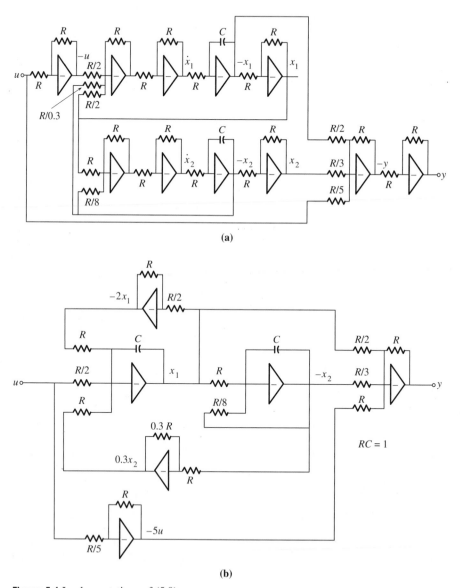

(a)

(b)

Figure 5.6 Implementations of (5.9).

$-(ax_1 + bx_2 + cx_3)$. The circuit in Figure 5.4(f) can act as an integrator, an adder, and multipliers. If we assign its output as x, then we have $-\dot{x} = av_1 + bv_2 + cv_3$. If we assign the output as $-x$, then we have $\dot{x} = av_1 + bv_2 + cv_3$. It is important to mention that we can assign the output either as x or $-x$, but we cannot alter its relationship with the inputs.

Figure 5.6(b) shows an op-amp circuit implementation of (5.9) by using the elements in Figure 5.4(e) and (f). It has two integrators. The output of one integrator is assigned as x_1; therefore, its input equals $-\dot{x}_1 = -2x_1 + 0.3x_2 + 2u$ as shown. The output of the second integrator is assigned as $-x_2$; therefore its input should equal $\dot{x}_2 = x_1 - 8x_2$ as shown. The rest is self-explanatory. Although the numbers of capacitors used in Figure 5.6(a) and (b) are the same, the numbers of operational amplifiers and resistors in Figure 5.6(b) are considerably smaller.

In actual operational amplifier circuits, the range of signals is limited by the supplied voltages, for example ± 15 volts. Therefore, a state-variable equation may have to be scaled before implementation. Otherwise, saturation may occur. This and other technical details are outside the scope of this text, and the interested reader is referred to References [24, 50].

5.5 REALIZATION PROBLEM

Consider a transfer function $G(s)$. To compute the response of $G(s)$ due to an input, we may find the Laplace transform of the input and then expand $G(s)U(s)$ into partial fraction expansion. From the expansion, we can obtain the response. This procedure requires the computation of all poles of $G(s)U(s)$, or all roots of a polynomial. This can be easily done by using MATLAB. A polynomial in MATLAB is represented by a row vector with coefficients ordered in descending powers. For example, the polynomial $(s + 1)^3(s + 1.001) = s^4 + 4.001s^3 + 6.003s^2 + 4.003s + 1.001$ is represented as

p = [1,4.001,6.003,4.003,1.001]

or

p = [1 4.001 6.003 4.003 1.001]

The entries are separated by commas or spaces. The command

roots(p)

will generate

-1.001
-1.0001
$-1 + 0.0001i$
$-1 - 0.0001i$

We see that the results differ slightly from the exact roots $-1, -1, -1$, and -1.001. If we perturb the polynomial to $s^4 + 4.002s^3 + 0.6002s^2 + 4.002s + 1$, then the command

roots([1 4.002 6.002 4.002 1])

will generate

-1.2379
$-9.9781+0.208i$
$-9.9781-0.208i$
-0.8078

The roots are quite different from those of the original polynomial, even though the two polynomials differ by less than 0.03%. Thus, roots of polynomials are very sensitive to their coefficients. One may argue that the roots of the two polynomials are repeated or clustered together; therefore, the roots are sensitive to coefficients. In fact, even if the roots are well spread, as in the polynomial

$$p(s) = (s + 1)(s + 2)(s + 3) \cdots (s + 19)(s + 20)$$

the roots are still very sensitive to the coefficients. See Reference [15, p. 219]. Furthermore, to develop a computer program to carry out partial fraction expansion is not simple. On the other hand, the response of state-variable equations is easy to program, as is shown in Section 5.2. Its computation does not require the computation of roots or eigenvalues, therefore it is less sensitive to parameter variations. For these reasons, it is desirable to compute the response of $G(s)$ through state-variable equations.

Consider a transfer function $G(s)$. If we can find a state-variable equation

$$\dot{\mathbf{x}}(t) = \mathbf{A}\mathbf{x}(t) + \mathbf{b}u(t) \tag{5.10a}$$

$$y(t) = \mathbf{c}\mathbf{x}(t) + du(t) \tag{5.10b}$$

such that the transfer function from u to y in (5.10) equals $G(s)$ or, from (2.75),

$$G(s) = \mathbf{c}(s\mathbf{I} - \mathbf{A})^{-1}\mathbf{b} + d$$

then $G(s)$ is said to be *realizable* and (5.10) is called a *realization* of $G(s)$. The term *realization* is well justified, for $G(s)$ can then be built or implemented using op-amp circuits through the state-variable equation. It turns out that $G(s)$ is realizable if and only if $G(s)$ is a proper rational function. If $G(s)$ is an improper rational function, its realization will assume the form

$$\dot{\mathbf{x}}(t) = \mathbf{A}\mathbf{x}(t) + \mathbf{b}u(t) \tag{5.11a}$$

$$y(t) = \mathbf{c}\mathbf{x}(t) + du(t) + d_1\dot{u}(t) + d_2\ddot{u}(t) + \cdots \tag{5.11b}$$

where $\dot{u}(t) = du(t)/dt$ and $\ddot{u}(t) = d^2u(t)/dt^2$. In this case, differentiators are needed to generate $\dot{u}(t)$ and $\ddot{u}(t)$. Note that differentiators are not included in Figure 5.3 and are difficult to build in practice. See Reference [18, p. 456]. Therefore, the state-variable equation in (5.11) is not used in practice. Thus, we study only the realization of proper rational transfer functions in the remainder of this chapter.

5.5.1 Realizations of $N(s)/D(s)$

Instead of discussing the general case, we use a transfer function of degree 4 to illustrate the realization procedure. Consider

$$G(s) = \frac{Y(s)}{U(s)} = \frac{b_4 s^4 + b_3 s^3 + b_2 s^2 + b_1 s + b_0}{a_4 s^4 + a_3 s^3 + a_2 s^2 + a_1 s + a_0} =: \frac{N(s)}{D(s)} \qquad (5.12)$$

where a_i and b_i are real constants and $a_4 \neq 0$. If $b_4 \neq 0$, the transfer function is biproper; if $b_4 = 0$, it is strictly proper. Before realization, we must carry out two preliminary steps: (i) decompose $G(s)$ into the sum of a constant and a strictly proper rational function, and (ii) carry out a normalization. This is illustrated by an example.

Example 5.5.1

Consider

$$G(s) = \frac{s^4 + 2s^3 - s^2 + 4s + 12}{2s^4 + 10s^3 + 20s^2 + 20s + 8} \qquad (5.13)$$

Clearly we have $G(\infty) = 1/2 = 0.5$; it is the ratio of the coefficients associated with s^4. We compute

$$G_s(s) := G(s) - G(\infty) = \frac{(s^4 + 2s^3 - s^2 + 4s + 12) -}{2s^4 + 10s^3 + 20s^2 + 20s + 8}$$
$$\frac{0.5(2s^4 + 10s^3 + 20s^2 + 20s + 8)}{2s^4 + 10s^3 + 20s^2 + 20s + 8} \qquad (5.14)$$
$$= \frac{-3s^3 - 11s^2 - 6s + 8}{2s^4 + 10s^3 + 20s^2 + 20s + 8}$$

It is strictly proper. Next we divide its numerator and denominator by 2 to yield

$$G_s(s) = \frac{-1.5s^3 - 5.5s^2 - 3s + 4}{s^4 + 5s^3 + 10s^2 + 10s + 4} =: \frac{\bar{N}(s)}{\bar{D}(s)}$$

This step normalizes the leading coefficient of $\bar{D}(s)$ to 1. Thus (5.13) can be written as

$$G(s) = \frac{s^4 + 2s^3 - s^2 + 4s + 12}{2s^4 + 10s^3 + 20s^2 + 20s + 8} \qquad (5.15)$$
$$= 0.5 + \frac{-1.5s^3 - 5.5s^2 - 3s + 4}{s^4 + 5s^3 + 10s^2 + 10s + 4}$$

This completes the preliminary steps of realization. We mention that (5.14) can also be obtained by direct division as

$$
\begin{array}{r}
0.5 \\
2s^4 + 10s^3 + 20s^2 + 20s + 8 \overline{)\, s^4 + 2s^3 - s^2 + 4s + 12} \\
\underline{s^4 + 5s^3 + 10s^2 + 10s + 4} \\
-3s^3 - 11s^2 - 6s + 8
\end{array}
$$

Using the preceding procedures, we can write $G(s)$ in (5.12) as

$$G(s) = \frac{Y(s)}{U(s)} = G(\infty) + \frac{\bar{b}_3 s^3 + \bar{b}_2 s^2 + \bar{b}_1 s + \bar{b}_0}{s^4 + \bar{a}_3 s^3 + \bar{a}_2 s^2 + \bar{a}_1 s + \bar{a}_0} =: d + \frac{\bar{N}(s)}{\bar{D}(s)} \quad (5.16)$$

Now we claim that the following state-variable equation

$$\dot{\mathbf{x}}(t) = \begin{bmatrix} -\bar{a}_3 & -\bar{a}_2 & -\bar{a}_1 & -\bar{a}_0 \\ 1 & 0 & 0 & 0 \\ 0 & 1 & 0 & 0 \\ 0 & 0 & 1 & 0 \end{bmatrix} \mathbf{x}(t) + \begin{bmatrix} 1 \\ 0 \\ 0 \\ 0 \end{bmatrix} u(t) \quad (5.17a)$$

$$y(t) = [\bar{b}_3 \quad \bar{b}_2 \quad \bar{b}_1 \quad \bar{b}_0] \mathbf{x}(t) + du(t) \quad (5.17b)$$

with $d = G(\infty)$, is a realization of (5.12) or (5.16). The dimension of this equation equals 4, the degree of the denominator of $G(s)$. The value of $G(\infty)$ yields the direct transmission part d. The first row of **A** consists of the coefficients, except the leading coefficients, of $\bar{D}(s)$ with sign changed. The pattern of the remainder of **A** is fixed; if we delete the first row and the last column, then it reduces to a unit matrix. The pattern of **b** is again fixed. It is the same for all $G(s)$; the first entry is 1 and the rest zero. The row vector **c** is formed from the coefficients of $\bar{N}(s)$. Thus the state-variable equation in (5.17) can be easily formed from the coefficients in (5.16). Given a $G(s)$, there are generally many (in fact, infinitely many) realizations. For convenience, we call (5.17) the *controllable-form* realization for reasons to be discussed in Chapter 12. See also Problem 5.18.

Now we use Mason's formula to show that the transfer function of (5.17) from u to y equals (5.16). Figure 5.7 shows a basic block diagram of (5.17). It has 4 integrators with state variables chosen as shown. The first equation of (5.17a) yields the lower part of Figure 5.7. The second equation of (5.17a) is $\dot{x}_2 = x_1$; thus, we simply connect the output of the first integrator, counting from the left, to the input of the second integrator. The third equation of (5.17a) is $\dot{x}_3 = x_2$; thus, we connect

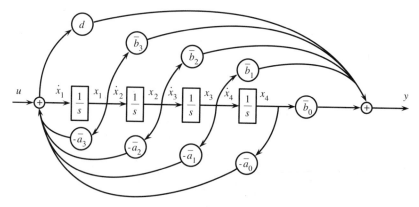

Figure 5.7 Basic block diagram of (5.17).

the output of the second integrator to the input of the third integrator. A similar remark applies to the fourth equation of (5.17a). From (5.17b), we can readily draw the upper part of Figure 5.7. The basic block diagram has four loops with loop gains $-\bar{a}_3/s$, $-\bar{a}_2/s^2$, $-\bar{a}_1/s^3$ and $-\bar{a}_0/s^4$. They touch each other. Thus, we have

$$\Delta = 1 - \left(\frac{-\bar{a}_3}{s} + \frac{-\bar{a}_2}{s^2} + \frac{-\bar{a}_1}{s^3} + \frac{-\bar{a}_0}{s^4} \right) = 1 + \frac{\bar{a}_3}{s} + \frac{\bar{a}_2}{s^2} + \frac{\bar{a}_1}{s^3} + \frac{\bar{a}_0}{s^4}$$

There are five forward paths from u to y with path gains:

$$P_1 = \frac{\bar{b}_0}{s^4} \qquad P_2 = \frac{\bar{b}_1}{s^3} \qquad P_3 = \frac{\bar{b}_2}{s^2} \qquad P_4 = \frac{\bar{b}_3}{s}$$

and

$$P_5 = d$$

Note that the first four paths touch all the loops; whereas P_5, the direct transmission path, does not touch any loop. Thus we have

$$\Delta_1 = \Delta_2 = \Delta_3 = \Delta_4 = 1$$

and

$$\Delta_5 = \Delta$$

Therefore the transfer function from u to y is

$$G(s) = \frac{\sum_{i=1}^{5} P_i \Delta_i}{\Delta} = \frac{P_1 + P_2 + P_3 + P_4 + d\Delta}{\Delta}$$

$$= d + \frac{\dfrac{\bar{b}_3}{s} + \dfrac{\bar{b}_2}{s^2} + \dfrac{\bar{b}_1}{s^3} + \dfrac{\bar{b}_0}{s^4}}{1 + \dfrac{\bar{a}_3}{s} + \dfrac{\bar{a}_2}{s^2} + \dfrac{\bar{a}_1}{s^3} + \dfrac{\bar{a}_0}{s^4}}$$

$$= d + \frac{\bar{b}_3 s^3 + \bar{b}_2 s^2 + \bar{b}_1 s + \bar{b}_0}{s^4 + \bar{a}_3 s^3 + \bar{a}_2 s^2 + \bar{a}_1 s + \bar{a}_0}$$

This is the same as (5.16). Thus (5.17) is a realization of (5.12) or (5.16). This can also be shown by computing $c(sI - A)^{-1}b + d$ in (5.17). This is more tedious and is skipped.

Because $G(s)$ is scalar—that is, $G(s) = G'(s)$—we have

$$G(s) = c(sI - A)^{-1}b + d = b'(sI - A')^{-1}c' + d$$

Thus, the following state-variable equation

$$\dot{x}(t) = \begin{bmatrix} -\bar{a}_3 & 1 & 0 & 0 \\ -\bar{a}_2 & 0 & 1 & 0 \\ -\bar{a}_1 & 0 & 0 & 1 \\ -\bar{a}_0 & 0 & 0 & 0 \end{bmatrix} x(t) + \begin{bmatrix} \bar{b}_3 \\ \bar{b}_2 \\ \bar{b}_1 \\ \bar{b}_0 \end{bmatrix} u(t) \qquad (5.18a)$$

$$y(t) = [1 \quad 0 \quad 0 \quad 0]\mathbf{x}(t) + du(t) \tag{5.18b}$$

is also a realization of (5.16). It is obtained from (5.17) by taking the transpose of **A** and interchanging **b** and **c**. This is called the *observable-form* realization of (5.16) for reasons to be discussed in Chapter 12. Although we use the same **x**, the **x** in (5.17) and the **x** in (5.18) denote entirely different variables.

Example 5.5.2

Consider the $G(s)$ in (5.13). It can be written as, as discussed in (5.15),

$$G(s) = \frac{s^4 + 2s^3 - s^2 + 4s + 12}{2s^4 + 10s^3 + 20s^2 + 20s + 8} \tag{5.19}$$

$$= 0.5 + \frac{-1.5s^3 - 5.5s^2 - 3s + 4}{s^4 + 5s^3 + 10s^2 + 10s + 4}$$

The denominator of $G(s)$ has degree 4, therefore the realization of $G(s)$ has dimension 4. Its controllable-form realization is

$$\dot{\mathbf{x}} = \begin{bmatrix} -5 & -10 & -10 & -4 \\ 1 & 0 & 0 & 0 \\ 0 & 1 & 0 & 0 \\ 0 & 0 & 1 & 0 \end{bmatrix} \mathbf{x} + \begin{bmatrix} 1 \\ 0 \\ 0 \\ 0 \end{bmatrix} u \tag{5.20a}$$

$$y = [-1.5 \quad -5.5 \quad -3 \quad 4]\mathbf{x} + 0.5u \tag{5.20b}$$

Its observable-form realization is

$$\dot{\mathbf{x}} = \begin{bmatrix} -5 & 1 & 0 & 0 \\ -10 & 0 & 1 & 0 \\ -10 & 0 & 0 & 1 \\ -4 & 0 & 0 & 0 \end{bmatrix} \mathbf{x} + \begin{bmatrix} -1.5 \\ -5.5 \\ -3 \\ 4 \end{bmatrix} u$$

$$y = [1 \quad 0 \quad 0 \quad 0]\mathbf{x} + 0.5u$$

Example 5.5.3

Consider

$$G(s) = \frac{1}{2s^3} = \frac{0.5}{s^3} = \frac{0 \cdot s^2 + 0 \cdot s + 0.5}{s^3 + 0 \cdot s^2 + 0 \cdot s + 0}$$

The denominator of $G(s)$ has degree 3, therefore the realization of $G(s)$ has dimension 3. Because $G(s)$ is strictly proper or $G(\infty) = 0$, the realization has no direct transmission part. The controllable-form realization of $G(s)$ is

$$\dot{\mathbf{x}} = \begin{bmatrix} 0 & 0 & 0 \\ 1 & 0 & 0 \\ 0 & 1 & 0 \end{bmatrix} \mathbf{x} + \begin{bmatrix} 1 \\ 0 \\ 0 \end{bmatrix} u$$

$$y = [0 \quad 0 \quad 0.5]\mathbf{x}$$

Exercise 5.5.1

Find controllable- and observable-form realizations of

a. $\dfrac{2s + 10}{s + 2}$

b. $\dfrac{3s^2 - s + 2}{s^3 + 2s^2 + 1}$

c. $\dfrac{4s^3 + 2s + 1}{2s^3 + 3s^2 + 2}$

Exercise 5.5.2

Find realizations of

a. $\dfrac{5}{2s^2 + 4s + 3}$

b. $\dfrac{2}{s^3 + s - 1}$

c. $\dfrac{3}{2s^5 + 1}$

To conclude this section, we mention that realizations can also be generated using MATLAB. The command tf2ss stands for transfer function to state space or state-variable equation. The $G(s)$ in (5.19) can be represented by

num = [1 2 −1 4 12]; den = [2 10 20 20 8];

where num stands for numerator and den, denominator. Then the command

[a,b,c,d] = tf2ss(num,den)

will generate the controllable-form realization in (5.20). To find the step response of $G(s)$, we type

   step(num,den)

if we use version 3.5 or the Student Edition of MATLAB. Then the response will appear on the monitor. If we use version 3.1 of MATLAB to find the step response of $G(s)$ from 0 to 20 seconds and with print-out interval 0.1, we type

   t=0:0.1:20;
   y=step(num,den,t);
   plot(t,y)

Then the response will appear on the monitor. Inside MATLAB, the transfer function is first realized as a state-variable equation by calling **tf2ss**, and then discretized by calling **c2d**. The response is then computed and plotted.

5.5.2 Tandem and Parallel Realizations

The realizations in the preceding section are easy to obtain, but they are sensitive to parameter variations in computer computation and op-amp circuit implementations. In this subsection, we discuss two different types of realizations. They are to be obtained from block diagrams. We use examples to illustrate the realization procedures.

Example 5.5.4

Consider the transfer function

$$G(s) = \frac{20}{(s+2)^2(2s+3)} = \frac{2}{s+2} \cdot \frac{1}{s+2} \cdot \frac{10}{2s+3} \qquad (5.21)$$

with grouping shown; the grouping is quite arbitrary. It is plotted in Figure 5.8(a). Now we assign the output of each block as a state variable as shown. From the first

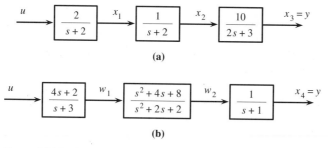

Figure 5.8 Tandem connections.

block from the left, we have

$$X_1(s) = \frac{2}{s + 2} U(s) \quad \text{or} \quad (s + 2)X_1(s) = 2U(s)$$

which becomes, in the time domain,

$$\dot{x}_1 + 2x_1 = 2u \quad \text{or} \quad \dot{x}_1 = -2x_1 + 2u$$

From the third block, we have

$$X_3(s) = \frac{10}{2s + 3} X_2(s) \quad \text{or} \quad (s + 1.5)X_3(s) = 5X_2(s)$$

which becomes

$$\dot{x}_3 + 1.5x_3 = 5x_2 \quad \text{or} \quad \dot{x}_3 = -1.5x_3 + 5x_2$$

Similarly, the second block implies $\dot{x}_2 = -2x_2 + x_1$. These equations and $y = x_3$ can be arranged as

$$\begin{bmatrix} \dot{x}_1 \\ \dot{x}_2 \\ \dot{x}_3 \end{bmatrix} = \begin{bmatrix} -2 & 0 & 0 \\ 1 & -2 & 0 \\ 0 & 5 & -1.5 \end{bmatrix} \begin{bmatrix} x_1 \\ x_2 \\ x_3 \end{bmatrix} + \begin{bmatrix} 2 \\ 0 \\ 0 \end{bmatrix} u \qquad (5.22a)$$

$$y = [0 \quad 0 \quad 1]\mathbf{x} \qquad (5.22b)$$

This is called a *tandem* or *cascade realization* because it is obtained from the tandem connection of blocks.

Example 5.5.5

Consider

$$G(s) = \frac{(4s + 2)(s^2 + 4s + 8)}{(s + 3)(s^2 + 2s + 2)(s + 1)} \qquad (5.23)$$

It has pairs of complex conjugate poles and zeros. If complex numbers are not permitted in realization, then these poles and zeros cannot be further factored into first-order factors. We group (5.23) and then plot it in Figure 5.8(b). The first block has the biproper transfer function $(4s + 2)/(s + 3)$. If we assign its output as a state variable as x_1, then we have

$$X_1(s) = \frac{4s + 2}{s + 3} U(s) \quad \text{or} \quad (s + 3)X_1(s) = (4s + 2)U(s)$$

which implies

$$\dot{x}_1 + 3x_1 = 4\dot{u} + 2u \quad \text{or} \quad \dot{x}_1 = -3x_1 + 2u + 4\dot{u}$$

This equation contains the derivative of u, thus we cannot assign the output of the first block as a state variable. Now we assign the outputs of the first and second blocks as w_1 and w_2 as shown in Figure 5.8(b). Then we have

$$W_1(s) = \frac{4s + 2}{s + 3} U(s) = \left(4 + \frac{-10}{s + 3}\right) U(s)$$

and

$$W_2(s) = \frac{s^2 + 4s + 8}{s^2 + 2s + 2} W_1(s) = \left(1 + \frac{2s + 6}{s^2 + 2s + 2}\right) W_1(s)$$

The first transfer function is realized in the observable form as

$$\dot{x}_1 = -3x_1 - 10u \qquad w_1 = x_1 + 4u \tag{5.24}$$

The second transfer function is realized in the controllable form as

$$\begin{bmatrix} \dot{x}_2 \\ \dot{x}_3 \end{bmatrix} = \begin{bmatrix} -2 & -2 \\ 1 & 0 \end{bmatrix} \begin{bmatrix} x_2 \\ x_3 \end{bmatrix} + \begin{bmatrix} 1 \\ 0 \end{bmatrix} w_1$$

$$w_2 = [2 \quad 6] \begin{bmatrix} x_1 \\ x_3 \end{bmatrix} + w_1 \tag{5.25}$$

The substitution of w_1 in (5.24) into (5.25) yields

$$\dot{x}_2 = -2x_2 - 2x_3 + x_1 + 4u \qquad \dot{x}_3 = x_2 \tag{5.26a}$$

$$w_2 = 2x_2 + 6x_3 + x_1 + 4u \tag{5.26b}$$

If we assign the output of the third block in Figure 5.8(b) as x_4, then we have, using (5.26b),

$$\dot{x}_4 = -x_4 + w_2 = -x_4 + 2x_2 + 6x_3 + x_1 + 4u \tag{5.27}$$

These equations and $y = x_4$ can be arranged as

$$\begin{bmatrix} \dot{x}_1 \\ \dot{x}_2 \\ \dot{x}_3 \\ \dot{x}_4 \end{bmatrix} = \begin{bmatrix} -3 & 0 & 0 & 0 \\ 1 & -2 & -2 & 0 \\ 0 & 1 & 0 & 0 \\ 1 & 2 & 6 & -1 \end{bmatrix} \begin{bmatrix} x_1 \\ x_2 \\ x_3 \\ x_4 \end{bmatrix} + \begin{bmatrix} -10 \\ 4 \\ 0 \\ 4 \end{bmatrix} u \tag{5.28a}$$

$$y = [0 \quad 0 \quad 0 \quad 1]\mathbf{x} \tag{5.28b}$$

This is a tandem realization of the transfer function in (5.23).

If a transfer function is broken into a product of transfer functions of degree 1 or 2, then we can obtain a tandem realization of the transfer function. The realization is not unique. Different grouping and different ordering yield different tandem realizations. The outputs of blocks with transfer function $b/(s + a)$ can be assigned as state variables (see also Problem 5.9). The outputs of blocks with transfer function $(s + b)/(s + a)$ or of degree 2 cannot be so assigned.

In the following, we discuss a different type of realization, called *parallel realization*.

Example 5.5.6

Consider the transfer function in (5.21). We use partial fraction expansion to expand it as

$$G(s) = \frac{20}{(s + 2)^2(2s + 3)} = \frac{-40}{s + 2} + \frac{-20}{(s + 2)^2} + \frac{40}{s + 1.5} \qquad (5.29)$$

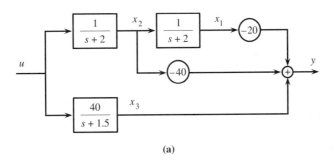

(a)

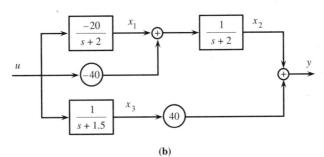

(b)

Figure 5.9 Parallel connections.

Figure 5.9 shows two different plots of (5.29). If we assign the output of each block as a state variable as shown, then from Figure 5.9(a), we can readily obtain

$$\dot{x}_1 = -2x_1 + x_2 \qquad \dot{x}_2 = -2x_2 + u$$

and

$$\dot{x}_3 = -1.5x_3 + 40u \qquad y = -20x_1 - 40x_2 + x_3$$

They can be expressed in matrix form as

$$\begin{bmatrix} \dot{x}_1 \\ \dot{x}_2 \\ \dot{x}_3 \end{bmatrix} = \begin{bmatrix} -2 & 1 & 0 \\ 0 & -2 & 0 \\ 0 & 0 & -1.5 \end{bmatrix} \begin{bmatrix} x_1 \\ x_2 \\ x_3 \end{bmatrix} + \begin{bmatrix} 0 \\ 1 \\ 40 \end{bmatrix} u \tag{5.30a}$$

$$y = [-20 \quad -40 \quad 1]\mathbf{x} \tag{5.30b}$$

This is a parallel realization of $G(s)$ in (5.29). The matrix $\mathbf{A}$ in (5.30a) is said to be in a *Jordan form*; it is a generalization of diagonal matrices.

Exercise 5.5.3

Show that the block diagram in Figure 5.9(b) with state variables chosen as shown can be described by

$$\begin{bmatrix} \dot{x}_1 \\ \dot{x}_2 \\ \dot{x}_3 \end{bmatrix} = \begin{bmatrix} -2 & 0 & 0 \\ 1 & -2 & 0 \\ 0 & 0 & -1.5 \end{bmatrix} \begin{bmatrix} x_1 \\ x_2 \\ x_3 \end{bmatrix} + \begin{bmatrix} -20 \\ -40 \\ 1 \end{bmatrix} u \tag{5.31a}$$

$$y = [0 \quad 1 \quad 40]\mathbf{x} \tag{5.31b}$$

Example 5.5.7

Consider the transfer function in (5.23). Using partial fraction expansion, we expand it as

$$G(s) = \frac{(4s + 2)(s^2 + 4s + 8)}{(s + 3)(s^2 + 2s + 2)(s + 1)} \tag{5.32}$$

$$= \frac{-5}{s + 1} + \frac{5}{s + 3} + \frac{4s + 12}{s^2 + 2s + 2}$$

It is plotted in Figure 5.10. With the variables chosen as shown, we have

$$\dot{x}_1 = -x_1 - 5u \qquad \dot{x}_2 = -3x_2 + 5u$$

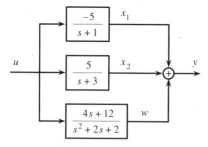

Figure 5.10 Parallel connections.

and

$$\begin{bmatrix} \dot{x}_3 \\ \dot{x}_4 \end{bmatrix} = \begin{bmatrix} -2 & -2 \\ 1 & 0 \end{bmatrix} \begin{bmatrix} x_3 \\ x_4 \end{bmatrix} + \begin{bmatrix} 1 \\ 0 \end{bmatrix} u$$

$$w = [4 \quad 12] \begin{bmatrix} x_3 \\ x_4 \end{bmatrix}$$

They can be combined to yield

$$\begin{bmatrix} \dot{x}_1 \\ \dot{x}_2 \\ \dot{x}_3 \\ \dot{x}_4 \end{bmatrix} = \begin{bmatrix} -1 & 0 & 0 & 0 \\ 0 & -3 & 0 & 0 \\ 0 & 0 & -2 & -2 \\ 0 & 0 & 1 & 0 \end{bmatrix} \begin{bmatrix} x_1 \\ x_2 \\ x_3 \\ x_4 \end{bmatrix} + \begin{bmatrix} -5 \\ 5 \\ 1 \\ 0 \end{bmatrix} u \qquad (5.33a)$$

$$y = [1 \quad 1 \quad 4 \quad 12]\mathbf{x} \qquad (5.33b)$$

This is a parallel realization of the transfer function in (5.32).

Tandem and parallel realizations are less sensitive to parameter variations than the controllable- and observable-form realizations are. For a comparison, see Reference [13].

5.6 MINIMAL REALIZATIONS

Consider the transfer function

$$G(s) = \frac{s^3 + 4s^2 + 7s + 6}{s^4 + 5s^3 + 10s^2 + 11s + 3} \qquad (5.34)$$

It is strictly proper and has 1 as the leading coefficient of its denominator. Therefore, its realization can be read from its coefficients as

$$\dot{\mathbf{x}} = \begin{bmatrix} -5 & -10 & -11 & -3 \\ 1 & 0 & 0 & 0 \\ 0 & 1 & 0 & 0 \\ 0 & 0 & 1 & 0 \end{bmatrix} \mathbf{x} + \begin{bmatrix} 1 \\ 0 \\ 0 \\ 0 \end{bmatrix} u \qquad (5.35a)$$

$$y = [1 \quad 4 \quad 7 \quad 6]\mathbf{x} \qquad (5.35b)$$

This is the controllable-form realization. Its observable-form realization is

$$\dot{\mathbf{x}} = \begin{bmatrix} -5 & 1 & 0 & 0 \\ -10 & 0 & 1 & 0 \\ -11 & 0 & 0 & 1 \\ -3 & 0 & 0 & 0 \end{bmatrix} \mathbf{x} + \begin{bmatrix} 1 \\ 4 \\ 7 \\ 6 \end{bmatrix} u \qquad (5.36a)$$

$$y = [1 \quad 0 \quad 0 \quad 0]\mathbf{x} \tag{5.36b}$$

Both realizations have dimension 4. If we use either realization to simulate or implement $G(s)$, then we need 4 integrators. It happens that the numerator and denominator of $G(s)$ have the common factor $s^2 + 2s + 3$. If we cancel this common factor, then $G(s)$ in (5.34) can be reduced to

$$G(s) = \frac{s + 2}{s^2 + 3s + 1} \tag{5.37}$$

This can be realized as

$$\dot{\mathbf{x}} = \begin{bmatrix} -3 & -1 \\ 1 & 0 \end{bmatrix} \mathbf{x} + \begin{bmatrix} 1 \\ 0 \end{bmatrix} u \tag{5.38a}$$

$$y = [1 \quad 2]\mathbf{x} \tag{5.38b}$$

or

$$\dot{\mathbf{x}} = \begin{bmatrix} -3 & 1 \\ -1 & 0 \end{bmatrix} \mathbf{x} + \begin{bmatrix} 1 \\ 2 \end{bmatrix} u \tag{5.39a}$$

$$y = [1 \quad 0]\mathbf{x} \tag{5.39b}$$

These realizations have dimension 2. They are called *minimal realizations* of $G(s)$ in (5.34) or (5.37), because they have the smallest number of state variables among all possible realizations of $G(s)$. The realizations in (5.35) and (5.36) are called *nonminimal realizations*. We mention that minimal realizations are minimal equations discussed in Section 2.8. Nonminimal realizations are not minimal equations.

If $D(s)$ and $N(s)$ have no common factor, then the controllable-form and observable-form realizations of $G(s) = N(s)/D(s)$ are minimal realizations. On the other hand, if we introduce a common factor into $N(s)$ and $D(s)$ such as

$$G(s) = \frac{s + 2}{s^2 + 3s + 1} = \frac{(s + 2)P(s)}{(s^2 + 3s + 1)P(s)}$$

then depending on the polynomial $P(s)$, we can find many nonminimal realizations. For example, if the degree of $P(s)$ is 5, then we can find a 7-dimensional realization. Such a nonminimal realization uses unnecessarily large numbers of components and is not desirable in practice.

Exercise 5.6.1

Find two minimal realizations for

$$G(s) = \frac{s^2 - 1}{s^3 + s^2 + s - 3}$$

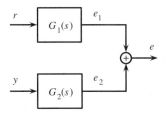

Figure 5.11 Two-input, one-output system.

Exercise 5.6.2

Find three different nonminimal realizations of dimension 3 for the transfer function $G(s) = 1/(s + 1)$.

5.6.1 Minimal Realization of Vector Transfer Functions[1]

Consider the connection of two transfer functions shown in Figure 5.11. It has two inputs r and y and one output e, and is called a *two-input, one-output system*. Using vector notation, the output and inputs can be expressed as

$$E(s) = G_1(s)R(s) + G_2(s)Y(s) = [G_1(s) \quad G_2(s)] \begin{bmatrix} R(s) \\ Y(s) \end{bmatrix} \quad (5.40)$$

where $G_1(s)$ and $G_2(s)$ are proper rational functions. We assume that both $G_1(s)$ and $G_2(s)$ are irreducible—that is, the numerator and denominator of each transfer function have no common factor. Certainly, we can find a minimal realization for $G_1(s)$ and that for $G_2(s)$. By connecting these two realizations, we can obtain a realization for the two-input, one-output system. This realization, however, may not be a minimal realization.

We now discuss a method of finding a minimal realization for the two-input, one-output system in Figure 5.11. First, the system is considered to have the following 1×2 vector transfer function

$$\mathbf{G}(s) = [G_1(s) \quad G_2(s)] \quad (5.41)$$

We expand $\mathbf{G}(s)$ as

$$\mathbf{G}(s) = [d_1 \quad d_2] + [G_{s1}(s) \quad G_{s2}(s)] =: [d_1 \quad d_2] + \begin{bmatrix} \dfrac{N_1(s)}{D_1(s)} & \dfrac{N_2(s)}{D_2(s)} \end{bmatrix} \quad (5.42)$$

where $d_i = G_i(\infty)$ and $G_{si}(s)$ are strictly proper, that is, deg $N_i(s) <$ deg $D_i(s)$. Let $\overline{D}(s)$ be the least common multiplier of $D_1(s)$ and $D_2(s)$ and have 1 as its leading

---

[1]The material in this section is used only in Chapter 10 and its study may be postponed.

coefficient. We then write (5.42) as

$$\mathbf{G}(s) = [d_1 \quad d_2] + \frac{1}{\overline{D}(s)}[\overline{N}_1(s) \quad \overline{N}_2(s)]$$

Note that deg $\overline{N}_i(s) < $ deg $\overline{D}(s)$, for $i = 1, 2$. For convenience of discussion, we assume

$$\overline{D}(s) = s^4 + \overline{a}_3 s^3 + \overline{a}_2 s^2 + \overline{a}_1 s + \overline{a}_0$$
$$\overline{N}_1(s) = \overline{b}_{31} s^3 + \overline{b}_{21} s^2 + \overline{b}_{11} s + \overline{b}_{01}$$

(5.43)

and

$$\overline{N}_2(s) = \overline{b}_{32} s^3 + \overline{b}_{22} s^2 + \overline{b}_{12} s + \overline{b}_{02}$$

Then a minimal realization of (5.42) is

$$\dot{\mathbf{x}}(t) = \begin{bmatrix} -\overline{a}_3 & 1 & 0 & 0 \\ -\overline{a}_2 & 0 & 1 & 0 \\ -\overline{a}_1 & 0 & 0 & 1 \\ -\overline{a}_0 & 0 & 0 & 0 \end{bmatrix} \mathbf{x}(t) + \begin{bmatrix} \overline{b}_{31} & \overline{b}_{32} \\ \overline{b}_{21} & \overline{b}_{22} \\ \overline{b}_{11} & \overline{b}_{12} \\ \overline{b}_{01} & \overline{b}_{02} \end{bmatrix} \begin{bmatrix} r(t) \\ y(t) \end{bmatrix}$$

(5.44a)

$$e(t) = [1 \quad 0 \quad 0 \quad 0]\mathbf{x}(t) + [d_1 \quad d_2]\begin{bmatrix} r(t) \\ y(t) \end{bmatrix}$$

(5.44b)

This is essentially a combination of the two observable-form realizations of $G_1(s)$ and $G_2(s)$.

Example 5.6.1

Consider $G_1(s) = (s + 3)/(s + 1)(s + 2)$ and $G_2(s) = (-s + 2)/2(s + 1)$ connected as shown in Figure 5.11. Let the input of $G_1(s)$ be r and the output be e_1. We expand $G_1(s)$ as

$$G_1(s) = \frac{E_1(s)}{R(s)} = \frac{s + 3}{(s + 1)(s + 2)} = \frac{s + 3}{s^2 + 3s + 2}$$

From its coefficients, we can obtain the following minimal realization:

$$\begin{bmatrix} \dot{x}_1(t) \\ \dot{x}_2(t) \end{bmatrix} = \begin{bmatrix} -3 & -2 \\ 1 & 0 \end{bmatrix}\begin{bmatrix} x_1(t) \\ x_2(t) \end{bmatrix} + \begin{bmatrix} 1 \\ 0 \end{bmatrix} r(t)$$

$$e_1(t) = [1 \quad 3]\begin{bmatrix} x_1(t) \\ x_2(t) \end{bmatrix}$$

To realize $G_2(s)$ with input y and output e_2, we expand it as

$$G_2(s) = \frac{E_2(s)}{Y(s)} = \frac{(-s + 2)}{2(s + 1)} = -0.5 + \frac{1.5}{s + 1}$$

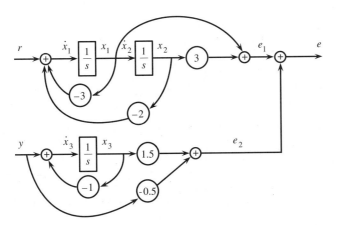

Figure 5.12 Nonminimum realization.

Then the following one-dimensional state-variable equation

$$\dot{x}_3(t) = -x_3(t) + y(t)$$

$$e_2(t) = 1.5x_3(t) - 0.5y(t)$$

is a minimal realization of $G_2(s)$. The addition of $e_1(t)$ and $e_2(t)$ yields $e(t)$. The basic block diagram of the connection is shown in Figure 5.12 and the combination of the two realizations yields

$$\begin{bmatrix} \dot{x}_1(t) \\ \dot{x}_2(t) \\ \dot{x}_3(t) \end{bmatrix} = \begin{bmatrix} -3 & -2 & 0 \\ 1 & 0 & 0 \\ 0 & 0 & -1 \end{bmatrix} \begin{bmatrix} x_1(t) \\ x_2(t) \\ x_3(t) \end{bmatrix} + \begin{bmatrix} 1 & 0 \\ 0 & 0 \\ 0 & 1 \end{bmatrix} \begin{bmatrix} r(t) \\ y(t) \end{bmatrix}$$

$$e(t) = e_1(t) + e_2(t) = \begin{bmatrix} 1 & 3 & 1.5 \end{bmatrix} \begin{bmatrix} x_1(t) \\ x_2(t) \\ x_3(t) \end{bmatrix} + \begin{bmatrix} 0 & -0.5 \end{bmatrix} \begin{bmatrix} r(t) \\ y(t) \end{bmatrix}$$

This is a three-dimensional realization.

Now we develop a minimal realization. The transfer functions $G_1(s)$ and $G_2(s)$ are combined to form a vector as

$$\mathbf{G}(s) = \begin{bmatrix} \dfrac{s + 3}{(s + 1)(s + 2)} & \dfrac{-s + 2}{2(s + 1)} \end{bmatrix}$$

$$= \begin{bmatrix} 0 & -0.5 \end{bmatrix} + \begin{bmatrix} \dfrac{s + 3}{(s + 1)(s + 2)} & \dfrac{3}{2(s + 1)} \end{bmatrix}$$

(5.45)

The least common multiplier, with leading coefficient 1, of D_1 and D_2 is

$$\overline{D}(s) = (s + 1)(s + 2) = s^2 + 3s + 2$$

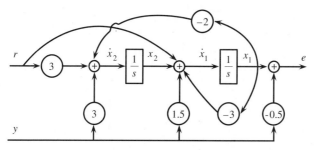

Figure 5.13 Minimal realization.

We then write (5.45) as

$$\mathbf{G}(s) = [0 \quad -0.5] + \frac{1}{s^2 + 3s + 2} [s + 3 \quad 1.5(s + 2)]$$

Thus a minimal realization is

$$\dot{\mathbf{x}}(t) = \begin{bmatrix} -3 & 1 \\ -2 & 0 \end{bmatrix} \mathbf{x}(t) + \begin{bmatrix} 1 & 1.5 \\ 3 & 3 \end{bmatrix} \begin{bmatrix} r(t) \\ y(t) \end{bmatrix} \tag{5.46a}$$

$$e(t) = [1 \quad 0]\mathbf{x}(t) + [0 \quad -0.5] \begin{bmatrix} r(t) \\ y(t) \end{bmatrix} \tag{5.46b}$$

Its basic block diagram is shown in Figure 5.13. It has two integrators, one less than the block diagram in Figure 5.12. Thus, the realization in Figure 5.12 is not a minimal realization. Note that the summer in Figure 5.11 is easily identifiable with the rightmost summer in Figure 5.12. This is not possible in Figure 5.13. The summer in Figure 5.11 is imbedded and cannot be identified with any summer in Figure 5.13.

Exercise 5.6.3

Find minimal realizations of the following 1×2 proper rational functions with inputs r and y, and output e:

a. $\left[\dfrac{s - 1}{s(s + 1)} \quad \dfrac{s + 1}{s} \right]$

b. $\left[\dfrac{2s^2 + 1}{s^2 + 3s + 1} \quad \dfrac{-s^2 + s - 2}{s^2 + 3s + 1} \right]$

PROBLEMS

5.1. Write a program to compute the output y of the following state-variable equations due to a unit-step input. The initial conditions are assumed to be zero. What integration step sizes will you choose?

a. $\dot{\mathbf{x}} = \begin{bmatrix} -1 & -2 \\ 81 & -0.9 \end{bmatrix} \mathbf{x} + \begin{bmatrix} 1.5 \\ 1.1 \end{bmatrix} u$

$y = [0.7 \quad 2.1]\mathbf{x}$

b. $\dot{\mathbf{x}} = \begin{bmatrix} 9 & -0.1 & 1 \\ 1.9 & 0 & 4.5 \\ 1 & 2 & 5 \end{bmatrix} \mathbf{x} + \begin{bmatrix} 1.1 \\ 0 \\ 2 \end{bmatrix} u$

$y = [2.5 \quad 1 \quad 1.2]\mathbf{x}$

5.2. Repeat Problem 5.1 using a commercially available software. Compare the results with those obtained in Problem 5.1.

5.3. Draw basic block diagrams for the equations in Problem 5.1.

5.4. Draw operational amplifier circuits for the two state-variable equations in Problem 5.1. Use the elements in Figure 5.4(e) and (f).

5.5. Develop a state-variable equation for the basic block diagram shown in Figure P5.5.

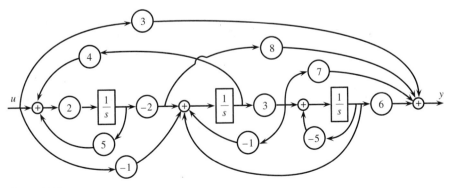

Figure P5.5

5.6. Develop a state-variable equation for the op-amp circuit shown in Figure P5.6.

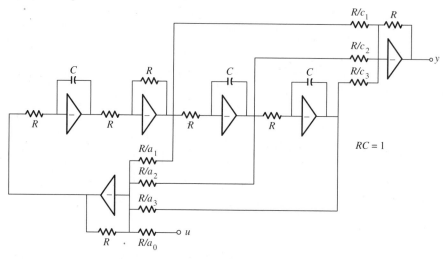

Figure P5.6

5.7. Draw a basic block diagram for the observable-form equation in (5.18) and then use Mason's formula to compute its transfer function.

5.8. Find realizations for the following transfer functions and draw their basic block diagrams.

a. $G_1(s) = \dfrac{s^2 + 2}{4s^3}$

b. $G_2(s) = \dfrac{3s^4 + 1}{2s^4 + 3s^3 + 4s^2 + s + 5}$

c. $G_3(s) = \dfrac{(s + 3)^2}{(s + 1)^2(s + 2)}$

5.9. Show the equivalence of the blocks in Figure P5.9.

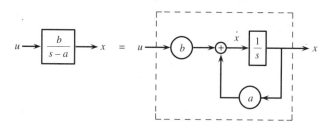

Figure P5.9

5.10. Use Problem 5.9 to change every block of Figure P5.10 into a basic block diagram and then develop state-variable equations to describe the two systems.

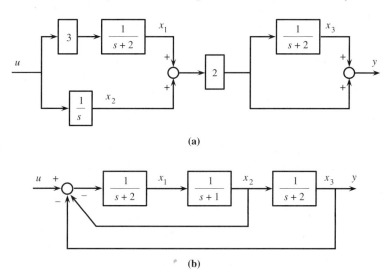

(a)

(b)

Figure P5.10

5.11. Find tandem and parallel realizations for the following transfer functions:

a. $\dfrac{(s + 3)^2}{(s + 1)^2(s + 2)}$

b. $\dfrac{(s + 3)^2}{(s + 2)^2(s^2 + 4s + 6)}$

5.12. a. Find a minimum realization of

$$G(s) = \frac{s^2 - 4}{(s - 2)(2s^2 + 3s + 4)}$$

b. Find realizations of dimension 3 and 4 for the $G(s)$ in (a).

5.13. a. Consider the armature-controlled dc motor in Figure 3.1. Show that if the state variables are chosen as $x_1 = \theta$, $x_2 = \dot{\theta}$ and $x_3 = i_a$, then its state-variable description is given by

$$\begin{bmatrix} \dot{x}_1 \\ \dot{x}_2 \\ \dot{x}_3 \end{bmatrix} = \begin{bmatrix} 0 & 1 & 0 \\ 0 & -f/J & k_t/J \\ 0 & -k_b/L_a & -R_a/L_a \end{bmatrix} \begin{bmatrix} x_1 \\ x_2 \\ x_3 \end{bmatrix} + \begin{bmatrix} 0 \\ 0 \\ 1/L_a \end{bmatrix} u$$

$$\theta = [1 \quad 0 \quad 0]\mathbf{x}$$

b. The transfer function of the motor is, as computed in (3.14),

$$G(s) = \frac{\Theta(s)}{U(s)} = \frac{k_t}{s[(Js + f)(R_a + L_a s) + k_t k_b]}$$

Find a realization of $G(s)$. [Although the realization is different from the state-variable equation in (a), they are equivalent. See Section 11.3.]

c. Every state variable in (a) is associated with a physical quantity. Can you associate every state variable in (b) with a physical quantity?

5.14. Consider the block diagram shown in Figure P5.14. (a) Compute its overall transfer function, realize it, and then draw a basic block diagram. (b) Draw a basic block diagram for each block and then connect them to yield the overall system. (c) If k is to be varied over a range, which basic block diagram, (a) or (b), is more convenient?

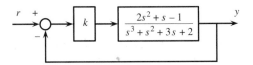

Figure P5.14

5.15. Consider the block diagram shown in Figure P5.15. It has a tachometer feedback with transfer function $2s$. Differentiators are generally not built using operational amplifier circuits. Therefore, the diagram cannot be directly simulated using operational amplifier circuits. Can you modify the diagram so that it can be so simulated?

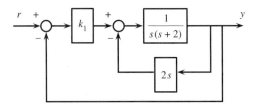

Figure P5.15

5.16. Find minimal realizations for the following 1×2 vector transfer functions:

a. $\left[\dfrac{s + 2}{(s + 1)^2} \quad \dfrac{s + 1}{s + 2} \right]$

b. $\left[\dfrac{s - 2}{(s + 1)(s + 2)} \quad \dfrac{s + 1}{s + 2} \right]$

c. $\left[\dfrac{s^2 + s + 1}{s^3 + 2s^2 + 1} \quad \dfrac{s^3 - 1}{s^3 + 2s^2 + 1} \right]$

5.17. Consider the block diagram shown in Figure 5.11. Suppose $G_1(s)$ and $G_2(s)$ are given as in Problem 5.16(c). Draw a basic block diagram. Can you identify the summer in Figure 5.11 with a summer in your diagram?

5.18. a. Consider the block diagram in Figure 5.7. Show that if the state variables x_1, x_2, x_3, and x_4 in Figure 5.7 are renamed as x_4, x_3, x_2, and x_1, then the block diagram can be described by

$$\dot{\mathbf{x}}(t) = \begin{bmatrix} 0 & 1 & 0 & 0 \\ 0 & 0 & 1 & 0 \\ 0 & 0 & 0 & 1 \\ -\bar{a}_0 & -\bar{a}_1 & -\bar{a}_2 & -\bar{a}_3 \end{bmatrix} \mathbf{x}(t) + \begin{bmatrix} 0 \\ 0 \\ 0 \\ 1 \end{bmatrix} u(t)$$

$$y(t) = [\bar{b}_0 \quad \bar{b}_1 \quad \bar{b}_2 \quad \bar{b}_3]\mathbf{x}(t) + du(t)$$

This is also called a controllable-form realization of (5.16). See References [15, 16]. This is an alternative controllable form of (5.17).

b. Show that an alternative observable form of (5.18) is

$$\dot{\mathbf{x}}(t) = \begin{bmatrix} 0 & 0 & 0 & -\bar{a}_0 \\ 1 & 0 & 0 & -\bar{a}_1 \\ 0 & 1 & 0 & -\bar{a}_2 \\ 0 & 0 & 1 & -\bar{a}_3 \end{bmatrix} \mathbf{x}(t) + \begin{bmatrix} \bar{b}_0 \\ \bar{b}_1 \\ \bar{b}_2 \\ \bar{b}_3 \end{bmatrix} u(t)$$

$$y(t) = [0 \quad 0 \quad 0 \quad 1]\mathbf{x}(t) + du(t)$$

6 Design Criteria, Constraints, and Feedback

6.1 INTRODUCTION

With the background introduced in the preceding chapters, we are now ready to study the design of control systems. Before introducing specific design techniques, it is important to obtain a total picture of the design problem. In this chapter, we first discuss the choice of plants and design criteria and then discuss noise and disturbance problems encountered in practice. These problems impose constraints on control systems, which lead to the concepts of well-posedness and total stability. We also discuss the reason for imposing constraints on actuating signals. Feedback systems are then shown to be less sensitive to plant perturbations and external disturbances then are open loop systems. Finally, we discuss two general approaches in the design of control systems.

6.2 CHOICE OF A PLANT

We use an example to illustrate the formulation of the design problem. Suppose we are given an antenna. The antenna's specifications, such as weight and moment of inertia, and its operational range, such as average and maximum speed and acceleration, are also given. We are asked to design a control system to drive the antenna. The first step in the design is to choose a motor. Before choosing the type of motor, we must estimate the required horsepower. Clearly if a motor is too small, it will not be powerful enough to drive the antenna; if it is too large, the cost will be unnecessarily high. The torque required to drive the antenna is

$$\text{Torque} = J\ddot{\theta}(t)$$

where J is the moment of inertia of the antenna and $\theta(t)$ is its angular displacement. The power needed to drive the antenna is

$$\text{Power} = \text{Torque} \times \text{Velocity} = J\ddot{\theta}(t)\dot{\theta}(t)$$

This equation shows that the larger the acceleration and the velocity, the larger the power needed. Let $\ddot{\theta}_{max}$ and $\dot{\theta}_{max}$ be the maximum acceleration and velocity. Then the required power is

$$\text{Power} = J(\ddot{\theta}_{max})(\dot{\theta}_{max}) \tag{6.1}$$

In this computation, the moments of inertia of motor and gear trains, which are not yet determined, are not included. Also, we consider neither the power required to overcome static, Coulomb, and viscous frictions nor disturbances due to gusting. Therefore, the horsepower of the motor should be larger than the one computed in (6.1). After the size of a motor is determined, we must select the type of motor: dc, ac, or hydraulic. The choice may depend on availability at the time of design, cost, reliability, and other considerations. Past experience may also be used in this choice. For convenience of discussion, we choose an armature-controlled dc motor to drive the antenna. A dc generator is also chosen as a power amplifier, as shown in Figure 6.1. This collection of devices, including the load, is called the *plant* of the control system. We see from the foregoing discussion that the choice of a plant is not unique.

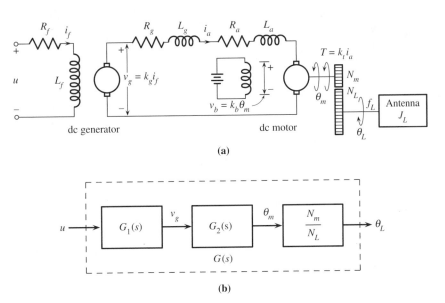

Figure 6.1 (a) Plant. (b) Its block diagram.

Different designers often choose different plants; even the same designer may choose different plants at different times.

Once a plant is chosen, the design problem is concerned with how to make the best use of the plant. Generally, the plant alone cannot meet the design objective. We must introduce compensators and transducers and hope that the resulting overall system will meet the design objective. If after trying all available design methods, we are still unable to design a good control system, then either the design objective is too stringent or the plant is not properly chosen. If the design objective can be relaxed, then we can complete the design. If not, then we must choose a new plant and repeat the design. Thus a change of the plant occurs only as a last resort. Otherwise, the plant remains fixed throughout the design.

6.3   PERFORMANCE CRITERIA

Once a plant with transfer function $G(s)$ is chosen, the design problem is to design an overall system, as shown in Figure 6.2, to meet design specifications. Different applications require different specifications. For example, in designing a control system to aim an astronomical telescope at a distant star, perfect aiming or accuracy is most important; how fast the system achieves the aiming is not critical. On the other hand, for a control system that drives guided missiles to aim at incoming enemy missiles, speed of response is as critical as accuracy. In general, the performance of control systems is divided into two parts: steady-state performance, which specifies accuracy, and transient performance, which specifies the speed of response. The steady-state performance may be defined for a step, ramp, or acceleration reference input. The transient response, however, is defined only for a step reference input.

Before proceeding, it is important to point out that the behavior of a control system depends only on its overall transfer function $G_o(s)$ from the reference input to the plant output y. It does not depend explicitly on the plant transfer function $G(s)$. Thus, the design problem is essentially the search for a $G_o(s)$ to meet design specifications. Let $G_o(s)$ be of the form

$$G_o(s) = \frac{B(s)}{A(s)} = \frac{\beta_0 + \beta_1 s + \beta_2 s^2 + \cdots + \beta_m s^m}{\alpha_0 + \alpha_1 s + \alpha_2 s^2 + \cdots + \alpha_n s^n} \tag{6.2}$$

with $n \geq m$. If $G_o(s)$ is not stable, the output $y(t)$ will approach infinity for almost any reference input and the system will break down or burn out. Thus, every $G_o(s)$

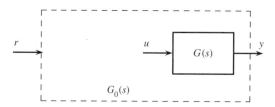

Figure 6.2 The design problem.

to be designed must be stable. The design problem then becomes the search for a *stable* $G_o(s)$ to meet design specifications.

6.3.1 Steady-State Performance—Accuracy

The steady-state performance is concerned with the response of $y(t)$ as t approaches infinity. It is defined for step, ramp, and acceleration inputs.

Step Reference Inputs

If the reference signal $r(t)$ is a step function with amplitude a, that is, $r(t) = a$, for $t \geq 0$, then the steady-state output $y_s(t)$ of the overall system, as derived in (4.25) or by using the final-value theorem, is

$$y_s(t) = \lim_{t \to \infty} y(t) = \lim_{s \to 0} sY(s) = \lim_{s \to 0} sG_o(s) \frac{a}{s} = aG_o(0) = a\frac{\beta_0}{\alpha_0}$$

In this derivation, the stability condition of $G_o(s)$ is essential. The percentage steady-state error is then defined as

$$\text{Position error} := e_p(t) := \lim_{t \to \infty} \left| \frac{a - y(t)}{a} \right| \tag{6.3}$$

$$= \left| \frac{a - a\frac{\beta_0}{\alpha_0}}{a} \right| = \left| \frac{\alpha_0 - \beta_0}{\alpha_0} \right| = |1 - G_o(0)|$$

Because step functions correspond to positions, this error is called the *position error*. Clearly if $G_o(0) = 1$ or $\beta_0 = \alpha_0$ then the position error is zero. If we require the position error to be smaller than γ or 100γ percent, then

$$\left| \frac{\alpha_0 - \beta_0}{\alpha_0} \right| < \gamma$$

which, using $\alpha_0 > 0$ because of the stability assumption of $G_o(s)$, implies

$$-\alpha_0\gamma < \beta_0 - \alpha_0 < \alpha_0\gamma$$

or

$$(1 - \gamma)\alpha_0 < \beta_0 < (1 + \gamma)\alpha_0 \tag{6.4}$$

Thus, the specification on the position error can easily be translated into the constant coefficients α_0 and β_0 of $G_o(s)$. Note that the position error is independent of α_i and β_i, for $i \geq 1$.

Exercise 6.3.1

Find the range of β_0 so that the position error of the following transfer function is smaller than 5%.

$$G(s) = \frac{\beta_0}{s^2 + 2s + 2}$$

[**Answer:** $1.9 < \beta_0 < 2.1$.]

Ramp Reference Inputs

If the reference signal is a ramp function or $r(t) = at$, for $t \geq 0$ and $a > 0$, then the steady-state output $y_s(t)$ of the overall system, as computed in (4.26), is

$$y_s(t) = G_o(0) \, at + G_o'(0) \, a = \frac{\beta_0}{\alpha_0} \, at + \frac{\alpha_0 \beta_1 - \beta_0 \alpha_1}{\alpha_0^2} \cdot a \tag{6.5}$$

where $G_o'(0) = dG_o(s)/ds|_{s=0}$. The percentage steady-state error due to a ramp function is defined as

$$
\begin{aligned}
\text{Velocity error} := e_v(t) := \lim_{t \to \infty} \left| \frac{r(t) - y(t)}{a} \right| &= \left| \frac{at - y_s(t)}{a} \right| \\
&= |(1 - G_o(0))t - G_o'(0)| \tag{6.6} \\
&= \left| \left(\frac{\alpha_0 - \beta_0}{\alpha_0} \right) t - \frac{\alpha_0 \beta_1 - \beta_0 \alpha_1}{\alpha_0^2} \right|
\end{aligned}
$$

This error will be called the *velocity error*, because ramp functions correspond to velocities. We see that if $G_o(0) = \beta_0/\alpha_0 \neq 1$, then $r(t)$ and $y_s(t)$ have different slopes, as shown in Figure 6.3(a), and their difference approaches infinity as $t \to \infty$. Thus the velocity error is infinity. Therefore, in order to have a finite velocity error, we must have $G_o(0) = 1$ or $\beta_0 = \alpha_0$. In this case, $r(t)$ and $y_s(t)$ have the same

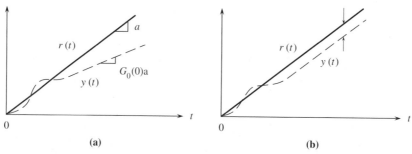

(a)   (b)

Figure 6.3 Velocity errors.

slope, as is shown in Figure 6.3(b), and the velocity error becomes finite and equals

$$\text{Velocity error} = e_v(t) = \left| \frac{\alpha_0 \beta_1 - \beta_0 \alpha_1}{\alpha_0^2} \right| = \left| \frac{\beta_1 - \alpha_1}{\alpha_0} \right| \qquad (6.7)$$

Thus the conditions for having a zero velocity error are $\alpha_0 = \beta_0$ and $\alpha_1 = \beta_1$, or $G_o(0) = 1$ and $G_o'(0) = 0$. They are independent of α_i and β_i, for $i \geq 2$.

The preceding analysis can be extended to acceleration reference inputs or any inputs that are polynomials of t. We will not do so here. We mention that, in addition to the steady-state performances defined for step, ramp, and acceleration functions, there is another type of steady-state performance, defined for sinusoidal functions. This specification is used in frequency-domain design and will be discussed in Chapter 8.

The plant output $y(t)$ is said to track *asymptotically* the reference input $r(t)$ if

$$\lim_{t \to \infty} |y(t) - r(t)| = 0$$

If the position error is zero, then the plant output will track asymptotically any step reference input. For easy reference, we recapitulate the preceding discussion in the following. Consider the design problem in Figure 6.2. No matter how the system is designed or what configuration is used, if its overall transfer function $G_o(s)$ in (6.2) is stable, then the overall system has the following properties:

1. If $G_o(0) = 1$ or $\alpha_0 = \beta_0$, then the position error is zero, and the plant output will track asymptotically any step reference input.
2. If $G_o(0) = 1$ and $G_o'(0) = 0$, or $\alpha_0 = \beta_0$ and $\alpha_1 = \beta_1$, then the velocity error is zero, and the plant output will track asymptotically any ramp reference input.
3. If $G_o(0) = 1$, $G_o'(0) = 0$, and $G_o''(0) = 0$, or $\alpha_0 = \beta_0$, $\alpha_1 = \beta_1$, and $\alpha_2 = \beta_2$, then the acceleration error is zero, and the plant output will track asymptotically any acceleration input.

Thus, the specifications on the steady-state performance can be directly translated into the coefficients of $G_o(s)$ and can be easily incorporated into the design.

The problem of designing a system to track asymptotically a *nonzero* reference signal is called the *tracking problem*. The more complex the reference signal, the more complex the overall system. For example, tracking ramp reference inputs imposes conditions on α_0, α_1, β_0, and β_1; tracking step reference inputs imposes conditions only on α_0 and β_0. If the reference signal is zero, then the problem is called the *regulating problem*. The response of regulating systems is excited by nonzero initial conditions or disturbances, and the objective of the design is to bring such nonzero responses to zero. This objective can be achieved if the system is designed to be stable; no other condition such as $G_o(0) = 1$ is needed. Thus, if a system is designed to track a step reference input or any nonzero reference input, then the system can also achieve regulation. In this sense, the regulating problem is a special case of the tracking problem.

6.3.2 System Types—Unity-Feedback Configuration

The conditions for the steady-state performance in the preceding section are stated for overall closed-loop transfer functions. Therefore, they are applicable to any control configuration once its overall transfer function is computed. In this subsection, we discuss a special case in which the conditions can be stated in terms of open-loop transfer functions. Consider the unity-feedback configuration shown in Figure 6.4(a), where $G(s)$ is the plant transfer function and $C(s)$ is a compensator. We define

$$G_l(s) = G(s)C(s)$$

and call it the *loop transfer function*. A transfer function is called a type i transfer function if it has i poles at $s = 0$. Thus, if $G_l(s)$ is of type i, then it can be expressed as

$$G_l(s) = \frac{N_l(s)}{s^i D_l(s)}$$

with $N_l(0) \neq 0$ and $D_l(0) \neq 0$. Now we claim that if $G_l(s)$ is of type 1, and if the unity-feedback system is stable, then the position error of the unity-feedback system is 0. Indeed, if $G_l(s)$ is of type 1, then the overall transfer function is

$$G_o(s) = \frac{\dfrac{N_l(s)}{sD_l(s)}}{1 + \dfrac{N_l(s)}{sD_l(s)}} = \frac{N_l(s)}{sD_l(s) + N_l(s)}$$

Therefore we have

$$G_o(0) = \frac{N_l(0)}{0 \times D_l(0) + N_l(0)} = \frac{N_l(0)}{N_l(0)} = 1$$

which implies that the position error is zero. Thus, the plant output will track asymptotically any step reference input. Furthermore, even if there are variations of the parameters of $N_l(s)$ and $D_l(s)$, the plant output will still track any step reference input so long as the overall system remains stable. Therefore, the tracking property is said to be *robust*. Using the same argument, we can show that if the loop transfer function is of type 2 and if the unity-feedback system is stable, then the plant output will track asymptotically and robustly any ramp reference input (Problem 6.7).

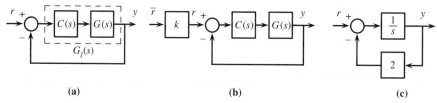

| (a) | (b) | (c) |

Figure 6.4 (a) Unity-feedback system. (b) Unity-feedback system with a forward gain. (c) Nonunity-feedback system.

If $G_l(s)$ is of type 0, that is, $G_l(s) = N_l(s)/D_l(s)$ with $N_l(0) \neq 0$ and $D_l(0) \neq 0$, then we have

$$G_o(s) = \frac{N_l(s)}{D_l(s) + N_l(s)} \quad \text{and} \quad G_o(0) = \frac{N_l(0)}{D_l(0) + N_l(0)} \neq 1$$

and the position error is different from zero. Thus, if the loop transfer function in Figure 6.4(a) is of type 0, then the plant output will not track asymptotically any step reference input. This problem can be resolved, however, by introducing the forward gain

$$k = \frac{D_l(0) + N_l(0)}{N_l(0)}$$

as shown in Figure 6.4(b). Then the transfer function $\overline{G}_o(s)$ from $\bar{r}$ to y has the property $\overline{G}_o(0) = 1$, and the plant output will track asymptotically any step reference input $\bar{r}$. In practice, there is no need to implement gain k. By a proper calibration or setting of r, it is possible for the plant output to approach asymptotically any desired value.

There is one problem with this design, however. If the parameters of $G_l(s) = G(s)C(s)$ change, then we must recalibrate or reset the reference input r. Therefore, this design is not robust. On the other hand, if $G_l(s)$ is of type 1, then the tracking property of Figure 6.4(a) is robust, and there is no need to reset the reference input. Therefore, in the design, it is often desirable to have type 1 loop transfer functions.

We mention that the preceding discussion holds only for unity-feedback systems. If a configuration is not unity feedback, such as the one shown in Figure 6.4(c), even if the plant is of type 1 and the feedback system is stable, the position error is not necessarily zero. For example, the transfer function of Figure 6.4(c) is

$$G_o(s) = \frac{\dfrac{1}{s}}{1 + \dfrac{2}{s}} = \frac{1}{s + 2}$$

Its position error, using (6.3), is

$$e_p = \left| \frac{2 - 1}{2} \right| = 0.5 = 50\%$$

The position error is not zero even though the plant is of type 1. Therefore, system types are useful in determining position or velocity errors in the unity-feedback configuration, but not necessarily useful in other configurations.

6.3.3 Transient Performance—Speed of Response

Transient performance is concerned with the speed of response or the speed at which the system reaches the steady state. Although the steady-state performance is defined for step, ramp, or acceleration reference inputs, the transient performance is defined

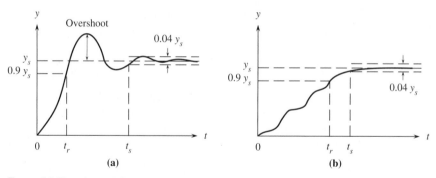

Figure 6.5 Transient performance.

only for step reference inputs. Consider the outputs due to a unit-step reference input shown in Figure 6.5, in which y_s denotes the steady state of the output. The transient response is generally specified in terms of the rise time, settling time, and overshoot. The *rise time* can be defined in many ways. We define it as the time required for the response to rise from 0 to 90% of its steady-state value, as shown in Figure 6.5. In other words, it is the smallest t_r such that

$$y(t_r) = 0.9y_s$$

The time denoted by t_s in Figure 6.5 is called the *settling time*. It is the time for the response to reach and remain inside $\pm 2\%$ of its steady-state value, or it is the smallest t_s such that

$$|y(t) - y_s| \leq 0.02y_s \qquad \text{for all } t \geq t_s$$

Let y_{max} be the maximum value of $|y(t)|$, for $t \geq 0$, or

$$y_{max} := \max |y(t)|$$

Then the *overshoot* is defined as

$$\text{Overshoot} := \frac{y_{max} - y_s}{y_s} \times 100\%$$

For the response in Figure 6.5(a), if $y_{max} = 1.3y_s$, then the overshoot is 30%. For the response in Figure 6.5(b), because $y_{max} = y_s$, the overshoot is zero or there is no overshoot.

Control systems are inherently time-domain systems, so the introduced specifications are natural and have simple physical interpretations. For example, in pointing a telescope at a star, the steady-state performance (accuracy) is the main concern; the specifications on the rise time, overshoot, and settling time are not critical. However, in aiming missiles at an aircraft, both accuracy and speed of response are important. In the design of an aircraft, the specification is often given as shown in Figure 6.6. It is required that the step response of the system be confined to the region shown. This region is obtained by a compromise between the comfort or

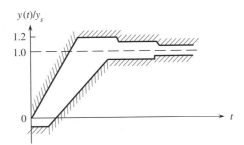

Figure 6.6 Allowable step response.

physical limitations of the pilot and the maneuverability of the aircraft. In the design of an elevator, any appreciable overshoot is undesirable. Different applications have different specifications.

A system is said to be *sluggish* if its rise time and settling time are large. If a system is designed for a fast response, or to have a small rise time and a small settling time, then the system may exhibit a large overshoot, as can be seen from Figure 4.7. Thus, the requirements on the rise time and overshoot are often conflicting and must be reached by compromise.

The steady-state response of $G_o(s)$ depends only on a number of coefficients of $G_o(s)$; thus the steady-state performance can easily be incorporated into the design. The transient response of $G_o(s)$ depends on both its poles and zeros. Except for some special cases, no simple relationship exists between the specifications and pole-zero locations. Therefore, designing a control system to meet transient specifications is not as simple as designing one to meet steady-state specifications.

6.4 NOISE AND DISTURBANCES

Noise and disturbances often arise in control systems. For example, if a potentiometer is used as a transducer, noise will be generated (because of brush jumps, wire irregularity, or variations of contact resistance). Motors and generators also generate noise because of irregularity of contact between carbon brushes and commutators. Shot noise and thermal noise are always present in electronic circuits. Therefore, noise, usually high-frequency noise, exists everywhere in control systems.

Most control systems will also encounter external disturbances. A cruising aircraft may encounter air turbulence or air pockets. A huge antenna may encounter strong or gusting winds. Fluctuations in power supply, mechanical vibrations, and hydraulic or pneumatic pressure will also disturb control systems.

Variation of load is also common in control systems. For example, consider a motor driving an audio or video tape. At the beginning and end, the amounts of tape on the reel are quite different; consequently, the moments of inertia of the load are not the same. As a result, the transfer function of the plant, as can be seen from (3.17), is not the same at all times. One way to deal with this problem is to choose

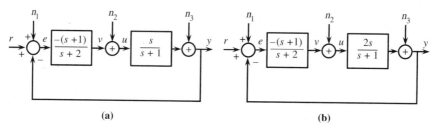

Figure 6.7 Systems with noise or disturbance entering at every block.

the average moment of inertia or the largest moment of inertia (the worst case), compute the transfer function, and use it in the design. This transfer function is called the *nominal* transfer function. The actual transfer function may differ from the nominal one. This is called *plant perturbation*. Aging may also change plant transfer functions. Plant perturbations are indeed inevitable in practice.

One way to deal with plant perturbation is to use the nominal transfer function in the design. The difference between actual transfer function and nominal transfer function is then considered as an external disturbance. Thus, disturbances may arise from external sources or internal load variations. To simplify discussion, we assume that noise and/or disturbance will enter at the input and output terminals of every block, as shown in Figure 6.7. These inputs also generate some responses at the plant output. These outputs are undesirable and should be suppressed or, if possible, eliminated. Therefore, a good control system should be able to track reference inputs and to reject the effects of noise and disturbances.

6.5   PROPER COMPENSATORS AND WELL-POSEDNESS

In this and the following sections we discuss some physical constraints in the design of control systems. Without these constraints, design would become purely a mathematical exercise and would have no relation to reality. The first constraint is that compensators used in the design must have proper transfer functions. As discussed in the preceding chapter, every proper transfer function can be realized as a state-variable equation and then built using operational amplifier circuits. If the transfer function of a compensator is improper, then its construction requires the use of pure differentiators. Pure differentiators built by using operational amplifiers may be unstable. See Reference [18]. Thus compensators with improper transfer functions cannot easily be built in practice. For this reason, all compensators used in the design will be required to have proper transfer functions.

In industry, proportional-integral-derivative (PID) controllers or compensators are widely used. The transfer functions of proportional and integral controllers are k_p and k_i/s; they are proper transfer functions. The transfer function of derivative controllers is $k_d s$, which is improper. However, in practice, derivative controllers are realized as

$$\frac{k_d s}{1 + \dfrac{k_d}{N} s}$$

for some constant N. This is a proper transfer function, and therefore does not violate the requirement that all compensators have proper transfer functions. In the remainder of this chapter, we assume that every component of a control system has a proper transfer function. If we encounter a tachometer with improper transfer function ks, we shall remodel it as shown in Figure 3.10(c). Therefore, the assumption remains valid.

Even though all components have proper transfer functions, a control system so built may not have a proper transfer function. This is illustrated by an example.

Example 6.5.1

Consider the system shown in Figure 6.7(a). The transfer functions of the plant and the compensator are all proper. Now we compute the transfer function $G_{yr}(s)$ from r to y. Because the system is linear, in computing $G_{yr}(s)$, all other inputs shown (n_i, $i = 1, 2$, and 3) can be assumed zero or disregarded. Clearly we have

$$G_{yr}(s) = \frac{\dfrac{-(s+1)}{s+2} \cdot \dfrac{s}{s+1}}{1 + \dfrac{-(s+1)}{s+2} \cdot \dfrac{s}{s+1}} = \frac{\dfrac{-s}{s+2}}{1 - \dfrac{s}{s+2}} = \frac{-s}{s+2-s} = -0.5s$$

It is improper! Thus the properness of all component transfer functions does not guarantee the properness of an overall transfer function.

Now we discuss the implication of improper overall transfer functions. As discussed in the preceding section, in addition to the reference input, noise and disturbance may enter a control system as shown in Figure 6.7(a). Suppose $r(t) = \sin t$ and $n_1(t) = 0.01 \sin 10{,}000t$ where $r(t)$ denotes a desired signal and $n_1(t)$ denotes a high-frequency noise. Because the transfer function $G_{yr}(s)$ is $-0.5s$, the plant output is simply the derivative of $r(t)$ and $n_1(t)$, scaled by -0.5. Therefore, we have

$$y(t) = (-0.5) \frac{d}{dt} (\sin t + 0.01 \sin 10{,}000t)$$

$$= -0.5 \cos t - 0.5(0.01) \times 10{,}000 \times \cos 10{,}000t$$

$$= -0.5 \cos t - 50 \cos 10{,}000t$$

Although the magnitude of the noise is one-hundredth of that of the desired signal at the input, it is one hundred times larger at the plant output. Therefore, the plant output is completely dominated by the noise and the system cannot be used in practice.

In conclusion, if a control system has an improper closed-loop transfer function, then high-frequency noise will be greatly amplified and the system cannot be used. Thus, a workable control system should not contain any improper closed-loop transfer function. This motivates the following definition.

□ Definition 6.1

A system is said to be *well-posed* or *closed-loop proper* if the closed-loop transfer function of *every* possible input/output pair of the system is proper. ■

We have assumed that noise and disturbance may enter a control system at the input and output terminals of each block. Therefore, we shall consider not only the transfer function from r to y, but also transfer functions from those inputs to all variables. Let G_{pq} denote the transfer function from input q to output p. Then the system in Figure 6.7(a) or (b) is well posed if the transfer functions G_{er}, G_{vr}, G_{ur}, G_{yr}, G_{en_1}, G_{un_1}, G_{yn_1}, G_{en_2}, G_{un_2}, G_{yn_2}, G_{en_3}, G_{un_3}, G_{yn_3}, are all proper. These transfer functions are clearly all closed-loop transfer functions and, strictly speaking, the adjective *closed-loop* is redundant. It is kept in Definition 6.1 to stress their difference from open-loop transfer functions.

The number of possible input/output pairs is quite large even for the simple systems in Figure 6.7. Therefore, it appears to be difficult to check the well-posedness of systems. Fortunately, this is not the case. In fact, the condition for a feedback system to be well posed is very simple. A system that is built with blocks with proper transfer functions is well posed if and only if

$$\Delta(\infty) \neq 0 \tag{6.8}$$

where Δ is the characteristic function defined in (3.37). For the feedback systems in Figure 6.7, the condition becomes

$$\Delta(\infty) = 1 + C(\infty)G(\infty) \neq 0 \tag{6.9}$$

For the system in Figure 6.7(a), we have $C(s) = -(s + 1)/(s + 2)$ and $G(s) = s/(s + 1)$ which imply $C(\infty) = -1, G(\infty) = 1$ and $1 + C(\infty)G(\infty) = 0$. Thus the system is not well posed. For the system in Figure 6.7(b), we have

$$1 + C(s)G(s)\big|_{s=\infty} = 1 + \frac{-(s + 1)}{s + 2} \cdot \frac{2s}{s + 1}\bigg|_{s=\infty} = 1 + (-1)(2) = -1 \neq 0$$

Thus the system is well posed. As a check we compute the closed-loop transfer functions from n_2 to u, y, e, and v in Figure 6.7(b). In this computation, all other inputs are assumed zero. The application of Mason's formula yields

$$G_{un_2}(s) = \frac{1}{1 + \dfrac{2s}{s + 1} \cdot \dfrac{-(s + 1)}{(s + 2)}} = \frac{1}{1 - \dfrac{2s}{s + 2}} = \frac{1}{\dfrac{s + 2 - 2s}{s + 2}} = \frac{s + 2}{-s + 2}$$

$$G_{yn_2}(s) = \frac{\dfrac{2s}{(s+1)}}{1 + \dfrac{2s}{s+1} \cdot \dfrac{-(s+1)}{s+2}} = \frac{2s(s+2)}{(s+1)(-s+2)}$$

$$G_{en_2}(s) = -G_{yn_2}(s) = \frac{-2s(s+2)}{(s+1)(-s+2)}$$

and

$$G_{vn_2}(s) = \frac{-\dfrac{2s}{s+1} \cdot \dfrac{-(s+1)}{s+2}}{1 + \dfrac{2s}{s+1} \cdot \dfrac{-(s+1)}{s+2}} = \frac{\dfrac{2s}{s+2}}{\dfrac{s+2-2s}{s+2}} = \frac{2s}{-s+2} \qquad (6.10)$$

They are indeed all proper. Because the condition is (6.8) can easily be met, a control system can easily be designed to be well posed. We remark that if a plant transfer function $G(s)$ is strictly proper and if $C(s)$ is proper, then the condition in (6.9) is automatically satisfied. Note that the conditions in (6.8) and (6.9) hold only if the transfer function of every block is proper. If any one of them is improper, then the conditions cannot be used.

To conclude this section, we discuss the relationship between well-posedness and properness of compensators. Properness of compensators is concerned with open-loop properness, whereas well-posedness is concerned with closed-loop properness. Open-loop properness does not imply closed-loop properness, as is demonstrated in the system in Figure 6.7(a). It can be verified, by computing all possible closed-loop transfer functions, that the system in Figure 6.8 is well posed. However, the system contains one improper compensator. Thus, well-posedness does not imply properness of compensators. In conclusion, open-loop properness and closed-loop properness are two independent concepts. They are also introduced for different reasons. The former is needed to avoid the use of differentiators in realizing compensators; the latter is needed to avoid amplification of high-frequency noise in overall systems.

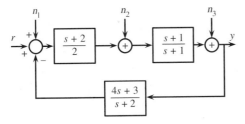

Figure 6.8 Well-posed system with an improper compensator.

6.6   TOTAL STABILITY

In the design of control systems, the first requirement is always the stability of transfer functions, $G_o(s)$, from the reference input r to the plant output y. However, this may not guarantee that systems will work properly. This is illustrated by an example.

Example 6.6.1

Consider the system shown in Figure 6.9. The transfer function from r to y is

$$G_o(s) = 2 \times \frac{\dfrac{s-1}{s+1} \cdot \dfrac{1}{s-1}}{1 + \dfrac{s-1}{s+1} \cdot \dfrac{1}{s-1}} = 2 \times \frac{\dfrac{1}{s+1}}{1 + \dfrac{1}{s+1}} = \frac{2}{s+2} \qquad (6.11)$$

It is stable. Because $G_o(0) = 1$, the position error is zero. The time constant of the system is $1/2 = 0.5$. Therefore, the plant output will track any step reference input in about $5 \times 0.5 = 2.5$ seconds. Thus the system appears to be a good control system.

A close examination of the system in Figure 6.9 reveals that there is a pole-zero cancellation between $C(s)$ and $G(s)$. Will this cause any problem? As was discussed earlier, noise or disturbance may enter a control system. We compute the transfer function from n to y in Figure 6.9:

$$G_{yn}(s) = \frac{\dfrac{1}{s-1}}{1 + \dfrac{s-1}{s+1} \cdot \dfrac{1}{s-1}} = \frac{\dfrac{1}{s-1}}{1 + \dfrac{1}{s+1}} = \frac{s+1}{(s-1)(s+2)} \qquad (6.12)$$

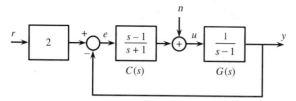

Figure 6.9 Feedback system with pole-zero cancellation.

It is unstable! Thus any nonzero noise, no matter how small, will excite an unbounded plant output and the system will burn out. Therefore the system cannot be used in practice, even though its transfer function from r to y is stable. This motivates the following definition.

□ **Definition   6.2**

A system is said to be *totally stable* if the closed-loop transfer function of every possible input-output pair of the system is stable. ■

Because of the presence of noise, every control system should be required to be totally stable. Otherwise, noise will drive some variable of the system to infinity and the system will disintegrate or burn out. From Example 6.6.1, we see that if a plant has an unstable pole, it is useless to eliminate it by direct cancellation. Although the canceled pole does not appear in the transfer function $G_o(s)$ from r to y, it appears in the transfer function from n to y. We call the pole a *missing pole* or a *hidden pole* from $G_o(s)$. Any unstable pole-zero cancellation will not actually eliminate the unstable pole, only make it hidden from some closed-loop transfer functions.

We now discuss the condition for a system to be totally stable. Consider a system that consists of a number of subsystems. Every subsystem is assumed to be completely characterized by its proper transfer function. See Section 2.5. Let $G_o(s)$ be the overall transfer function. If the number of poles of $G_o(s)$ equals the total number of poles of all subsystems, then the system is completely characterized by $G_o(s)$. If not, the system is not completely characterized by $G_o(s)$ and $G_o(s)$ is said to have *missing poles*. Missing poles arise from pole-zero cancellation[1] and their number equals

Number of poles of $G_o(s)$ $-$ [Total number of poles of all subsystems]

With this preliminary, we are ready to state the condition for a system to be totally stable. The condition can be stated for *any* closed-loop transfer function and its missing poles. A system is totally stable if and only if the poles of any overall transfer function and its missing poles are all stable poles. For example, consider the system shown in Figure 6.9. Its overall transfer function $G_o(s)$ from r to y is computed in (6.11). It has 1 pole, which is less than the total number of poles of $G(s)$ and $C(s)$. Therefore, there is one missing pole. The pole of $G_o(s)$ is stable, but the missing pole $s - 1$ is unstable. Therefore, the system is not totally stable. The same conclusion can also be reached by using the transfer function G_{yn} from n to y computed in (6.12). The number of poles of G_{yn} equals the total number of poles of $G(s)$ and $C(s)$. Therefore, G_{yn} has no missing pole. The system in Figure 6.9 is totally stable if and only if all poles of G_{yn} are stable poles. This is not the case. Therefore, the system is not totally stable.

6.6.1 Imperfect Cancellations

As discussed in the preceding section, a system cannot be totally stable if there is any unstable pole-zero cancellation. In fact, it is impossible to achieve exact pole-

---

[1]In addition to pole-zero cancellations, missing poles may also arise from parallel connection and other situations. See Reference [15, pp. 436–437]. In this text, it suffices to consider only pole-zero cancellations.

zero cancellation in practice. For example, suppose we need a 10-kΩ resistor to realize $C(s) = (s - 1)/(s + 1)$ in Figure 6.9. However, the resistor we use may not have a resistance exactly equal to 10 kΩ. Furthermore, because of change in temperature or aging, the resistance may also change. Thus, the compensator we realize may become $C'(s) = (s - 0.9)/(s + 1.1)$, rather than the intended $(s - 1)/(s + 1)$, as shown in Figure 6.10, and the unstable pole $(s - 1)$ will not be canceled. Even if we can realize $C(s) = (s - 1)/(s + 1)$ exactly, the plant transfer function $1/(s - 1)$ may change due to load variations or aging. Again, exact pole-zero cancellation cannot be achieved. In conclusion, because of inexact implementation of compensators and plant perturbations due to load variations and aging, all we can achieve is imperfect pole-zero cancellation. In this section, we study the effect of imperfect cancellations. The transfer function from r to y in Figure 6.10 is

$$G'_o(s) = 2 \times \frac{\dfrac{s - 0.9}{s + 1.1} \cdot \dfrac{1}{s - 1}}{1 + \dfrac{s - 0.9}{s + 1.1} \cdot \dfrac{1}{s - 1}} = \frac{2(s - 0.9)}{s^2 + 0.1s - 1.1 + s - 0.9}$$

$$= \frac{2(s - 0.9)}{s^2 + 1.1s - 2} = \frac{2(s - 0.9)}{(s + 2.0674)(s - 0.9674)}$$

$$(6.13)$$

It is unstable! Thus, its step response $y(t)$ will approach infinity and is entirely different from the one in Figure 6.9. In conclusion, unstable pole-zero cancellations are permitted neither in theory nor in practice.

Stable pole-zero cancellations, however, are an entirely different matter. Consider the system shown in Figure 6.11(a). The plant transfer function is $G(s) = 3/(s^2 + 0.1s + 100)$, and the compensator transfer function is $C(s) = (s^2 + 0.1s + 100)/s(s + 2)$. Note that $C(s)$ is of type 1, therefore the position error of the unity feedback system is zero. The overall transfer function from r to y is

$$G_o(s) = \frac{\dfrac{s^2 + 0.1s + 100}{s(s + 2)} \cdot \dfrac{3}{s^2 + 0.1s + 100}}{1 + \dfrac{s^2 + 0.1s + 100}{s(s + 2)} \cdot \dfrac{3}{s^2 + 0.1s + 100}}$$

$$= \frac{\dfrac{3}{s(s + 2)}}{1 + \dfrac{3}{s(s + 2)}} = \frac{3}{s^2 + 2s + 3}$$

$$(6.14)$$

The number of the poles of $G_o(s)$ is 2, which is 2 less than the total number of poles of $G(s)$ and $C(s)$. Thus, $G_o(s)$ has two missing poles; they are the roots of $s^2 + 0.1s + 100$. Because the poles of $G_o(s)$ and the two missing poles are stable, the system is totally stable. The unit-step response of $G_o(s)$ is computed, using MATLAB, and plotted in Figure 6.11(b) with the solid line.

Now we study the effect of imperfect pole-zero cancellations. Suppose the compensator becomes

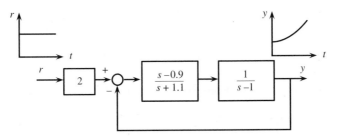

Figure 6.10 Feedback system with an imperfect cancellation.

$$C'(s) = \frac{s^2 + 0.09s + 99}{s(s + 2)}$$

due to aging or inexact realization. With this $C'(s)$, the transfer function of Figure 6.11(a) becomes

$$
\begin{aligned}
G'_o(s) &= \frac{\dfrac{s^2 + 0.09s + 99}{s(s + 2)} \cdot \dfrac{3}{s^2 + 0.1s + 100}}{1 + \dfrac{s^2 + 0.09s + 99}{s(s + 2)} \cdot \dfrac{3}{s^2 + 0.1s + 100}} \\[2mm]
&= \frac{3(s^2 + 0.09s + 99)}{s(s + 2)(s^2 + 0.1s + 100) + 3(s^2 + 0.09s + 99)} \\[2mm]
&= \frac{3s^2 + 0.27s + 297}{s^4 + 2.1s^3 + 103.2s^2 + 200.27s + 297}
\end{aligned}
$$

(6.15)

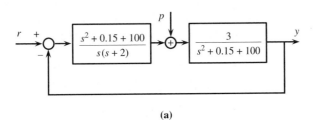

(a)

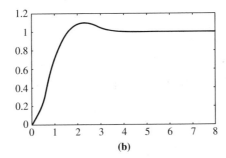

(b)

Figure 6.11 (a) Feedback system. (b) Its step response.

Its unit-step response is plotted in Figure 6.11(b) with the dotted line. We see that the two responses in Figure 6.11(b) are hardly distinguishable. Therefore, unlike unstable pole-zero cancellation, imperfect stable pole-zero cancellations may not cause any serious problem in control systems.

6.6.2 Design Involving Pole-Zero Cancellations

Unstable pole-zero cancellations are not permitted in the design of control systems. How about stable pole-zero cancellations? As was discussed earlier, pole-zero cancellations do not really cancel the poles. The poles become hidden in some closed-loop transfer functions, but may appear in other closed-loop transfer functions. For example, the transfer function from disturbance p to the plant output y in Figure 6.11(a) is given by

$$G_{yp} = \frac{\dfrac{3}{s^2 + 0.1s + 100}}{1 + \dfrac{s^2 + 0.1s + 100}{s(s+2)} \cdot \dfrac{3}{s^2 + 0.1s + 100}} \tag{6.16}$$

$$= \frac{\dfrac{3}{s^2 + 0.1s + 100}}{1 + \dfrac{3}{s(s+2)}} = \frac{3s(s+2)}{(s^2 + 2s + 3)(s^2 + 0.1s + 100)}$$

We see that the complex-conjugate poles $s^2 + 0.1s + 100 = (s + 0.05 + j9.999)$ $(s + 0.05 - j9.999)$, which do not appear in $G_o(s)$, do appear in G_{yp}. Because the poles have a very small real part and large imaginary parts, they will introduce high-frequency oscillation and slow decay. Indeed, if the disturbance is modeled as a unit-step function, then the excited output is as shown in Figure 6.12. Even though the response eventually approaches zero, it introduces too much oscillation and takes too long to decay to zero. Therefore, the system is not good, and the stable pole-zero cancellation should be avoided.

Can pole-zero cancellations be used in the design? Unstable pole-zero cancellations are not permitted, because the resulting system can never be totally stable.

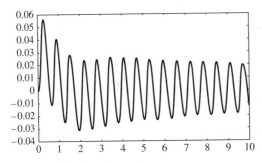

Figure 6.12 Effect of step disturbance.

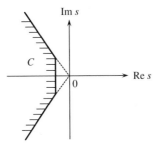

Figure 6.13 Permissible pole cancellation region.

A system can be totally stable with stable pole-zero cancellations. However, if canceled stable poles are close to the imaginary axis or have large imaginary parts, then disturbance or noise may excite a plant output that is oscillatory and slow decaying. If poles lie inside the region C shown in Figure 6.13, then their cancellations will not cause any problems. Therefore, perfect or imperfect cancellation of poles lying inside the region C is permitted in theory and in practice. The exact boundary of the region C depends on each control system and performance specifications, and will be discussed in the next chapter.

There are two reasons for using pole-zero cancellation in design. One is to simplify design, as will be seen in the next chapter. The other reason is due to necessity. In model matching, we may have to introduce pole-zero cancellations to insure that the required compensators are proper. This is discussed in Chapter 10.

6.7 SATURATION—CONSTRAINT ON ACTUATING SIGNALS

In the preceding sections, we introduced the requirements of well-posedness and total stability in the design of control systems. In this section, we introduce one more constraint. This constraint arises from modeling and from devices used, and is probably the most difficult to meet in design.

Strictly speaking, most physical systems are nonlinear. However, they can often be modeled as linear systems within some limited operational ranges. If signals remain inside the ranges, the linear models can be used. If not, the linear models do not hold, and the results of linear analyses may not be applicable. This is demonstrated by an example.

Example 6.7.1

Consider the system shown in Figure 6.14. The element A is an amplifier with gain 2. The overall transfer function is

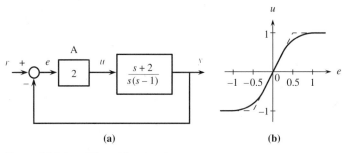

(a)   **(b)**

Figure 6.14 Amplifier with saturation.

$$G_o(s) = \frac{2 \cdot \dfrac{s + 2}{s(s - 1)}}{1 + 2 \cdot \dfrac{s + 2}{s(s - 1)}} = \frac{2(s + 2)}{s^2 - s + 2s + 4} = \frac{2(s + 2)}{s^2 + s + 4} \qquad (6.17)$$

Because $G_o(0) = 4/4 = 1$, the system has zero position error and the plant output will track any step reference input without an error. The plant outputs excited by $r = 0.3$, 1.1, and 1.15 are shown in Figure 6.15 with the solid lines.

In reality, the amplifier may have the characteristic shown in Figure 6.14(b). For ease of simulation, the saturation is approximated by the dashed lines shown. The responses of the system due to $r = 0.3$, 1.1, and 1.15 are shown in Figure 6.15 with the dashed lines. These responses are obtained by computer simulations. If $r = 0.3$, the amplifier will not saturate and the response is identical to the one obtained by using the linear model. If $r = 1.1$, the amplifier saturates and the response differs from the one obtained by using the linear model. If $r = 1.15$, the

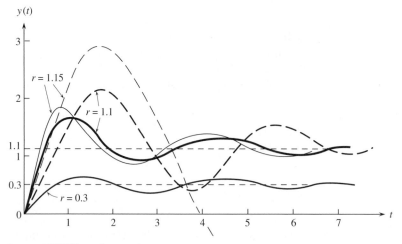

Figure 6.15 Effect of saturation.

response approaches infinity oscillatorily and the system is not stable, although the linear model is always stable for any r. This example shows that linear analysis cannot be used if signals run outside the linear range.

This example is intentionally chosen to dramatize the effect of saturation. In most cases, saturation will make systems only more sluggish, rather than unstable. In any case, if a control system saturates, then the system will not function as designed. Therefore, in design, we often require

$$|u(t)| \leq M \qquad (6.18)$$

for all $t \geq 0$, where $u(t)$ is the actuating signal and M is a constant. This constraint arises naturally if valves are used to generate actuating signals. The actuating signals reach their maximum values when the valves are fully open. In a ship steering system, the constraint exists because the rudder can turn only a finite number of degrees. In electric motors, because of saturation of the magnetic field, the constraint also exists. In hydraulic motors, the movement of pistons in the pilot cylinder is limited. Thus, the constraint in (6.18) exists in most plants. Strictly speaking, similar constraints should also be imposed upon compensators. If we were to include all these constraints, the design would become very complicated. Besides, compared with plants, compensators are rather inexpensive, and hence, if saturated, can be replaced by ones with larger linear ranges. Therefore, the saturation constraint is generally imposed only on the plant.

Actuating signals depend on reference input signals. If the amplitude of a reference signal is doubled, so is that of the actuating signal. Therefore, in checking whether or not the constraint in (6.18) is met, we shall use the largest reference input signal. However, for convenience, the constraint M in (6.18) will be normalized to correspond to unit-step reference inputs. Therefore, in design, we often require the actuating signal due to a unit-step reference input to have a magnitude less than a certain value.

To keep a plant from saturating is not a simple problem, because in the process of design, we don't know what the exact response of the resulting system will be. Hence, the saturation problem can be checked only after the completion of the design. If saturation does occur, the system may have to be redesigned to improve its performance.

6.8   OPEN-LOOP AND CLOSED-LOOP CONFIGURATIONS

In this section we discuss the configuration of control systems. Given a control problem, it is generally possible to use many different configurations to achieve the design. It is therefore natural to compare the relative merits of various configurations. To simplify the discussion, we compare only the two configurations shown in Figure 6.16. Figure 6.16(a) is an open-loop configuration, Figure 6.16(b) a closed-loop or feedback configuration.

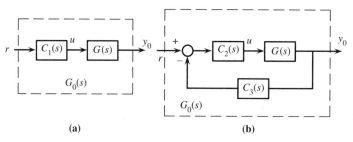

Figure 6.16 Open- and closed-loop systems.

In order to compare the two configurations, we must consider noise and disturbances as discussed in Section 6.4. If there were no noise and disturbances in control systems, then there would be no difference between open-loop and closed-loop configurations. In fact, the open-loop configuration may sometimes be preferable because it is simpler and less expensive. Unfortunately, noise and disturbances are unavoidable in control systems.

The major difference between the open-loop and feedback configurations is that the actuating signal of the former does not depend on the plant output. It is predetermined, and will not change even if the actual plant output is quite different from the desired value. The actuating signal of a feedback system depends on the reference signal and the plant output. Therefore, if the plant output deviates from the desired value due to noise, external disturbance, or plant perturbations, the deviation will reflect on the actuating signal. Thus, a properly designed feedback system should perform better than an open-loop system. This will be substantiated by examples.

Example 6.8.1

Consider the two amplifiers shown in Figure 6.17. Figure 6.17(a) is an open-loop amplifier. The amplifier in Figure 6.17(b) is built by connecting three identical open-loop amplifiers and then introducing a feedback from the output to the input as shown, and is called a *feedback amplifier*. Their block diagrams are also shown in Figure 6.17. The gain of the open-loop amplifier is $A = -10R/R = -10$. From Figure 6.17(b), we have

$$e = -\left(\frac{10R}{R}r + \frac{10R}{R_f}y\right) = -10\left(r + \frac{R}{R_f}y\right) = A\left(r + \frac{R}{R_f}y\right)$$

Thus the constant β in the feedback loop equals R/R_f. Note that the feedback is positive feedback. The transfer function or the gain of the feedback amplifier is

$$A_o = \frac{A^3}{1 - \beta A^3} \tag{6.19}$$

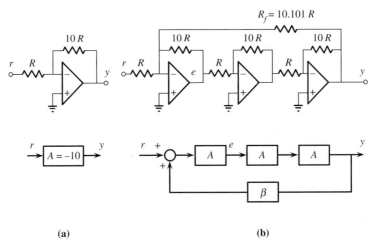

(a)   **(b)**

Figure 6.17 Open- and closed-loop amplifiers.

In order to make a fair comparison, we shall require both amplifiers to have the same gain—that is, $A_o = A = -10$. To achieve this, β can readily be computed as $\beta = 0.099$, which implies $R_f = 10.101R$.

The feedback amplifier needs three times more hardware and still has the same gain as the open-loop amplifier. Therefore, there seems no reason to use the former. Indeed, this is the case if there is no perturbation in gain A.

Now suppose gain A decreases by 10% each year due to aging. In other words, the gain becomes -9 in the second year, -8.1 in the third year, and so forth. We compute

$$A_o = \frac{(-9)^3}{1 - 0.099(-9)^3} = -9.96$$

$$A_o = \frac{(-8.1)^3}{1 - 0.099(-8.1)^3} = -9.91$$

and so forth. The results are listed in the following:

	1	2	3	4	5	6	7	8	9	10
$-A$	10	9.0	8.1	7.29	6.56	5.9	5.3	4.78	4.3	3.87
$-A_o$	10	9.96	9.92	9.84	9.75	9.63	9.46	9.25	8.96	8.6

We see that although the open-loop gain A decreases by 10% each year, the closed-loop gain decreases by from 0.4% in the first year to 4.1% in the tenth year. Thus the feedback system is much less sensitive to plant perturbation.

If an amplifier is to be taken out of service when its gain falls below 9, then the open-loop amplifier can serve only one year, whereas the closed-loop amplifier can last almost nine years. Therefore, even though the feedback amplifier uses three times more hardware, it is actually three times more economical than the open-loop amplifier. Furthermore, the labor cost of yearly replacement of open-loop amplifiers can be saved.

Example 6.8.2

Consider a speed control problem in an aluminum factory. Heated aluminum ingots are pressed into sheets through rollers as shown in Figure 6.18(a). The rollers are driven by armature-controlled dc motors; their speeds are to be kept constant in order to maintain uniform thickness of aluminum sheets. Let the transfer function of the motor and rollers be

$$G(s) = \frac{10}{5s + 1} \tag{6.20}$$

Its time constant is 5; therefore it will take 25 seconds (5 × time constant) for the rollers to reach the final speed. This is very slow, and we decide to design an overall system with transfer function

$$G_o(s) = \frac{2}{s + 2} \tag{6.21}$$

Its time constant is $1/2 = 0.5$; therefore the speed of response of this system is much faster. Because $G_o(0) = 2/2 = 1$, the plant output of $G_o(s)$ will track any step reference input without an error.

Now we shall implement $G_o(s)$ in the open-loop and closed-loop configurations shown in Figure 6.18(b) and (c). For the open-loop configuration, we require

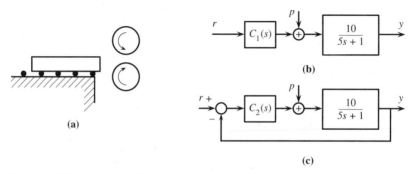

Figure 6.18 Speed control of rollers.

$G_o(s) = G(s)C_1(s)$. Thus the open-loop compensator $C_1(s)$ is given by

$$C_1(s) = \frac{G_o(s)}{G(s)} = \frac{2}{s + 2} \cdot \frac{5s + 1}{10} = \frac{5s + 1}{5(s + 2)} \tag{6.22}$$

For the closed-loop configuration, we require

$$G_o(s) = \frac{C_2(s)G(s)}{1 + C_2(s)G(s)} \tag{6.23}$$

or

$$G_o(s) + G_o(s)C_2(s)G(s) = C_2(s)G(s)$$

Thus the closed-loop compensator $C_2(s)$ is given by

$$C_2(s) = \frac{G_o(s)}{G(s)(1 - G_o(s))} = \frac{2/(s + 2)}{\dfrac{10}{5s + 1}\left(1 - \dfrac{2}{s + 2}\right)} = \frac{2/(s + 2)}{\dfrac{10}{5s + 1} \cdot \dfrac{s}{s + 2}}$$

$$= \frac{5s + 1}{5s} = 1 + \frac{1}{5s} \tag{6.24}$$

It consists of a proportional compensator with gain 1 and an integral compensator with transfer function $1/5s$; therefore it is called a *PI compensator* or *controller*. We see that it is a type 1 transfer function.

If there are no disturbances and plant perturbation, the open-loop and closed-loop systems should behave identically, because they have the same overall transfer function. Because $G_o(0) = 1$, they both track asymptotically any step reference input without an error. In practice, the load of the rollers is not constant. From Figure 6.18(a), we see that before and after the engagement of an ingot with the rollers, the load is quite different. Even after the engagement, the load varies because of nonuniform thickness of ingots. We study the effect of this load variation in the following.

Plant Perturbation

The transfer function of the motor and rollers is assumed as $G(s) = 10/(5s + 1)$ in (6.20). Because the transfer function depends on the load, if the load changes, so does the transfer function. Now we assume that, after the design, the transfer function changes to

$$\overline{G}(s) = \frac{9}{(4.5s + 1)} \tag{6.25}$$

This is called *plant perturbation*. We now study its effect on the open-loop and closed-loop systems.

After plant perturbation, the open-loop overall transfer function becomes

$$\overline{G}_{oo}(s) = C_1(s)\overline{G}(s) = \frac{5s + 1}{5(s + 2)} \cdot \frac{9}{(4.5s + 1)} \tag{6.26}$$

Because $\overline{G}_{oo}(0) = 9/10 \neq 1$, this perturbed system will not track asymptotically any step reference input. Thus the tracking property of the open-loop system is lost after plant perturbation.

Now we compute the overall transfer function of the closed-loop system with perturbed $\overline{G}(s)$ in (6.25). Clearly, we have

$$\overline{G}_{oc}(s) = \frac{C_2(s)\overline{G}(s)}{1 + C_2(s)\overline{G}(s)} = \frac{\dfrac{5s + 1}{5s} \cdot \dfrac{9}{4.5s + 1}}{1 + \dfrac{5s + 1}{5s} \cdot \dfrac{9}{4.5s + 1}}$$

$$= \frac{9(5s + 1)}{5s(4.5s + 1) + 9(5s + 1)} = \frac{45s + 9}{22.5s^2 + 50s + 9}$$

(6.27)

Because $\overline{G}_{oc}(0) = 1$, this perturbed overall closed-loop system still track any step reference input without an error. In fact, because the compensator is of type 1, no matter how large the plant perturbation is, the system will always track asymptotically any step reference input, so long as the overall system remains stable. This is called *robust* tracking. In conclusion, the tracking property is destroyed by plant perturbation in the open-loop system but is preserved in the closed-loop system.

Disturbance Rejection

One way to study the effect of load variations is to introduce plant perturbations as in the preceding paragraphs. Another way is to introduce a disturbance $p(t)$ into the plant input as shown in Figure 6.18(b) and (c). Now we study the effect of this disturbance in the open-loop and closed-loop systems. From Figure 6.18(b), we see that the transfer function from p to y is not affected by the open-loop compensator $C_1(s)$. If the disturbance is modeled as a step function of magnitude a, then it will excite the following plant output

$$Y_p(s) = \frac{10}{5s + 1} \cdot \frac{a}{s}$$

(6.28)

Its steady-state output, using the final-value theorem, is

$$y_p(\infty) = \lim_{t \to \infty} y_p(t) = \lim_{s \to 0} sY_p(s) = \lim_{s \to 0} s \cdot \frac{10}{5s + 1} \cdot \frac{a}{s} = 10a$$

In other words, in the open-loop configuration, the step disturbance will excite a nonzero plant output; therefore, the speed of the rollers will differ from the desired speed. For example, if the disturbance is as shown in Figure 6.19(a), then the speed will be as shown in Figure 6.19(b). This differs from the desired speed and will cause unevenness in thickness of aluminum sheets. Thus, the open-loop system is not satisfactory.

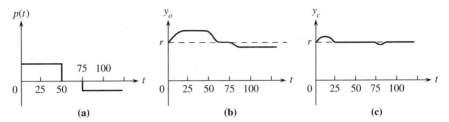

Figure 6.19 Effects of disturbance.

Now we study the closed-loop system. The transfer function from p to y is, using Mason's formula,

$$G_{yp}(s) = \frac{G(s)}{1 + G(s)C_2(s)} = \frac{10/(5s + 1)}{1 + \dfrac{10}{5s + 1} \cdot \dfrac{5s + 1}{5s}} = \frac{10s}{(5s + 1)(s + 2)} \quad (6.29)$$

Now if the disturbance is $P(s) = a/s$, the steady-state output y_p due to the disturbance is

$$y_p(\infty) = \lim_{t \to \infty} y_p(t) = \lim_{s \to 0} sY_p(s) = \lim_{s \to 0} sG_{yp}(s) \cdot \frac{a}{s}$$

$$= \lim_{s \to 0} \frac{10sa}{(5s + 1)(s + 2)} = 0$$

This means that the effect of the disturbance on the plant output eventually vanishes. Thus, the speed of the rollers is completely controlled by the reference input, and thus, in the feedback configuration, even if there are disturbances, the speed will return, after the transient dies out, to the desired speed, as shown in Figure 6.19(c). Consequently, evenness in the thickness of aluminum sheets can be better maintained.

We remark that in the closed-loop system in Figure 6.18(c), there is a pole-zero cancellation. The canceled pole is $-1/5$, which is stable but quite close to the $j\omega$-axis. Although this pole does not appear in $G_o(s)$ in (6.21), it appears in $G_{yd}(s)$ in (6.29). Because of this pole (its time constant is 5 seconds), it will take roughly 25 seconds (5 × time constant) for the effect of disturbances to vanish, as is shown in Figure 6.19(c). It is possible to use different feedback configurations to avoid this pole-zero cancellation. This is discussed in Chapter 10. See also Problem 6.14.

From the preceding two examples, we conclude that the closed-loop or feedback configuration is less sensitive to plant perturbation and disturbances than the open-loop configuration. Therefore, in the remainder of this text, we use only closed-loop configurations in design.

6.9   TWO BASIC APPROACHES IN DESIGN

With the preceding discussion, the design of control systems can now be stated as follows: Given a plant, design an overall system to meet a given set of specifications. We use only feedback configurations because they are less sensitive to disturbances and plant perturbation than open-loop configurations are. Because improper compensators cannot easily be built in practice, we use only compensators with proper transfer functions. The resulting system is required to be well posed so that high-frequency noise will not be unduly amplified. The design cannot have unstable pole-zero cancellation, otherwise the resulting system cannot be totally stable. Because of the limitation of linear models and devices used, a constraint must generally be imposed on the magnitude of actuating signals. The following two approaches are available to carry out this design:

1.   We first choose a feedback configuration and a compensator with open parameters. We then adjust the parameters so that the resulting feedback system will hopefully meet the specifications.
2.   We first search for an overall transfer function $G_o(s)$ to meet the specifications. We then choose a feedback configuration and compute the required compensator.

These two approaches are quite different in philosophy. The first approach starts from internal compensators and works toward external overall transfer functions. Thus, it is called the *outward* approach. This approach is basically a trial-and-error method. The root-locus and frequency-domain methods discussed in Chapters 7 and 8 take this approach. The second approach starts from external overall transfer functions and then computes internal compensators, and is called the *inward* approach. This approach is studied in Chapters 9 and 10. These two approaches are independent and can be studied in either order. In other words, we may study Chapters 7 and 8, and then 9 and 10, or study first Chapters 9 and 10, and then Chapters 7 and 8.

To conclude this chapter, we mention a very important fact of feedback. Consider a plant with transfer function $G(s) = N(s)/D(s)$ and consider the feedback configuration shown in Figure 6.20. Suppose the transfer function of the compensator is $C(s) = B(s)/A(s)$. Then the overall transfer function is given by

$$G_o(s) = \frac{C(s)G(s)}{1 + C(s)G(s)} = \frac{\dfrac{B(s)}{A(s)} \cdot \dfrac{N(s)}{D(s)}}{1 + \dfrac{B(s)}{A(s)} \cdot \dfrac{N(s)}{D(s)}} = \frac{B(s)N(s)}{A(s)D(s) + B(s)N(s)}$$

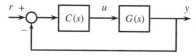

Figure 6.20 Feedback system.

The zeros of $G(s)$ and $C(s)$ are the roots of $N(s)$ and $B(s)$; they remain to be the zeros of $G_o(s)$. In other words, feedback does not affect the zeros of $G(s)$ and $C(s)$. The poles of $G(s)$ and $C(s)$ are the roots of $D(s)$ and $A(s)$; after feedback, the poles of $G_o(s)$ become the roots of $A(s)D(s) + B(s)N(s)$. The total numbers of poles before feedback and after are the same, but their positions have now been *shifted* from $D(s)$ and $A(s)$ to $A(s)D(s) + B(s)N(s)$. Therefore, feedback affects the poles but not the zeros of the plant transfer function. The given plant can be stable or unstable, but we can always introduce feedback and compensators to shift the poles of $G(s)$ to desired position. Therefore feedback can make a good overall system out of a bad plant. In the outward approach, we choose a $C(s)$ and hope that $G_o(s)$ will be a good overall transfer function. In the inward approach, we choose a good $G_o(s)$ and then compute $C(s)$.

PROBLEMS

6.1. Find the ranges of β_i so that the following transfer functions have position errors smaller than 10%.

    a. $\dfrac{\beta_1 s + \beta_0}{s^2 + 2s + 2}$

    b. $\dfrac{\beta_2 s^2 + \beta_1 s + \beta_0}{s^3 + 3s^2 + 2s + 3}$

    c. $\dfrac{\beta_2 s^2 + \beta_1 s + \beta_0}{s^3 + 2s^2 + 9s + 68}$

6.2. Find the ranges of β_i so that the transfer functions in Problem 5.1 have velocity errors smaller than 10%.

6.3. Consider the three systems shown in Figure P6.3. Find the ranges of k so that the systems are stable and have position errors smaller than 10%.

6.4. Repeat Problem 6.3 so that the systems have velocity errors smaller than 10%.

6.5. **a.** Find the range of k_0 such that the system in Figure P6.5 is stable. Find the value of k_0 such that the system has a zero position error or, equivalently, such that y will track asymptotically a step reference input.

    b. If the plant transfer function in Figure P6.5 becomes $5.1/(s - 0.9)$ due to aging, will the output still track asymptotically any step reference input? If not, such a tracking is said to be *not robust*.

6.6. Consider the unity feedback system shown in Figure 6.4(a). We showed there that if the loop transfer function $G_l(s) = C(s)G(s)$ is of type 1 or, equivalently, can be expressed as

$$G_l(s) = \frac{N_l(s)}{s\overline{D}_l(s)}$$

where $N_l(0) \neq 0$ and $\overline{D}_l(0) \neq 0$, and if the feedback system is stable, then the

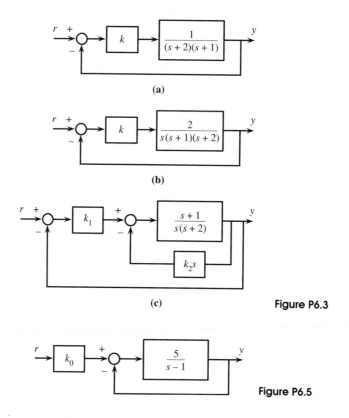

(a)

(b)

(c)

Figure P6.3

Figure P6.5

plant output will track asymptotically any step reference input. Now show that the tracking is *robust* in the sense that, even if there are perturbations in $N(s)$ and $\overline{D}(s)$, the position error is still zero as long as the system remains stable.

6.7. **a.** Consider the unity feedback system shown in Figure 6.4(a). Show that if $G_l(s)$ is of type 2 or, equivalently, can be expressed as

$$G_l(s) = \frac{N_l(s)}{s^2 \overline{D}_l(s)}$$

with $N_l(0) \neq 0$ and $\overline{D}_l(0) \neq 0$, and if the unity feedback system is stable, then its velocity error is zero. In other words, the plant output will track asymptotically any ramp reference input.

b. Show that the tracking of a ramp reference input is robust even if there are perturbations in $N_l(s)$ and $\overline{D}_l(s)$ as long as the system remains stable. Note that $G_l(s)$ contains $1/s^2$, which is the Laplace transform of the ramp reference input. This is a special case of the *internal model principle*, which states that if $G_l(s)$ contains $R(s)$, then $y(t)$ will track $r(t)$ asymptotically and the tracking is robust. See Reference [15].

6.8. Consider the system shown in Figure P6.8. Show that the system is stable. The plant transfer function $G(s)$ is of type 1, is the position error of the feedback system zero? In the unity feedback system, the position and velocity error can be determined by system type. Is this true in nonunity feedback or other configurations?

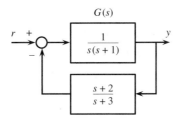

$$G(s)$$

Figure P6.8

6.9. Show that if a system is designed to track t^2, then the system will track any reference input of the form $r_0 + r_1 t + r_2 t^2$.

6.10. The movement of a recorder's pen can be controlled as shown in Figure P6.10(a). Its block diagram is shown in Figure P6.10(b). Find the range of k such that the position error is smaller than 1%.

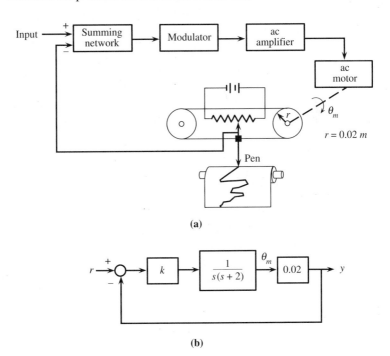

Figure P6.10

6.11. Consider the systems shown in Figure P6.11. Which of the systems are not well posed? If not, find the input-output pair that has an improper closed-loop transfer function.

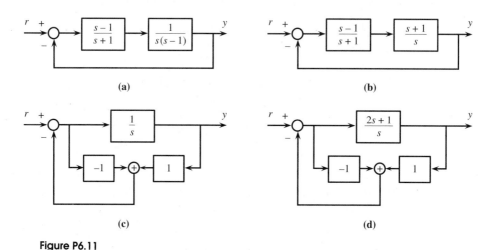

(a)   (b)

(c)   (d)

Figure P6.11

6.12. Discuss the total stability of the systems shown in Figure P6.11.

6.13. Consider the speed control of rollers discussed in Figure 6.18. We now model the plant transfer function as $10/(\tau s + 1)$, with τ ranging between 4 and 6. Use the compensators computed in Figure 6.18 to compute the steady-state outputs of the open-loop and feedback systems due to a unit-step reference input for the following three cases: (a) τ equals the nominal value 5, (b) $\tau = 4$, and (c) $\tau = 6$. Which system, open-loop or feedback system, is less sensitive to parameter variations?

6.14. a. Consider the plant transfer function shown in Figure 6.18. Find a k in Figure P6.14, if it exists, such that the overall transfer function in Figure P6.14 equals $2/(s + 2)$.

b. If the plant has a disturbance as shown in Figure 6.18, find the steady-state output of the overall system in (a) due to a unit-step disturbance input.

c. Which feedback system, Figure 6.18(c) or Figure P6.14, is less sensitive to plant perturbations? The loop transfer function in Figure 6.18(c) is of type 1. Is the loop transfer function in Figure P6.14 of type 1?

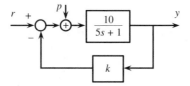

Figure P6.14

6.15. Consider the system shown in Figure P6.15. The noise generated by the amplifier is represented by n. If $r = \sin t$ and $n = 0.1 \sin 10t$, what are the steady-state outputs due to r and n? What is the ratio of the amplitudes of the outputs excited by r and n?

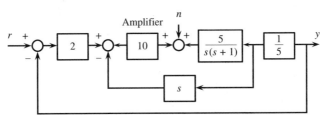

Figure P6.15

6.16. Consider the systems shown in Figure P6.16. (a) If the plant, denoted by P, has the following nominal transfer function

$$\frac{1}{s(s^2 + 2s + 3)}$$

show that the two systems have the same steady-state output due to $r(t) = \sin 0.1t$. (b) If, due to aging, the plant transfer function becomes

$$\frac{1}{s(s^2 + 2.1s + 3.06)}$$

what are the steady-state outputs of the two systems due to the same r? (c) Which system has a steady-state output closer to the one in (a)?

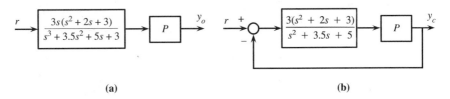

(a)   (b)

Figure P6.16

6.17. The comparison in Problem 6.16 between the open-loop and closed-loop systems does not consider the noise due to the transducer (which is used to introduce feedback). Now the noise is modeled as shown in Figure P6.17.

a. Compute the steady-state y_c due to $n(t) = 0.1 \sin 10t$.

b. What is the steady-state y_c due to $r(t) = \sin 0.1t$ and $n(t) = 0.1 \sin 10t$?

c. Compare the steady-state error in the open-loop system in Figure P6.16(a) with the one in the closed-loop system in Figure P6.17. Is the reduction in the steady-state error due to the feedback large enough to offset the increase of the steady-state error due to the noise of the transducer?

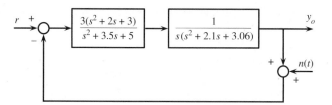

Figure P6.17

6.18. a. Consider the feedback system shown in Figure P6.18. The nominal values of all k_i are assumed to be 1. What is its position error?

b. Compute the position error if $k_1 = 2$ and $k_2 = k_3 = 1$. Compute the position error if $k_2 = 2$ and $k_1 = k_3 = 1$. Compute the position error if $k_3 = 2$ and $k_1 = k_2 = 1$.

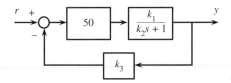

Figure P6.18

7 *The Root-Locus Method*

7.1 INTRODUCTION

As was discussed in the preceding chapter, inward and outward approaches are available for designing control systems. There are two methods in the outward approach: the root-locus method and the frequency-domain method. In this chapter, we study the root-locus method.

In the root-locus method, we first choose a configuration, usually the unity-feedback configuration, and a gain, a compensator of degree 0. We then search the gain and hope that a good control system can be obtained. If not, we then choose a different configuration and/or a more complex compensator and repeat the design. Because the method can handle only one parameter at a time, the form of compensators must be restricted. This is basically a trial-and-error method. We first use an example to illustrate the basic idea.

7.2 QUADRATIC TRANSFER FUNCTIONS WITH A CONSTANT NUMERATOR

Consider a plant with transfer function

$$G(s) = \frac{1}{s(s + 2)} \tag{7.1}$$

It could be the transfer function of a motor driving a load. The problem is to design an overall system to meet the following specifications:

1. Position error $= 0$
2. Overshoot $\leq 5\%$
3. Settling time ≤ 9 seconds
4. Rise time as small as possible.

Before carrying out the design, we must first choose a configuration and a compensator with one open parameter. The simplest possible feedback configuration and compensator are shown in Figure 7.1. They can be implemented using a pair of potentiometers and an amplifier. The overall transfer function is

$$G_o(s) = \frac{k \cdot \dfrac{1}{s(s+2)}}{1 + k \cdot \dfrac{1}{s(s+2)}} = \frac{k}{s^2 + 2s + k} \tag{7.2}$$

The first requirement in the design is the stability of $G_o(s)$. Clearly, $G_o(s)$ is stable if and only if $k > 0$. Because $G_o(0) = k/k = 1$, the system has zero position error for every $k > 0$. Thus the design reduces to the search for a positive k to meet requirements (2) through (4). Arbitrarily, we choose $k = 0.36$. Then $G_o(s)$ becomes

$$G_o(s) = \frac{0.36}{s^2 + 2s + 0.36} = \frac{0.36}{(s+0.2)(s+1.8)} \tag{7.3}$$

One way to find out whether or not $G_o(s)$ will meet (2) and (3) is to compute analytically the unit-step response of (7.3). A simpler method is to carry out computer simulation. If the system does not meet (2) or (3), then $k = 0.36$ is not acceptable. If the system meets (2) and (3), then $k = 0.36$ is a possible candidate. We then choose a different k and repeat the process. Finally, we choose from those k meeting (2) and (3) the one that has the smallest rise time. This completes the design.

From the preceding discussion, we see that the design procedure is very tedious and must rely heavily on computer simulation. The major difficulty arises from the fact that the specifications are given in the time domain, whereas the design is carried out using transfer functions, or in the s-plane. Therefore, if we can translate the time-domain specifications into the s-domain, the design can be considerably simplified. This is possible for a special class of transfer functions and will be discussed in the next subsection.

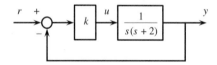

Figure 7.1 Unity-feedback system.

7.2.1 Desired Pole Region

Consider a control system with transfer function

$$G_o(s) = \frac{\omega_n^2}{s^2 + 2\zeta\omega_n s + \omega_n^2} \qquad (7.4)$$

where ζ is the damping ratio and ω_n the natural frequency. It is a quadratic transfer function with a constant numerator. This system was studied in Section 4.3. The poles of $G_o(s)$ are

$$-\zeta\omega_n \pm \omega_n\sqrt{\zeta^2 - 1}$$

If $\zeta < 1$, the poles are complex conjugate as shown in Figure 7.2(a). For $\zeta < 1$, the unit-step response of $G_o(s)$, as derived in (4.10), is

$$y(t) = \mathcal{L}^{-1}\left[\frac{\omega_n^2}{s^2 + 2\zeta\omega_n s + \omega_n^2} \cdot \frac{1}{s}\right] \qquad (7.5)$$

$$= 1 - \frac{\omega_n}{\omega_d} e^{-\sigma t} \sin(\omega_d t + \theta)$$

where $\omega_d = \omega_n(1 - \zeta^2)^{1/2}$, $\sigma = \zeta\omega_n$, and $\theta = \cos^{-1}\zeta$. The steady-state response of $y(t)$ is $y_s = 1$, and the maximum value, as computed in (4.13), is

$$y_{max} = \max |y(t)| = 1 + e^{-\pi\zeta(1-\zeta^2)^{-1/2}}$$

Thus the overshoot, as defined in Section 6.3.3, is

$$\text{Overshoot} = \left|\frac{y_{max} - 1}{1}\right| = e^{-\pi\zeta(1-\zeta^2)^{-1/2}}$$

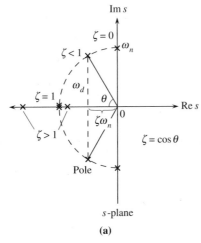

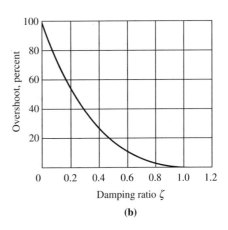

(a)

(b)

Figure 7.2 Damping ratio and overshoot.

We see that the overshoot depends only on the damping ratio. The relationship is plotted in Figure 7.2(b). From the plot, the range of ζ for a given overshoot can be obtained. For example, if the overshoot is required to be less than 20%, then the damping ratio must be larger than 0.45, as can be read from Figure 7.2(b). Now we translate this relationship into a pole region. Because

$$\zeta = \cos \theta$$

where θ is defined in Figure 7.2(a), and because $\cos \theta$ is a decreasing function of θ in $0 \le \theta \le 90°$, if $\zeta \ge \zeta_o$, then $\theta \le \theta_o = \cos^{-1}\zeta_o$. Using this fact, the specification on overshoot can easily be translated into a pole region in the s-plane. For example, we have

$$\text{Overshoot} \le 10\% \rightarrow \zeta \ge 0.6 \rightarrow \theta \le \cos^{-1}0.6 = 53°$$

and

$$\text{Overshoot} \le 5\% \rightarrow \zeta \ge 0.7 \rightarrow \theta \le \cos^{-1}0.7 = 45°$$

In other words, for the system in (7.4), if the overshoot is required to be less than 5%, then the poles of $G_o(s)$ must lie inside the sector bounded by 45°, as is shown in Figure 7.3. This translates the specification on overshoots into a desired pole region.

Next we consider the settling time. As defined in Section 6.3.3, the settling time is the time needed for the step response to reach and stay within 2% of its steady-state value. The difference between $y(t)$ in (7.5) and its steady state $y_s = 1$ is

$$D := |y(t) - 1| = \left| -\frac{\omega_n e^{-\sigma t}}{\omega_d} \sin(\omega_d t + \theta) \right| \le \frac{\omega_n}{\omega_d} e^{-\sigma t} = \frac{e^{-\sigma t}}{\sqrt{1 - \zeta^2}} \quad (7.6)$$

for $\zeta < 1$. Note that $\sigma = \zeta\omega_n$ is the magnitude of the real part of the poles (see Figure 7.2); it is the distance of the complex-conjugate poles from the imaginary

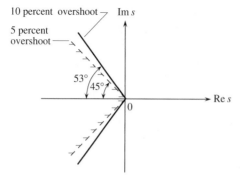

Figure 7.3 Overshoot and pole region.

axis. Thus the settling time t_s is the smallest t such that

$$\frac{e^{-\sigma t}}{\sqrt{1 - \zeta^2}} \le 0.02 \qquad \text{for } t > t_s \tag{7.7}$$

Clearly, for given ζ and ω_n, the settling time can be computed from (7.7). It is, however, desirable to develop a formula that is easier to employ. If $\zeta < 0.8$, then (7.7) becomes

$$D \le \frac{e^{-\sigma t}}{\sqrt{1 - \zeta^2}} \le 1.7 e^{-\sigma t}$$

This is smaller than 0.02 if $t \ge 4.5/\sigma$. Hence, given a settling time t_s, if $\sigma \ge 4.5/t_s$ or, equivalently,

$$-(\text{Real parts of the poles}) \ge \frac{4.5}{t_s} \tag{7.8}$$

then the specification on settling time can be met. Although (7.8) is developed for complex poles with damping ratio smaller than 0.8, it also holds for real poles with $\zeta > 1.05$. Thus, in general, if both poles of $G_o(s)$ in (7.4) lie on the left-hand side of the vertical line shown in Figure 7.4, then the specification on settling time can be met. The condition in (7.8) is consistent with the statement that the step response reaches and remains within 1% of its steady-state value in five time constants. The settling time is defined for 2% and equals $4.5 \times$ time constant.

Now we can combine the specifications on overshoot and settling time. The poles of (7.4) must lie inside the section as shown in Figure 7.3 to meet the overshoot specification and must lie on the left-hand side of the vertical line in Figure 7.4 to meet the settling time specification. Therefore, to meet both specifications, the poles must lie inside the region denoted by C in Figure 7.4. The exact boundary of C can be obtained from the specifications on overshoot and settling time.

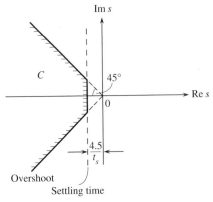

Figure 7.4 Desired pole region.

Exercise 7.2.1

Find the desired pole regions for the following specifications:

a. Overshoot < 20%, and settling time < 3 seconds.

b. Overshoot < 10%, and settling time < 10 seconds.

By definition, the rise time is the time required for the step response of $G_o(s)$ to rise from 0 to 90 percent of its steady-state value. The translation of the rise time into a pole region cannot be done quantitatively, as in the case of overshoot and settling time. All we can say is that, generally, the farther away the closest pole[1] from the origin of the s-plane, the smaller the rise time. Strictly speaking, this statement is not correct, as can be seen from Figure 4.7. The rise times of the responses in Figure 4.7 are all different, even though the distances of the corresponding complex poles from the origin all equal ω_n. On the other hand, because the time scale of Figure 4.7 is $\omega_n t$, as the distance ω_n increases, the rise time decreases. Thus, the assertion holds for a fixed ζ. Because there is no other better guideline, the assertion that the farther away the closest pole from the origin, the smaller the rise time will be used in the design.

We recapitulate the preceding discussion as follows:

Overshoot	Sector with $\theta = \cos \zeta$ and ζ is determined from Fig. 7.2(b).
Settling time	4.5/(shortest distance of poles from the $j\omega$-axis).
Rise time	Proportional to 1/(shortest distance of poles from the origin).

These simple rules, although not necessarily exact, are very convenient to use in design.

7.2.2 Design Using Desired Pole Region

Now we return to the design problem in Figure 7.1. As discussed earlier, if k is positive, then the system is stable and has zero position error. Hence, in the following, we shall find a positive k to meet transient specifications. The specification on overshoot requires that the poles of $G_o(s)$ lie inside the sector bounded by $\theta = 45°$, as shown in Figure 7.5. The settling time requires the poles to be located on the left-

---

[1]The system has two poles. If they are complex, the distances of the two complex-conjugate poles from the origin are the same. If they are real and distinct, then one pole is closer to the origin then the other. We consider only the distance to the closer pole. If a system has three or more poles, then we consider the pole closest to the origin.

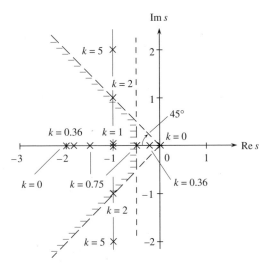

Figure 7.5 Poles of $G_o(s)$.

hand side of the vertical line passing through $-4.5/t_s = -0.5$. Hence, if all poles of $G_o(s)$ lie inside the shaded region in Figure 7.5, the overall system will meet the specifications on overshoot and settling time.

The poles of $G_o(s)$ in (7.2) for $k = 0.36, 0.75, 1, 2,$ and 5 are computed in the following:

Gain	Poles	Comments
$k = 0.36$	$-0.2, -1.8$	meet (2) but not (3)
$k = 0.75$	$-0.5, -1.5$	meet both (2) and (3)
$k = 1$	$-1, -1$	meet both (2) and (3)
$k = 2$	$-1 \pm j$	meet both (2) and (3)
$k = 5$	$-1 \pm j2$	meet (3) but not (2)

They are plotted in Figure 7.5. Note that there are two poles for each k. For $k = 0.36$, although one pole lies inside the region, the other is on the right-hand side of the vertical line. Hence if we choose $k = 0.36$, the system will meet the specification on overshoot but not that on settling time. If we choose $k = 5$, then the system will meet the specification on settling time but not that on overshoot. However for $k = 0.75, 1,$ and 2, all the poles are within the allowable region, and the system meets the specifications on overshoot and settling time. Now we discuss how to choose a k from 0.75, 1, and 2, so that the rise time will be the smallest. The poles corresponding to $k = 0.75$ are -0.5 and -1.5; therefore, the distance of the closer pole from the origin is 0.5. The poles corresponding to $k = 1$ are -1 and -1. Their distance from the origin is 1 and is larger than 0.5. Therefore, the system with $k = 1$ has a smaller rise time than the one with $k = 0.75$. The poles corresponding to $k = 2$ are $-1 \pm j1$. Their distance from the origin is $\sqrt{1 + 1} = 1.4$, which is the largest among $k = 0.75$, 1, and 2. Therefore the system with $k = 2$ has the smallest rise time or, equivalently, responds fastest. The unit-step responses of the system are shown in Figure 7.6. They bear out the preceding discussion.

For this example we are able to find a gain k to meet all the specifications. If some of the specifications are more stringent, then no k may exist. For example, if the settling time is required to be less than 2 seconds, then all poles of $G_o(s)$ must lie on the left-hand side of the vertical line passing through $-4.5/2 = -2.25$. From Figure 7.5, we see that no poles meet the requirement. Therefore, no k in Figure 7.1 can yield a system with settling time less than 2 seconds. In this case, we must choose a different configuration and/or a more complicated compensator and repeat the design.

Exercise 7.2.2

Consider a plant with transfer function $2/s(s + 4)$. (a) Find the range of k in Figure 7.1 such that the resulting system has overshoot less than 5%. (b) Find the range of k such that the system has settling time smaller than 4.5 seconds. (c) Find the range of k to meet both (a) and (b). (d) Find a value of k from (c) such that the system has the smallest rise time.

[**Answers:**   (a) $0 < k < 4$. (b) $1.5 < k < \infty$. (c) $1.5 < k < 4$. (d) $k = 4$.]

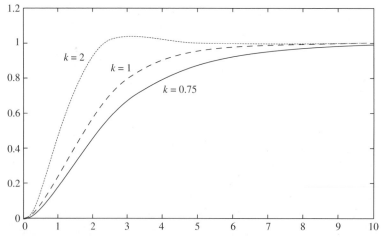

Figure 7.6 Unit-step responses.

7.3 MORE ON DESIRED POLE REGION

The example in the preceding sections illustrates the essential idea of the design method to be introduced in this chapter. The method consists of two major components:

1. The translation of the transient performance into a desired pole region. We then try to place the poles of the overall system inside the region by choosing a parameter.
2. In order to facilitate the choice of the parameter, the poles of the overall system as a function of the parameter will be plotted graphically. The method of plotting is called the *root-locus method*.

In this section we discuss further the desired pole region. The root-locus method is discussed in the next section.

The desired pole region in Figure 7.4 is developed from a quadratic transfer function with a constant numerator. We shall check whether it is applicable to other types of transfer functions. Consider

$$G_o(s) = \frac{1}{(s^2 + 1.2s + 1)\left(1 + \dfrac{s}{a}\right)} \tag{7.9}$$

In addition to the complex-conjugate poles $-0.6 \pm j0.8$, $G_o(s)$ has one plot at $-a$. If $a = \infty$, (7.9) reduces to (7.4) and its unit-step response, as shown in Figure 7.7, has an overshoot of 10% and a settling time of about 7 seconds, as predicted by the desired pole region developed in Figure 7.4. Consider now $a = 4$. The pole $s = -4$ is located quite far away from the pair of complex-conjugate poles. Because the response e^{-4t} of the pole approaches zero rapidly, the unit-step response of $G_o(s)$ with $a = 4$ is quite close to the one with $a = \infty$, as shown in Fig. 7.7. Thus the unit-step response of $G_o(s)$ is essentially determined by the pair of complex-conjugate poles $-0.6 \pm j0.8$. In this case the pair is called the *dominant poles* and $G_o(s)$

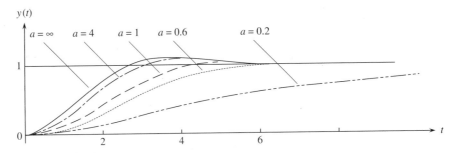

Figure 7.7 Unit-step responses of (7.9).

essentially reduces to a quadratic transfer function with a constant numerator. Thus the desired pole region in Figure 7.4 can still be employed.

The unit-step responses of $G_o(s)$ for $a = 1, 0.6,$ and 0.2 are also shown in Figure 7.7. We see that as the pole moves closer to the origin, the overshoot becomes smaller (eventually becoming zero) and the settling time becomes larger. In these cases, the quantitative specification for the overshoot in Figure 7.2 no longer holds. However, the specifications regarding the settling time and rise time appear to be still applicable.

Next we examine the effect of introducing a zero by considering

$$G_o(s) = \frac{1 + \dfrac{s}{a}}{s^2 + 1.2s + 1} \tag{7.10}$$

Figure 7.8(a) shows the unit-step responses of $G_o(s)$ for $a = 4, 1, 0.6,$ and 0.2, and Figure 7.8(b) shows the unit-step responses of $G_o(s)$ for $a = -4, -1, -0.6,$ and -0.2. The responses for $a = 4$ and -4 are quite similar to the one for $a = \infty$. In other words, if the zero is far away (either in the right half plane or in the left half plane) from the complex conjugate poles, the concept of dominant poles is still applicable. As the left-half-plane zero moves closer to the origin, the overshoot and settling time become larger. However, the rise time becomes smaller. If the zero is in the right half plane, or $a < 0$, the unit-step response will become negative and then positive. This is called *undershoot*.[2] For $a = -4$, the undershoot is hardly detectable. However as the right-half-plane zero moves closer to the origin, the undershoot becomes larger. The overshoot, settling time, and rise time also become larger. Thus, the quantitative specifications developed in Figure 7.4 are no longer applicable. This is not surprising, because the response of a system depends on its poles and zeros, whereas zeros are not considered in the development of the desired pole region.

Even for the simple systems in (7.9) and (7.10), the relationships between the specifications for the transient performance and the pole region are no longer as precise as for quadratic transfer functions with a constant numerator. However, because there is no other simple design guideline, the desired pole region developed in Figure 7.4 will be used for all overall transfer functions. Therefore, if an overall transfer function is not quadratic as in (7.4) and cannot be approximated by (7.4), then there is no guarantee that the resulting system will meet the transient specifications by placing all poles inside the desired pole region. It is therefore important to simulate the resulting system on a computer, to check whether or not it really meets the specifications. If it does not, the system must be redesigned.

---

[2]It was shown by Norimatsu and Ito [49] that if $G_o(s)$ has an odd number of open right-half-plane real zeros, then undershoots always occur in step responses of $G_o(s)$.

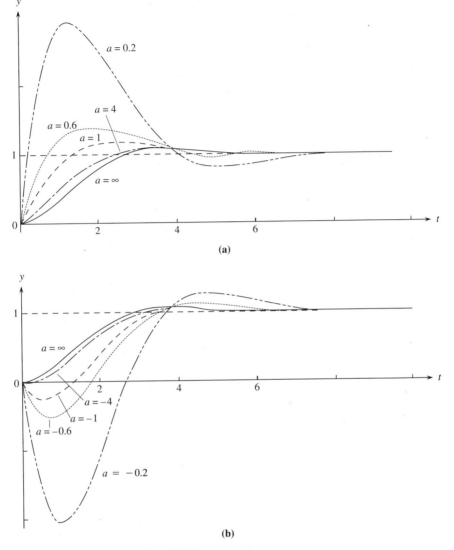

Figure 7.8 Unit-step responses of (7.10).

7.4 PLOT OF ROOT LOCI

From the example in Section 7.2, we see that the design requires the computation of the poles of $G_o(s)$ as a function of k. In this section, we shall discuss this problem.

Consider the unity-feedback system shown in Figure 7.9, where $G(s)$ is a proper rational function and k is a real constant. Let $G(s) = N(s)/D(s)$. Then the overall

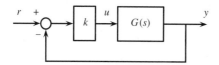

Figure 7.9 Unity-feedback system.

transfer function is

$$G_o(s) = \frac{kG(s)}{1 + kG(s)} = \frac{k\dfrac{N(s)}{D(s)}}{1 + k\dfrac{N(s)}{D(s)}} = \frac{kN(s)}{D(s) + kN(s)}$$

The poles of $G_o(s)$ are the *zeros* of the rational function

$$1 + kG(s) = 1 + k\frac{N(s)}{D(s)} \tag{7.11}$$

or the *roots* of the polynomial

$$D(s) + kN(s) \tag{7.12}$$

or the *solutions* of the equation

$$1 + kG(s) = 0 \tag{7.13}$$

The roots of $D(s) + k/N(s)$ or the poles of $G_o(s)$ as a function of a *real k* are called the *root loci*. Many software programs are available for computing root loci. For example, for $G(s) = 1/s(s + 2) = 1/(s^2 + 2s + 0)$, the following commands in version 3.1 of MATLAB

```
num=[1];den=[1 2 0];
k=0:0.5:10;
r=rlocus(num,den,k);
plot(r,'x')
```

will plot 21 sets of the poles of $kG(s)/(1 + kG(s))$ for $k = 0, 0.5, 1, 1.5, \ldots, 9.5$, and 10. If we use version 3.5 or the Student Edition of MATLAB, the command

```
rlocus(num,den)
```

will plot the complete root loci on the screen. Therefore, to use an existing computer program to compute root loci is very simple. Even so, it is useful to understand the general properties of root loci. From the properties, we can often obtain a rough plot of root loci even without any computation or measurement. This can then be used to check the correctness of computer printout.

To simplify discussion, we assume

$$G(s) = \frac{q(s + z_1)(s + z_2)}{(s + p_1)(s + p_2)(s + p_3)} \tag{7.14}$$

where $-z_i$ and $-p_i$ denote, respectively, zeros and poles, and q is a real constant, positive or negative. Because $G(s)$ is assumed to have real coefficients, complex-conjugate poles and zeros must appear in pairs. Now we shall write $1 + kG(s) = 0$ as

$$G(s) = \frac{q(s + z_1)(s + z_2)}{(s + p_1)(s + p_2)(s + p_3)} = -\frac{1}{k} \tag{7.15}$$

Then the roots of $D(s) + kN(s)$ are those s, real or complex, which satisfy (7.15) for some real k. Note that for each s, say s_1, each factor on the left-hand side of (7.15) is a vector *emitting* from a pole or zero to s_1, as shown in Figure 7.10. The magnitude $|\cdot|$ is the length of the vector. The phase $\sphericalangle$ is the angle measured from the direction of positive real axis; it is positive if measured counterclockwise, negative if measured clockwise. The substitution of

$$s_1 + z_i = |s_1 + z_i|\, e^{j\sphericalangle(s_1 + z_i)} = |s_1 + z_i|\, e^{j\theta_i}$$

and

$$s_1 + p_i = |s_1 + p_i|\, e^{j\sphericalangle(s_1 + p_i)} = |s_1 + p_i|\, e^{j\phi_i}$$

into (7.15) yields

$$\frac{|q|\, |s_1 + z_1|\, |s_1 + z_2|\, e^{j(\sphericalangle q + \theta_1 + \theta_2)}}{|s_1 + p_1|\, |s_1 + p_2|\, |s_1 + p_3|\, e^{j(\phi_1 + \phi_2 + \phi_3)}} = -\frac{1}{k} \tag{7.16}$$

This equation actually consists of two parts: the magnitude condition

$$\frac{|q|\, |s_1 + z_1|\, |s_1 + z_2|}{|s_1 + p_1|\, |s_1 + p_2|\, |s_1 + p_3|} = \left| -\frac{1}{k} \right| \tag{7.17}$$

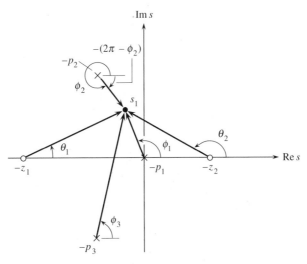

Figure 7.10 Vectors in s-plane.

and the phase condition

$$\measuredangle\, q + \theta_1 + \theta_2 - (\phi_1 + \phi_2 + \phi_3) = \measuredangle\left(-\frac{1}{k}\right) \tag{7.18}$$

Note that $\measuredangle\, q$ equals 0 if $q > 0$, π or $-\pi$ if $q < 0$, θ_i and ϕ_i can be positive (if measured counterclockwise) or negative (if measured clockwise). In the remainder of this and the next sections, we discuss only the phase condition. The magnitude condition will not arise until Section 7.4.3.

Because k is real, we have

$$\measuredangle\left(-\frac{1}{k}\right) = \begin{cases} \pm\pi,\ \pm3\pi,\ \pm5\pi,\ \dots & \text{if } k > 0 \\ 0,\ \pm2\pi,\ \pm4\pi,\ \dots & \text{if } k < 0 \end{cases}$$

Two angles will be considered the same if they differ by $\pm2\pi$ radians or $\pm360°$ or their multiples. Using this convention, the phase condition in (7.18) becomes

$$\text{Total phase} := \measuredangle\, q + \theta_1 + \theta_2 - (\phi_1 + \phi_2 + \phi_3) = \begin{cases} \pi & \text{if } k > 0 \\ 0 & \text{if } k < 0 \end{cases} \tag{7.19}$$

We see that the constant k does not appear explicitly in (7.19). Thus the search for the root loci becomes the search for all s_1 at which the total phase of $G(s_1)$ equals 0 or π. If s_1 satisfies (7.19), then there exists a *real* k_1 such that $D(s_1) + k_1N(s_1) = 0$. This k_1 can be computed from (7.17).

We recapitulate the preceding discussion in the following. The poles of $G_o(s)$ or, equivalently, the roots of $D(s) + kN(s)$ for some real k are those s_1 such that the total phase of $G(s_1)$ equals 0 or π. The way to search for those s_1 is as follows. First we choose an arbitrary s_1 and draw vectors from the poles and zeros of $G(s)$ to s_1 as shown in Figure 7.10. We then use a protractor to measure the phase of each vector. If the total phase is 0 or π, then s_1 is a point on the root loci. If the total phase is neither 0 nor π, then s_1 is not on the root loci. We then try a different point and repeat the process. This is a trial-and-error method and appears to be hopelessly complicated. However, using the properties to be discussed in the next subsection, we can often obtain a rough sketch of root loci without any measurement.

7.4.1 Properties of Root Loci—Phase Condition

Consider a transfer function with real coefficients expressed as

$$G(s) := \frac{q(s + z_1)(s + z_2) \cdots (s + z_m)}{(s + p_1)(s + p_2) \cdots (s + p_n)} =: \frac{N(s)}{D(s)} \tag{7.20}$$

This step can be achieved easily by using

tf2zp

in MATLAB. The transfer function has n poles and m zeros, with $n \geq m$. In discussing the root loci, it is important to express every term in the form of $(s + \alpha)$. Note that $(s + \alpha)$ is a vector extending from $-\alpha$ to s. If we use the form $1 + \tau s$, then its meaning is not apparent. The root loci will be developed graphically on the complex s-plane. We first plot all poles and zeros on the s-plane and then measure the angle from every pole and zero to a chosen s. Note that the constant q does not appear on the plot, but it still contributes a phase to (7.19). To simplify discussion, we assume in this section that $q > 0$. Then the phase of q is zero and the total phase of $G(s)$ will be contributed by the poles and zeros only. We discuss now the general properties of the roots of the polynomial

$$D(s) + kN(s) \qquad (7.21)$$

or the zeros of the rational function

$$1 + k\frac{N(s)}{D(s)} = 1 + kG(s)$$

or the solutions of the equation

$$\frac{N(s)}{D(s)} = \frac{q(s + z_1)(s + z_2) \cdots (s + z_m)}{(s + p_1)(s + p_2) \cdots (s + p_n)} = -\frac{1}{k} \qquad (7.22)$$

as a function of *real k*. To simplify discussion, we consider only $k \geq 0$. In this case, the root loci consist of *all* s at which $G(s)$ has a total phase of π radians or 180°. This is the phase condition in (7.19); the magnitude condition in (7.17) will not be used in this section.

PROPERTY 1

The root loci consist of n continuous trajectories as k varies continuously from 0 to ∞. The trajectories are symmetric with respect to the real axis. ■

The polynomial in (7.21) has degree n. Thus for each real k, there are n roots. Because the roots of a polynomial are continuous functions of its coefficients, the n roots form n continuous trajectories as k varies from 0 to ∞. Because the coefficients of $G(s)$ are real by assumption, complex-conjugate roots must appear in pairs. Therefore the trajectories are symmetric with respect to the real axis.

PROPERTY 2

Every section of the real axis with an odd number of *real* poles and zeros (counting together) on its right side is a part of the root loci for $k \geq 0$.[3] ■

If $k \geq 0$, the root loci consist of those s with total phases equal to 180°. Recall that we have assumed $q > 0$, thus the total phase of $G(s)$ is contributed by poles and zeros only. We use examples to establish this property. Consider

$$G_1(s) = \frac{(s + 4)}{(s - 1)(s + 2)} \qquad (7.23)$$

---

[3]More generally, if $q > 0$ and $k > 0$ or $q < 0$ and $k < 0$, then every section of the real axis whose right-hand side has an odd number of real poles and real zeros is part of the root loci. If $q < 0$ and $k > 0$ or $q > 0$ and $k < 0$, then every section of the real axis whose right-hand side has an even number of real poles and real zeros is part of the root loci.

Their poles and zeros are plotted in Figure 7.11(a). If we choose $s_1 = 2.5$ in Figure 7.11(a) and draw vectors from poles 1 and -2 to s_1' and from zero -4 to s_1, then the phase of every vector is zero. Therefore, the total phase of $G_1(s_1)$ is zero. Thus $s_1 = 2.5$ is not a zero of $1 + kG_1(s) = 0$ for any positive real k. If we choose $s_2 = 0$ and draw vectors as shown in Figure 7.11(a), then the total phase is

$$0 - 0 - \pi = -\pi$$

which equals π after the addition of 2π. Thus $s_2 = 0$ is on the root loci. In fact, every point in $[-2, 1]$ has a total phase of π, thus the entire section between $[-2, 1]$ is part of the root loci. The total phase of every point between $[-4, -2]$ can be shown to be 2π, therefore the section is not on the root loci. The total phase of every point in $(\infty, -4]$ is π, thus it is part of the root loci. The two sections $(\infty, -4]$ and $[-2, 1]$ have odd numbers of real poles and zeros on their right-hand sides.

The transfer function in (7.23) has only real poles and zeros. Now we consider

$$G_2(s) = \frac{2(s + 2)}{(s + 3)^2(s + 1 + j4)(s + 1 - j4)} \tag{7.24}$$

which has a pair of complex-conjugate poles. The net phase due to the pair to any point on the real axis equals 0 or 2π as shown in Figure 7.11(b). Therefore, in applying property 2, complex-conjugate poles and zeros can be disregarded. Thus for $k > 0$, the sections $(-\infty, -3]$ and $[-3, -2]$ are part of the root loci.

Exercise 7.4.1

Consider the transfer function

$$G_3(s) = \frac{s + 4}{(s - 1)(s + 1)^2} \tag{7.25}$$

Find the root loci on the real axis for $k \geq 0$.

Exercise 7.4.2

Consider the transfer function

$$G_4(s) = \frac{(s + 2 + j2)(s + 2 - j2)}{(s - 1)(s + 2)(s + 3)} \tag{7.26}$$

Find the root loci on the real axis for positive k.

PROPERTY 3

The n trajectories migrate from the poles of $G(s)$ to the zeros of $G(s)$ as k increases from 0 to ∞.   ∎

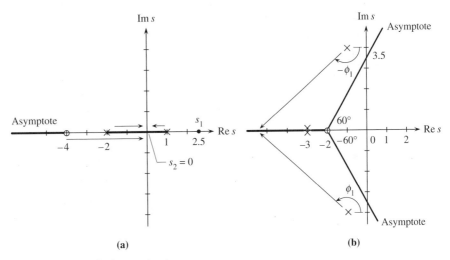

Figure 7.11 Root loci on real axis.

The roots of (7.21) are simply the roots of $D(s)$ if $k = 0$. The roots of $D(s) + kN(s) = 0$ are the same as the roots of

$$\frac{D(s)}{k} + N(s) = 0$$

Thus its roots approach those of $N(s)$ as $k \to \infty$. Therefore, as k increases from 0 to ∞, the root loci exit from the poles of $G(s)$ and enter the zeros of $G(s)$. There is one problem, however. The number of poles and the number of zeros may not be the same. If n (the number of poles) $> m$ (the number of zeros), then m trajectories will enter the m zeros. The remaining $(n - m)$ trajectories will approach $(n - m)$ asymptotes, as will be discussed in the next property.

PROPERTY 4

For large s, the root loci will approach $(n - m)$ number of straight lines, called *asymptotes*, emitting from

$$\left(\frac{\Sigma \text{ Poles } - \Sigma \text{ Zeros}}{\text{No. of poles } - \text{ No. of zeros}}, 0 \right) \tag{7.27a}$$

called the *centroid*.[4] These $(n - m)$ asymptotes have angles

$$\frac{\pm \pi}{n - m} \qquad \frac{\pm 3\pi}{n - m} \qquad \frac{\pm 5\pi}{n - m} \qquad \cdots \tag{7.27b}$$

---

[4]If $G(s)$ has no zeros, then the centroid equals the center of gravity of all poles.

These formulas will give only $(n - m)$ distinct angles. We list some of the angles in the following table.

$n - m$	Angles of asymptotes
1	$180°$
2	$\pm 90°$
3	$\pm 60°,\ 180°$
4	$\pm 45°,\ \pm 135°$
5	$\pm 36°,\ \pm 108°,\ 180°$

■

We justify the property by using the pole-zero pattern shown in Figure 7.12(a). For s_1 very large, the poles and zeros can be considered to cluster at the same point—say, a—as shown in Figure 7.12(b). Note the units of the scales in Figure 7.12(a) and (b). Consequently the transfer function in (7.22) can be approximated by

$$\frac{q(s + z_1) \cdots (s + z_m)}{(s + p_1) \cdots (s + p_n)} \approx \frac{q}{(s - a)^{n-m}} \qquad \text{for } s \text{ very large} \qquad (7.28)$$

In other words, all m zeros are canceled by poles, and only $(n - m)$ poles are left at a. Now we compute the relationship among z_i, p_i, and a. After canceling q, we turn (7.28) upside down and then expand it as

$$\frac{(s + p_1) \cdots (s + p_n)}{(s + z_1) \cdots (s + z_m)} = \frac{s^n + (\Sigma p_i)s^{n-1} + \cdots}{s^m + (\Sigma z_i)s^{m-1} + \cdots} \approx (s - a)^{n-m}$$

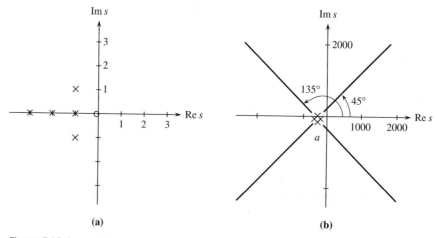

(a)   **(b)**

Figure 7.12 Asymptotes.

which implies, by direct division and expansion,

$$s^{n-m} + [(\Sigma p_i) - (\Sigma z_i)]s^{n-m-1} + \cdots \approx s^{n-m} - (n-m)as^{n-m-1} + \cdots$$

Equating the coefficients of s^{n-m-1} yields

$$(n-m)a = -[(\Sigma p_i) - (\Sigma z_i)] = \Sigma(-p_i) - (\Sigma - z_i)$$

or

$$a = \frac{\Sigma(-p_i) - (\Sigma - z_i)}{n-m} = \frac{(\Sigma \text{ Poles}) - (\Sigma \text{ zeros})}{(\text{No. of poles}) - (\text{No. of zeros})}$$

This establishes (7.27a).

With all $(n-m)$ number of real poles located at a, it becomes simple to find all s_1 with a total phase of π, or, more generally, $\pm\pi, \pm 3\pi, \pm 5\pi, \ldots$ Thus each pole must contribute $\pm\pi/(n-m), \pm 3\pi/(n-m), \pm 5\pi/(n-m), \ldots$ This establishes (7.27b). We mention that $(n-m)$ asymptotes divide 360° equally and are symmetric with respect to the real axis.

Now we shall use this property to find the asymptotes for $G_1(s)$ in (7.23) and $G_2(s)$ in (7.24). The difference between the numbers of poles and zeros of $G_1(s)$ is 1; therefore, there is only one asymptote in the root loci of $G_1(s)$. Its degree is $\pi/1 = 180°$; it coincides with the negative real axis. In this case, it is unnecessary to compute the centroid. For the transfer function $G_2(s)$ in (7.24), the difference between the numbers of poles and zeros is 3; therefore, there are three asymptotes in the root loci of $G_2(s)$. Using (7.27a), the centroid is

$$\frac{-3 - 3 - 1 - j4 - 1 + j4 - (-2)}{3} = \frac{-6}{3} = -2$$

Thus the three asymptotes emit from $(-2, 0)$. Their angles are $\pm 60°$ and 180°. Note that the asymptotes are developed for large s, thus the root loci will *approach* them for large s or large k.

Now we shall combine Properties (3) and (4) as follows: If $G(s)$ has n poles and m zeros, as k increases from 0 to ∞, n trajectories will emit from the n poles. Among the n trajectories, m of them will approach the m zeros; the remaining $(n-m)$ trajectories will approach the $(n-m)$ asymptotes.[5]

Exercise 7.4.3

Find the centroids and asymptotes for $G_3(s)$ in (7.25) and $G_4(s)$ in (7.26).

[**Answers:** $(1.5, 0)$, $\pm 90°$; no need to compute centroid, 180°.]

---

[5]The $G(s)$ in (7.20) can be, for s very large, approximated by q/s^{n-m}. Because it equals zero at $s = \infty$, $G(s)$ can be considered to have $n-m$ number of zeros at $s = \infty$. These zeros are located at the end of the $(n-m)$ asymptotes. If these infinite zeros are included, then the number of zeros equals the number of poles, and the n trajectories will emit from the n poles and approach the n finite and infinite zeros.

PROPERTY 5

Breakaway points—Solutions of $D(s)N'(s) - D'(s)N(s) = 0$. ■

Consider the transfer function $G(s) = N(s)/D(s) = (s + 4)/(s - 1)(s + 2)$. Part of the root loci of $1 + kG(s)$ is shown in Figure 7.11(a), repeated in Figure 7.13. As k increases, the two roots of $D(s) + kN(s)$ move away from poles 1 and -2 and move toward each other inside the section $[-2, 1]$. As k continues to increase, the two roots will eventually collide and split or break away. Such a point is called a *breakaway point*. Similarly, as k approaches infinity, one root will approach zero -4 and another will approach $-\infty$ along the asymptote that coincides with the negative real axis. Because the root loci are continuous, the two roots must come in or break in somewhere in the section $(-\infty, -4]$ as shown in Figure 7.13. Such a point is also called a *breakaway point*. Breakaway points can be computed analytically.

A breakaway point is where two roots collide and break away; therefore, there are at least two roots at every breakaway point. Let s_o be a breakaway point of $D(s) + kN(s)$. Then it is a repeated root of $D(s) + kN(s)$. Consequently, we have

$$D(s_o) + kN(s_o) = 0 \tag{7.29a}$$

and

$$\frac{d}{ds}[D(s) + kN(s)]\bigg|_{s=s_o} = D'(s_o) + kN'(s_o) = 0 \tag{7.29b}$$

where the prime denotes differentiation with respect to s. The elimination of k from (7.29) yields

$$D(s_o) - \frac{D'(s_o)}{N'(s_o)} N(s_o) = 0$$

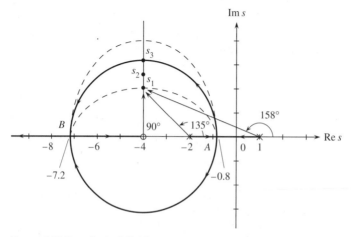

Figure 7.13 Root loci of $G_1(s)$.

which implies

$$D(s_o)N'(s_o) - D'(s_o)N(s_o) = 0 \qquad (7.30)$$

Thus a breakaway point s_o must satisfy (7.30) and can be obtained by solving the equation. For example, if $G(s) = (s + 4)/(s - 1)(s + 2)$, then

$$D(s) = s^2 + s - 2 \qquad D'(s) = 2s + 1$$

$$N(s) = s + 4 \qquad N'(s) = 1$$

and

$$D(s)N'(s) - D'(s)N(s) = s^2 + s - 2 - (2s^2 + 9s + 4) = 0 \qquad (7.31)$$

or

$$s^2 + 8s + 6 = 0$$

Its roots are -0.8 and -7.2. Thus the root loci have two breakaway points at A $= -0.8$ and B $= -7.2$ as shown in Figure 7.13. For this example, the two solutions yield two breakaway points. In general, not every solution of (7.30) is necessarily a breakaway point for $k \geq 0$. Although breakaway points occur mostly on the real axis, they may appear elsewhere, as shown in Figure 7.14(a). If two loci break away from a breakaway point as shown in Figure 7.13 and Figure 7.14(a), then their tangents will be $180°$ apart. If four loci break away from a breakaway point (it has four repeated roots) as shown in Figure 7.14(b), then their tangents will equally divide $360°$.

With the preceding properties, we are ready to complete the root loci in Figure 7.13 or, equivalently, the solutions of

$$\frac{s + 4}{(s - 1)(s + 2)} = -\frac{1}{k}$$

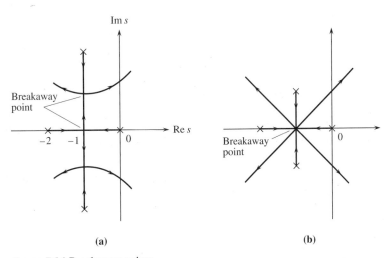

(a)   **(b)**

Figure 7.14 Breakaway points.

for $k \geq 0$. As discussed earlier, the sections $(-\infty, -4]$ and $[-2, 1]$ are parts of the root loci. There is one asymptote that coincides with the negative real part. There are two breakaway points as shown. Because the root loci are continuous, the root loci must assume the *form* indicated by the dotted line shown in Figure 7.13. The exact loci, however, must be obtained by measurement. Arbitrarily we choose an s_1 and draw vectors from zero -4 and poles -2 and 1 to s_1 as shown in Figure 7.13. The phase of each vector is measured using a protractor. The total phase is

$$90° - 135° - 158° = -203°$$

which is different from $\pm 180°$. Thus s_1 is not on the root loci. We then try s_2, and the total phase is measured as $-190°$. It is not on the root loci. We then try s_3, and the total phase roughly equals $-180°$. Thus s_3 is on the root loci. From the fact that they break away at point A, pass through s_3, and come in at point B, we can obtain the root loci as shown. Clearly the more points we find on the root loci, the more accurate the plot. The root loci in Figure 7.13 happens to be a circle with radius 3.2 and centered at -4. This completes the plot of the root loci of $G_1(s)$ in (7.23).

Exercise 7.4.4

Find the breakaway points for $G_3(s)$ in (7.25) and $G_4(s)$ in (7.26). Also complete the root loci of $G_3(s)$.

PROPERTY 6

Angle of departure or arrival.   ■

Every trajectory will depart from a pole. If the pole is real and distinct, the direction of the departure is usually 0° or 180°. If the pole is complex, then the direction of the departure may assume any degree between 0° and 360°. Fortunately this angle can be measured in one step. Similarly the angle for a trajectory to arrive at a zero can also be measured in one step. We now discuss their measurement.

Consider the transfer function $G_2(s)$ in (7.24). Its partial root loci are obtained in Figure 7.11(b) and repeated in Figure 7.15(a). There are four poles, so there are four trajectories. One departs from the pole at -3 and enters the zero at -2. One departs from another pole at -3 and moves along the asymptote on the negative real axis. The last two trajectories will depart from the complex-conjugate poles and move toward the asymptotes with angles $\pm 60°$. To find the angle of departure, we draw a small circle around pole $-1 + j4$ as shown in Figure 7.15(b). We then find a point s_1 on the circle with a total phase equal to π. Let s_1 be an arbitrary point on the circle and let the phase from pole $-1 + j4$ to s_1 be denoted by θ_1. If the radius of the circle is very small, then the vectors drawn from the zero and all other poles to s_1 are the same as those drawn to the pole at $-1 + j4$. Their angles can be measured, using a protractor, as 76°, 63°, and 90°. Therefore, the total phase of $G_2(s)$

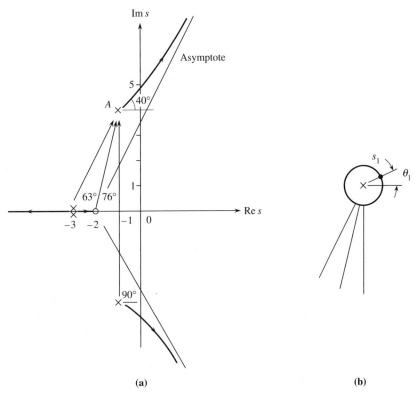

Figure 7.15 Root loci of $G_2(s)$.

at s_1 is

$$76° - (63° + 63° + 90° + \theta_1) = -140° - \theta_1$$

Note that there are two poles at -3, therefore there are two $63°$ in the phase equation. In order for s_1 to be on the root loci, the total phase must be $\pm 180°$. Thus we have $\theta_1 = 40°$. This is the angle of departure.

Once we have the asymptote and the angle of departure, we can draw a rough trajectory as shown in Figure 7.15. Certainly, if we find a point, say $A = j5$ shown in the figure, with total phase $180°$, then the plot will be more accurate. In conclusion, using the properties discussed in this section, we can often obtain a rough sketch of root loci with a minimum amount of measurement.

Exercise 7.4.5

Compute the angle of arrival for $G_4(s)$ in (7.26) and then complete its root loci.

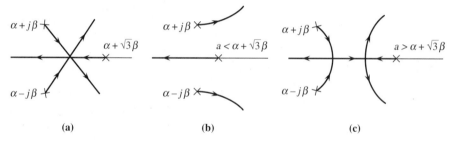

Figure 7.16 Root loci of (7.32).

7.4.2 Complexities of Root Loci

Using the properties discussed in the preceding subsection, we can often obtain a rough sketch of root loci. However, because the roots of polynomials are very sensitive to coefficients, small changes in coefficients may yield entirely different root loci. For example, consider the transfer function[6]

$$G(s) = \frac{1}{(s - \alpha - j\beta)(s - \alpha + j\beta)(s - a)} \tag{7.32}$$

It has three poles. If $a = \alpha + \sqrt{3}\beta$ or, equivalently, if the three poles form an equilateral triangle, then the root loci are all straight lines as shown in Figure 7.16(a). If $a < \alpha + \sqrt{3}\beta$, then the root loci are as shown in Figure 7.16(b). If $a > \alpha + \sqrt{3}\beta$, then the root loci have two breakaway points as shown in Figure 7.16(c). Although the relative positions of the three poles are the same for the three cases, their root loci have entirely different patterns.

As an another example, consider

$$G(s) = \frac{s + 1}{s^2(s + a)} \tag{7.33}$$

Its approximate root loci for $a = 3, 7, 9$, and 11 are shown in Figure 7.17. It has two asymptotes with degrees $\pm 90°$, emitting respectively from

$$\frac{0 + 0 + (-a) - (-1)}{3 - 1} = \frac{1 - a}{2} = -1, -3, -4, -5$$

As a moves away from the origin, the pattern of root loci changes drastically. Therefore to obtain exact root loci from the properties is not necessarily simple. On the other hand, none of the properties is violated in these plots. Therefore the properties can be used to check the correctness of root loci by a computer.

---

[6]This example was provided by Dr. Byunghak Seo.

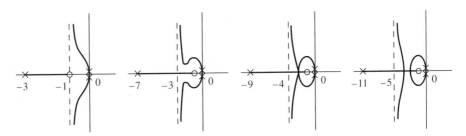

Figure 7.17 Root loci of (7.33).

7.4.3 Stability Range from Root Loci—Magnitude Condition

The plot of root loci up to this point used only the phase condition in (7.19). Now we discuss the use of the magnitude equation in (7.17). We use it to find the stability range of systems. Consider the unity-feedback system shown in Figure 7.18, where the plant transfer function is given by

$$G(s) = \frac{N(s)}{D(s)} = \frac{s^2 - 2s + 5}{s^3 + 5s^2 + 12s - 18}$$

$$= \frac{(s - 1 + j2)(s - 1 - j2)}{(s - 1)(s + 3 + j3)(s + 3 - j3)}$$

(7.34)

The system was studied in (4.22) and the stability range of k was computed, using the Routh test, as

$$3.6 < k < 5.54$$

Now we shall recompute it using the root-locus method. First we plot the root loci of

$$\frac{(s - 1 + j2)(s - 1 - j2)}{(s - 1)(s + 3 + j3)(s + 3 - j3)} = \frac{-1}{k}$$

(7.35)

for $k > 0$. The section $(-\infty, 1]$, plotted with the heavy line in Figure 7.19, is part of the root loci because its right-hand side has one real pole. The difference between the numbers of poles and zeros is 1; therefore, there is one asymptote with degree 180°, which coincides with the negative real axis. The angle of departure at pole $-3 + j3$ is measured as 242°; the angle of arrival at zero $1 + j2$ is measured as

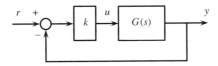

Figure 7.18 Unity-feedback system.

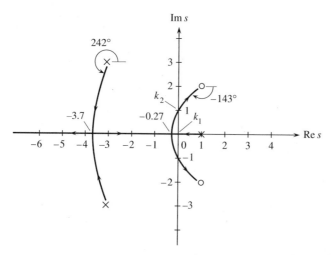

Figure 7.19 Stability range from root loci.

$-143°$. We compute

$$D(s)N'(s) - D'(s)N(s) = (s^3 + 5s^2 + 12s - 18)(2s - 2)$$
$$-(3s^2 + 10s + 12)(s^2 - 2s + 5) = -s^4 + 4s^3 + 7s^2 - 86s - 24$$

Its roots are computed, using MATLAB, as $3.9870 \pm j2.789$, -3.7, and -0.274. Clearly, -3.7 and -0.274 are breakaway points, but not the complex-conjugate roots. Using the breakaway points and departure angles, we can readily obtain a rough sketch of root loci as shown in Figure 7.19 with heavy lines. As k increases from 0 to ∞, the three roots of $D(s) + kN(s)$ or, equivalently, the three poles of the unity-feedback system in Figure 7.18 will move along the trajectories as indicated by arrows.

The rates of emigration of the three closed-loop poles are not necessarily the same. To see this, we list the poles for $k = 0$, 1 and 2:

$k = 0$: $1, -3 \pm j3$
$k = 1$: $0.83, -3.4 \pm j2$
$k = 2$: $0.62, -5.1, -2.9$

We see that, at $k = 1$, the complex-conjugate poles have moved quite far away from $-3 \pm j3$, but the real pole is still very close to 1. As k continues to increase, the complex-conjugate poles collide at $s = -3.7$, and then one moves to the left and the other to the right on the real axis. The pole moving to the left approaches $-\infty$ as k approaches infinity. The one moving to the right collides with the real pole emitting from 1 at $s = -0.274$. They split and then enter the complex-conjugate zeros of $G(s)$ with angles $\pm 143°$. They cross the imaginary axis roughly at $s = \pm j1$.

From the preceding discussion, we can now determine the stability range of k. At $k = 0$, the unity-feedback system has one unstable pole at $s = 1$ and one pair of stable complex-conjugate poles at $-3 \pm j3$. As k increases, the unstable closed-loop pole moves from 1 into the left half plane. It is on the imaginary axis at $k = k_1$, and then becomes stable for $k > k_1$. Note that the complex-conjugate closed-loop poles remain inside the open left half plane and are stable as k increases from 0 to k_1. Therefore, the unity-feedback system is stable if $k > k_1$. The three closed-loop poles remain inside the open left half plane until $k = k_2$, where the root loci intersect with the imaginary axis. Then the closed-loop complex-conjugate poles move into the right half plane and become unstable. Therefore, the unity-feedback system is stable if $k_1 < k < k_2$.

To compute k_1 and k_2, we must use the magnitude equation. The magnitude of (7.35) is

$$\frac{|s - 1 + j2|\,|s - 1 - j2|}{|s - 1|\,|s + 3 + j3|\,|s + 3 - j3|} = \left|\frac{-1}{k}\right| = \frac{1}{k} \tag{7.36}$$

where we have used the fact that $k > 0$. Note that k_1 is the gain of the root loci at $s = 0$ and k_2 the gain at $s = j1$. To compute k_1, we set $s = 0$ in (7.36) and compute

$$\frac{1}{k_1} = \frac{|-1 + j2|\,|-1 - j2|}{|3 + j3|\,|3 - j3|} = \frac{\sqrt{5}\,\sqrt{5}}{\sqrt{18}\,\sqrt{18}} = \frac{1}{3.6} \tag{7.37}$$

which implies $k_1 = 3.6$. This step can also be carried out by measurement. We draw vectors from all the poles and zeros to $s = 0$ and then measure their magnitudes. Certainly, excluding possible measurement errors, the result should be the same as (7.37). To compute k_2, we draw vectors from all the poles and zeros to $s = j1$ and measure their magnitudes to yield

$$\frac{1.4 \times 3.2}{1.4 \times 3.6 \times 5} = \frac{1}{k_2}$$

which implies

$$k_2 = 5.6$$

Thus we conclude that the overall system is stable in the range

$$3.6 = k_1 < k < k_2 = 5.6$$

This result is the same as the one obtained by using the Routh test.

Exercise 7.4.6

Consider the $G_2(s)$ in (7.24) with its root loci plotted in Figure 7.15. Find the range of positive k in which the system is stable.

[**Answer:** $0 \le k < 38$.]

7.5   DESIGN USING THE ROOT-LOCUS METHOD

In this section we discuss the design using the root-locus method. We use the example in (7.1) to develop a design procedure. The procedure, however, is applicable to the general case. It consists of the following steps:

Step 1: Choose a configuration and a compensator with one open parameter k such as the one in Figure 7.1.

Step 2: Compute the overall transfer function and then find the range of k for the system to be stable and to meet steady-state specifications. If no such k exists, go back to Step 1.

Step 3: Plot root loci that yield the poles of the overall system as a function of the parameter.

Step 4: Find the desired pole region from the specifications on overshoot and settling time as shown in Figure 7.4.

Step 5: Find the range of k in which the root loci lie inside the desired pole region. If no such k exists, go to Step 1 and choose a more complicated compensator or a different configuration.

Step 6: Find the range of k that meets 2 and 5. If no such k exists, go to Step 1.

Step 7: From the range of k in Step 6, find a k to meet the remaining specifications, such as the rise time or the constraint on the actuating signal. This step may require computer simulation of the system.

We remark that in Step 2, the check of stability may be skipped because the stability of the system is automatically met in Step 5 when all poles lie inside the desired pole region. Therefore, in Step 2, we may simply find the range of k to meet the specifications on steady-state performance.

Example 7.5.1

We use an example to illustrate the design procedure. Consider a plant with transfer function

$$G(s) = \frac{s + 4}{(s + 2)(s - 1)} \tag{7.38}$$

This plant has two poles and one zero. Design an overall system to meet the following specifications:

1. Position error $\leq 10\%$
2. Overshoot $\leq 5\%$
3. Settling time ≤ 4.5 seconds
4. Rise time as small as possible.

Step 1: We try the unity-feedback configuration shown in Figure 7.20.

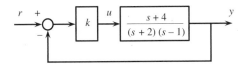

Figure 7.20 Unity-feedback system.

Step 2: The overall transfer function is

$$G_o(s) = \frac{k \cdot \dfrac{s + 4}{(s + 2)(s - 1)}}{1 + k \cdot \dfrac{s + 4}{(s + 2)(s - 1)}} = \frac{k(s + 4)}{s^2 + (k + 1)s + 4k - 2} \qquad (7.39)$$

The conditions for $G_o(s)$ to be stable are $4k - 2 > 0$ and $k + 1 > 0$, which imply

$$k > \frac{2}{4} = 0.5$$

Thus the system is stable for $k > 0.5$. Next we find the range of k to have position error less than 10%. The specification requires, using (6.3),

$$e_p(t) = \left| \frac{4k - 2 - 4k}{4k - 2} \right| = \left| \frac{-2}{4k - 2} \right| = \frac{1}{2k - 1} \le 0.1 \qquad (7.40)$$

where we have used the fact that $k > 0.5$, otherwise the absolute value sign cannot be removed. The inequality in (7.40) implies

$$10 \le 2k - 1$$

or

$$k \ge \frac{11}{2} = 5.5 \qquad (7.41)$$

Thus, if $k \ge 5.5$, then the system in Figure 7.20 is stable and meets specification (1). The larger k is, the smaller the position error.

Steps 3 and 4: Using the procedure in Section 7.4.1, we plot the root loci of $1 + kG(s) = 0$ in Figure 7.21. For convenience of discussion, the poles corresponding to $k = 0.5, 0.7, 1, 5, \ldots$ are also indicated. They are actually obtained by using MATLAB. Note that for each k, there are two poles, but only one is indicated. The specification on overshoot requires all poles to lie inside the sector bounded by 45°. The specification on settling time requires all poles to lie on the left-hand side of the vertical line passing through $-4.5/t_s = -1$. The sector and the vertical line are also plotted in Figure 7.21.

Step 5: Now we shall find the ranges of k to meet the specifications on overshoot and settling time. From Figure 7.21, we see that if $0.5 < k < 1$, the two

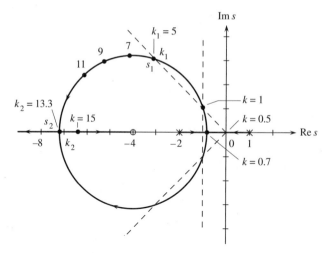

Figure 7.21 Root loci of (7.38).

poles lie inside the sector bounded by 45°. If $1 < k < 5$, the two poles move outside the sector. They again move inside the sector for $k > 5$. Thus if $0.5 < k < 1$ or $5 < k$, the overall system meets the specification on overshoot. If $k < 1$, although one pole of $G_o(s)$ is on the left-hand side of the vertical line passing through -1, one pole is on the right-hand side. If $k > 1$, then both poles are on the left-hand side. Thus if $k > 1$, the system meets the specification on settling time.

Step 6: The preceding discussion is summarized in the following:

$k > 0.5$: stable

$k > 5.5$: meets specification (1). The larger k is, the smaller the position error.

$k > 5$ or $1 > k > 0.5$: meets specification (2)

$k > 1$: meets specification (3).

Clearly in order to meet (1), (2), and (3), k must be larger than 5.5.

Step 7: The last step of the design is to find a k in $k > 5.5$ such that the system has the smallest rise time. To achieve this, we choose a k such that the closest pole is farthest away from the origin. From the plot we see that as k increases, the two complex-conjugate poles of $G_o(s)$ move away from the origin. At $k = 13.3$, the two complex poles become repeated poles at $s = -7.2$. At $k = 15$, the poles are -10.4 and -6.4; one pole moves away from the origin, but the other moves closer to the origin. Thus, at $k = 13.3$, the poles of $G_o(s)$ are farthest away from the origin and the system has the smallest rise time. This completes the design.

It is important to stress once again that the desired pole region in Figure 7.4 is developed for quadratic transfer functions with a constant numerator. The $G_o(s)$ in (7.39) is not such a transfer function. Therefore, it is advisable to simulate the resulting system. Figure 7.22 shows the unit-step responses of the system in (7.39) for $k = 13.3$ (dashed line) and $k = 5.5$ (solid line). The system with $k = 13.3$ is better than the one with $k = 5.5$. Its position error, settling time, and overshoot are roughly 4%, 1.5 seconds, and 10%. The system meets the specifications on position error and settling time, but not on overshoot. This system will be acceptable if the requirement on overshoot can be relaxed. Otherwise, we must redesign the system.

The root loci in Figure 7.21 are obtained by using a personal computer; therefore, the gain k is also available on the plot. If the root loci are obtained by hand, then the value of k is not available on the plot. In this case, we must use the magnitude equation

$$\left| \frac{(s + 4)}{(s + 2)(s - 1)} \right| = \left| \frac{-1}{k} \right|$$

to compute k. For example, to find the value of k_1 shown in Figure 7.21, we draw vectors from all poles and zeros to s_1 and then *measure* their magnitudes to yield

$$\left| \frac{(s + 4)}{(s + 2)(s - 1)} \right|_{s=s_1} = \frac{3.2}{3.2 \times 5} = \left| \frac{-1}{k_1} \right|$$

which implies $k_1 = 5$. To compute k_2, we draw vectors from all poles and zeros to s_2 and measure their magnitudes to yield

$$\left| \frac{(s + 4)}{(s + 2)(s - 1)} \right|_{s=s_2} = \frac{3.2}{5.2 \times 8.2} = \left| \frac{-1}{k_2} \right|$$

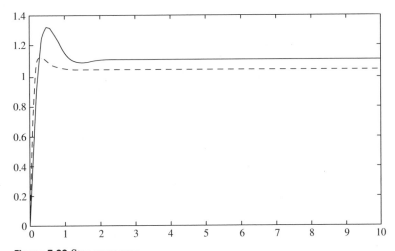

Figure 7.22 Step responses.

which implies $k_2 = 13.3$. Thus, the gain can be obtained from the magnitude equation.

7.5.1 Discussion

1. Although we studied only the unity-feedback configuration in the preceding section, the root-locus method is actually applicable to any configuration as long as its overall transfer function can be expressed as

$$G_o(s) = \frac{N_o(s, k)}{p(s) + kq(s)} \tag{7.42}$$

where $p(s)$ and $q(s)$ are polynomials, independent of k, and k is a real parameter to be adjusted. Since the root-locus method is concerned only with the poles of $G_o(s)$, we plot the roots of

$$p(s) + kq(s) \tag{7.43a}$$

or the solutions of

$$\frac{q(s)}{p(s)} = -\frac{1}{k} \tag{7.43b}$$

as a function of real k. We see that (7.43a) and (7.43b) are the same as (7.12) and (7.13), thus all discussion in the preceding sections is directly applicable to (7.42). For example, consider the system shown in Figure 7.23. Its overall transfer function is

$$G_o(s) = \frac{k_1 \cdot \dfrac{s + k_2}{s + 2} \cdot \dfrac{10}{s(s^2 + 2s + 2)}}{1 + k_1 \cdot \dfrac{s + k_2}{s + 2} \cdot \dfrac{10}{s(s^2 + 2s + 2)}}$$

$$= \frac{10k_1(s + k_2)}{s(s + 2)(s^2 + 2s + 2) + 10k_1(s + k_2)}$$

It has two parameters, k_1 and k_2. If we use a digital computer to plot the root loci, it makes no difference whether the equation has one, two, or more parameters. Once the root loci are obtained, the design procedure is identical to the one discussed in the preceding sections. If the root loci are to be plotted by hand, we are able to handle only one parameter at a time. Arbitrarily, we choose

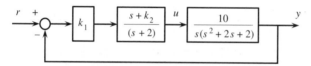

Figure 7.23 System with two parameters.

$k_1 = 5$. Then $G_o(s)$ becomes

$$G_o(s) = \frac{50(s + k_2)}{[s(s + 2)(s^2 + 2s + 2) + 50s] + k_2 \cdot 50}$$

This is in the form of (7.42). Thus the root-locus method is applicable. In this case, the root loci are a function of k_2.

2. The root-locus method considers only the poles. The zeros are not considered, as can be seen from (7.42). Thus the method is essentially a pole-placement problem. The poles, however, cannot be arbitrarily assigned; they can be assigned only along the root loci.

3. The desired pole region in Figure 7.4 is developed for quadratic transfer functions with a constant numerator. When it is used to design other types of transfer functions, it is advisable to simulate resulting systems to check whether they really meet the given specifications.

7.6   PROPORTIONAL-DERIVATIVE (PD) CONTROLLER

In this section we give an example that uses a proportional-derivative (PD) controller. Consider a plant with transfer function

$$G(s) = \frac{2}{s(s + 1)(s + 5)} \tag{7.44}$$

Design an overall system to meet the specifications:

1. Velocity error as small as possible
2. Overshoot $\leq 5\%$
3. Settling time < 5 seconds
4. Rise time as small as possible.

As a first try, we choose the unity-feedback system shown in Figure 7.24. The overall transfer function is

$$G_o(s) = \frac{2k}{s(s + 1)(s + 5) + 2k} = \frac{2k}{s^3 + 6s^2 + 5s + 2k}$$

A necessary condition for $G_o(s)$ to be stable is $k > 0$. Thus we plot the root loci of

$$\frac{2}{s(s + 1)(s + 5)} = -\frac{1}{k}$$

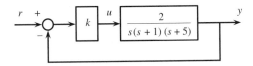

Figure 7.24 Unity-feedback system.

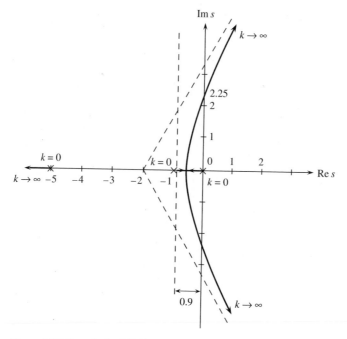

Figure 7.25 Root loci of (7.44).

for $k > 0$. The root loci are shown in Figure 7.25. There are three asymptotes with centroid at

$$\frac{0 - 1 - 5}{3} = -2$$

and with angles $\pm 60°$ and $180°$. The breakaway point can also be computed analytically by solving

$$D(s)N'(s) - D'(s)N(s) = -(3s^2 + 12s + 5) = -3(s + 0.47)(s + 3.5)$$

Its solutions are -0.47 and -3.5. Clearly -0.47 is a breakaway point, but -3.5 is not.[7]

In order for the resulting system to have settling time less than 5 seconds, all the poles of $G_o(s)$ must lie on the left-hand side of the vertical line passing through the point $-4.5/t_s = -0.9$. From the root loci in Figure 7.25 we see that this is not possible for any $k > 0$. Therefore, the configuration in Figure 7.24 cannot meet the specifications.

As a next try, we introduce an additional tachometer feedback as shown in Figure 7.26. Now the compensator consists of a proportional compensator with gain

---

[7]It is a breakaway point of the root loci for $k < 0$.

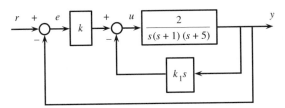

Figure 7.26 PD controller.

k and a derivative compensator with transfer function $k_1 s$, thus it is called a *PD compensator* or *controller*.[8] It has two parameters, k and k_1. Because we can handle only one parameter at a time, we shall choose a value for k. First we choose $k = 1$ and carry out the design. It is found that the design is not possible for any k_1. Next we choose $k = 5$. Then the overall transfer function of Figure 7.26 becomes

$$G_o(s) = \frac{\dfrac{2k}{s(s+1)(s+5)}}{1 + \dfrac{2k}{s(s+1)(s+5)} + \dfrac{2k_1 s}{s(s+1)(s+5)}}$$

$$= \frac{2k}{s(s+1)(s+5) + 2k + 2k_1 s} \tag{7.45}$$

$$= \frac{10}{s^3 + 6s^2 + 5s + 2k_1 s + 10}$$

The root loci of $(s^3 + 6s^2 + 5s + 10) + k_1(2s)$ or of

$$-\frac{1}{k_1} = \frac{2s}{s^3 + 6s^2 + 5s + 10} \tag{7.46}$$

$$= \frac{2s}{(s+5.42)(s+0.29+j1.33)(s+0.29-j1.33)}$$

are plotted in Figure 7.27. There are three trajectories. One moves from pole -5.4 to the zero at $s = 0$ along the negative real axis; the other two are complex conjugates and approach the two asymptotes with centroid at

$$a = \frac{(-5.42 - 0.29 + j1.33 - 0.29 - j1.33) - (0)}{3 - 1} = -3$$

and angles $\pm 90°$. Some of k_1 are also indicated on the plot.

---

[8]A different arrangement of PD controllers is $U(s) = (k + k_1 s)E(s)$. See Chapter 11. The arrangement in Figure 7.26, that is, $U(s) = kE(s) + k_1 sY(s)$, is preferable, because it differentiates $y(t)$ rather than $e(t)$, which often contains discontinuity at $t = 0$. Therefore, the chance for the actuating signal in Figure 7.26 to become saturated is less.

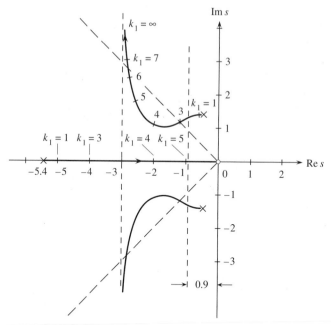

Figure 7.27 Root loci of (7.46).

Because $G_o(0) = 1$, the system in (7.45) has zero position error and its velocity error, using (6.7), is

$$e_v(t) = \left| \frac{5 + 2k_1 - 0}{10} \right| = \frac{2k_1 + 5}{10}$$

Thus the smaller k_1, the smaller the error. To meet the specification on overshoot, all poles must lie in the sector bounded by 45°, as shown in Figure 7.27. The real pole lies inside the sector for all $k_1 > 0$. The complex poles move into the section at about $k_1 = 3$ and move out at about $k_1 = 6.5$. Therefore, if $3 < k_1 < 6.5$, then all three closed-loop poles lie inside the sector and the system meets the specification on overshoot. To meet the specification on settling time, all poles must lie on the left-hand side of the vertical line passing through $-4.5/5 = -0.9$. The real pole moves into the right-hand side at about $k_1 = 5$; the complex poles move into the left-hand side at about $k_1 = 2.5$. Therefore if $2.5 < k_1 < 5$, then all poles lie on the left-hand side of the vertical line and the system meets the specification on settling time. Combining the preceding two conditions, we conclude that if $3 < k_1 < 5$, then the system meets the specifications on overshoot and settling time.

The condition for the system to have the smallest rise time is that the closest pole be as far away as possible from the origin. Note that for each k_1, $G_o(s)$ in (7.45) has one real pole and one pair of complex-conjugate poles. We list in the following the poles and their shortest distance from the origin for $k_1 = 3, 4,$ and 5:

k_1	Poles	Shortest Distance
3	$-3.8, -1.1 \pm j1.2$	1.63
4	$-2, -2 \pm j1$	2
5	$-1, -2.5 \pm j1.94$	1

Because the system corresponding to $k_1 = 4$ has the largest shortest distance, it has the smallest rise time among $k_1 = 3, 4,$ and 5. Recall that the velocity error is smaller if k_1 is smaller. Therefore, if the requirement on velocity error is more important, then we choose $k_1 = 3$. If the requirement on rise time is more important, than we choose $k_1 = 4$. This completes the design.

The overall transfer function in (7.45) is not quadratic; therefore, the preceding design may not meet the design specifications. Figure 7.28 shows the unit-step responses of (7.45) for $k_1 = 4$ (solid line) and 3 (dashed line). The overshoot, settling, and rise times of the system with $k_1 = 4$ are, respectively, 0, 3.1 and 2.2 seconds. The system meets all design specifications. The overshoot, settling, and rise times of the system with $k_1 = 3$ are, respectively, 4.8%, 6.1, and 1.9 seconds. The system does not meet the specification on settling time but meets the specification on overshoot. Note that the system with $k_1 = 3$ has a smaller rise time than the system with $k_1 = 4$, although the distance of its closest poles from the origin for $k_1 = 3$ is

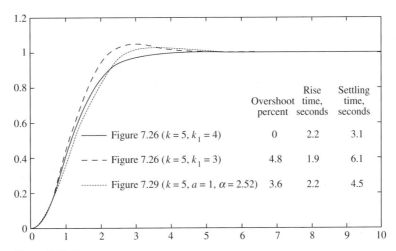

Figure 7.28 Step responses.

smaller than that for $k_1 = 4$. Therefore the rule that the farther away the closest pole from the origin, the smaller the rise time, is not applicable for this system. In conclusion, the system in Figure 7.26 with $k = 5$ and $k_1 = 4$ meets all design requirements and the design is completed.

7.7   PHASE-LEAD AND PHASE-LAG NETWORKS

Consider again the design problem studied in Section 7.6. As shown there, the design cannot be achieved by using the configuration in Figure 7.24. However, if we introduce an additional tachometer feedback or, equivalently, if we use a PD controller, then the design is possible. The use of tachometer feedback, however, is not the only way to achieve the design. In this section, we discuss a different design by using a compensating network as shown in Figure 7.29. The transfer function of the compensating network is chosen as

$$C(s) := \frac{s + a}{s + \alpha a} \tag{7.47}$$

It is called a *phase-lead network*, if $\alpha > 1$; a *phase-lag network*, if $\alpha < 1$. The reason for calling it phase-lead or phase-lag will be given in the next chapter. See also Problem 7.9.

The transfer function of the system in Figure 7.29 is

$$
\begin{aligned}
G_o(s) &= \frac{k \dfrac{s + a}{s + \alpha a} \dfrac{2}{s(s + 1)(s + 5)}}{1 + k \dfrac{s + a}{s + \alpha a} \dfrac{2}{s(s + 1)(s + 5)}} \\[2mm]
&= \frac{2k(s + a)}{s(s + 1)(s + 5)(s + \alpha a) + 2k(s + a)}
\end{aligned}
\tag{7.48}
$$

Its denominator has degree 4 and the design using (7.48) will be comparatively complex. To simplify design, we shall introduce a stable pole-zero cancellation. Because both -1 and -5 lie inside the desired pole region, either one can be canceled. Arbitrarily, we choose to cancel the pole at -1. Thus we choose $a = 1$ in (7.47) and the overall transfer function in (7.48) reduces to

$$G_o(s) = \frac{2k}{s(s + 5)(s + \alpha) + 2k} \tag{7.49}$$

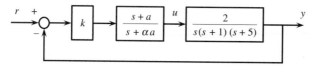

Figure 7.29 Unity-feedback system with compensating network.

If k is chosen as $k = 5$, then (7.49) becomes

$$G_o(s) = \frac{10}{s^3 + (5 + \alpha)s^2 + 5\alpha s + 10}$$

$$= \frac{10}{(s^3 + 5s^2 + 10) + \alpha s(s + 5)}$$

(7.50)

The root loci of $(s^3 + 5s^2 + 10) + \alpha s(s + 5)$ or, equivalently, of

$$\frac{-1}{\alpha} = \frac{s(s + 5)}{s^3 + 5s^2 + 10}$$

$$= \frac{s(s + 5)}{(s + 5.35)(s - 0.18 + j1.36)(s - 0.18 - j1.36)}$$

(7.51)

as a function of α are plotted in Figure 7.30.

Now the specification on overshoot requires that the roots be inside the sector bounded by 45° as shown. The settling time requires that the roots be on the left-hand side of the vertical line passing through point $(-0.9, 0)$. We see from Figure 7.30 that if $\alpha_1 \leq \alpha \leq \alpha_3$, then these two specifications are satisfied. In order to have the rise time be as small as possible, the pole closest to the origin should be as far away as possible from the origin. Again from Figure 7.30 we rule out the range from α_2 to α_3. The roots corresponding to any α in (α_1, α_2) are roughly the same distance from the origin, thus we may pick an α from this range. The last specification is that the velocity error be as small as possible. The velocity error of $G_o(s)$ in (7.50) is,

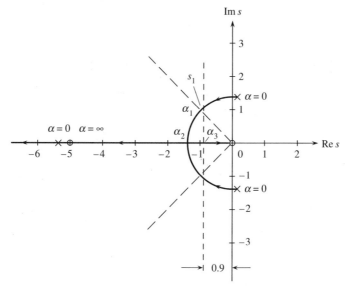

Figure 7.30 Root loci of (7.51).

from (6.7),

$$e_v(t) = \left| \frac{5\alpha - 0}{10} \right| = \left| \frac{\alpha}{2} \right| \times 100\%$$

Hence in order to have the smallest possible velocity error, we choose α to be α_1. The parameter α_1 can be obtained from (7.51) by measurement as

$$\left| \frac{-1}{\alpha_1} \right| = \left| \frac{s(s + 5)}{(s + 5.35)(s - 0.18 + j1.36)(s - 0.18 - j1.36)} \right|_{s=s_1}$$

$$= \frac{1.4 \times 4.1}{4.45 \times 1.2 \times 2.7} = \frac{1}{2.52}$$

which implies $\alpha_1 = 2.52$. Hence by choosing $k = 5$, $a = 1$, and $\alpha = 2.52$, the system in Figure 7.29 may meet all the design specifications, and the design is completed. The total compensator is

$$\frac{k(s + a)}{s + \alpha a} = \frac{5(s + 1)}{s + 2.52} \tag{7.52}$$

It is a phase-lead network.

The unit-step response of the system in Figure 7.29 with (7.52) as its compensator is plotted in Figure 7.28 with the dotted line. Its overshoot is about 3.6%; its settling time is 4.5 seconds. It also responds very fast. Thus the design is satisfactory.

7.8   CONCLUDING REMARKS

We give a number of remarks to conclude this chapter.

1. The root-locus method is basically a graphical method of plotting the roots of a polynomial as a function of a real parameter. In the plotting, we use only the phase condition in (7.19). From the properties of roots of polynomials, we can often obtain a rough sketch of root loci without any measurement or computation. Plotting exact root loci is best done on a personal computer.

2. The root-locus design method tries to choose a parameter such that the poles are in a nice location. To guide the choice, we develop a desired pole region from quadratic transfer functions with a constant numerator. If an overall transfer function is not quadratic, there is no guarantee that the resulting system will meet the given specifications.

3. The method is a trial-and-error method. Therefore, it may take us several trials before we succeed in designing an acceptable system.

4. In the root-locus design method, the constraint on actuating signals is not considered. The constraint can be checked only after the completion of the design. If the constraint is not met, then we may have to redesign the system.

5. The design method considers only the poles of overall transfer functions. The zeros are not considered. Therefore, the method is a special case of the pole-

placement problem in Chapter 10. The pole-placement problem in Chapter 10 can assign poles in any location; the root-locus method can assign poles only along the root loci.

PROBLEMS

7.1. Sketch the root loci for the unity-feedback system shown in Figure 7.1 with

a. $G(s) = \dfrac{s + 4}{s^2(s + 1)}$

b. $G(s) = \dfrac{(s + 4)(s + 6)}{(s - 1)(s + 1)}$

c. $G(s) = \dfrac{s^2 + 2s + 2}{(s + 1)^2(s^2 + 4s + 6)}$

7.2. Sketch the root loci of the polynomials

a. $s^3 + 2s^2 + 3s + ks + 2k$

b. $s^3(1 + 0.001s)(1 + 0.002s) + k(1 + 0.1s)(1 + 0.25s)$

7.3. Use the root-locus method to show that

a. The polynomial $s^3 + s^2 + s + 2$ has one real root in $(-2, -1)$ and a pair of complex-conjugate roots with real part in $(0, 1)$. [*Hint:* Write the polynomial as $s^2(s + 1) + k(s + 2)$ with $k = 1$.]

b. The polynomial

$$s^5 + 2s^4 - 15s^3 + s^2 - 2s - 15$$

has three real roots and a pair of complex-conjugate roots. Also show that the three real roots lie in $(5, 3)$, $(0, -3)$, and $(-5, -\infty)$.

7.4. The root loci of the system shown in Figure P7.4(a) are given in Figure P7.4(b). Find the following directly from measurement on the graph.

a. The stability range of k.

b. The real pole that has the same value of k as the pair of pure imaginary poles.

c. The k that meets (i) overshoot $\leq 20\%$, (ii) settling time ≤ 10 seconds, and (iii) smallest possible position error.

7.5. Consider the feedback system shown in Figure P7.5. Sketch root loci, as a function of positive real k, for the following:

a. $G(s) = \dfrac{1}{s(s + 1)}$, $\quad H(s) = \dfrac{4(s + 2)}{s + 4}$

b. $G(s) = \dfrac{s^2 + 4s + 4}{s(s - 1)}$, $\quad H(s) = \dfrac{s + 5}{s^2 + 2s + 2}$

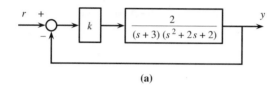

(a)

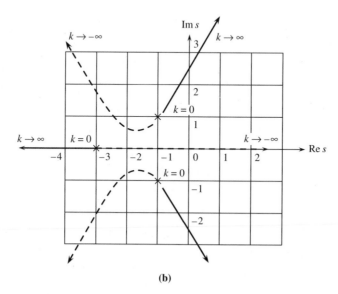

(b)

Figure P7.4

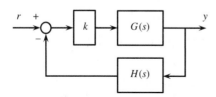

Figure P7.5

7.6. Consider the unity-feedback system shown in Figure 7.1. Let the plant transfer function be

$$G(s) = \frac{s + 1}{(s - 0.2 + j2)(s - 0.2 - j2)}$$

Find the ranges of k to meet the following

a. Position error $< 10\%$

b. a. and overshoot $< 15\%$

c. a., b., and settling time < 4.5 seconds

d. a., b., c., and the smallest possible rise time.

7.7. A machine tool can be automatically controlled by a punched tape, as shown in Figure P7.7(a). This type of control is called *numerical control*. By neglecting the quantization problem, the system can be modeled as shown in Figure P7.7(b). Find a gain k such that the system has zero position error and zero overshoot. Numerical control cannot have overshoot, otherwise it will overcut or the tool will break. No constraints are imposed on the settling and rise times.

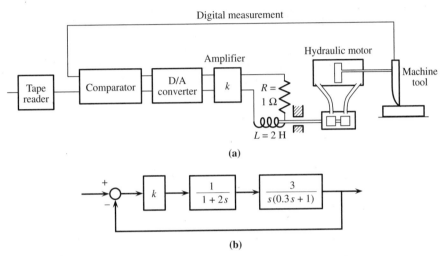

(a)

(b)

Figure P7.7

7.8. The depth below sea level of a submarine can be maintained by the control system shown in Figure P7.8. The transfer function from the stern plane angle θ to the actual depth y of the submarine can be modeled as

$$G(s) = \frac{10(s + 2)^2}{(s + 10)(s^2 + 0.1)}$$

The depth of the submarine is measured by a pressure transducer, which is assumed to have a transfer function of 1. Find the smallest k such that the

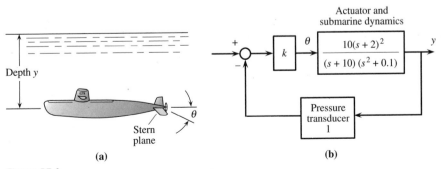

(a)   **(b)**

Figure P7.8

position error is less than 5%, the settling time is less than 10 seconds, and the overshoot is less than 2%.

7.9. **a.** Consider $C(s) = (s + 2)/(s + 1)$. Compute its phase at $s = j1$. Is it positive or negative?

   b. Consider $C(s) = (s + a)/(s + b)$. Show that the phase of $G(j\omega)$ for every $\omega > 0$ is positive for $0 < a < b$ and negative for $0 < b < a$. (Thus, the transfer function is called a *phase-lead* network if $b > a$ and a *phase-lag* network if $a > b$.)

7.10. Consider the unity-feedback system shown in Figure P7.10. Use the Routh test to find the range of real a for the system to be stable. Verify the result by using the root-locus method. Find the a such that the system has the smallest settling time and overshoot. Is it a phase-lead or phase-lag network?

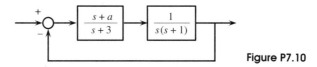

Figure P7.10

7.11. The speed of a motor shaft can be controlled accurately using a phase-locked loop [39]. The schematic diagram of such a system and its block diagram are shown in Figure P7.11. The desired speed is transformed into a pulse sequence with a fixed frequency. The encoder at the motor shaft generates a pulse stream whose frequency is proportional to the motor speed. The phase comparator generates a voltage proportional to the difference in phase and frequency. Sketch the root loci of the system. Does there exist a k such that the settling time of the system is smaller than 1 second and the overshoot is smaller than 10 percent?

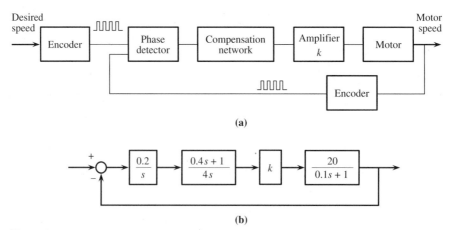

Figure P7.11

7.12. The transfer function from the thrust deflection angle u to the pitch angle θ of a guided missile is found to be

$$G(s) = \frac{4(s + 0.05)}{s(s + 2)(s - 1.2)}$$

The configuration of the compensator is chosen as shown in Figure P7.12. The transfer function of the actuator is $G_1(s) = 1/(s + 6.1)$. If $k_1 = 2k_2$, find a k_1, if it exists, such that the position error is less than 10%, the overshoot is less than 15%, and the settling time is less than 10 seconds.

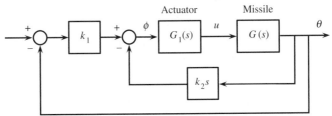

Figure P7.12

7.13. Consider the control system shown in Figure P7.13. Such a system may be used to drive potentiometers, dials, and other devices. Find k_1 and k_2 such that the position error is zero, the settling time is less than 1 second, and the overshoot is less than 5%. Can you achieve the design without plotting root loci?

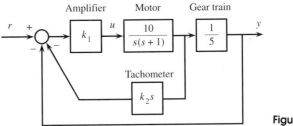

Figure P7.13

7.14. One way to stabilize an ocean liner, for passengers' comfort, is to use a pair of fins as shown in Figure P7.14(a). The fins are controlled by an actuator, which is itself a feedback system consisting of a hydraulic motor. The transfer function of the actuator, compared with the dynamics of the liner, may be simplified as a constant k. The equation governing the roll motion of the liner is

$$J\ddot{\theta}(t) + \eta\dot{\theta}(t) + \alpha\theta(t) = ku(t)$$

where θ is the roll angle, and $ku(t)$ is the roll moment generated by the fins. The block diagram of the linear and actuator is shown in Figure P7.14(b). It is assumed that $\alpha/J = 0.3$, $\eta/2\sqrt{\alpha J} = 0.1$, and $k/\alpha = 0.05$. A possible configuration is shown in Figure P7.14(c). If $k_1 = 5$, find a k_2, if it exists, such that (1) position error $\leq 15\%$, (2) overshoot $\leq 5\%$, and (3) settling time ≤ 30 seconds. If no such k_2 exists, choose a different k_1 and repeat the design.

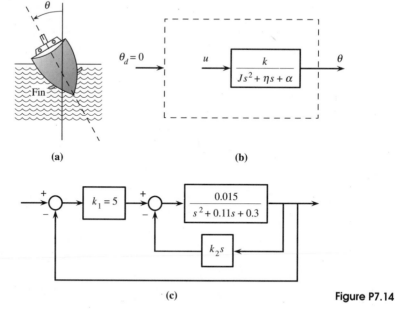

(a)   **(b)**

(c)   **Figure P7.14**

7.15. A highly simplified model for controlling the yaw of an aircraft is shown in Figure P7.15(a), where θ is the yaw error and ϕ is the rudder deflection. The rudder is controlled by an actuator whose transfer function can be approximated as a constant k. Let J be the moment of inertia of the aircraft with respect to the yaw axis. For simplicity, it is assumed that the restoring torque is proportional to the rudder deflection $\phi(t)$; that is,

$$J\ddot{\theta}(t) = -k\phi(t)$$

The configurations of compensators are chosen as shown in Figure P7.15(b), (c), and (d), where $G(s) = -k/Js^2 = -2/s^2$. We are required to design an overall system such that (1) velocity error $\leq 10\%$, (2) overshoot $\leq 10\%$, (3) settling time ≤ 5 seconds, and (4) rise time is as small as possible. Is it possible to achieve the design using configuration (b)? How about (c) and (d)? In using (c) and (d), do you have to plot the root loci? In this problem, we assume that the saturation of the actuating signal will not occur.

7.16. Consider the plant discussed in Section 6.2 and shown in Figure 6.1. Its transfer function is computed as

$$G(s) = \frac{300}{s(s^3 + 184s^2 + 760.5s + 162)}$$

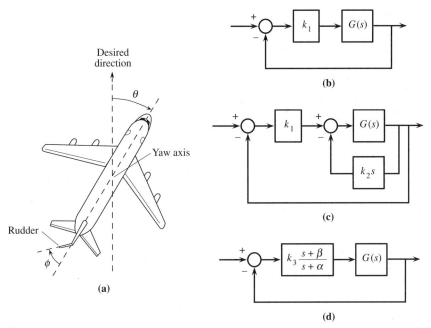

Figure P7.15

Design an overall system such that (1) position error $\leq 10\%$, (2) settling time ≤ 5 seconds, and (3) overshoot is as small as possible.

7.17. Consider the system shown in Figure 7.26. Let $k = 10$. Use the root-locus method to find a k_1 so that the system meets the specifications listed in Section 7.6.

7.18. Consider the system shown in Figure 7.26. Let $k_1 = 4$. Use the root-locus method to find a k so that the system meets the specifications listed in Section 7.6.

7.19. In Figure 7.29, if we choose $a = 5$, then the system involves a stable pole-zero cancellation at $s = -5$. Is it possible to find k and α in Figure 7.29 so that the system meets the specifications listed in Section 7.6? Compare your design with the one in Section 7.7 which has a stable pole-zero cancellation at $s = -1$.

8 *Frequency-Domain Techniques*

8.1 INTRODUCTION

In this chapter we introduce a design method that, like the root-locus method, takes the outward approach. In this approach, we first choose a configuration, then search a compensator and hope that the resulting overall system will meet design specifications. The method is mainly limited to the unity-feedback configuration shown in Figure 8.1, however. Because of this, it is possible to translate the design specifications for the overall system into specifications for the plant transfer function $G(s)$. If $G(s)$ does not meet the specifications, we then search for a compensator $C(s)$ so that $C(s)G(s)$ will meet the specifications and hope that the resulting unity-feedback configuration in Figure 8.1 will perform satisfactorily. Thus, in this method we work directly on $G(s)$ and $C(s)$. However, the objective is still the overall system $G_o(s) = G(s)C(s)/(1 + G(s)C(s))$. This feature is not shared by any other design method.

The method has another important feature; it uses only the information of $G(s)$ along the positive imaginary axis, that is, $G(j\omega)$ for all $\omega \geq 0$. Thus the method is called the *frequency-domain method*. As discussed in Chapter 4, $G(j\omega)$ can be obtained by direct measurement. Once $G(j\omega)$ is measured, we may proceed directly to the design without computing the transfer function $G(s)$. On the other hand, if we are given a transfer function $G(s)$, we must first compute $G(j\omega)$ before carrying out the design. Thus we discuss first the plotting of $G(j\omega)$.

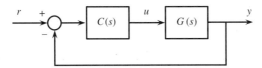

Figure 8.1 Unity-feedback system.

8.2 FREQUENCY-DOMAIN PLOTS

We use a simple example to illustrate the basic concept. Consider

$$G(s) = \frac{1}{s + 0.5} \tag{8.1}$$

or

$$G(j\omega) = \frac{1}{j\omega + 0.5}$$

We discuss the plot of $G(j\omega)$ as a function of real $\omega \geq 0$. Although ω is real, $G(j\omega)$ is, in general, complex. If $\omega = 0$, then $G(0) = 2$. If $\omega = 0.2$, then

$$G(j0.2) = \frac{1}{0.5 + j0.2} = \frac{1}{\sqrt{0.29}e^{\tan^{-1}(0.2/0.5)}} = \frac{1}{0.53e^{j22°}} = 1.9e^{-j22°}$$

Similarly, we can compute

$$G(j0.5) = 1.4e^{-j45°} \qquad G(j2) = 0.5e^{-j76°} \qquad G(j10) = 0.1e^{-j87°}$$

Using these data we can plot a number of $G(j\omega)$ by using different sets of coordinates. The plot in Figure 8.2(a) is called the *polar plot*. Its horizontal and vertical axes are, respectively, Re $G(j\omega)$ and Im $G(j\omega)$, where Re and Im stand for the real part and imaginary part. Thus a point in the plane is a vector with magnitude or gain $|G(j\omega)|$ and phase $\sphericalangle G(j\omega)$. For example, if $\omega = 0$, it is the vector or point A shown in Figure 8.2(a); it has magnitude 2 and phase 0°. Point B, corresponding to $\omega = 0.5$, has magnitude 1.4 and phase $-45°$. The plot of $G(s)$ in (8.1) happens to be a semicircle as shown.

The plot in Figure 8.2(b) is called the *log magnitude-phase* plot. It is a plot of gain versus phase on rectangular coordinates as shown. The gain on the vertical coordinate is expressed in decibels (dB), defined as

$$\text{dB} = 20 \log |G(j\omega)|$$

For example, we have

$$20 \log 2 = 6 \text{ dB}$$

$$20 \log 1.9 = 5.6 \text{ dB}$$

$$20 \log 1.4 = 2.9 \text{ dB}$$

$$20 \log 0.5 = -6 \text{ dB}$$

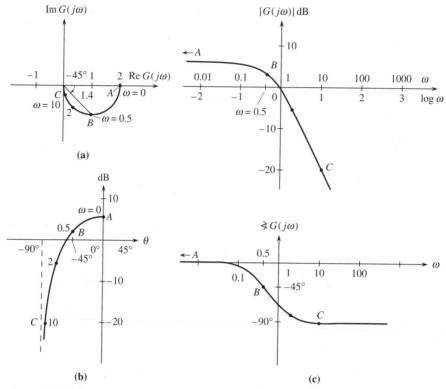

Figure 8.2 Frequency of plots of $G(s)$.

and

$$20 \log 0.1 = -20 \text{ dB}$$

Note that the decibel gain is positive if $|G(j\omega)| > 1$, and negative if $|Gj\omega| < 1$. The phase, on the horizontal coordinate, is in linear angles. The log magnitude-phase plot of $G(s)$ in (8.1) is shown in Figure 8.2(b). For example, point A, corresponding to $\omega = 0$, has magnitude 6 dB and phase 0°; point B, corresponding to $\omega = 0.5$, has magnitude 2.9 dB and phase $-45°$ and so forth. The plot is quite different from the one in Figure 8.2(a).

The plot in Figure 8.2(c) is called the *Bode plot*. It actually consists of two plots: gain versus frequency, and phase versus frequency. The gain is expressed in decibels, the phase in degrees. The frequency on the horizontal coordinate is expressed in logarithmic units as shown. Thus $\omega = 1$ corresponds to 0; $\omega = 10$ corresponds to 1; $\omega = 100$ corresponds to 2, and so forth. Note that $\omega = 0$ appears at $-\infty$. Thus point A should appear at $-\infty$ and has 6 dB and zero degree. The complete Bode plot of $G(s)$ in (8.1) is plotted in Figure 8.2(c). We remark that ω appears as a variable on the plots in Figure 8.2(a) and (b) whereas it appears as coordinates in

Figure 8.2(c). Although the three plots in Figure 8.2 look entirely different, they are plots of the same $G(j\omega)$. It is clear that if any plot is available, the other two plots can be obtained by change of coordinates.

   With digital computers, the computation and plotting of $G(j\omega)$ become very simple. Even so, it is useful to be able to estimate a rough sketch of $G(j\omega)$. This is illustrated by an example.

Example 8.2.1

   Plot the polar plot of

$$G(s) = \frac{2}{s(s + 1)(s + 2)} \tag{8.2}$$

Before computing $G(j\omega)$ for any ω, we shall estimate first the values of $G(j\omega)$ as $\omega \to 0$ and $\omega \to \infty$. Clearly we have

$$s \to 0 \text{ or } \omega \to 0: \quad G(s) \approx \frac{2}{2s} \text{ or } G(j\omega) \approx \frac{1}{j\omega} \Rightarrow |G(j\omega)| \to \infty, \sphericalangle\, G(j\omega) = -90°$$

and

$$s \to \infty \text{ or } \omega \to \infty: \quad G(s) \approx \frac{2}{s^3} \Rightarrow |G(j\omega)| \to 0, \sphericalangle\, G(j\omega) = -270°$$

They imply that for ω very small, the phase is $-90°$ and the amplitude is very large. Thus the plot will start somewhere in the region denoted by A shown in Figure 8.3(a). As ω increases to infinity, the plot will approach zero or the origin with phase

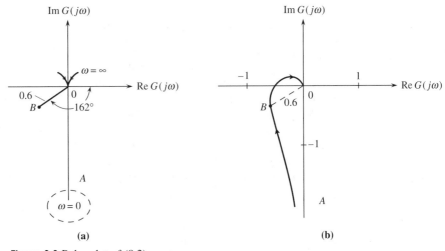

| (a) | (b) |

Figure 8.3 Polar plot of (8.2).

$-270°$ or $+90°$ as shown in Figure 8.3(a). Recall that a phase is positive if measured clockwise, negative if measured counterclockwise. Now we compute $G(j\omega)$ at $\omega = 1$:

$$G(j1) = \frac{2}{j1(j1 + 1)(j1 + 2)} = \frac{2}{e^{j90°} \cdot 1.4e^{j45°} \cdot 2.2e^{j27°}} = 0.6e^{-j162°}$$

It is plotted in Figure 8.3(a) as point B. In other words, as ω increases from 0 to ∞, the plot will start from region A, pass through point B, and approach the origin along the positive imaginary axis. Thus the plot may assume the form shown in Figure 8.3(b). Thus a *rough* polar plot of $G(s)$ in (8.2) can be easily obtained as shown. Clearly the more points of $G(j\omega)$ we compute, the more accurate the plot.

We stress once again that in plotting $G(j\omega)$, it is useful to estimate first the values of $G(s)$ at $\omega = 0$ and $\omega = \infty$. These values can be used to check the correctness of a digital computer computation.

Exercise 8.2.1

Plot the polar, log magnitude-phase, and Bode plots of $G(s) = 1/(s + 2)$.

To conclude this section, we discuss the plot of $G(s) = 1/(s + 0.5)$ using MATLAB. We first list the commands for version 3.1 of MATLAB:

```
n=[1];d=[1 0.5];
w=logspace(-1,2) or w=logspace(-1,2,200);
[re,im]=nyquist(n,d,w);
plot(re,im),title('Polar plot')
[mag,pha]=bode(n,d,w);
db=20*log10(mag);
plot(pha,db),title('Log magnitude-phase plot')
semilogx(w,db),title('Bode gain plot')
semilogx(w,pha),title('Bode phase plot')
```

The numerator and denominator of $G(s)$ are represented by the row vectors n and d, with coefficients arranged in descending power of s, separated by spaces or commas. Command logspace($-1,2,200$) generates 200 equally spaced frequencies in logarithmic scale between $10^{-1} = 0.1$ and $10^2 = 100$ radians per second. If 200 is not typed, the default is 50. Thus logspace($-1,2$) generates 50 equally spaced frequencies between 0.1 and 100. Command nyquist(n,d,w)[1] computes the real part and imaginary part of $G(j\omega)$ at w. Thus, plot(re,im) generates a polar plot. Command bode(n,d,w) computes the magnitude and phase of $G(j\omega)$ at w. The magni-

---

[1]The name Nyquist will be introduced in a later section. The Nyquist plot of $G(s)$ is defined as $G(j\omega)$ for $\omega \geq 0$ and $\omega < 0$, whereas the polar plot is defined for only $\omega \geq 0$. Because command nyquist(n,d,w) in version 3.1 of MATLAB computes only positive ω, a better name would be polar(n,d,w).

tude is converted into decibels by db=20*log10(mag). Command plot(pha,db) plots the phase on the *x*- or horizontal axis with linear scale and the gain in decibels on the vertical axis. Thus the plot is a log magnitude-phase plot. Command semilogx(w,db) plots ω on the horizontal axis with logarithmic scale and the gain in dB on the vertical axis. Thus the plot is a Bode gain plot. Similarly, semilogx(w,pha) generates a Bode phase plot. For version 3.5 or the Student Edition of MATLAB, the commands

```
n=[1];d=[1 0.5];
bode(n,d)
```

will plot the Bode gain and phase plots on the screen. The command

```
nyquist(n,d)
```

will plot the polar plot ($G(j\omega)$, $\omega \geq 0$) with the solid line and its mirror image ($G(j\omega)$, $\omega < 0$) with the dashed line on the screen. Thus, the use of MATLAB to generate frequency plots is very simple.

8.3 PLOTTING BODE PLOTS

In this section, we discuss the plot of Bode plots by hand. One may wonder why we bother to study this when the plot can be easily obtained on a personal computer. Indeed, one can argue strongly for not studying this section. But in the study, we can learn the following: the reason for using logarithmic scales for frequency and magnitude, the mechanism for identifying a system from its Bode plot, and the reason for using the Bode plot, rather than the polar or log magnitude-phase plot, in the design. Besides, the plot of Bode plots by hand is quite simple; it does not require much computation.

We use an example to discuss the basic procedure of plotting Bode plots. Consider

$$G(s) = \frac{5s + 50}{s^2 + 99.8s - 20} = \frac{5(s + 10)}{(s - 0.2)(s + 100)} \tag{8.3}$$

First we write it as

$$G(s) = \frac{-5 \times 10\left(1 + \frac{1}{10}s\right)}{0.2 \times 100\left(1 - \frac{1}{0.2}s\right)\left(1 + \frac{1}{100}s\right)}$$

$$= \frac{-2.5\left(1 + \frac{1}{10}s\right)}{\left(1 - \frac{1}{0.2}s\right)\left(1 + \frac{1}{100}s\right)} \tag{8.4}$$

It is important to express every term in the form of $1 + \tau s$. The gain of $G(s)$ in decibels is $20 \log |G(j\omega)|$ or

$$
\begin{aligned}
20 \log |G(s)| = {} & 20 \log |-2.5| + 20 \log \left| 1 + \frac{1}{10} s \right| \\
& -20 \log \left| 1 - \frac{1}{0.2} s \right| - 20 \log \left| 1 + \frac{1}{100} s \right|
\end{aligned} \tag{8.5}
$$

and the phase of $G(s)$ is

$$
\begin{aligned}
\sphericalangle G(s) = {} & \sphericalangle (-2.5) + \sphericalangle \left(1 + \frac{1}{10} s \right) \\
& - \sphericalangle \left(1 - \frac{1}{0.2} s \right) - \sphericalangle \left(1 + \frac{1}{100} s \right)
\end{aligned} \tag{8.6}
$$

We see that the decibel gain of $G(s)$ is simply the algebraic sum of the decibel gain of each term of $G(s)$. Adding all gains of the terms in the numerator and then subtracting those in the denominator yields the gain of $G(s)$. Similar remarks apply to the phase of $G(s)$. Other than the constant term, (8.4) consists only of linear factors of form $(1 + \tau s)$. Therefore we discuss these linear factors first.

Gain of Linear Factors

Consider $20 \log |1 \pm \tau j\omega|$, for $\tau > 0$ and $\omega \geq 0$. We use the following approximations:

ω very small or $\omega \leq 0.1/\tau$: $20 \log |1 \pm j\omega\tau| \approx 20 \log 1 = 0$ dB

ω very large or $\omega \geq 10/\tau$: $20 \log |1 \pm j\omega\tau| \approx 20 \log \omega\tau$

In other words, for ω very small, $20 \log |1 \pm j\omega\tau|$ can be approximated by a horizontal line passing through 0 dB. For ω very large, $20 \log |1 \pm j\omega\tau|$ can be approximated by the straight line $20 \log \omega\tau$. If $\omega = 1/\tau$, then $20 \log \omega\tau = 20 \log 1 = 0$, thus the straight line $20 \log \omega\tau$ intersects the 0 dB line at $\omega = 1/\tau$. The gain of $20 \log \omega\tau$ increases 20 dB whenever ω increases 10 times, thus the straight line has a slope of 20 dB/decade as shown in Figure 8.4. The point $1/\tau$ where the straight line intersects with the 0-dB line is called the *corner frequency*. Once the corner frequency is determined, the two straight lines can be readily drawn. The two straight lines are called the *asymptotes*.

Now the question is: How good is the approximation? For $\omega \leq 0.1/\tau$, the difference between $20 \log |1 \pm j\omega\tau|$ and the 0-dB line is less than 0.04 dB. Similarly, if $\omega \geq 10/\tau$, the difference between $20 \log |1 \pm j\omega\tau|$ and $20 \log \omega\tau$ is again less than 0.04 dB. Thus, for $\omega \leq 0.1/\tau$ and $\omega \geq 10/\tau$, or for ω less than one-tenth of the corner frequency and larger than ten times the corner frequency, the Bode gain plot almost coincides with the asymptotes. For ω in the frequency interval $(0.1/\tau, 10/\tau)$, the exact plot differs from the asymptotes. The largest difference occurs at corner frequency $1/\tau$, and equals

$$20 \log |1 \pm j\omega\tau| = 20 \log |1 \pm j1| = 20 \log \sqrt{2} = 3 \text{ dB}$$

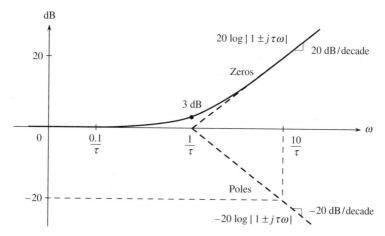

Figure 8.4 Bode gain plot of $\pm 20 \log |1 \pm j\tau\omega|$ with $\tau > 0$.

If we pick the 3-dB point and draw a smooth curve as shown in Figure 8.4, then the curve will be a very good approximation of the Bode gain plot of $(1 \pm \tau s)$. Note that the Bode gain plot of $(1 + \tau s)$ is identical to that of $(1 - \tau s)$. If they appear in the numerator, their asymptotes go up with slope 20 dB/decade. If they appear in the denominator, their asymptotes go down with slope -20 dB/decade. In other words, the asymptotes of $20 \log |1 \pm j\tau\omega|$ go up with slope 20 dB/decade; the asymptotes of $-20 \log |1 \pm j\tau\omega|$ go down with slope -20 dB/decade.

Now we plot the Bode gain plot of (8.4) or (8.5). The corner frequency of zero $(1 + s/10)$ is 10, and its asymptote goes up with slope 20 dB/decade as shown in Figure 8.5 with the dashed lines. The corner frequency of pole $(1 - s/0.2)$ is 0.2, and its asymptote goes down with slope -20 dB/decade as shown with the dotted

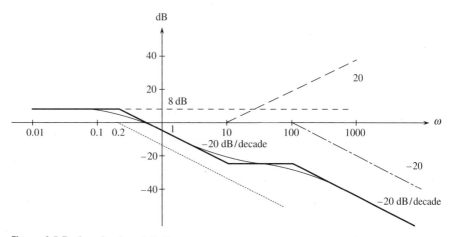

Figure 8.5 Bode gain plot of (8.4).

lines. Note that the negative sign makes no difference in the gain plot. The corner frequency of pole $(1 + s/100)$ is 100, and its asymptote goes down with slope -20 dB/decade as shown with the dashed-and-dotted lines. Thus, there are three pairs of asymptotes. Now we consider gain -2.5 in (8.4). Its decibel gain is 20 log $|-2.5| = 8$ dB; it is a horizontal line, independent of ω, as shown. The sum of the horizontal line and the three pairs of asymptotes is shown by the heavy solid line. It is obtained by adding them point by point. Because the plot consists of only straight lines, we need to compute the sums only at $\omega = 0.2$, 10, 100, and at a point larger than $\omega = 100$. The sum can also be obtained as follows. From $\omega = 0$ to $\omega = 0.2$, the sum of an 8-dB line and three 0-dB lines clearly equals 8 dB. Between [0.2, 10], there is only one asymptote with slope -20 dB/decade. Thus we draw from $\omega = 0.2$ a line with slope -20 dB/decade up to $\omega = 10$ as shown. Between [10, 100], there is one asymptote with slope 20 dB/decade and one with slope -20 dB/decade, thus the net is 0 dB/decade. Therefore, we draw a horizontal line from $\omega = 10$ to $\omega = 100$ as shown. For $\omega \geq 100$, two asymptotes have slope -20 dB/decade and one has slope 20 dB/decade. Thus, the net is a straight line with slope -20 dB/decade. There are three corner frequencies, at $\omega = 0.2$, 10, and 100. Because they are far apart, the effects of their Bode gain plots on each other are small. Thus the difference between the Bode plot and the asymptotes at every corner frequency roughly equals 3 dB. Using this fact, the Bode gain plot can then be obtained by drawing a smooth curve as shown. This completes the plotting of the Bode gain plot of (8.4).

Phase of Linear Factors

Consider the phase of $(1 \pm \tau s)$ or, more precisely, of $(1 \pm j\omega\tau)$ with $\tau > 0$ and $\omega \geq 0$. We shall use the following approximations:

$$\omega \text{ very small or } \omega \leq 0.1/\tau: \quad \sphericalangle (1 \pm j\omega\tau) \approx \sphericalangle (1) = 0°$$

and

$$\omega \text{ very large or } \omega \geq 10/\tau: \quad \sphericalangle (1 \pm j\omega\tau) \approx \begin{cases} \sphericalangle (+j\omega\tau) = 90° \\ \sphericalangle (-j\omega\tau) = -90° \end{cases}$$

In other words, for ω in $(-\infty, 0.1/\tau)$, the phase of $(1 \pm j\omega\tau)$ can be approximated by 0°; for ω in $(10/\tau, \infty)$, the phase of $(1 + j\omega\tau)$ can be approximated by 90° and the phase of $(1 - j\omega\tau)$, by $-90°$ as shown in Figure 8.6. We then connect the end points by a dashed straight line as shown. The exact phase of $(1 + j\omega\tau)$ is also plotted in Figure 8.6 using a solid line. We see that at $\omega = 0.1/\tau$ and $\omega = 10/\tau$, the differences are 5.7°. There is no difference at $\omega = 1/\tau$. The differences at the midpoints between $0.1/\tau$ and $1/\tau$ and between $1/\tau$ and $10/\tau$ are 3.4°. The differences are fairly small between the dashed straight line and the exact phase plot. Thus, the phase plot of $(1 + j\omega\tau)$ can be approximated by the straight lines. Note that the phase of $(1 - j\omega\tau)$ equals the reflection of the phase of $(1 + j\omega\tau)$ to negative angles. Thus the phase of $(1 - j\omega\tau)$ can also be approximated by the straight lines.

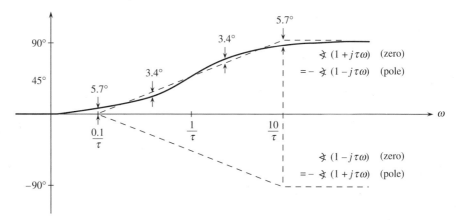

Figure 8.6 Bode phase plots of $1 \pm j\omega\tau$ with $\tau > 0$.

For the gain, we have

$$|1 + j\omega\tau| = |1 - j\omega\tau|$$

For the phase, we have

$$\sphericalangle (1 + j\omega\tau) = -\sphericalangle (1 - j\omega\tau) \qquad \sphericalangle (1 - j\omega\tau) = -\sphericalangle (1 + j\omega\tau)$$

Thus, the Bode gain plots of $(1 + \tau s)$ and $(1 - \tau s)$ are the same. If they appear as zeros, then their plots are as shown in the upper part of Figure 8.4; if they appear as poles, then their plots are as shown in the lower part of Figure 8.4. The situation in phase plots is different. If $(1 + \tau s)$ appears as a zero, or $(1 - \tau s)$ appears as a pole, then their phases are as shown in the upper part of Figure 8.6. If $(1 + \tau s)$ appears as a pole, or $(1 - \tau s)$ appears as a zero, then their phases are as shown in the lower part of Figure 8.6.

With the preceding discussion, we are ready to plot the Bode phase plot of (8.3) or (8.6). The phase of gain -2.5 is 180°. It is a horizontal line passing through 180° as shown in Figure 8.7. The corner frequency of zero $(1 + s/10)$ is 10; its phase for ω smaller than one-tenth of 10 is 0°; its phase for ω larger than ten times 10 is $+90°$. The phase for ω in $(1, 100)$ is approximated by the dashed line shown in Figure 8.7. The phase for pole $(1 - s/0.2)$ is $-\sphericalangle (1 - j\omega/0.2)$. Its phase is 0°, for $\omega < 0.02$, and $-(-90°) = +90°$, for $\omega > 2$. The approximated phase is plotted with dotted lines. Similarly the phase of pole $(1 + s/100)$ or $-\sphericalangle (1 + j\omega/100)$ is plotted with dashed-and-dotted lines. Their sum is denoted by the solid line; it is obtained using the procedures discussed for the gain plot. By smoothing the straight lines, the Bode phase plot can then be obtained (not shown).

Once a Bode phase plot is completed, it is always advisable to check the plot for $\omega \to 0$ and $\omega \to \infty$. If $\omega \to 0$, (8.3) reduces to -2.5 and its phase is 180°. If $\omega \to \infty$, (8.3) reduces to $5s/s^2 = 5/s$ and its phase is $-90°$ which also equals 270°. Thus the plot in Figure 8.7 checks with the two extreme cases. Note that phases are considered the same if they differ by 360° or its multiples. For example, if the phase

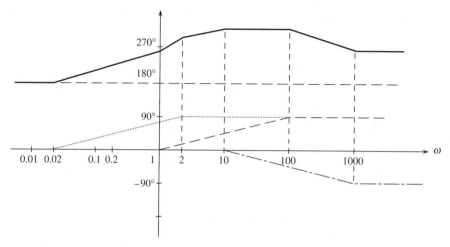

Figure 8.7 Bode phase plot of (8.3).

of -2.5 is plotted as $-180°$, then the phase plot in Figure 8.7 will be shifted down by $360°$, which is the one generated by calling **bode** in MATLAB.

Exercise 8.3.1

Plot the Bode plots of

a. $G(s) = \dfrac{10}{(s + 5)(s + 100)}$

b. $G(s) = \dfrac{2(s - 5)}{(s + 5)^2(s + 10)}$

Up to this point, we have considered transfer functions that contain only linear factors. Now we discuss transfer functions that also contain quadratic factors and poles or zeros at the origin.

Poles or Zeros at the Origin

The repeated pole $1/s^i$ has a logarithmic magnitude of $-20 \log |j\omega|^i = -20i \log \omega$ dB and a phase of $-i \times 90°$. If $i = 1$, the gain plot is a straight line passing through the 0-dB line at $\omega = 1$ and with slope -20 dB/decade, as shown in Figure 8.8. Its phase is a horizontal line passing through $-90°$. In Figure 8.8, we also plot the Bode plot of $1/s^i$ for $i = 2, -1$, and -2. The plots are very simple.

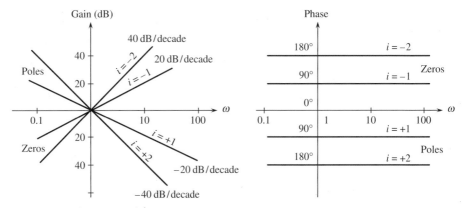

Figure 8.8 Bode plot of $1/s^i$.

Quadratic Factors

Consider the complex-conjugate poles[2]

$$E(s) := \frac{\omega_n^2}{s^2 + 2\zeta\omega_n s + \omega_n^2} = \frac{1}{1 + \dfrac{2\zeta}{\omega_n}s + \left(\dfrac{s}{\omega_n}\right)^2} \tag{8.7}$$

with $0 \leq \zeta < 1$. The logarithmic magnitude for $s = j\omega$ is

$$20 \log |E(j\omega)| = -20 \log \left|1 + \frac{2\zeta}{\omega_n}(j\omega) - \left(\frac{\omega}{\omega_n}\right)^2\right|$$

As for linear factors, we use the following approximations:

ω very small or $\omega \leq 0.1\omega_n$: $\quad 20 \log |E(j\omega)| \approx -20 \log |1| = 0$ dB
ω very large or $\omega \geq 10\omega_n$: $\quad 20 \log |E(j\omega)| \approx -20 \log |(\omega/\omega_n)^2|$
$$= -40 \log \omega/\omega_n$$

They are two straight lines, called *asymptotes*. One asymptote has slope zero, the other has slope -40 dB/decade. They intersect at corner frequency $\omega = \omega_n$ as shown in Figure 8.9. The Bode gain plot for ω in $(0.1\omega_n, 10\omega_n)$ depends highly on damping ratio ζ and differs greatly from the asymptotes as shown in Figure 8.9. If $\zeta = 0.5$, the gain plot is very close to the asymptotes. If $\zeta = 0.7$, the difference at the corner frequency is 3 dB. The smaller ζ, the larger the overshoot. Thus in plotting the Bode gain plot of the quadratic factor, we first compute the corner frequency and draw the two asymptotes, one with slope zero, the other with slope

---

[2]The Bode gain plot of $1/(1 - (2\zeta/\omega_n)s + (s/\omega_n)^2)$ equals that of (8.7), and its Bode phase plot equals that of (8.7) reflected to positive angles. To simplify discussion, we study only (8.7).

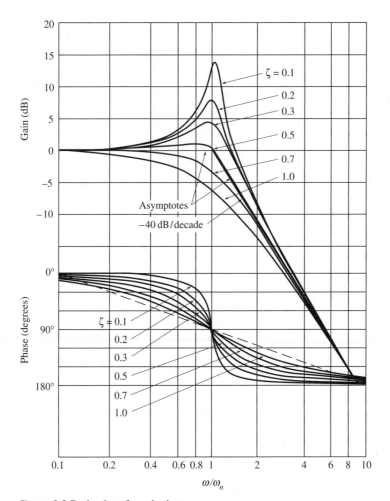

Figure 8.9 Bode plot of quadratic term.

-40 dB/decade. We then compute the damping ratio and use Figure 8.9 to draw an approximate Bode gain plot. Note that if the quadratic factor appears as zeros, then one asymptote will go up with slope $+40$ dB/decade, and the plots in Figure 8.9 are reversed.

The phase plot of the quadratic factor in (8.7) can be approximated by straight lines as follows:

ω very small or $\omega \leq 0.1\omega_n$: $\measuredangle E(j\omega) \approx -\measuredangle 1 = 0°$

ω very large or $\omega \geq 10\omega_n$: $\measuredangle E(j\omega) \approx -\measuredangle (j\omega/\omega_n)^2 = -180°$

The exact phases for ω in $(0.1\omega_n, 10\omega_n)$ are plotted in Figure 8.9 for various ζ. For small ζ, they are quite different from the dashed straight line shown. Thus in plotting the Bode phase plot of (8.7), we must compute the corner frequency ω_n as well as the damping ratio ζ and then use Figure 8.9.

Example 8.3.1

Plot the Bode plot of the transfer function

$$G(s) = \frac{50(s + 2)}{s(s^2 + 4s + 100)}$$

To compute ω_n and ζ of the quadratic term, we equate

$$\omega_n^2 = 100 \qquad 2\zeta\omega_n = 4$$

which imply $\omega_n = 10$ and $\zeta = 0.2$. We then express every term of $G(s)$ in the form

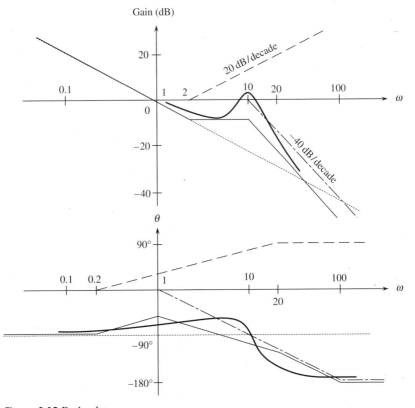

Figure 8.10 Bode plot.

of $1 + (\cdot)$ as in (8.4) or (8.7):

$$G(s) = \frac{50 \times 2 \left(1 + \dfrac{1}{2}s\right)}{100s \left(1 + \dfrac{4}{100}s + \dfrac{s^2}{100}\right)} = \frac{1 + \dfrac{1}{2}s}{s\left[1 + 2\left(\dfrac{0.2}{10}\right)s + \left(\dfrac{s}{10}\right)^2\right]}$$

The asymptotes of the zero, the pole at the origin, and the quadratic term are plotted in Figure 8.10 using, respectively, the dashed, dotted, and dashed-and-dotted lines. The sums of these asymptotes are denoted by the thin solid lines. Because the damping ratio is 0.2, the Bode plot is quite different from the asymptotes at $\omega = \omega_n = 10$. Using the plot in Figure 8.9, we can obtain the Bode gain and phase plots of $G(s)$ as shown in Figure 8.10 with the heavy solid lines.

Exercise 8.3.2

Plot the Bode plots of

a. $G(s) = \dfrac{1}{s^2(s + 1)}$

b. $G(s) = \dfrac{100}{s(s^2 + 4s + 100)}$

8.3.1 Non-Minimum-Phase Transfer Functions

Consider the following two transfer functions

$$G_1(s) = \frac{s - z}{s + p} \quad \text{and} \quad G_2(s) = \frac{s + z}{s + p}$$

with their poles and zeros plotted in Figure 8.11. The transfer function $G_1(s)$ has one zero $s = z$ in the open right half plane. If we reflect the zero into the left half plane,

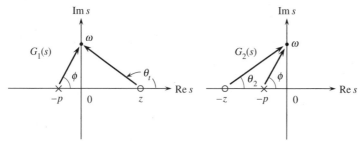

Figure 8.11 Non-minimum-phase transfer function.

we obtain $G_2(s)$. Now we compare the gains and phases of $G_1(s)$ and $G_2(s)$ on the positive $j\omega$-axis. From Figure 8.11, we can see that the vectors from zero z to any point on the $j\omega$-axis and from zero $-z$ to the same point have the same length; therefore, we have

$$|G_1(j\omega)| = |G_2(j\omega)|$$

for all $\omega \geq 0$. Actually, this fact has been used constantly in developing Bode gain plots. Although $G_1(s)$ and $G_2(s)$ have the same gain plots, their phases are quite different. From Figure 8.11, we have

$$\measuredangle\, G_1(j\omega) = \theta_1 - \phi \qquad \measuredangle\, G_2(j\omega) = \theta_2 - \phi$$

At $\omega = 0$, $\theta_1 = 180° > \theta_2 = 0°$. At $\omega = \infty$, $\theta_1 = \theta_2 = 90°$. In general, $\theta_1 \geq \theta_2$ for all $\omega \geq 0$. Thus we have

$$\measuredangle\, G_1(j\omega) \geq \measuredangle\, G_2(j\omega)$$

for all $\omega \geq 0$. Thus, if a transfer function has right-half-plane zeros, reflecting these zeros into the left half plane gives a transfer function with the same amplitude but a smaller phase than the original transfer function at every $\omega \geq 0$. This motivates the following definition:

□ Definition 8.1

A proper rational transfer function is called a *minimum-phase* transfer function if all its zeros lie inside the open left half s-plane. It is called a *non-minimum-phase* transfer function if it has zeros in the closed right half plane. Zeros in the closed right half plane are called *non-minimum-phase zeros*. Zeros in the open left half plane are called *minimum-phase zeros*. ■

We mention that there are two different definitions of minimum-phase transfer functions in the literature. One requires both poles and zeros to lie inside the open left half plane; the other requires only zeros. Both definitions are ambiguous about zeros on the $j\omega$-axis. Our definition of non-minimum-phase zeros includes zeros on the $j\omega$-axis.

Exercise 8.3.3

Plot the Bode plots of

$$\frac{(s - 2)}{s(s - 1)} \qquad \text{and} \qquad \frac{(s + 2)}{s(s - 1)}$$

Do they have the same amplitude plots? How about their phase plots? Phases are considered the same if they differ by $\pm 360°$ or their multiples. If their phases are set equal at $\omega = \infty$, which transfer function has a smaller phase at every $\omega \geq 0$?

Exercise 8.3.4

Which of the following are non-minimum-phase?

$$\frac{s-1}{(s+2)^2} \qquad \frac{s+1}{(s-2)^2} \qquad \frac{s(s+2)}{(s+1)^2}$$

[**Answers:**   1 and 3.]

8.3.2 Identification

Determination of the transfer function $G(s)$ of a system from measured data is an identification problem. As discussed in Section 4.7.1, if a system is stable, then $|G(j\omega)|$ and $\angle G(j\omega)$ can be obtained by measurement. This is also possible if $G(s)$ has only one unstable pole at $s = 0$ (see Problem 4.21). From $|G(j\omega)|$ and $\angle G(j\omega)$, we can readily obtain the Bode plot of $G(s)$. Now we discuss how to obtain $G(s)$ from its Bode plot. This is illustrated by examples.

Example 8.3.2

Find the transfer function of the Bode plot shown in Figure 8.12(a). First we approximate the gain plot by the three straight dashed lines shown. They intersect at $\omega = 1$ and $\omega = 10$. We begin with the leftmost part of the gain plot. There is a

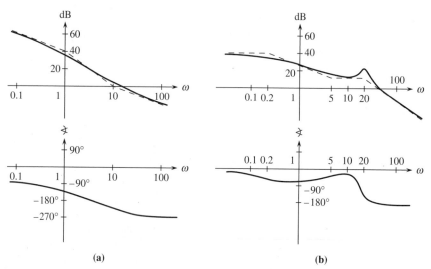

(a)   (b)

Figure 8.12 Bode plots.

straight line with slope -20 dB/decade, therefore the transfer function has one pole at $s = 0$. At $\omega = 1$, the slope becomes -40 dB/decade, a decrease of 20 dB/decade, thus there is a pole with corner frequency $\omega = 1$. At $\omega = 10$, the slope becomes -20 dB/decade, an increase of 20 dB/decade, therefore there is a zero with corner frequency $\omega = 10$. Thus, the transfer function must be of the form

$$G(s) = \frac{k\left(1 \pm \dfrac{s}{10}\right)}{s\left(1 \pm \dfrac{s}{1}\right)}$$

This form is very easy to obtain. Wherever there is a decrease of 20 dB/decade in slope, there must be a pole. Wherever there is an increase of 20 dB/decade, there must be a zero. The constant k can be determined from an arbitrary ω. For example, the gain is 37 dB at $\omega = 1$ or $s = j1$. Thus we have

$$37 = 20 \times \log \left| \frac{k\left(1 \pm \dfrac{j1}{10}\right)}{j1 \times \left(1 \pm \dfrac{j1}{1}\right)} \right| = 20 \times \log \left| \frac{k\sqrt{1 + 0.01}}{1 \times \sqrt{1 + 1}} \right|$$

$$= 20 \times \log \left| \frac{k \times 1.005}{1.414} \right|$$

which implies

$$|k| = \frac{1.414}{1.005} 10^{37/20} = 1.4 \times 10^{1.85} = 1.4 \cdot 70.79 = 99.6$$

Now we use the phase plot to determine the signs of each term. The phase of $G(s)$ for ω very small is determined by k/s. If k is negative, the phase of k/s is $+90°$; if k is positive, the phase is $-90°$. From the phase plot in Figure 8.12(a), we conclude that k is positive and equals 99.6. If the sign of pole $1 \pm s$ is negative, the pole will introduce positive phase into $G(s)$ or, equivalently, the phase of $G(s)$ will increase as ω passes through the corner frequency 1. This is not the case as shown in Figure 8.12(a), therefore we have $1 + s$. If the sign of zero $1 \pm 0.1s$ is positive, the zero will introduce positive phase into $G(s)$ or, equivalently, the phase of $G(s)$ will increase as ω passes through the corner frequency at 10. This is not the case as shown in Figure 8.12(a), thus we have $1 - 0.1s$ and the transfer function of the Bode plot is

$$G(s) = \frac{99.6(1 - 0.1s)}{s(1 + s)} = \frac{-9.96(s - 10)}{s(s + 1)}$$

Example 8.3.3

Find the transfer function of the Bode plot in Figure 8.12(b). The gain plot is first approximated by the straight lines shown. There are three corner frequencies: 0.2, 5, and 20. At $\omega = 0.2$, the slope becomes -20 dB/decade; therefore, there is one pole at 0.2 or $(1 \pm s/0.2)$. At $\omega = 5$, the slope changes from -20 dB/decade to 0; therefore, there is a zero at 5 or $(1 \pm s/5)$. At $\omega = 20$, the slope changes from 0 to -40 dB/decade; therefore, there is a repeated pole or a pair of complex-conjugate poles with corner frequency 20. Because of the small bump, it is a quadratic term. The bump is roughly 10 dB high, and we use Figure 8.9 to estimate its damping ratio ζ as 0.15. Therefore, the transfer function of the Bode plot is of the form

$$G(s) = \frac{k \left(1 \pm \dfrac{s}{5} \right)}{\left(1 \pm \dfrac{s}{0.2} \right)\left(1 \pm \dfrac{2 \times 0.15}{20} s + \left(\dfrac{s}{20} \right)^2 \right)}$$

The gain of $G(s)$ at $\omega \to 0$ or $s = 0$ is 40 dB. Thus we have

$$20 \times \log |G(0)| = 20 \times \log |k| = 40$$

which implies $|k| = 100$ or $k = \pm 100$. Using the identical argument as in the preceding example we can conclude from the phase plot that we must take the positive sign in all $\pm$. Thus the transfer function of the Bode plot is

$$G(s) = \frac{100 \left(1 + \dfrac{s}{5} \right)}{\left(1 + \dfrac{s}{0.2} \right)\left(1 + \dfrac{0.3}{20} s + \left(\dfrac{s}{20} \right)^2 \right)} = \frac{1600(s + 5)}{(s + 0.2)(s^2 + 6s + 400)}$$

From these examples, we see that if a Bode plot can be nicely approximated by straight lines with slope ± 20 dB/decade or its multiples, then its transfer function can be readily obtained. The Bode gain plot determines the form of the transfer function. Wherever the slope decreases by 20 dB/decade, there is a pole; wherever it increases by 20 dB/decade, there is a zero. Signs of poles or zeros are then determined from the Bode phase plot. If the Bode plot is obtained by measurement, then, except for a possible pole at $s = 0$, the system must be stable. Therefore, we can simply assign positive sign to the poles without checking the phase plot. If the transfer function is known to be of minimum phase, then we can assign positive sign to the zeros without checking the phase plot. In fact, if a transfer function is stable and of minimum phase, then there is a unique relationship between the gain plot and phase plot, and we can determine the transfer function from the gain plot alone. To conclude this section, we mention that devices, such as the HP3562A Dynamic System Analyzer, are available to measure Bode plots and then generate transfer functions. This facilitates considerably the identification of transfer functions.

8.4 STABILITY TEST IN THE FREQUENCY DOMAIN

Consider the unity-feedback system shown in Figure 8.1. We discuss in this section a method of checking the stability of the feedback system from its open-loop transfer function $G(s)C(s)$. The transfer function of the unity-feedback system is

$$G_o(s) = \frac{G(s)C(s)}{1 + G(s)C(s)} =: \frac{G_l(s)}{1 + G_l(s)} \tag{8.8}$$

where $G_l(s) := G(s)C(s)$ is called the *loop transfer function*. The stability of $G_o(s)$ is determined by the poles of $G_o(s)$ or the zeros of the rational function

$$1 + G_l(s)$$

Recall that we have introduced two methods of checking whether or not all zeros of $(1 + G_l(s))$ have negative real parts. If we write $G_l(s) = N_l(s)/D_l(s)$, then the zeros of $(1 + G_l(s))$ are the roots of the polynomial $D_l(s) + N_l(s)$ and we may apply the Routh test. Another method is to plot the root loci of $G_l(s) = -1/k$ with $k = 1$, as was discussed in Section 7.4.3. In this section, we shall introduce yet another method of checking whether or not all zeros of $(1 + G_l(s))$ lie inside the open left half plane. The method uses only the frequency plot of $G_l(s)$ and is based on the principle of argument in the theory of complex variables.

8.4.1 Principle of Argument

Consider a rational function $F(s)$. The substitution of a value s_i, a point in the s-plane as shown in Figure 8.13, into $F(s)$ yields a value $F(s_i)$, a point in the F-plane. In mathematical terminology, s_i in the s-plane is said to be *mapped* into $F(s_i)$ in the F-plane. In this sense, the polar plot in Figure 8.2(a) is the mapping of

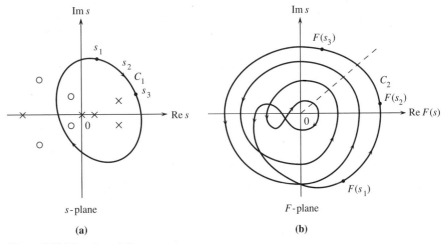

Figure 8.13 Mapping of C_1.

the positive $j\omega$-axis of the s-plane by $G(s)$ in (8.1) into the G-plane. A simple closed curve is defined as a curve that starts and ends at the same point without going through any point twice.

Principle of Argument

Let C_1 be a simple closed curve in the s-plane. Let $F(s)$ be a rational function of s that has neither pole nor zero on C_1. Let Z and P be the numbers of zeros and poles of $F(s)$ (counting the multiplicities) encircled by C_1. Let C_2 be the mapping of C_1 by $F(s)$ into the F-plane. Then C_2 will encircle the origin of the F-plane $(Z - P)$ times in the same direction as C_1. ■

Because $F(s)$ has no pole on C_1, the value of $F(s)$ at every s on C_1 is well defined. Because $F(s)$ is a continuous function of s, the mapping of C_1 by $F(s)$ is a continuous closed curve, denoted by C_2, in the F-plane as shown in Figure 8.13. The closed curve, however, is not necessarily simple—that is, it may intersect with itself. Furthermore C_2 will not pass through the origin, because $F(s)$ has no zero on C_1. Now if $Z - P$ is positive—that is, $F(s)$ has more zeros than poles encircled by C_1—then C_2 will travel in the same direction as C_1. If $Z - P$ is negative, then C_2 will travel in the opposite direction. In counting the encirclement, we count the *net* encirclement. For example, if C_2 encircles the origin once in the clockwise direction and three times in the counterclockwise direction, then C_2 is considered to encircle the origin two times in the counterclockwise direction. Consider the mapping shown in Figure 8.13. The contour C_1 is chosen to travel in the clockwise direction and encircles four poles and two zeros of $F(s)$. Thus we have $Z - P = -2$. Now if we plot $F(s)$ along C_1, the resulting contour C_2 must encircle the origin of the F-plane twice in the direction opposite to C_1 or in the counterclockwise direction. Indeed, the contour C_2 encircles the origin three times in the counterclockwise direction, once in the clockwise direction. Therefore the net encirclement is twice in the counterclockwise direction. One way to count the encirclements is to draw a straight line from the origin as shown in Figure 8.13; then the number of encirclements equals the number of crossings of the straight line by C_2 minus the number of reverse crossings by C_2.

8.4.2   The Nyquist Plot

Consider

$$G_o(s) = \frac{G_l(s)}{1 + G_l(s)} =: \frac{G_l(s)}{F(s)} \tag{8.9}$$

where $F(s) = 1 + G_l(s)$. The condition for $G_o(s)$ to be stable is that the numerator of $F(s)$ is a Hurwitz polynomial or, equivalently, $F(s)$ has no zero in the closed right half s-plane. In checking stability, the contour C_1 will be chosen to enclose the entire closed right half plane as shown in Figure 8.14(a), in which the radius R of the semicircle should be very large or infinity. The direction of C_1 is arbitrarily chosen to be clockwise. Now the mapping of C_1 by $F(s)$ into the F-plane is called the *Nyquist*

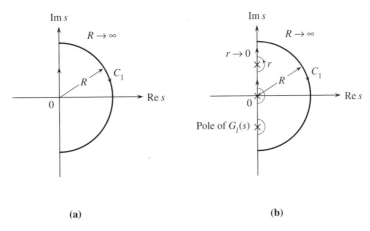

Figure 8.14 Contour of C_1.

plot of $F(s)$. Similarly, we may define the Nyquist plot of $G_l(s)$ as the mapping of C_1 by $G_l(s)$. We first use an example to illustrate the plotting of the Nyquist plots of $G_l(s)$ and $F(s)$.

Example 8.4.1

Consider

$$G_l(s) = \frac{8s}{(s-1)(s-2)} \tag{8.10}$$

We compute $G_l(j1) = 2.6e^{j161°}$, $G_l(j2) = 2.6e^{j198°}$, and $G_l(j10) = 0.8e^{-j107°}$. Using these, we can plot the polar plot of $G_l(s)$ or the plot of $G_l(j\omega)$ for $\omega \geq 0$ as shown in Figure 8.15 with the solid curve. It happens to be a circle. Because all coefficients of $G_l(s)$ are real, we have

$$G_l(-j\omega) = G_l^*(j\omega)$$

where the asterisk denotes the complex conjugate. Thus the mapping of $G_l(j\omega)$, for $\omega \leq 0$, is simply the reflection, with respect to the real axis, of the mapping of $G_l(j\omega)$, for $\omega \geq 0$, as shown with the dashed curve. Because $G_l(s)$ is strictly proper, every point of the semicircle in Figure 8.14(a) with $R \to \infty$ is mapped by $G_l(s)$ into 0. Thus, the complete Nyquist plot of $G_l(s)$ consists of the solid and dashed circles in Figure 8.15. As ω travels clockwise in Figure 8.14(a), the Nyquist plot of $G_l(s)$ travels counterclockwise as shown.

If $G_l(s)$ is proper, then the mapping of the infinite semicircle in Figure C_1 by $G_l(s)$ is simply a point in the G_l-plane. Thus the Nyquist plot of $G_l(s)$ consists mainly of $G_l(j\omega)$ for all ω. The polar plot of $G_l(s)$, however, is defined as $G_l(j\omega)$ for $\omega \geq 0$. Therefore the Nyquist plot of $G_l(s)$ consists of the polar plot of $G_l(s)$ and its reflection with respect to the real axis. Therefore, the Nyquist plot can be readily obtained from the polar plot.

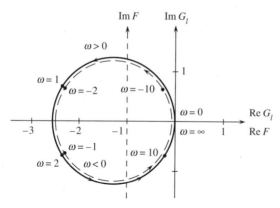

Figure 8.15 Nyquist plot of (8.10).

Because $F(s) = 1 + G_l(s)$, the Nyquist plot of $F(s)$ is simply the Nyquist plot of $G_l(s)$ shifted to the right by one unit. This can be more easily achieved by choosing the coordinates of the F-plane as shown in Figure 8.15. In other words, the origin of the F-plane is the point $(-1, 0)$ in the G_l-plane. Therefore, once the Nyquist plot of $G_l(s)$ is obtained, the Nyquist plot of $F(s)$ is already there.

Exercise 8.4.1

Plot the Nyquist plots of $1/(s + 1)$ and $2/(s - 1)$.

The transfer function in (8.10) has one zero and no pole on the imaginary axis, and its Nyquist plot can easily be obtained. Now if a transfer function $G_l(s)$ contains poles on the imaginary axis, then $G_l(s)$ is not defined at every point of C_1 in Figure 8.14(a) and its Nyquist plot cannot be completed. For this reason, if $G_l(s)$ contains poles on the imaginary axis as shown in Figure 8.14(b), then the contour C_1 must be modified as shown. That is, wherever there is a pole on the imaginary axis, the contour is indented by a very small semicircle with radius r. Ideally, the radius r should approach zero. With this modification, the Nyquist plot of $G_l(s)$ can then be completed. This is illustrated by an example. Before proceeding, we mention that the command nyquist in MATLAB will yield an incorrect or incomplete Nyquist plot if $G_l(s)$ has poles on the imaginary axis.

Example 8.4.2

Consider

$$G_l(s) = \frac{s + 1}{s^2(s - 2)} \tag{8.11}$$

Its poles and zero are plotted in Figure 8.16(a). Because $G_l(s)$ has poles at the origin, the contour C_1 at the origin is replaced by the semicircle $re^{j\theta}$, where θ varies from

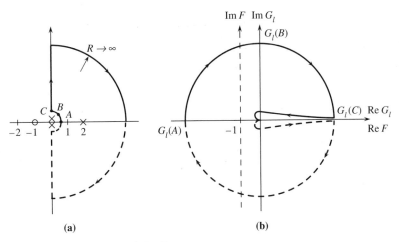

Figure 8.16 Nyquist plot of (8.11).

$-90°$ to $90°$ and r is very small. To compute the mapping of the small semicircle by $G_l(s)$, we use, for s very small,

$$G_l(s) = \frac{s+1}{s^2(s-2)} \approx \frac{1}{s^2(-2)} = \frac{-1}{2s^2} \tag{8.12}$$

Now consider point A or $s = re^{j0°}$ in Figure 8.16(a). It is mapped by (8.11) or (8.12) into

$$G_l(A) := \frac{-1}{2r^2 e^{j0°}} = \frac{e^{j180°}}{2r^2} = \frac{1}{2r^2} e^{j180°}$$

Its phase is $180°$ and its amplitude is very large because r is very small. Similarly, we have the following

$$B: \quad s = re^{j45°} \quad G_l(B) = \frac{e^{j180°}}{2r^2 e^{j90°}} = \frac{1}{2r^2} e^{j90°}$$

$$C: \quad s = re^{j90°} \quad G_l(C) = \frac{e^{j180°}}{2r^2 e^{j180°}} = \frac{1}{2r^2} e^{j0°}$$

We then compute $G_l(j0.1) = 50e^{j9°}$, $G_l(j1) = 0.6e^{j71°}$, and $G_l(j10) = 0.01e^{j162°}$. From these, we can plot $G_l(j\omega)$ for $\omega > 0$ (including section ABC) as shown in Figure 8.16(b) with the solid lines. The plot of $G_l(j\omega)$ for $\omega < 0$ is the reflection, with respect to the real axis, of the plot for $\omega > 0$ and is shown in Figure 8.15(b) with the dashed lines. Every point on the large semicircle with $R \to \infty$ is mapped by $G_l(s)$ into the origin of the G_l-plane. Thus the plot in Figure 8.16(b) is the conplete Nyquist plot of $G_l(s)$ in (8.11). If the Nyquist plot of $F(s) = 1 + G_l(s)$ is desired, we may simply add a set of coordinates as shown in Figure 8.16(b).

Exercise 8.4.2

Plot the Nyquist plots of

$$G_I(s) = \frac{1}{s(s + 1)}$$

and $F(s) = 1 + G_I(s)$.

8.4.3 Nyquist Stability Criterion

Consider the unity-feedback system shown in Figure 8.17(a). Its overall transfer function is

$$G_o(s) = \frac{G_I(s)}{1 + G_I(s)} =: \frac{G_I(s)}{F(s)} \tag{8.13}$$

THEOREM 8.1   (Nyquist Stability Criterion)

The $G_o(s)$ in (8.13) is stable if and only if the Nyquist plot of $G_I(s)$ does not pass through critical point $(-1, 0)$ and the number of counterclockwise encirclements of $(-1, 0)$ equals the number of *open* right-half-plane poles of $G_I(s)$. ∎

To prove this theorem, we first show that $G_o(s)$ is stable if and only if the Nyquist plot of $F(s)$ does not pass through the origin of the F-plane and the number of counterclockwise encirclements of the origin equals the number of open right-half-plane poles of $G_I(s)$. Clearly $G_o(s)$ is stable if and only if $F(s)$ has no closed right-half-plane zeros. If the Nyquist plot of $F(s)$ passes through the origin of the F-plane, then $F(s)$ has zeros on the imaginary axis and $G_o(s)$ is not stable. We assume in the following that $F(s)$ has no zeros on the imaginary axis. Let Z and P be, respectively, the numbers of open right-half-plane zeros and poles of $F(s)$ or, equivalently, the numbers of zeros and poles of $F(s)$ encircled by C_1. Because $F(s)$ and $G_I(s)$ have the same denominator, P also equals the number of open right-half-plane poles of $G_I(s)$. Now the principle of argument states that

$$N = Z - P$$

Clearly $F(s)$ has no open right-half-plane zeros, or $G_o(s)$ is stable if and only if $Z = 0$ or $N = -P$. Because C_1 is chosen to travel in the clockwise direction, the stability condition requires the encirclements to be in the counterclockwise direction. This establishes the assertion.

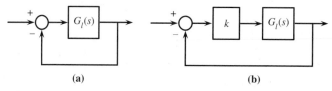

(a)   (b)

Figure 8.17 Unity-feedback systems.

From the discussion in the preceding subsection, the encirclement of the Nyquist plot of $F(s)$ around the origin of the F-plane is the same as the encirclement of the Nyquist plot of $G_l(s)$ around point $(-1, 0)$ on the G_l-plane. This establishes the theorem. We note that if the Nyquist plot of $G_l(s)$ passes through $(-1, 0)$, then $G_o(s)$ has at least one pole on the imaginary axis and $G_o(s)$ is not stable.

We discuss now the application of the theorem.

Example 8.4.3

Consider the unity-feedback system in Figure 8.17(a) with $G_l(s)$ given in (8.10). $G_l(s)$ has two poles in the open right half plane. The Nyquist plot of $G_l(s)$ is shown in Figure 8.15. It encircles $(-1, 0)$ twice in the counterclockwise direction. Thus, the unity-feedback system is stable. Certainly, this can also be checked by computing

$$G_o(s) = \frac{G_l(s)}{1 + G_l(s)} = \frac{\dfrac{8s}{(s-1)(s-2)}}{1 + \dfrac{8s}{(s-1)(s-2)}}$$

$$= \frac{8s}{(s-1)(s-2) + 8s} = \frac{8s}{s^2 + 5s + 2}$$

which is stable.

Example 8.4.4

Consider the unity-feedback system in Figure 8.17(a) with $G_l(s)$ given in (8.11). $G_l(s)$ has one open right-half-plane pole. Its Nyquist plot is shown in Figure 8.16; it encircles $(-1, 0)$ once in the clockwise direction. Although the number of encirclements is right, the direction is wrong. Thus the unity-feedback system is not stable. This can also be checked by computing

$$G_o(s) = \frac{G_l(s)}{1 + G_l(s)} = \frac{s + 1}{s^3 - 2s^2 + s + 1}$$

which is clearly not stable.

In application, we may encounter the problem of finding the range of k for the system in Figure 8.17(b) or

$$G_o(s) = \frac{kG_l(s)}{1 + kG_l(s)} \tag{8.14}$$

to be stable. Although Theorem 8.1 can be directly applied to solve the problem, it is more convenient to modify the Nyquist stability criterion as follows:

THEOREM 8.2

The $G_o(s)$ in (8.14) is stable if and only if the Nyquist plot of $G_l(s)$ does not pass through the critical point at $(-1/k, 0)$ and the number of counterclockwise encirclements of $(-1/k, 0)$ equals the number of open right-half-plane poles of $G_l(s)$. ∎

This theorem reduces to Theorem 8.1 if $k = 1$. The establishment of this theorem is similar to that of Theorem 8.1 and will not be repeated. We will now discuss its application.

Example 8.4.5

Consider the unity-feedback system shown in Figure 8.17(b) with

$$G_l(s) = \frac{8}{(s + 1)(s^2 + 2s + 2)} \tag{8.15}$$

Find the stability range of k.

The Nyquist plot of $G_l(s)$ is plotted in Figure 8.18. $G_l(s)$ has no open RHP pole. Thus the feedback system is stable if and only if the Nyquist plot does not encircle $(-1/k, 0)$. If $-1/k$ lies inside $[0, 4]$, the Nyquist plot encircles it once; if $-1/k$ lies inside $[-0.8, 0]$, the Nyquist plot encircles it twice. Thus if $-1/k$ lies inside $[0, 4]$ or $[-0.8, 0]$, the feedback system is not stable. On the other hand, if

$$-\infty \le -\frac{1}{k} < -0.8 \tag{8.16a}$$

or

$$4 < -\frac{1}{k} \le \infty \tag{8.16b}$$

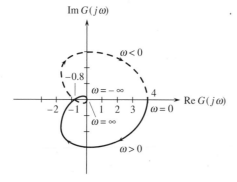

Figure 8.18 Nyquist plot of (8.15).

then the Nyquist plot does not encircle $(-1/k, 0)$ and the feedback system in Figure 8.17(b) is stable. Now (8.16a) implies $0 \leq k < 5/4$ and (8.16b) implies $0 \geq k > -1/4$. Thus the stability range of the system is

$$-\frac{1}{4} < k < \frac{5}{4} \tag{8.17}$$

This result is the same as (4.21) which is obtained by using the Routh test. We mention that the stability range can also be obtained by using the root-locus method. This will not be discussed.

We give a special case of Theorem 8.1 in which loop transfer functions have no open right-half-plane poles. Poles on the imaginary axis, however, are permitted.

COROLLARY 8.1

If the loop transfer function $G_l(s)$ in Figure 8.17(a) has no open right-half-plane poles, then $G_o(s)$ in (8.13) is stable if and only if the Nyquist plot of $G_l(s)$ does not encircle critical point $(-1, 0)$ nor passes through it. ■

Before proceeding, we mention that the Nyquist criterion is applicable to non-unity-feedback configurations such as the one shown in Figure 8.19. Its transfer function from r to y is

$$G_o(s) = \frac{C_1(s)G(s)}{1 + C_1(s)C_2(s)G(s)} \tag{8.18}$$

If we define $G_l(s) := C_1(s)C_2(s)G(s)$, then $G_o(s)$ in (8.18) is stable if and only if the Nyquist plot of $G_l(s)$ does not pass through $(-1, 0)$ and the number of counter-clockwise encirclements of $(-1, 0)$ equals the number of open right-half-plane poles of $G_l(s)$.

8.4.4 Relative Stability—Gain Margin and Phase Margin

The Routh test is probably the easiest method of checking stability. Therefore, the purpose of introducing the Nyquist criterion is not so much for checking stability, but because it will reveal the degree of stability and then provide a method of designing control systems. If the Nyquist plot of $G_l(s)$ passes through critical point

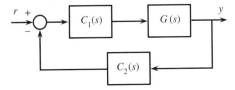

Figure 8.19 Non-unity-feedback system.

$(-1, 0)$ at, say, $s = j\omega_0$, then the numerator of $(1 + G_l(s))$ equals zero at $s = j\omega_0$. In other words, the numerator has a zero on the imaginary axis and is not Hurwitz. Consequently, $G_o(s)$ is not stable. Thus, the distance between the Nyquist plot of $G_l(s)$ and critical point $(-1, 0)$ can be used as a measure of stability of $G_o(s)$. Generally, the larger the distance, the more stable the system. The distance can be found by drawing a circle touching the Nyquist plot as shown in Figure 8.20(a). Such a distance, however, is not easy to measure and is not convenient for design. Therefore, it will be replaced by phase margin and gain margin.

Consider the Nyquist plot for $\omega \geq 0$ or, equivalently, the polar plot shown in Figure 8.20. Let $\omega_p > 0$ be the frequency at which $\measuredangle G_l(j\omega_p) = 180°$. This is called the *phase crossover frequency*; it is the frequency at which the polar plot of $G_l(s)$ passes through the negative real axis. The distance between -1 and $G_l(j\omega_p)$ as shown in Figure 8.20(b) is called the *gain margin*. The distance, however, is not measured on linear scale; it is measured in decibels defined as

$$\text{Gain margin} := 20 \log |-1| - 20 \log |G_l(j\omega_p)|$$
$$= -20 \log |G_l(j\omega_p)| \tag{8.19}$$

For example, if $G_l(j\omega_p) = -0.5$, then the gain margin is $+6$ dB. If $G_l(j\omega_p) = 0$, then the gain margin is ∞. Note that if $G_l(j\omega_p)$ lies between -1 and 0, then $|G_l(j\omega_p)| < 1$ and the gain margin is positive. If $|G_l(j\omega_p)| > 1$, or the polar plot of $G_l(j\omega_p)$ intersects the real axis on the left hand side of -1 as shown in Figure 8.20(c), then the gain margin is negative.

Let $\omega_g > 0$ be the frequency such that $|G_l(j\omega_g)| = 1$. It is the frequency at which the polar plot of $G_l(s)$ intersects with the unit circle as shown in Figure 8.20(b) and is called the *gain crossover frequency*. If we draw a straight line from the origin to $G_l(j\omega_g)$, then the angle between the straight line and negative real axis is called the *phase margin*. To be more precise, the phase margin is defined as

$$\text{Phase margin} := |\measuredangle (-1)| - |\measuredangle G_l(j\omega_g)| = 180° - |\measuredangle G_l(j\omega_g)| \tag{8.20}$$

where the phase of $G_l(j\omega_g)$ must be measured in the clockwise direction. Note that if the intersection with the unit circle occurs in the third quadrant as shown in Figure

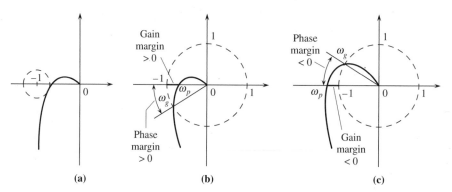

Figure 8.20 Relative stability.

8.20(b), then the phase margin is positive. If the intersection occurs in the second quadrant as shown in Figure 8.20(c), then the phase margin is negative.

The phase and gain margins can be much more easily obtained from the Bode plot. In fact, the definitions in (8.19) and (8.20) are developed from the Bode plot. Recall that the Bode plot and polar plot differ only in the coordinates and either one can be obtained from the other. Suppose the polar plots in Figure 8.20(b) and (c) are translated into the Bode plots shown in Figure 8.21(a) and (b). Then the gain crossover frequency ω_g is the frequency at which the gain plot crosses the 20 log $1 = 0$-dB line. We then draw a vertical line downward to the phase plot. The distance in degrees between the $-180°$ line and the phase plot is the phase margin. If the phase of $G_l(j\omega_g)$ lies above the $-180°$ line, the phase margin is positive, as shown in Figure 8.21(a). If it lies below, the phase margin is negative, as shown in Figure 8.21(b). The phase crossover frequency ω_p is the frequency at which the phase plot crosses the $-180°$ line. We then draw a vertical line upward to the gain plot. The distance in decibels between the 0-dB line and the gain plot at ω_p is the gain margin. If the gain at ω_p lies below the 0-dB line, as shown in Figure 8.21(a), the gain margin is positive. Otherwise it is negative, as shown in Figure 8.21(b). Thus, the phase and gain margins can readily be obtained from the Bode plot.

Although the phase and gain margins can be more easily obtained from the Bode plot, their physical meaning can be more easily visualized on the Nyquist plot. For example, if $G_l(s)$ has no pole in the open right half plane and has a polar plot roughly of the form shown in Figure 8.20, then its Nyquist plot (the polar plot plus its reflection) will not encircle $(-1, 0)$ if both the phase and gain margins are positive. Thus, the overall system $G_o(s) = G_l(s)/(1 + G_l(s))$ is stable. In conclusion,

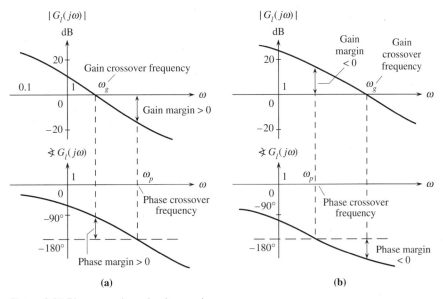

Figure 8.21 Phase margin and gain margin.

if $G_l(s)$ has no open right-half-plane pole, then *generally* the overall system $G_o(s) = G_l(s)/(1 + G_l(s))$ is stable if the phase margin and the gain margin of $G_l(s)$ are both positive. Furthermore, the larger the gain and phase margins, the more stable the system. In design we may require, for example, that the gain margin be larger than 6 dB and that the phase margin be larger than 30°. If either the phase margin or gain margin equals 0 or is negative, then the system $G_o(s)$ is generally not stable.

If a loop transfer function $G_l(s)$ has open right-half-plane poles, in order for $G_o(s)$ to be stable, the Nyquist plot must encircle the critical point. In this case, the polar plot may have a number of phase-crossover frequencies and a number of gain-crossover frequencies as shown in Figure 8.22(a), thus phase margins and gain margins are not unique. Furthermore, some phase margins must be positive and some negative in order for the system to be stable. Thus, the use of phase margins and gain margins becomes complicated. For this reason, if loop transfer functions have open right-half-plane poles, the concepts of phase and gain margins are less useful.

Even if loop transfer functions have no pole in the open right half plane, care must still be exercised in using the phase and gain margins. First, the polar plots of such transfer functions may not be of the form shown in Figure 8.20. They could have more than one phase margin and/or more than one gain margin as shown in Figure 8.22(b). Moreover, if a polar plot is as shown in Figure 8.22(c), even though $G_l(s)$ has a large phase margin and a large gain margin, the closed loop system $G_o(s) = G_l(s)/(1 + G_l(s))$ has a poor degree of stability. Therefore, the relationship between the degree of stability and phase and gain margins is not necessarily exact.

Exercise 8.4.3

Find the gain and phase margins of the following transfer functions:

a. $\dfrac{2}{s(s + 1)}$

b. $\dfrac{5}{s(s + 1)(s + 2)}$

c. $\dfrac{1}{s^2(s + 1)}$

8.5   FREQUENCY-DOMAIN SPECIFICATIONS FOR OVERALL SYSTEMS

With the preceding background, we are ready to discuss the design problem. The problem is: given a plant with transfer function $G(s)$, design a feedback system with transfer function $G_o(s)$ to meet a set of specifications. The specifications are generally stated in terms of position error, rise time, settling time, and overshoot. Because they

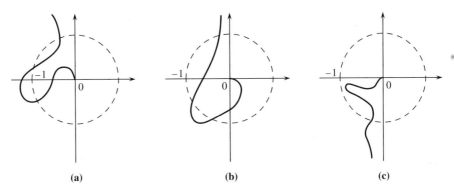

(a)   (b)   (c)

Figure 8.22 Applicability of phase and gain margins.

are defined for the time responses of $G_o(s)$, they are called *time-domain specifications*. Now, if the design is to be carried out using frequency plots, we must translate the time-domain specifications into the frequency domain. This will be carried out in this section. Recall that the specifications are stated for the overall transfer function $G_o(s)$, not for the plant transfer function $G(s)$.

Steady-State Performance

Let $G_o(s)$ be written as

$$G_o(j\omega) = |G_o(j\omega)|e^{j\theta(\omega)}$$

where $\theta(\omega) = \tan^{-1}[\text{Im } G_o(j\omega)/\text{Re } G_o(j\omega)]$. The plot of $|G_o(j\omega)|$ with respect to ω is called the *amplitude plot* and the plot of $\theta(\omega)$ with respect to ω is called the *phase plot* of $G_o(s)$. Typical amplitude and phase plots of control systems are shown in Figure 8.23. From the final-value theorem, we know that the steady-state performance (in the time domain) is determined by $G_o(s)$ as $s \to 0$ or $G_o(j\omega)$ as $\omega \to 0$ (in the frequency domain). Indeed, the position error or the steady-state error due to a step-reference input is, as derived in (6.3),

$$\text{Position error} = e_p(t) = |1 - G_o(0)| \times 100\% \qquad (8.21)$$

Thus from $G_o(0)$, the position error can immediately be determined. For example, if $G_o(0) = 1$, then $e_p = 0$; if $G_o(0) = 0.95$, then the position error is 5%. The velocity error or the steady-state error due to a ramp-reference input is, as derived in (6.6),

$$\text{Velocity error} = e_v(t) = |(1 - G_o(0))t - G_o'(0)| \times 100\% \qquad (8.22a)$$

In order to have finite velocity error, we require $G_o(0) = 1$. In this case, (8.22a) reduces to

$$\text{Velocity error} = v_p(t) = |G_o'(0)| \times 100\% \qquad (8.22b)$$

which implies that the velocity error depends only on the slope of $G_o(j\omega)$ at $\omega = 0$. If the slope is zero, velocity error is zero. Thus the steady-state performance can be easily translated into the values of $G_o(j\omega)$ and its derivatives at $\omega = 0$.

Transient Performance

The specifications on the transient performance consist of overshoot, settling time, and rise time. These specifications are closely related to M_p and bandwidth shown in Figure 8.23. The constant M_p, called the *peak resonance*, is defined as

$$M_p := \max \left| G_o(j\omega) \right| \tag{8.23}$$

for $\omega \geq 0$. It is the largest magnitude of $G_o(j\omega)$ in positive frequencies. The *bandwidth* is defined as the frequency range in which the magnitude of $G_o(j\omega)$ is equal to or larger than $0.707|G_o(0)|$. The frequency ω_c with the property $|G_o(j\omega_c)| = 0.707|G_o(0)|$ is called the *cutoff frequency*. If $G_o(0) = 1$ or 0 dB, the amplitude of $G_o(s)$ at ω_c is

$$20 \log 0.707 = -3 \text{ dB}$$

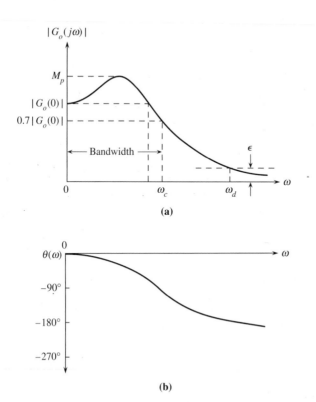

(a)

(b)

Figure 8.23 Typical frequency plot of control systems.

and, because the power is proportional to $|G_o(j\omega)|^2$, the power of $G_o(s)$ at ω_c is

$$(0.707)^2 = 0.5$$

thus ω_c is also called the -3 dB or *half-power point*. In conclusion, the cut-off frequency of $G_o(s)$ is defined as the frequency at which its amplitude is 70% or 3 dB below the level at $\omega = 0$, or at which the power is half of that at $\omega = 0$.

Now we discuss the relationship between the specifications on transient performance, and the peak resonance and bandwidth. Similar to the development of the desired pole region in Chapter 7, we consider the following quadratic transfer function

$$G_o(s) := \frac{\omega_n^2}{s^2 + 2\zeta\omega_n s + \omega_n^2} \qquad (8.24)$$

Clearly we have

$$G_o(j\omega) = \frac{\omega_n^2}{-\omega^2 + 2j\zeta\omega_n\omega + \omega_n^2}$$

and

$$|G_o(j\omega)|^2 = \frac{\omega_n^4}{(\omega_n^2 - \omega^2)^2 + (2\zeta\omega_n\omega)^2}$$

By algebraic manipulation, it can be shown that

$$M_p = \max |G_o(j\omega)| = \begin{cases} \dfrac{1}{2\zeta\sqrt{1 - \zeta^2}} & \text{for } \zeta < 0.707 \\ 1 & \text{for } \zeta \geq 0.707 \end{cases} \qquad (8.25)$$

Thus the peak resonance M_p depends only on the damping ratio. The same situation holds for the overshoot as shown in Figure 7.2(b). Now we plot (8.25) on Figure 7.2(b) to yield Figure 8.24. From Figure 8.24, the relationship between overshoot and peak resonance can easily be obtained. For example, if we require the overshoot to be less than 20%, then from Figure 8.24, we see that M_p must be less than 1.25. Similarly, if we require the overshoot to be less than 10%, then we require $M_p \leq$ 1.05. Thus, the peak resonance dictates the overshoot.

The bandwidth of $G_o(s)$ is related to the speed of response. The larger the bandwidth, the faster the response. This rule of thumb is widely accepted in engineering, even though the exact relationship is not known and the statement may not be true for every control system. The speed of response here may mean the rise time or settling time. We give a plausible argument of the statement by using the quadratic transfer function in (8.24). From the horizontal coordinate $\omega_n t$ in Figure 4.7, we argued in Section 7.2.1 that the rise time is inversely proportional to ω_n. The Bode plot of (8.24) is shown in Figure 8.9. The intersection of the -3-dB horizontal line with the gain plot yields the cutoff frequency and the bandwidth. Because the hor-

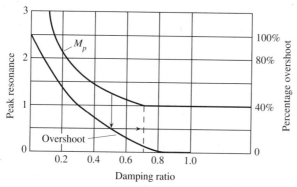

Figure 8.24 Peak resonance and overshoot.

izontal coordinate is ω/ω_n, the bandwidth is proportional to ω_n. Thus we conclude that the larger the bandwidth, the smaller the rise time or the faster the response.

In addition to bandwidth, we often impose constraint on the amplitude of the gain plot at high frequencies. For example, we may require the magnitude of $G_o(j\omega)$ to be less than, say, ϵ for all $\omega > \omega_d$ as shown in Figure 8.25. For the same bandwidth, if ω_d is closer to the cutoff frequency, then the cutoff rate of the gain plot will be larger as shown. The implication of this high-frequency specification on the time response is not clear. However, it has a very simple implication on noise rejection. Noise or unwanted signals with frequency spectra lying in the range $\omega > \omega_d$ will be greatly suppressed or eliminated.

The peak resonance and bandwidth are defined on the magnitude plot of $G_l(s)$. The phase plot of $G_l(s)$ is not used. If the phase plot is not a linear function of ω, then distortion will occur. Distortion of signals in control systems is not as critical as in signal processing and filter design; hence, in the design of control systems, generally no specification is imposed on the phase plot of $G_o(s)$.

We summarize the preceding discussion in Table 8.1. The second column lists the time-domain specifications, the third column lists the frequency-domain specifications. Both are specified for overall systems. The last column of the table will be developed in the next section.

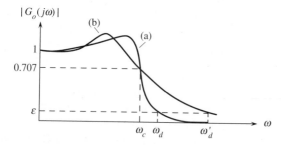

Figure 8.25 Two frequency plots with same bandwidth.

Table 8.1 Specifications of Control Systems

	Time Domain	Frequency Domain	
	Overall System	Overall System	Loop Transfer Function
Steady-State Performance	Position error Velocity error	$\|1 - G_o(0)\|$ $\|G'_o(0)\|$ if $G_o(0) = 1$	$\|1/(1 + K_p)\|$ $\|1/K_v\|$
Transient Performance	Overshoot Rise time Settling time	Peak resonance Bandwidth High-frequency gain	Gain and phase margins Gain crossover frequency High-frequency gain

8.6 FREQUENCY-DOMAIN SPECIFICATIONS FOR LOOP TRANSFER FUNCTIONS—UNITY-FEEDBACK CONFIGURATION

With the frequency-domain specifications discussed in the preceding section, the design problem now becomes: Given a plant with frequency-domain plot $G(j\omega)$, design a system such that the overall transfer function $G_o(j\omega)$ will meet the specifications. There are two possible approaches to carry out the design. The first approach is to compute $G_o(j\omega)$ from $G(j\omega)$ for a chosen configuration. If $G_o(j\omega)$ does not meet the specifications, we then introduce a compensator and recompute $G_o(j\omega)$. We repeat the process until $G_o(j\omega)$ meets the specifications. The second approach is to translate the specifications for $G_o(j\omega)$ into a set of specifications for $G(j\omega)$. If $G(j\omega)$ does not meet the specifications, we introduce a compensator $C(s)$ so that $C(j\omega)G(j\omega)$ will meet the specifications. Because the first approach is much more complicated than the second approach, it is rarely used in practice. The method introduced in the remainder of this chapter takes the second approach.

In order to take this approach, the specifications for the overall system must be translated into those for $G(j\omega)$. We shall do so for the unity-feedback configuration shown in Figure 8.1. Define $G_l(s) = C(s)G(s)$. Then the transfer function of the unity-feedback system in Figure 8.1 is

$$G_o(s) = \frac{C(s)G(s)}{1 + C(s)G(s)} = \frac{G_l(s)}{1 + G_l(s)} \tag{8.26}$$

We call $G_l(s)$ the *loop transfer function*.

Steady-State Performance

Consider the loop transfer function $G_l(s)$. We define

$$K_p := \lim_{s \to 0} G_l(s) = G_l(0) \tag{8.27a}$$

$$K_v := \lim_{s \to 0} sG_l(s) \tag{8.27b}$$

Now we compute steady-state errors in terms of K_p and K_v. The error $e = r - y$ of the unity-feedback system in Figure 8.1 can be computed in the Laplace-transform domain as

$$E(s) = R(s) - Y(s) = R(s) - G_o(s)R(s) \tag{8.28}$$

$$= \left(1 - \frac{G_l(s)}{1 + G_l(s)}\right) R(s) = \frac{1}{1 + G_l(s)} \cdot R(s)$$

If $r(t) = 1$ or $R(s) = 1/s$, then the steady-state error in (8.28) is the position error defined in (6.3).[3] Using the final-value theorem, we have

$$e_p(t) = \lim_{t \to \infty} |e(t)| = \lim_{s \to 0} |sE(s)| = \lim_{s \to 0} \left| s \left(\frac{1}{1 + G_l(s)}\right) \frac{1}{s} \right|$$

$$= \left| \frac{1}{1 + G_l(0)} \right| = \left| \frac{1}{1 + K_p} \right| \times 100\% \tag{8.29}$$

We see that the position error depends only on K_p, thus K_p is called the *position-error constant*.

If $r(t) = t$ or $R(s) = 1/s^2$, then the steady-state error in (8.28) is the velocity error defined in (6.6).[4] Using the final-value theorem, we have

$$e_v(t) = \lim_{t \to \infty} |e(t)| = \lim_{s \to 0} |sE(s)| = \lim_{s \to 0} \left| s \left(\frac{1}{1 + G_l(s)}\right) \frac{1}{s^2} \right|$$

$$= \lim_{s \to 0} \left| \frac{1}{s + sG_l(0)} \right| = \left| \frac{1}{K_v} \right| \times 100\% \tag{8.30}$$

Here we have implicitly assumed that $e_v(t)$ approaches a constant, otherwise the final-value theorem cannot be applied. We see that the velocity error depends only on K_v, thus K_v is called the *velocity-error constant*. Once the position or velocity error is specified, we can use (8.29) or (8.30) to find the range of K_p or K_v. This translates the steady-state specifications for overall systems into specifications for loop transfer functions.

As discussed in Section 6.3.2, the position or velocity error of unity-feedback systems can be determined from the system type of $G_l(s)$. The loop transfer function $G_l(s)$ is of type i if it has i poles at the origin. For example, $G_l(s) = 2/(s - 1)$ and $(s + 2)/(s^2 + s + 10)$ are of type 0, and their position-error constants are -2 and $2/10 = 0.2$. Their velocity-error constants are zero. The loop transfer function $G_l(s) = 1/s$ and $1/s(s + 2)$ are of type 1. Their position-error constants are infinity, and their velocity-error constants are 1 and $1/2 = 0.5$. These are summarized in Table 8.2 in which k denotes some nonzero constant and K_a is defined as $\lim_{s \to 0} s^2 G_l(s)$ and is called the *acceleration-error constant*.

---

[3]If $r(t) = a$, then the error must be divided by a. Because $a = 1$, this normalization is not needed.

[4]If $r(t) = at$, then the error must be divided by a. Because $a = 1$, this normalization is not needed.

Table 8.2 Error Constants

	K_p	K_v	K_a	
Type 0	k	0	0	
Type 1	∞	k	0	$e_p = \left\| \dfrac{1}{1 + K_p} \right\|$
Type 2	∞	∞	k	$e_v = \left\| \dfrac{1}{K_v} \right\|$

Now if $G_l(s)$ is of type 1, then $K_p = \infty$ and $e_p = 0$. Thus the unity-feedback system in Figure 8.1 will track any step-reference input without an error. If $G_l(s)$ is of type 2, then $K_v = \infty$ and $e_v = 0$. Thus the system will track any ramp-reference input without an error. These are consistent with the conclusions in Section 6.3.2. To conclude this part, we mention that (8.29) and (8.30) are established for unity-feedback systems. They are not necessarily applicable to other configurations.

Transient Performance[5]

The transient performance of $G_o(s)$ is specified in terms of the peak resonance M_p, bandwidth, and high-frequency gain. Now we shall translate these into a set of specifications for $G_l(s)$. To do so, we must first establish the relationship between $G_o(j\omega)$ and $G_l(j\omega) = G(j\omega)C(j\omega)$. Let the polar plot of $G_l(s)$ be as shown in Figure 8.26(a). Consider the vector $G_l(j\omega_1)$. Then the vector drawn from $(-1, 0)$ to $G_l(j\omega_1)$ equals $1 + G_l(j\omega_1)$. Their ratio

$$\frac{G_l(j\omega_1)}{1 + G_l(j\omega_1)} = G_o(j\omega_1)$$

yields $G_o(j\omega_1)$. Therefore it is possible to translate $G_l(j\omega)$ graphically into $G_o(j\omega)$.

To facilitate the translation, we first compute the loci on the G_l-plane that have constant $|G_o(j\omega)|$. Let $x + jy$ be a point of $G_l(j\omega)$ on the G_l-plane. Clearly we have

$$|G_o(j\omega)| = \left| \frac{G_l(j\omega)}{1 + G_l(j\omega)} \right| = \left| \frac{x + jy}{1 + x + jy} \right| = \left[\frac{x^2 + y^2}{(1 + x)^2 + y^2} \right]^{1/2} \tag{8.31}$$

Let $|G_o(j\omega)| = M$. Then (8.31) implies

$$M^2 = \frac{x^2 + y^2}{(1 + x)^2 + y^2}$$

or

$$(1 + x)^2 M^2 + y^2 M^2 = x^2 + y^2$$

[5]This subsection establishes the last column in Table 8.1. It may be skipped without loss of continuity.

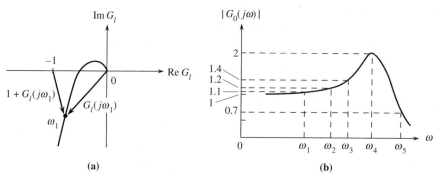

Figure 8.26 From $G_l(j\omega)$ to $G_o(j\omega)$.

which, after some simple algebraic manipulation, becomes

$$\left(x - \frac{M^2}{1 - M^2}\right)^2 + y^2 = \left(\frac{M}{1 - M^2}\right)^2$$

It is the equation of a circle with radius $M/(1 - M^2)$ and centered at $(M^2/(1 - M^2), 0)$. A family of such circles for various M are plotted in Figure 8.27. They are called *the constant M-loci*.[6] If the constant M-loci are superposed on the polar plot of $G_l(s)$ as shown, then $G_o(j\omega)$ can be read out directly as shown in Figure 8.26(b). The circle with the largest M which the polar plot of $G_l(s)$ touches tangentially yields the peak resonance M_p.

In order to better see the relationship between the peak resonance and phase and gain margins, we expand part of the plot of Figure 8.27 in Figure 8.28. From the intersections of the circle of $M = 1.2$ with the real axis and the unit circle, we may conclude that if the gain margin of $G_l(s)$ is smaller than 5.3 dB or the phase margin of $G_l(s)$ is smaller than 50°, then the peak resonance of $G_o(s)$ must be at least 1.2. Conversely, if a polar plot is roughly of the form shown in Figure 8.28, then we *may* have the following:

Gain margin $\geq$ 10 dB, Phase margin $\geq$ 45° $\Rightarrow$ $M_p \approx 1.3$

and

Gain margin $\geq$ 12 dB, Phase margin $\geq$ 60° $\Rightarrow$ $M_p \approx 1.0$

This establishes a relationship between the peak resonance of $G_o(s)$ and the gain and phase margins of $G_l(s)$.

Using Figure 8.28, we can also read the bandwidth of $G_o(s)$ from the polar plot of $G_l(s)$. If $G_o(0) = 1$, cutoff frequency ω_c is the frequency at which the polar plot intersects with the M-circle of 0.7. In the example in Figure 8.28, as ω increases, the polar plot first passes through the unit circle centered at the origin and then

---

[6]It is also possible to plot the loci of constant phases of $G_o(s)$ on the G_l-plane. The plot consists of a family of circles called *constant N-loci*. The plot of constant M- and N-loci on the log magnitude-phase plot is called the *Nichols chart*. They are not used in this text and will not be discussed.

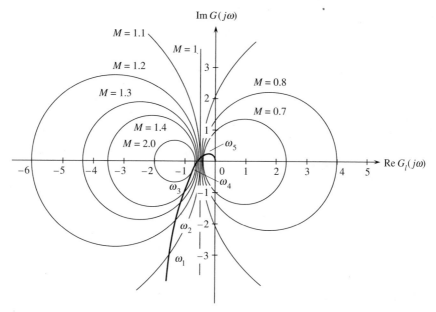

Figure 8.27 Constant M-loci.

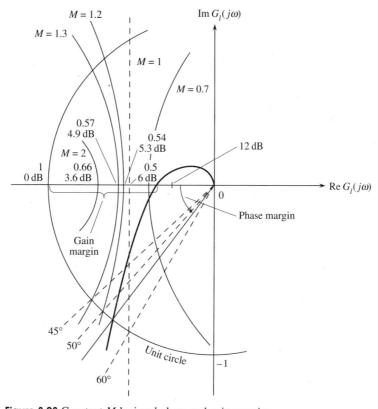

Figure 8.28 Constant M-loci and phase and gain margins.

through the M-circle of 0.7, thus we have

$$\omega_g \leq \omega_c \tag{8.32}$$

where ω_g is the gain-crossover frequency of $G_l(s)$. Note that ω_c is defined for $G_o(s)$, whereas ω_g is defined for $G_l(s)$. Although (8.32) is developed for the polar plot in Figure 8.28, it is true in general, and is often used in frequency-domain design. For example, if an overall system is required to respond as fast as possible or, equivalently, to have a bandwidth as large as possible, then we may search for a compensator $C(s)$ so that the loop transfer function $G_l(s) = C(s)G(s)$ has a gain-crossover frequency as large as possible.

The high-frequency specification on $G_o(s)$ can also be translated into that of $G_l(s) = C(s)G(s)$. If $G(s)$ is strictly proper and if $C(s)$ is proper, then $|G_l(j\omega)| << 1$, for large ω. Thus we have

$$|G_o(j\omega)| = \left| \frac{G_l(j\omega)}{1 + G_l(j\omega)} \right| \approx |G_l(j\omega)|$$

for large ω. Hence the specification of $G_o(j\omega) < \epsilon$, for $\omega \geq \omega_d$, can be translated to $|G_l(j\omega)| < \epsilon$, for $\omega \geq \omega_d$.

The preceding discussion is tabulated in the last column of Table 8.1. We see that there are three sets of specifications on accuracy (steady-state performance) and speed of response (transient performance). The first set is given in the time domain, the other two are given in the frequency domain. The first two sets are specified for overall systems, the last one is specified for loop transfer functions in the unity feedback configuration. It is important to mention that even though specifications are stated for loop transfer functions, the objective is still to design a good overall system.

8.6.1   Why Use Bode Plots?

With the preceding discussion, the design problem becomes as follows: Given a plant with proper transfer function $G(s)$, find a compensator $C(s)$ in the unity-feedback configuration in Figure 8.1 such that the loop transfer function $G_l(s) = C(s)G(s)$ will meet the specifications on K_p, or K_v, the phase margin, gain margin, and gain-crossover frequency. The design is to be carried out using only frequency plots. This chapter introduced three frequency plots—namely, the polar plot, log magnitude-phase plot, and Bode plot. These three plots are all equivalent, and theoretically any one can be used in the design. In practice, however, the Bode plot is used almost exclusively, for the following two reasons. First, the Bode plot can be drawn from its asymptotes; thus the Bode plot is the easiest to draw among the three plots. Second and more important, we have

$$\text{Bode plot of } C(s)G(s) = \text{Bode plot of } G(s) + \text{Bode plot of } C(s)$$

This applies to the magnitude plot as well as to the phase plot, that is,

$$20 \log |C(j\omega)G(j\omega)| = 20 \log |C(j\omega)| + 20 \log |G(j\omega)|$$

and

$$\angle C(j\omega)G(j\omega) = \angle C(j\omega) + \angle G(j\omega)$$

Thus in the design, if the Bode plot of $G(s)$ does not meet the specifications, we simply add the Bode plot of $C(s)$ to it until the sum meets the specifications. On the other hand, if we use the polar plot of $G(s)$ to carry out the design, the polar plot of $G(s)$ is of no use in subsequent design because the polar plot of $C(s)G(s)$ cannot easily be drawn from the polar plot of $G(s)$. Therefore the polar plot is less often used. In the log magnitude-phase plot, the frequency appears as a parameter on the plot, thus the summation of $C(j\omega)$ and $G(j\omega)$ involves summations of vectors and is not as convenient as in the Bode plot. Thus, the Bode plot is most often used in the frequency-domain design.

8.6.2 Design from Measured Data

If the transfer function of a plant is known, we can compute its K_p or K_v, plot its Bode plot, and then proceed to the design. Even if the transfer function is not known, we can still carry out the design from measured data. This is a unique feature of the frequency-domain method. Instruments are available to measure the frequency plot, in particular, the Bode plot of plants. Once a Bode plot is obtained by measurement, we can then read out its phase margin and gain margin. Its position- or velocity-error constants can also be read out from the plot. At very low frequencies, if the slope of the gain plot is zero, as shown in Figure 8.29(a), the transfer function is of type 0; if the slope is -20 dB/decade, it is of type 1; if the slope is -40 dB/decade, it is of type 2; and so forth. For a type 0 transfer function, we have, at very low frequencies or at $\omega = 0$,

$$20 \log |G(j\omega)| = 20 \log |K_p|$$

Therefore, if we extend the horizontal line to intersect with the vertical coordinate at, say, a dB, then we have

$$20 \log |K_p| = a$$

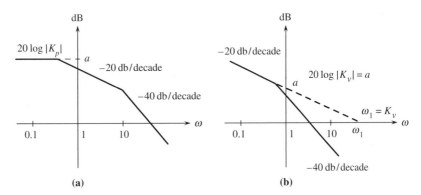

Figure 8.29 Bode plots.

which implies $K_p = 10^{a/20}$. Thus the position-error constant can be easily obtained from the plot.

Every type 1 transfer function can be expressed as

$$G(s) = \frac{k(1 + b_1 s)(1 + b_2 s) \cdots}{s(1 + a_1 s)(1 + a_2 s)(1 + a_3 s) \cdots}$$

Clearly, its position-error constant is infinity. At very low frequencies, the Bode gain plot is governed by

$$20 \log |G(j\omega)| \approx 20 \log \left|\frac{k}{j\omega}\right| = 20 \log \left|\frac{k}{\omega}\right| = 20 \log \left|\frac{K_v}{\omega}\right| \tag{8.33}$$

This is a straight line with slope -20 dB/decade. We extend the straight line to intersect with the vertical axis at, say, a dB and intersect with the horizontal axis at, say, ω_1 radians per second. The vertical axis passes through $\omega = 1$, thus (8.33) becomes

$$20 \log \left|\frac{K_v}{1}\right| = a$$

which implies $K_v = 10^{a/20}$. The gain of (8.33) is 0 dB at $\omega = \omega_1$, thus we have

$$0 = 20 \log \left|\frac{K_v}{\omega_1}\right|$$

which implies $K_v = \omega_1$. Thus, the velocity-error constant of type 1 transfer functions can also be easily obtained from the Bode gain plot. In conclusion, from the leftmost asymptote of Bode gain plots, the constants K_p and K_v can be easily obtained. Once K_p, K_v, the phase margin, and the gain margin are read out from the measured data, we can then proceed to the design.

8.7   DESIGN ON BODE PLOTS

Before discussing specific design techniques, we review the problem once again. Given a plant with transfer function $G(s)$, the objective is to design an overall system to meet a set of specifications in the time domain. Because the design will be carried out by using frequency plots, the specifications are translated into the frequency domain for the overall transfer function $G_o(s)$, as shown in Table 8.1. For the unity-feedback configuration shown in Figure 8.1, the specifications for $G_o(j\omega)$ can be further translated into those for the loop transfer function $G_l(s) = G(s)C(s)$ as shown in Table 8.1. Therefore the design problem now becomes: Given a plant with Bode plot $G(j\omega)$, find a compensator $C(s)$ in Figure 8.1 such that the Bode plot of $C(s)G(s)$ will meet the specifications on position- or velocity-error constant, phase margin, gain margin, gain-crossover frequency, and high-frequency gain. If this is successful, all we can hope for is that the resulting overall system $G_o(s) = G_l(s)/(1 + G_l(s))$ would be a good control system. Recall that the translations of the specifications in Table 8.1 are developed mainly from quadratic transfer functions; they may not hold in general. Therefore, it is always advisable to simulate the resulting overall system to check whether it really meets the design specifications.

The search for $C(s)$ is essentially a trial-and-error process. Therefore we always start from a simple compensator and, if we are not successful, move to a more complicated one. The compensators used in this design are mainly of the following four types: (a) gain adjustment (amplification or attenuation), (b) phase-lag compensation, (c) phase-lead compensation, and (d) lag-lead compensation. Before proceeding, we mention a useful property. Consider the Bode plots shown in Figure 8.30. It is assumed that the plant has no open right-half-plane poles nor open right-half-plane zeros. Under this assumption, the phase can be estimated from the slopes of the asymptotes of the gain plot. If a slope is -20 dB/decade, the phase will approach $-90°$. If a slope is -40 dB/decade, the phase will approach $-180°$. If a slope is -60 dB/decade, the phase will approach $-270°$. Because of this property, if the slope of the asymptote at the gain-crossover frequency is -60 dB/decade, as shown in Figure 8.30(a), then the phase will approach $-270°$ and the phase margin will be negative. Consequently the feedback system will be unstable. On the other hand, if the slope of the asymptote at the gain-crossover frequency is -20 dB/decade as shown in Figure 8.30(b), then the phase margin is positive. If the slope of the asymptote at the gain-crossover frequency is -40 dB/decade, the phase margin can be positive or negative. For this reason, if it is possible, the asymptote at the gain-crossover frequency is designed to have slope -20 dB/decade. This is the case in almost every design in the remainder of this chapter.

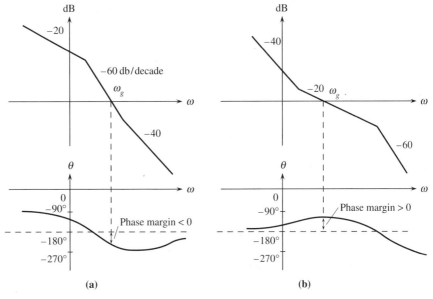

Figure 8.30 Bode plots.

8.7.1 Gain Adjustment

The simplest possible compensator $C(s)$ is a gain k. The Bode gain plot of $C(s)G(s) = kG(s)$ is $20 \log |k| + 20 \log |G(j\omega)|$. Thus, the introduction of gain k will simply shift up the Bode gain plot of $G(s)$ if $|k| > 1$, and shift it down if $|k| < 1$. If gain k is positive, its introduction will not affect the phase plot of $G(s)$. For some problems, it is possible to achieve a design by simply shifting a Bode gain plot up or down.

Example 8.7.1

Consider the unity-feedback system shown in Figure 8.31. Let the plant transfer function be $G(s) = 1/s(s + 2)$ and let the compensator $C(s)$ be simply a constant k. Find a gain k such that the loop transfer function $kG(s)$ will meet the following:

1. Position error $\leq 10\%$.
2. Phase margin $\geq 60°$, gain margin ≥ 12 dB.
3. Gain-crossover frequency as large as possible.

The plant is of type 1, thus its position-error constant K_p is infinity and its position error $|1/(1 + K_p)|$ is zero for any k. The Bode plot of $G(s) = 1/s(s + 2)$ or

$$G(s) = \frac{1}{2s\left(1 + \frac{1}{2}s\right)} = \frac{0.5}{s\left(1 + \frac{1}{2}s\right)}$$

is shown in Figure 8.32 with the solid lines. The gain plot crosses the 0-dB line roughly at 0.5; thus the gain-crossover frequency is $\omega_g = 0.5$ rad/s. The phase margin is then measured from the plot as 76°. To find the gain margin, we must first find the phase-crossover frequency. Because the phase approaches the $-180°$ line asymptotically as shown, it intersects the line at $\omega = \infty$, and the phase-crossover frequency ω_p is infinity. The gain plot goes down to $-\infty$ dB with slope -40 dB/decade as $\omega \to \infty$. Thus, the gain margin at $\omega_p = \infty$ is infinity. Thus, the Bode plot of $G(s)$ meets the specifications in (1) and (2). If we do not require the specification in (3), then there is no need to introduce any compensator and the design is completed. It is important to point out that no compensator means $C(s) = k = 1$ and that the unity feedback is still needed as shown in Figure 8.31.

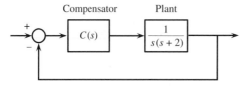

Figure 8.31 Unity-feedback system.

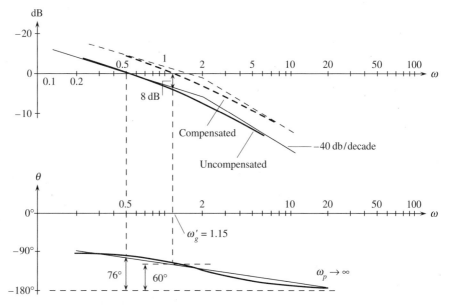

Figure 8.32 Bode plot of $1/s(s + 2)$.

Now we shall adjust k to make the gain-crossover frequency as large as possible. If we increase k from 1, the gain plot will move upward and the gain-crossover frequency will increase (shift to the right). This will cause the phase margin to decrease, as can be seen from Figure 8.32. In other words, as the gain-crossover frequency increases, the phase margin will decrease. To find the largest permissible k, we draw a horizontal line with 60° phase margin. Its intersection with the phase plot yields the largest permissible gain-crossover frequency, which is read from the plot as $\omega'_g = 1.15$. If we draw a vertical line upward from $\omega'_g = 1.15$, then we can read from the gain plot that if we add 8 dB to the gain plot or shift the gain plot up 8 dB, then the new gain-crossover frequency will be shifted to $\omega'_g = 1.15$. Thus the required gain k is

$$20 \log k = 8$$

which implies $k = 2.5$. Note that the gain margin remains infinity for $k = 2.5$. Thus the design is completed by introducing $k = 2.5$.

This problem was also designed using the root-locus technique in Section 7.2. The result was $k = 2$. For this problem, the root-locus and frequency-domain methods yield comparable results.

8.8 PHASE-LAG COMPENSATION

First we use the example in Figure 8.31 to show that adjustment of a gain alone sometimes cannot achieve a design.

Example 8.8.1

Consider a plant with transfer function $G(s) = 1/s(s + 2)$. Find a compensator $C(s)$ in Figure 8.1 such that $C(s)G(s)$ will meet the following: (1) velocity error $\leq 10\%$, (2) phase margin $\geq 60°$, and (3) gain margin ≥ 12 dB.

First we choose $C(s) = k$ and see whether or not the design is possible. The loop transfer function is

$$G_l(s) = kG(s) = \frac{k}{s(s + 2)} \tag{8.34}$$

It is of type 1 and its velocity-error constant is

$$K_v = \lim_{s \to 0} sG_l(s) = \frac{k}{2} \tag{8.35}$$

In order to meet (1), we require

$$\left| \frac{1}{K_v} \right| = \left| \frac{2}{k} \right| \leq 0.1 \tag{8.36}$$

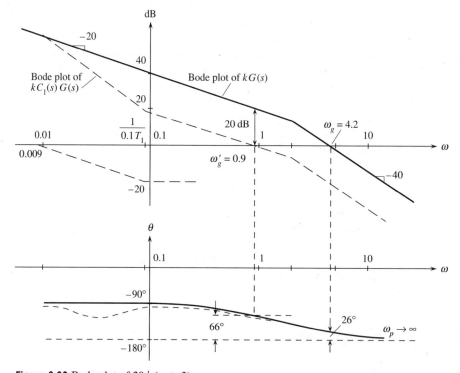

Figure 8.33 Bode plot of $20/s(s + 2)$.

which implies $k \geq 20$. We choose $k = 20$ and plot the Bode plot of

$$kG(s) = 20 \cdot \frac{1}{s(s + 2)} = \frac{10}{s\left(1 + \frac{1}{2}s\right)} \tag{8.37}$$

in Figure 8.33. We plot only the asymptotes. We see that the gain-crossover frequency roughly equals $\omega_g = 4.2$ rad/s, and the phase margin roughly equals 26°. The phase-crossover frequency is $\omega_p = \infty$, and the gain margin is infinity. Although the gain margin meets the specifications, the phase margin does not. If we restrict $C(s)$ to be k, the only way to increase the phase margin is to decrease k. This, however, will violate (1). Thus for this problem, it is not possible to achieve the design by adjusting k alone.

Now we introduce a compensating network that is more complex than a gain. Consider the network shown in Figure 8.34(a). Its transfer function is

$$C_1(s) = \frac{E_2(s)}{E_1(s)} = \frac{R_2 + \frac{1}{Cs}}{R_1 + R_2 + \frac{1}{Cs}} = \frac{1 + R_2Cs}{1 + (R_1 + R_2)Cs} = \frac{1 + aT_1s}{1 + T_1s} \tag{8.38}$$

where

$$0 < a := \frac{R_2}{R_1 + R_2} < 1 \tag{8.39}$$

and

$$T_1 := (R_1 + R_2)C \tag{8.40}$$

The pole and zero of $C_1(s)$ are $-1/T_1$ and $-1/aT_1$. Because $a < 1$, the zero is farther away from the origin than the pole, as shown in Figure 8.34(b). For any $\omega \geq 0$, the phase of $C_1(s)$ equals $\theta - \phi$. Because $\phi \geq \theta$, the phase of $C_1(s)$ is negative for all $\omega \geq 0$. Thus the network is called a *phase-lag network*.

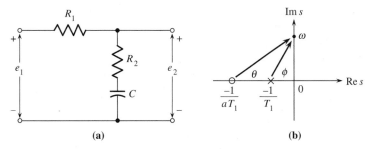

(a) (b)

Figure 8.34 Phase-lag network.

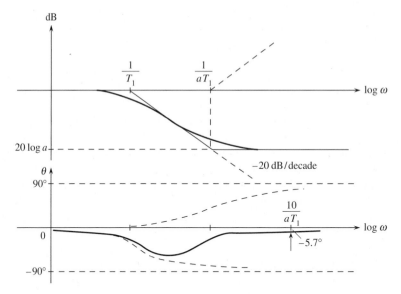

Figure 8.35 Bode plot of (8.38).

Now we plot the Bode plot of $C_1(s)$. The corner frequency of the pole is $1/T_1$. We draw two asymptotes from $1/T_1$, one with slope -20 dB/decade. The corner frequency of the zero is $1/aT_1$. We draw two asymptotes from $1/aT_1$, one with slope 20 dB/decade. Note that the corner frequency $1/aT_1$ is on the right-hand side of the corner frequency $1/T_1$, because $a < 1$. The summation of these yields the Bode gain plot of $C_1(s)$ as shown in Figure 8.35. For ω very small, the Bode gain plot is a horizontal line with gain 1 or 0 dB. For ω very large, $C_1(s)$ can be approximated by $(1 + aT_1s)/(1 + T_1s) \approx aT_1s/T_1s = a$; thus its Bode gain plot is a horizontal line with gain a or $20 \log a$ dB. Because $a < 1$, $20 \log a$ is a negative number. Thus the phase-lag network introduces an attenuation of $20 \, |\log a|$. The Bode phase plot of $C_1(s)$ can be similarly obtained as shown in Figure 8.35. We see that the phase is negative for all ω as expected.

A phase-lag network has two parameters a and T_1. The amount of attenuation introduced by the network is determined by a. In employing a phase-lag network, we use mainly its attenuation property. The phase characteristic will not be used except the phase at $10/aT_1$, ten times the right-hand-side corner frequency. From Figure 8.6, we see that the phase at $10/aT_1$ is at most $5.7°$. Now we shall redesign the problem in Example 8.8.1.

Example 8.8.1 (continued)

Consider a plant with transfer function $G(s) = 1/s(s + 2)$. Find a compensator $C(s)$ in Figure 8.1 such that $C(s)G(s)$ will meet (1) velocity error $\leq 10\%$, (2) phase margin $\geq 60°$, and (3) gain margin ≥ 12 dB.

Now we choose $C(s)$ as $kC_1(s)$, where $C_1(s)$ is given in (8.38). Because $C_1(0) = 1$, the velocity-error constant K_v of $C(s)G(s)$ is

$$K_v = \lim_{s \to 0} sG_l(s) = \lim_{s \to 0} skC_1(s)G(s) = \frac{k}{2}$$

Thus, in order to meet (1), we require, as in (8.36), $k \geq 20$. The Bode plot of

$$20 \cdot \frac{1}{s(s + 2)}$$

is plotted in Figure 8.33 with the solid lines. Its gain-crossover frequency is $\omega_g = 4.5$ rad/s and its phase margin is $25°$. The phase margin, however, is required to be at least $60°$. This will be achieved by introducing a phase-lag network. First we search for a new gain-crossover frequency that has a phase margin of $60°$ plus $6°$ (this will be explained later), or $66°$. This can be found by drawing a horizontal line with $66°$ phase margin. Its intersection with the phase plot yields the new gain-crossover frequency. It is read from Figure 8.33 as $\omega_g' = 0.9$. We then draw a vertical line upward to the gain plot. We see that if the gain is attenuated by 20 dB at $\omega \ll \omega_g'$, then the new gain plot will pass through $\omega_g' = 0.9$ at the 0-dB line. A phase-lag network will introduce an attenuation of $20 |\log a|$, thus we set

$$20 \log a = -20$$

which implies $a = 0.1$.

The remaining problem is to determine T_1 in (8.38). If the right-hand-side corner frequency $1/aT_1$ is close to ω_g', the phase margin at ω_g' will be greatly reduced by the phase of the network. If the corner frequency is very far away from ω_g' (on the left-hand side), then the phase margin will be hardly affected by the phase of the network. Therefore it seems that we should choose T_1 so that $1/aT_1$ is as small as possible. However, this will cause a different problem. If $1/aT_1$ is very small, then $C(s)$ has a pole very close to the origin, as shown in Figure 8.34(b). If we plot the root locus of $C(s)G(s)$, then we can see that the closed-loop system will have a pole very close to the origin and the corresponding time constant will be very large. Therefore, if $1/aT_1$ is very small, the speed of response of the resulting unity-feedback system may become slow. For this reason, we do not want to place $1/aT_1$ too far away from ω_g'. In practice, $1/aT_1$ is often placed at one decade below ω_g'; that is,

$$\frac{1}{aT_1} = \frac{\omega_g'}{10}$$

In this case, the phase-lag network has at most a phase lag of $5.7°$ at ω_g', as shown in Figure 8.35. Thus the phase of $G(s)$ at ω_g' will be reduced by roughly $6°$ after introducing $C(s)$. This is the reason for adding $6°$ to the required phase margin in determining ω_g'.[7] For this problem, we have

$$aT_1 = \frac{10}{\omega_g'} = \frac{10}{0.9} = 11.1$$

and

$$T_1 = \frac{11.1}{a} = \frac{11.1}{0.1} = 111$$

Thus the compensator is

$$C(s) = kC_1(s) = 20 \cdot \frac{1 + aT_1s}{1 + T_1s} = 20 \cdot \frac{1 + 11.1s}{1 + 111s} \qquad (8.41)$$

With this compensator, the Bode plot of $C(s)G(s)$ will meet the specifications on velocity error and phase margin. Now we check whether it will meet the specification on gain margin. From Figure 8.33, we see that the phase-crossover frequency is $\omega_p = \infty$ and the gain margin is infinity. Thus, $C(s)G(s)$ meets all specifications, and the design is completed.

We summarize the procedure of designing phase-lag networks in the following.

Step 1: Compute the position- or velocity-error constant from the specification on steady-state error.

Step 2: Plot the Bode plot of the plant with the required error constant. Measure the phase margin and gain margin from the plot.

Step 3: If we decide to use a phase-lag compensator, determine the frequency at which the Bode plot in Step 2 has the required phase margin plus 6°. This frequency is designated as the new gain-crossover frequency ω_g'.

Step 4: Measure the attenuation needed to bring the gain plot down to ω_g'. This attenuation will be provided by a phase-lag network. Let this attenuation be α. Compute a from $\alpha = -20 \log a$ or $a = 10^{-\alpha/20}$

Step 5: Compute T_1 from

$$\frac{1}{aT_1} = \frac{\omega_g'}{10} \qquad (8.42)$$

that is, the corner frequency $1/aT_1$ is placed one decade below the new gain-crossover frequency ω_g'.

Step 6: If the resulting system satisfies all other specifications, then the design is completed. The compensating network can be realized as shown in Figure 8.34(a).

---

[7]The design on Bode plots is carried out mostly by measurements. It is difficult to differentiate between 5.7° and 6° on a plot. Therefore we need not be concerned too much about accuracy in this method.

To conclude this section, we remark that the use of a phase-lag network will reduce the gain-crossover frequency. Consequently the bandwidth of the unity-feedback system may be reduced. For this reason, this type of compensation will make a system more sluggish.

8.9 PHASE-LEAD COMPENSATION

In this section we introduce a network that has a positive phase for every $\omega > 0$. Consider the network shown in Figure 8.36(a). It is built by using two resistors, one capacitor, and an amplifier. The impedance of the parallel connection of R_1 and the capacitor with impedance $1/Cs$ is

$$\frac{R_1 \cdot \dfrac{1}{Cs}}{R_1 + \dfrac{1}{Cs}} = \frac{R_1}{R_1 Cs + 1}$$

Thus the transfer function from e_1 to e_2 in Figure 8.36(a) is

$$C_2(s) := \frac{E_2(s)}{E_1(s)} = \frac{R_1 + R_2}{R_2} \cdot \frac{R_2}{R_2 + \dfrac{R_1}{R_1 Cs + 1}}$$

$$= \frac{R_1 + R_2}{R_2} \cdot \frac{R_2 + R_1 R_2 Cs}{R_1 + R_2 + R_1 R_2 Cs} =: \frac{1 + bT_2 s}{1 + T_2 s} \qquad (8.43)$$

where

$$b = \frac{R_1 + R_2}{R_2} > 1 \qquad (8.44)$$

and

$$T_2 = \frac{R_1 R_2}{R_1 + R_2} \cdot C \qquad (8.45)$$

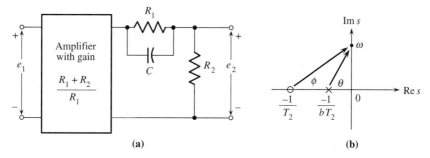

(a)   (b)

Figure 8.36 Phase-lead network.

The pole and zero of $C_2(s)$ are plotted in Figure 8.36(b). The phase of $C_2(s)$ equals $\theta - \phi$ as shown. Because $\theta > \phi$ for all positive ω, the phase of $C_2(s)$ is positive. Thus it is called a *phase-lead network*. Note that $C_2(0) = 1$.

The corner frequency of the zero is $1/bT_2$; the corner frequency of the pole is $1/T_2$. Because $b > 1$, $1/bT_2 < 1/T_2$ and the corner frequency of the zero is on the left-hand side of that of the pole. Thus the Bode gain plot of $C_2(s)$ is as shown in Figure 8.37. The gain at low frequencies is 1 or 0 dB. For large ω, we have $C_2(s) \approx bT_2 s/T_2 s = b$, thus the gain is b or $20 \log b$ dB as shown. Unlike the phase-lag network, the phase of the phase-lead network is essential in the design, therefore we must compute its phase. The phase of $(1 + bT_2 s)/(1 + T_2 s)$ at $s = j\omega$ equals

$$\phi(\omega) = \tan^{-1} bT_2\omega - \tan^{-1} T_2\omega = \tan^{-1} \frac{bT_2\omega - T_2\omega}{1 + bT_2^2\omega^2}$$

Thus we have

$$\tan \phi(\omega) = \frac{bT_2\omega - T_2\omega}{1 + bT_2^2\omega^2} \tag{8.46}$$

Since the phase plot is symmetric with respect to the midpoint of $1/bT_2$ and $1/T_2$, as shown in Figure 8.37, the maximum phase occurs at the midpoint. Because of the logarithmic scale, the midpoint ω_m is given by

$$\log \omega_m = \frac{1}{2}\left(\log \frac{1}{bT_2} + \log \frac{1}{T_2}\right) = \log \frac{1}{\sqrt{b}T_2}$$

or

$$\omega_m = \frac{1}{\sqrt{b}T_2} \tag{8.47}$$

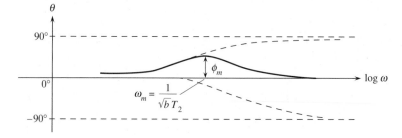

Figure 8.37 Bode plot of (8.43).

Thus the maximum phase at ω_m equals, substituting (8.47) into (8.46),

$$\tan \phi_m = \frac{(b-1)T_2\omega_m}{1 + bT_2^2\omega_m^2} = \frac{\dfrac{(b-1)}{\sqrt{b}}}{1 + 1} = \frac{b-1}{2\sqrt{b}}$$

which implies

$$\sin \phi_m = \frac{b-1}{b+1} \qquad \text{and} \qquad b = \frac{1 + \sin \phi_m}{1 - \sin \phi_m} \tag{8.48}$$

We see that the larger the constant b, the larger the maximum phase ϕ_m. However, the network in Figure 8.36 also requires a larger amplification. In practice, constant b is seldom chosen to be greater than 15. We mention that the gain equals $10 \log b$ at $\omega = \omega_m$, as shown in Figure 8.37.

The philosophy of using a phase-lead network is entirely different from that of using a phase-lag network. A phase-lag network is placed far away from the new gain-crossover frequency so that its phase will not affect seriously the phase margin. A phase-lead network, on the other hand, must be placed so that its maximum phase will contribute wholly to the phase margin. Therefore, ω_m should be placed at the new gain-crossover frequency. To achieve this, however, is not as simple as in the design of phase-lag networks. The procedure of designing phase-lead networks is explained in the following:

Step 1: Compute the position-error or velocity-error constant from the specification on steady-state error.

Step 2: Plot the Bode plot of $kG(s)$, the plant with the required position- or velocity-error constant. Determine the gain-crossover frequency ω_g and phase-crossover frequency ω_p. Measure the phase margin ϕ_1 and gain margin from the plot.

Step 3: If we decide to use a phase-lead compensator, calculate $\psi =$ (required phase margin) $- \phi_1$. The introduction of a phase-lead compensator will shift the gain-crossover frequency to the right and, consequently, decrease the phase margin. To compensate for this reduction, we add θ, say $5°$, to ψ. Compute $\phi_m = \psi + \theta$.

Step 4: Compute constant b from (8.48), which yields phase ϕ_m.

Step 5: If we place this maximum phase at ω_g or, equivalently, set $\omega_m = \omega_g$, because the network has positive gain, the gain-crossover frequency of $C_2(s)G(s)$ will be shifted to the right and the maximum phase will not appear at the new gain-crossover frequency. For this reason, we must compute first the new gain-crossover frequency before placing ω_m. We draw a horizontal line with gain $-10 \log b$. Its intersection with the Bode gain plot of $kG(s)$ yields the new gain-crossover frequency, denoted by ω_g'. Measure the phase margin ϕ_2 of $kG(s)$ at this frequency. If $\phi_1 - \phi_2 > \theta$, choose a larger θ in Step 3 and repeat Steps 4 and 5. If $\phi_1 - \phi_2 < \theta$, go to the next step.

Step 6: Set $\omega_m = \omega'_g$ and compute T_2 from (8.47). If the resulting system satisfies all other specifications, the design is completed. The network can then be realized as shown in Figure 8.36(a).

Example 8.9.1

We shall redesign the system discussed in the preceding section by using a phase-lead network. Consider a plant with transfer function $G(s) = 1/s(s + 2)$. Find a compensator $C(s)$ in Figure 8.1 such that $C(s)G(s)$ will meet (1) velocity error $\le$ 10%, (2) phase margin $\ge 60°$, and (3) gain margin ≥ 12 dB.

Now we shall choose $C(s)$ as $kC_2(s)$, where $C_2(s)$ is given in (8.43). Because $C_2(0) = 1$, the velocity-error constant K_v of $C(s)G(s)$ is

$$K_v = \lim_{s \to 0} sG_l(s) = \lim_{s \to 0} skC_2(s)G(s) = \frac{k}{2}$$

Thus we require $k \ge 20$ in order to meet the specification in (1). The Bode plot of $kG(s) = 20/s(s + 2)$ is plotted in Figure 8.38 with the solid lines. The phase margin ϕ_1 is 26°. The required phase margin is 60°. Thus we have $\psi = 60 - 26 = 34°$. If we introduce a phase-lead network, the gain-crossover frequency will increase and the corresponding phase margin will decrease. In order to compensate for this reduction, we choose arbitrarily $\theta = 5°$. Then we have $\phi_m = 34° + 5° = 39°$. This is the total phase needed from a phase-lead network.

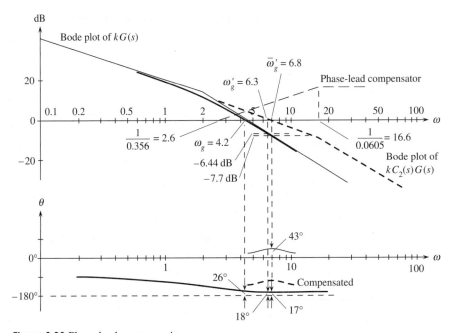

Figure 8.38 Phase-lead compensation.

Next we use (8.48) to compute b as

$$b = \frac{1 + \sin 39°}{1 - \sin 39°} = \frac{1.63}{0.37} = 4.4$$

We then draw a horizontal line with $-10 \log b = -10 \log 4.4 = -6.44$ dB as shown in Figure 8.38. Its intersection with the gain plot yields the new gain-crossover frequency as $\omega'_g = 6.3$. The corresponding phase margin is then read from the plot as $\phi_2 = 18°$. Because $\phi_1 - \phi_2 = 26° - 18° = 8°$, which is larger than $\theta = 5°$, the choice of $\theta = 5°$ is not satisfactory. As a next try, we choose $\theta = 9°$. Then we have $\phi_m = 34° + 9° = 43°$ and the corresponding b is computed from (8.48) as 5.9. Next we draw a horizontal line with $-10 \log b = -10 \log 5.9 = -7.7$ dB. Its intersection with the gain plot yields $\overline{\omega}'_g = 6.8$ rad/s. The phase margin at $\overline{\omega}'_g$ is 17°. The reduction of the phase margin is $26° - 17° = 9°$ which equals θ. Thus the choice of $\theta = 9°$ is satisfactory and the corresponding $b = 5.9$ can be used in the design. Now we set $\omega_m = \overline{\omega}'_g$ or, using (8.47),

$$\omega_m = \frac{1}{\sqrt{bT_2}} = \overline{\omega}'_g = 6.8$$

which implies

$$T_2 = \frac{1}{\sqrt{b}\,\overline{\omega}'_g} = \frac{1}{\sqrt{5.9} \times 6.8} = \frac{1}{16.5} = 0.06$$

Thus the required phase-lead compensator is

$$C_2(s) = \frac{1 + bT_2 s}{1 + T_2 s} = \frac{1 + 0.36s}{1 + 0.06s}$$

and the total compensator is

$$C(s) = kC_2(s) = 20 \cdot \frac{0.36(s + 2.8)}{0.06(s + 16.9)} = 120 \cdot \frac{s + 2.8}{s + 16.7} \qquad \text{(8.49)}$$

This completes the design.

The introduction of phase-lead compensators will increase the gain-crossover frequency and, consequently, the bandwidth of the overall system. This, in turn, will increase the speed of response of the resulting system. Before proceeding, we compute the root loci and step responses of the preceding three designs in Figure 8.39. The design using the gain adjustment alone (shown with the solid line) has a very large overshoot, although its phase margin is 60° and gain margin is infinity. This is not entirely surprising, because the relationships among the overshoot, damping factor, peak resonance, phase margin, and gain margin are only approximate. The design using a phase-lag network (shown with the dashed line) is rather sluggish; the settling time of the system is very large. The design using a phase-lead network

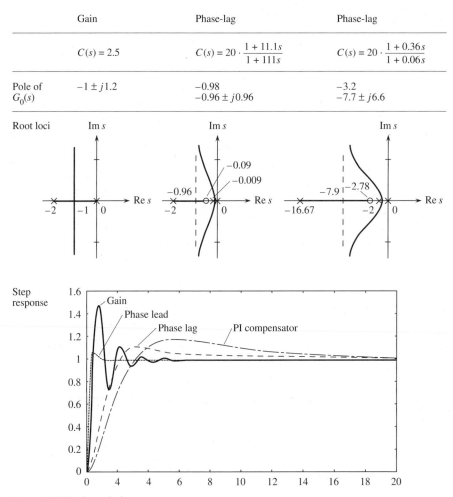

Figure 8.39 Various designs.

(shown with the dotted line) has the smallest overshoot among the three designs; its rise time and settling time are also smallest. Therefore it is the best design among the three.

The phase-lead compensation does not always yield a good design. For example, if the phase in the neighborhood of the gain-crossover frequency decreases rapidly, then the reduction of the phase margin due to a phase-lead compensator will offset the phase introduced by the compensator. In this case, the specification on phase margin cannot be met. Thus, the phase-lead compensation is not effective if the phase at the gain-crossover frequency decreases rapidly. To conclude this section, we compare phase-lag and phase-lead networks in Table 8.3.

Table 8.3 Comparisons of Phase-Lead and Phase-Lag Networks

Phase-Lag Network	Phase-Lead Network
1. The pole is closer to the origin than the zero. Its phase is negative for every positive ω.	The zero is closer to the origin than the pole. Its phase is positive for every positive ω.
2. Shifts down the gain-crossover frequency; consequently, decreases the bandwidth and the speed of response.	Shifts up the gain-crossover frequency; consequently, increases the bandwidth and the speed of response.
3. Placed one decade below the new gain-crossover frequency to reduce the effect of the network on the phase margin.	Placed over the new gain-crossover frequency to add the maximum phase on the phase margin.
4. No additional gain is needed.	Additional gain is needed.
5. Design can be achieved in one step.	Design may require trial and error.

In addition to the phase-lag and phase-lead networks, we may also use a network with transfer function

$$C_3(s) = \frac{1 + aT_1 s}{1 + T_1 s} \cdot \frac{1 + bT_2 s}{1 + T_2 s}$$

in the design. The transfer function is the product of the transfer function of a phase-lag network and that of a phase-lead network. Thus it is called a *lag-lead network*. In the design, we use the attenuation property of the phase-lag part and the positive phase of the phase-lead part. The basic idea is identical to those discussed in the preceding two sections and will not be repeated.

8.10 PROPORTIONAL-INTEGRAL (PI) COMPENSATORS

A compensator that consists of a gain (also called a *proportional compensator*) and an integral compensator is called a *PI compensator* or *controller*. The transfer function of an integral compensator with gain k_i is k_i/s. Thus, the transfer function of PI compensators is

$$C_3(s) = k + \frac{k_i}{s} = \frac{k_i + ks}{s} = \frac{k_i(1 + cs)}{s} \qquad (8.50)$$

where $c = k/k_i$. This is a special case of the phase-lag network shown in Figure 8.34(b) with the pole located at the origin. The phase of $C_3(s)$ is clearly negative for all positive ω. We shall use this compensator to redesign the problem in the preceding two sections. Although PI controllers are a special case of phase-lag networks, the procedure for designing phase-lag networks cannot be used here. Instead we will use the idea in designing phase-lead networks in this problem.

Example 8.10.1

Consider a plant with transfer function $G(s) = 1/s(s + 2)$. Find a PI compensator $C_3(s)$ in Figure 8.1 such that $C_3(s)G(s)$ will meet (1) velocity error $\leq 10\%$, (2) phase margin $\geq 60°$, and (3) gain margin ≥ 12 dB.

The transfer function of $C_3(s)G(s)$ is

$$C_3(s)G(s) = \frac{k_i(1 + cs)}{s} \cdot \frac{1}{s(s + 2)} = \frac{k_i(1 + cs)}{2s^2(1 + 0.5s)} \tag{8.51}$$

which has two poles at $s = 0$. Thus it is of type 2, and the velocity error is zero for any k_i (see Section 6.3.2). The same conclusion can also be reached by using Table 8.2. For a type 2 transfer function, the velocity-error constant K_v is infinity. Thus, its velocity error $|1/K_v|$ is zero for any k_i.

We first assume $c = 0$ and plot in Figure 8.40 the Bode plot of

$$\frac{k_i}{2s^2\left(1 + \frac{1}{2}s\right)} \tag{8.52}$$

with $k_i/2 = 1$. The gain-crossover frequency is $\omega_g = 1$ and the phase margin can be read from the plot as $-27°$. Because the phase plot approaches the $-180°$ line at $\omega = 0$, the phase-crossover frequency is $\omega_p = 0$ and the gain margin is negative infinity. Changing k_i will shift the gain plot up or down and, simultaneously, shift the gain-crossover frequency to the right or left. However, the phase margin is always

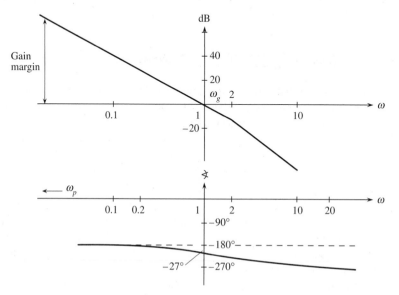

Figure 8.40 Bode plot of $1/s^2(1 + 0.5s)$.

negative and the gain margin remains negative infinity. Thus, if $c = 0$, the unity-feedback system is unstable for any k_i and the design is not possible.

If $k_i > 0$ and $c > 0$, the Bode phase plot of $C_3(s)G(s)$ is given by

$$
\sphericalangle \, (C_3(s)G(s)) = \sphericalangle \left(\frac{1}{s^2} \right) + \sphericalangle \left(\frac{1 + cs}{1 + 0.5s} \right)
$$

$$
= -180° + \sphericalangle \left(\frac{1 + cs}{1 + 0.5s} \right) \tag{8.53}
$$

If $c < 0.5$, then $(1 + cs)/(1 + 0.5s)$ is a phase-lag network and its phase is negative for all positive ω. In this case, the phase in (8.53) is always smaller than $-180°$. Thus, the phase margin is negative, and the design is not possible. If $c > 0.5$, then $(1 + cs)/(1 + 0.5s)$ is a phase-lead network, and it will introduce positive phases into (8.53). Thus the design is possible. We write

$$
\frac{1 + cs}{1 + 0.5s} = \frac{1 + \bar{c} \times 0.5s}{1 + 0.5s} \tag{8.54}
$$

with $\bar{c} > 1$ and compare it with (8.43), then $\bar{c} = b$ and (8.48) can be used to compute $\bar{c}$. Now the phase margin of $C_3(s)G(s)$ is required to be at least 60°, therefore we have

$$
\bar{c} = \frac{1 + \sin 60°}{1 - \sin 60°} = \frac{1 + 0.866}{1 - 0.866} = 13.9
$$

Substituting this $\bar{c}$ into (8.54) yields

$$
\frac{1 + cs}{1 + 0.5s} = \frac{1 + 13.9 \times 0.5s}{1 + 0.5s} = \frac{1 + 6.95s}{1 + 0.5s}
$$

and $C_3(s)G(s)$ in (8.51) becomes

$$
C_3(s)G(s) = \frac{k_i(1 + 6.95s)}{2s^2(1 + 0.5s)} \tag{8.55}
$$

The corner frequencies of the pole and zero of the phase-lead network are, respectively, $1/0.5 = 2$ and $1/6.95 = 0.14$. Thus the maximum phase of the network occurs at

$$
\omega_m = \sqrt{0.14 \times 2} = \sqrt{0.28} = 0.53 \tag{8.56}
$$

and equals 60°. Figure 8.41 shows the Bode plot of (8.55) with $k_i/2 = 1$. We plot only the asymptotes for the gain plot. Because the phase of $1/s^2$ is $-180°$, the phase of (8.55) is simply the phase of $(1 + 6.95s)/(1 + 0.5s)$ shifted down by 180°. From the plot, we see that the gain-crossover frequency is roughly 3 radians per second. If we draw a vertical line down to the phase plot, the phase margin can be read out as 25°. This is less than the required 60°. Thus if $k_i/2 = 1$ or $k_i = 2$, $C_3(s)G(s)$ in (8.55) is not acceptable.

If we increase k_i, the gain plot will move up, the gain-crossover frequency will shift to the right, and the corresponding phase margin will decrease. On the other hand, decreasing k_i will move down the gain plot and shift the gain-crossover frequency to the left. Thus the phase margin will increase. Now c in the phase-lead network $(1 + cs)/(1 + 0.5s)$ is chosen so that the phase is 60° at $\omega_m = 0.53$. Thus, we shall choose k_i so that the gain-crossover frequency equals ω_m. The gain at ω_m is roughly 22 dB, so we set

$$20 \log \left(\frac{k_i}{2} \right) = -22$$

which implies

$$\frac{k_i}{2} = 10^{-22/20} = 0.08$$

If we choose $k_i = 2 \times 0.08 = 0.16$, then $C_3(s)G(s)$ in (8.55) will meet the specification on phase margin.

The phase plot approaches the $-180°$ line at $\omega = 0$ and $\omega = \infty$, as shown in Figure 8.41. Thus, there are two phase-crossover frequencies. Their corresponding gain margins are, respectively, $-\infty$ and ∞. It is difficult to see the physical meaning of these gain margins from Figure 8.41, so we shall plot the Nyquist plot of $C_3(s)G(s)$, where the phase and gain margins are originally defined. Figure 8.42

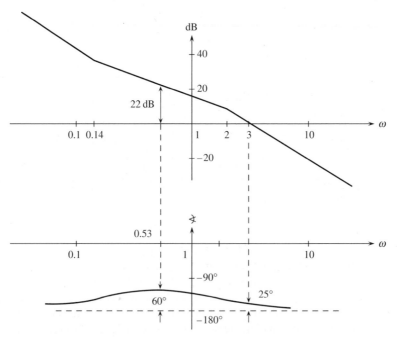

Figure 8.41 Bode plot of $(1 + 6.95s)/s^2(1 + 0.5s)$.

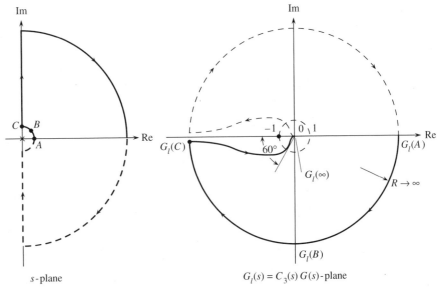

Figure 8.42 Nyquist plot of $C_3(s)G(s)$ in (8.57).

shows the Nyquist plot of (8.55) with $k_i = 0.16$ or

$$C_3(s)G(s) = \frac{0.16(1 + 6.95s)}{2s^2(1 + 0.5s)} \tag{8.57}$$

We see that the Nyquist plot encircles the critical point $(-1, 0)$ once in the counterclockwise direction and once in the clockwise direction. Therefore, the number of net encirclements is zero. The loop transfer function in (8.57) has no open right-half-plane pole, so we conclude from the Nyquist stability criterion that the unity-feedback system in Figure 8.1 is stable. From the Nyquist plot, we see that the phase margin is 60°. There are two gain-crossover frequencies with gain margins ∞ and $-\infty$. If there are two or more phase-crossover frequencies, we shall consider the one that is closest to $(-1, 0)$. In this case, we consider the phase-crossover frequency at $\omega = \infty$; its gain margin is ∞. Thus $C_3(s)G(s)$ also meets the requirement on gain margin and the design is completed. In conclusion, if we introduce the following PI compensator

$$C_3(s) = \frac{0.16(1 + 6.95s)}{s}$$

the unity-feedback system in Figure 8.1 will meet all specifications. The response of this system is shown in Figure 8.39 with the dashed-and-dotted lines. Its response is slower than the one using the phase-lag compensator and much slower than the one using the phase-lead network. Therefore, there is no reason to restrict compensators to PI form for this problem.

8.11 CONCLUDING REMARKS

We give a number of remarks to conclude this chapter.

1. An important feature of the Bode-plot design method is that it can be carried out from measured data without knowing transfer functions. All other methods discussed in this text require the knowledge of transfer functions to carry out design.

2. The design is carried out on loop transfer functions. In order to do so, specifications for overall systems are translated into those for loop transfer functions in the unity-feedback configuration. Thus the method is applicable only to this configuration.

3. The relationships between the specifications such as rise time and overshoot for overall systems and the specifications such as phase and gain margins for loop transfer functions are not exact. Therefore, it is advisable to simulate the resulting systems after the design.

4. If a plant transfer function has open right-half-plane poles, then its Bode plot may have two or more phase and gain margins. In this case, the use of phase and gain margins becomes complex. For this reason, the Bode-plot design method is usually limited to plants without open right-half-plane poles.

5. In this chapter, we often use asymptotes of Bode gain plots to carry out the design. This is done purposely because the reader can see better the plots and the design ideas. In actual design, one should use more accurate plots. On the other hand, because the relationships between phase and gain margins and time responses are not exact, design results using asymptotes may not differ very much from those using accurate plots.

6. The method is a trial-and-error method. Therefore, a number of trials may be needed to design an acceptable system.

7. In the Bode-plot design method, the constraint on actuating signals is not considered. The constraint can be checked only after the completion of the design. If the constraint is not met, we may have to redesign the system.

8. The method is essentially developed from the Nyquist plot, which checks whether or not a polynomial is Hurwitz. In this sense, the design method considers only poles of overall systems. Zeros are not considered.

PROBLEMS

8.1. Plot the polar plots, log magnitude-phase plots, and Bode plots of

$$\frac{10}{(s - 2)} \quad \text{and} \quad \frac{20}{s(s + 1)}$$

8.2. Plot the Bode plots of the following transfer functions:

a. $G(s) = \dfrac{s + 5}{s(s + 1)(s + 10)}$

b. $G(s) = \dfrac{s - 5}{s(s + 1)(s + 10)}$

c. $G(s) = \dfrac{10}{(s + 2)(s^2 + 8s + 25)}$

8.3. **a.** Consider the Bode gain plot shown in Figure P8.3. Find all transfer functions that have the gain plot.

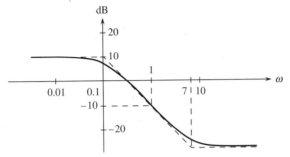

Figure P8.3

b. If the transfer functions are known to be stable, find all transfer functions that have the gain plot.

c. If the transfer functions are known to be minimum phase, find all transfer functions that have the gain plot.

d. If the transfer functions are known to be stable and of minimum phase, find all transfer functions that have the gain plot. Is this transfer function unique?

8.4. Consider the three Bode plots shown in Figure P8.4. What are their transfer functions?

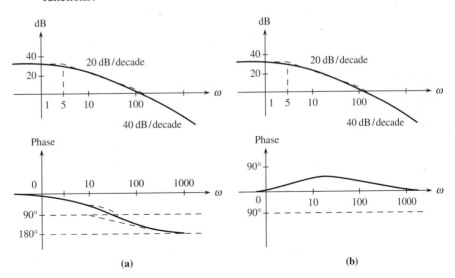

Figure P8.4

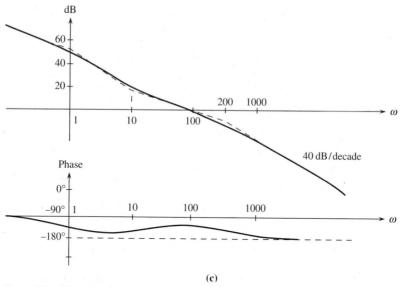

(c)

Figure P8.4 (Continued)

8.5. Consider

$$G(s) = \frac{k(1 + 0.5s)}{s(bs + 1)(s + 10)}$$

Its Bode plot is plotted in Figure P8.5. What are k and b?

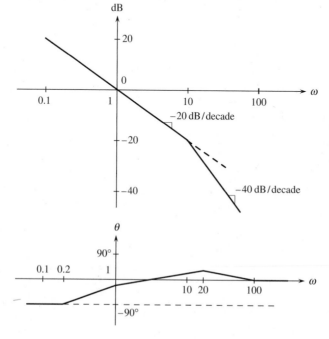

Figure P8.5

8.6. A typical frequency response of a Fairchild μA741 operational amplifier is shown in Figure P8.6. What is its transfer function?

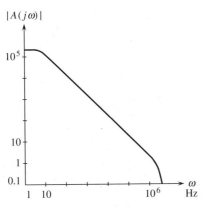
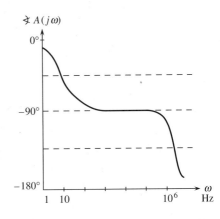

Figure P8.6

8.7. Use the Nyquist criterion to determine the stability of the system shown in Figure 8.17(a) with

a. $G_l(s) = \dfrac{2s + 10}{(s + 1)(s - 1)}$

b. $G_l(s) = \dfrac{2s + 1}{s(s - 1)}$

c. $G_l(s) = \dfrac{100(s + 1)}{s(s - 1)(s + 10)}$

8.8. Consider the unity-feedback system shown in Figure 8.17(b). If the polar plot of $G_l(s)$ is of the form shown in Figure P8.8, find the stability range for the following cases:

a. $G_l(s)$ has no open right-half-plane (RHP) pole and zero.

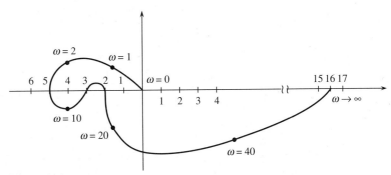

Figure P8.8

b. $G_l(s)$ has one open RHP zero, no open RHP pole.

c. $G_l(s)$ has one open RHP pole.

d. $G_l(s)$ has two open RHP poles, one open RHP zero.

e. $G_l(s)$ has three open RHP poles.

8.9. Find the stability range of the system in Figure 8.17(b) with

$$G_l(s) = \frac{s + 2}{(s^2 + 3s + 6.25)(s - 1)}$$

using (a) the Routh test, (b) the root-locus method and (c) the Nyquist stability criterion.

8.10. What are the gain-crossover frequency, phase-crossover frequency, gain margin, phase margin, position-error constant, and velocity-error constant for each of the transfer functions in Problem 8.2?

8.11. Repeat Problem 8.10 for the Bode plots shown in Fig. P8.4.

8.12. Consider the unity-feedback system shown in Figure 8.1. The Bode plot of the plant is shown in Figure P8.12. Let the compensator $C(s)$ be a gain k. (a) Find the largest k such that the phase margin is 45 degrees. (b) Find a k such that the gain margin is 20 dB.

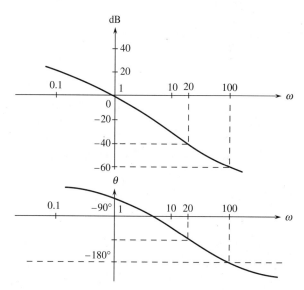

Figure P8.12

8.13. Consider the system shown in Figure 8.1. The Bode plot of the plant is shown in Figure P8.13. The compensator $C(s)$ is chosen as a gain k. Find k to meet (1) phase margin $\geq 60°$, (2) gain margin ≥ 10 dB, and (3) position error $\leq 25\%$.

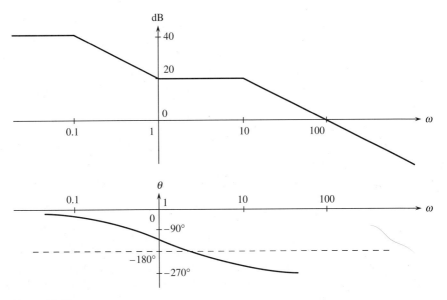

Figure P8.13

8.14. The Bode plot of the plant in Figure 8.1 is shown in Figure P8.14. Find a phase-lag network to meet (1) phase margin $\geq 60°$ and (2) gain margin ≥ 10 dB.

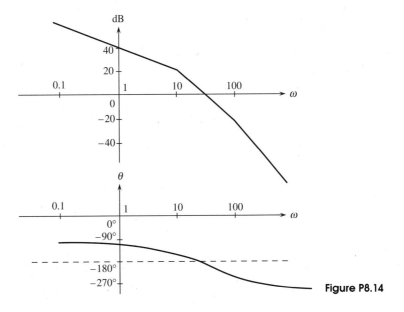

Figure P8.14

8.15. Consider the control of the depth of a submarine discussed in Problem 7.8. Design an overall system to meet (1) position error $\leq 10\%$, (2) phase margin

$\geq 60°$, and (3) gain margin ≥ 10 dB. Compare the design with the one in Problem 7.8.

8.16. Consider the ship stabilization problem discussed in Problem 7.14. Design an overall system to meet (1) position error $\leq 15\%$, (2) phase margin $\geq 60°$, (3) gain margin ≥ 10 dB, and (4) gain-crossover frequency ≥ 10 rad/s.

8.17. Consider the problem of controlling the yaw of the airplane discussed in Problem 7.15. Design an overall system to meet (1) velocity error $\leq 10\%$, (2) phase margin $\geq 30\%$, (3) gain margin ≥ 6 dB, and (4) gain-crossover frequency as large as possible.

8.18. Consider the plant transfer function given in Problem 7.16, which is factored as

$$G(s) = \frac{300}{s(s + 0.225)(s + 3.997)(s + 179.8)}$$

Design an overall system to meet (1) position error $= 0$, (2) phase margin $\geq 55°$, (3) gain margin ≥ 6 dB, and (4) gain-crossover frequency is not smaller than that of the uncompensated plant. [*Hint:* Use a lag-lead network.]

8.19. a. Plot the Bode plot of

$$G(s) = \frac{s - 1}{s^2(s + 10)}$$

What are its phase margin and gain margin?

b. Compute $G_o(s) = G(s)/(1 + G(s))$. Is $G_o(s)$ stable?

c. Is it always true that the unity-feedback system is stable if the phase and gain margins of its loop transfer function are both positive?

9
The Inward Approach—Choice of Overall Transfer Functions

9.1 INTRODUCTION

In the design of control systems using the root-locus method or the frequency-domain method, we first choose a configuration and a compensator with open parameters. We then search for parameters such that the resulting overall system will meet design specifications. This approach is essentially a trial-and-error method; therefore, we usually choose the simplest possible feedback configuration (namely, a unity-feedback configuration) and start from the simplest possible compensator—namely, a gain (a compensator of degree 0). If the design objective cannot be met by searching the gain, we then choose a different configuration or a compensator of degree 1 (phase-lead or phase-lag network) and repeat the search. This approach starts from internal compensators and then designs an overall system to meet design specifications; therefore, it may be called the *outward* approach.

In this and the following chapters we shall introduce a different approach, called the *inward* approach. In this approach, we first search for an overall transfer function to meet design specifications, and then choose a configuration and compute the required compensators. Choice of overall transfer functions will be discussed in this chapter. The implementation problem—namely, choosing a configuration and computing the required compensators—will be discussed in the next chapter.

Consider a plant with proper transfer function $G(s) = N(s)/D(s)$ as shown in Figure 9.1. In the inward approach, the first step is to choose an overall transfer function $G_o(s)$ from the reference input r to the plant output y to meet a set of

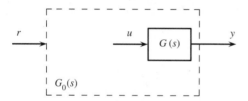

Figure 9.1 Design of control systems.

specifications. We claim that

$$G_o(s) = 1$$

is the best possible system we can design. Indeed if $G_o(s) = 1$, then $y(t) = r(t)$, for $t \geq 0$. Thus the position and velocity errors are zero; the rise time, settling time, and overshoot are all zero. Thus no other $G_o(s)$ can perform better than $G_o(s) = 1$. Note that although $r(t) = y(t)$, the power levels at the reference input and plant output are different. The reference signal may be provided by turning a knob by hand; the plant output $y(t)$ may be the angular position of an antenna with weight over several tons.

Although $G_o(s) = 1$ is the best system, we may not be able to implement it in practice. Recall from Chapter 6 that practical constraints, such as proper compensators, well-posedness, and total stability, do exist in the design of control systems. These constraints impose some limitations in choosing $G_o(s)$. We first discuss this problem.

9.2 IMPLEMENTABLE TRANSFER FUNCTIONS

Consider a plant with transfer function

$$G(s) = \frac{N(s)}{D(s)}$$

where $N(s)$ and $D(s)$ are two polynomials and are assumed to have no common factors. We assume $n = \deg D(s) \geq \deg N(s)$, that is, $G(s)$ is proper and has degree n. An overall transfer function $G_o(s)$ is said to be *implementable* if there exists a configuration such that the transfer function from the reference input r to the plant output y in Figure 9.1 equals $G_o(s)$ and the design meets the following four constraints:

1. All compensators used have proper rational transfer functions.
2. The resulting system is well-posed.
3. The resulting system is totally stable.
4. There is no plant leakage in the sense that all forward paths from r to y pass through the plant.

The first constraint is needed, as discussed in Section 5.4, for building compensators using operational amplifier circuits. If a compensator has an improper transfer function, then it cannot be easily built in practice. The second and third constraints are needed, as discussed in Chapter 6, to avoid amplification of high-frequency noise and to avoid unstable pole-zero cancellations. The fourth constraint implies that all power must pass through the plant and that no compensator be introduced in parallel with the plant. This constraint appears to be reasonable and seems to be met by every configuration in the literature. This constraint is called ''no plant leakage'' by Horowitz [35].

If an overall transfer function $G_o(s)$ is not implementable, then no matter what configuration is used to implement it, the design will violate at least one of the preceding four constraints. Therefore, in the inward approach, the $G_o(s)$ we choose must be implementable.

The question then is how to tell whether or not a $G_o(s)$ is implementable. It turns out that the answer is very simple.

THEOREM 9.1

Consider a plant with proper transfer function $G(s) = N(s)/D(s)$. Then $G_o(s)$ is implementable if and only if $G_o(s)$ and

$$T(s) := \frac{G_o(s)}{G(s)}$$

are proper and stable. ∎

We discuss first the necessity of the theorem. Consider, for example, the configuration shown in Figure 9.2. Noise, which may enter into the intput and output terminals of each block, is not shown. If the closed-loop transfer function from r to y is $G_o(s)$ and if there is no plant leakage, then the *closed-loop* transfer function from r to u is $T(s)$. Well-posedness requires every closed-loop transfer function to be proper, thus $T(s)$ and $G_o(s)$ must be proper. Total stability requires every closed-loop transfer function to be stable, thus $G_o(s)$ and $T(s)$ must be stable. This establishes the necessity of the theorem. The sufficiency of the theorem will be established constructively in the next chapter.

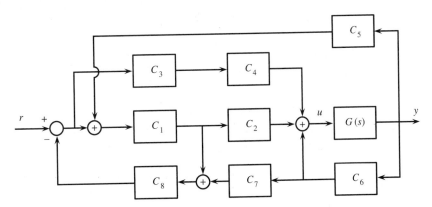

Figure 9.2 Feedback system without plant leakage.

We discuss now the implication of Theorem 9.1. Let us write

$$G(s) = \frac{N(s)}{D(s)} \qquad G_o(s) = \frac{N_o(s)}{D_o(s)} \qquad T(s) = \frac{N_t(s)}{D_t(s)}$$

We assume that the numerator and denominator of each transfer function have no common factors. The equality $G_o(s) = G(s)T(s)$ or

$$\frac{N_o(s)}{D_o(s)} = \frac{N(s)}{D(s)} \cdot \frac{N_t(s)}{D_t(s)}$$

implies

$$\deg D_o(s) - \deg N_o(s) = \deg D(s) - \deg N(s) + (\deg D_t(s) - \deg N_t(s))$$

Thus if $T(s)$ is proper, that is, $\deg D_t(s) \geq \deg N_t(s)$, then we have

$$\deg D_o(s) - \deg N_o(s) \geq \deg D(s) - \deg N(s) \qquad (9.1)$$

Conversely, if (9.1) holds, then $\deg D_t(s) \geq \deg N_t(s)$, and $T(s)$ is proper.

Stability of $G_o(s)$ and $T(s)$ requires both $D_o(s)$ and $D_t(s)$ to be Hurwitz. From

$$T(s) = \frac{N_t(s)}{D_t(s)} = \frac{G_o(s)}{G(s)} = \frac{N_o(s)}{D_o(s)} \cdot \frac{D(s)}{N(s)}$$

we see that if $N(s)$ has closed right-half-plane (RHP) roots, and if these roots are not canceled by $N_o(s)$, then $D_t(s)$ cannot be Hurwitz. Therefore, in order for $T(s)$ to be stable, all the closed RHP roots of $N(s)$ must be contained in $N_o(s)$. This establishes the following corollary.

COROLLARY 9.1

Consider a plant with proper transfer function $G(s) = N(s)/D(s)$. Then $G_o(s) = N_o(s)/D_o(s)$ is implementable if and only if

(a) $\deg D_o(s) - \deg N_o(s) \geq \deg D(s) - \deg N(s)$ (pole-zero excess inequality).
(b) All closed RHP zeros of $N(s)$ are retained in $N_o(s)$ (retainment of non-minimum-phase zeros).
(c) $D_o(s)$ is Hurwitz. ∎

As was defined in Section 8.3.1, zeros in the closed RHP are called non-minimum-phase zeros. Zeros in the open left half plane are called minimum-phase zeros. Poles in the closed RHP are called *unstable poles*. We see that the non-minimum-phase zeros of $G(s)$ impose constraints on implementable $G_o(s)$ but the unstable poles of $G(s)$ do not. This can be easily explained from the unity-feedback configuration shown in Figure 9.3. Let

$$G(s) = \frac{N(s)}{D(s)} \qquad C(s) = \frac{N_c(s)}{D_c(s)}$$

be respectively the plant transfer function and compensator transfer function. Let

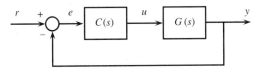

Figure 9.3 Unity-feedback configuration.

$G_o(s) = N_o(s)/D_o(s)$ be the overall transfer function from the reference input r to the plant output y. Then we have

$$G_o(s) = \frac{N_o(s)}{D_o(s)} = \frac{C(s)G(s)}{1 + C(s)G(s)} = \frac{N(s)N_c(s)}{D(s)D_c(s) + N(s)N_c(s)} \tag{9.2}$$

We see that $N(s)$ appears directly as a factor of $N_o(s)$. If a root of $N(s)$ does not appear in $N_o(s)$, the only way to achieve this is to introduce the same root in $D(s)D_c(s) + N(s)N_c(s)$ to cancel it. This cancellation is an unstable pole-zero cancellation if the root of $N(s)$ is in the closed right half s-plane. In this case, the system cannot be totally stable and the cancellation is not permitted. Therefore all non-minimum-phase zeros of $G(s)$ must appear in $N_o(s)$. The poles of $G(s)$ or the roots of $D(s)$ are *shifted* to $D(s)D_c(s) + N(s)N_c(s)$ by feedback, and it is immaterial whether $D(s)$ is Hurwitz or not. Therefore, unstable poles of $G(s)$ do not impose any constraint on $G_o(s)$, but non-minimum-phase zeros of $G(s)$ do. Although the preceding assertion is developed for the unity-feedback system shown in Figure 9.3, it is generally true that, in any feedback configuration without plant leakage, feedback will *shift* the poles of the plant transfer function to new locations but will not affect its zeros. Therefore the non-minimum-phase zeros of $G(s)$ impose constraints on $G_o(s)$ but the unstable poles of $G(s)$ do not.

Example 9.2.1

Consider

$$G(s) = \frac{(s + 2)(s - 1)}{s(s^2 - 2s + 2)}$$

Then we have

$G_o(s) = 1$ Not implementable, because it violates (a) and (b) in Corollary 9.1.

$G_o(s) = \dfrac{s + 2}{(s + 3)(s + 1)}$ Not implementable, meets (a) and (c) but violates (b).

$G_o(s) = \dfrac{s - 1}{s(s + 2)}$ Not implementable, meets (a) and (b), violates (c).

$G_o(s) = \dfrac{s - 1}{(s + 3)(s + 1)}$ Implementable.

$$G_o(s) = \frac{s - 1}{(s + 3)(s + 1)^2} \qquad \text{Implementable.}$$

$$G_o(s) = \frac{(2s - 3)(s - 1)}{(s + 2)^3} \qquad \text{Implementable.}$$

$$G_o(s) = \frac{(2s - 3)(s - 1)(s + 1)}{(s + 2)^5} \qquad \text{Implementable.}$$

Exercise 9.2.1

Given $G(s) = (s - 2)/(s - 3)^2$, are the following implementable?

a. $\dfrac{1}{s + 1}$ **b.** $\dfrac{s - 2}{s + 1}$ **c.** $\dfrac{s - 2}{(s + 1)^2}$ **d.** $\dfrac{(s - 2)(s - 3)}{(s + 1)^3}$

[**Answers:** No, no, yes, yes.]

Exercise 9.2.2

Given $G(s) = (s + 1)/(s - 3)^2$, are the following implementable?

a. $\dfrac{1}{s + 1}$ **b.** $\dfrac{s - 2}{s + 1}$ **c.** $\dfrac{s - 2}{(s + 1)^2}$ **d.** $\dfrac{(s - 2)s^4}{(s + 2)^6}$

[**Answers:** Yes, no, yes, yes.]

From the preceding examples, we see that if the pole-zero excess inequality is met, then all poles and all minimum-phase zeros of $G_o(s)$ can be arbitrarily assigned. To be precise, all poles of $G_o(s)$ can be assigned anywhere inside the open left half s-plane (to insure stability). Other than retaining all non-minimum-phase zeros of $G(s)$, all minimum-phase zeros of $G_o(s)$ can be assigned anywhere in the entire s-plane. In the assignment, if a complex number is assigned as a zero or pole, its complex conjugate must also be assigned. Otherwise, the coefficients of $G_o(s)$ will be complex, and $G_o(s)$ cannot be realized in the real world. Therefore, roughly speaking, if $G_o(s)$ meets the pole-zero excess inequality, its poles and zeros can be arbitrarily assigned.

Consider a plant with transfer function $G(s)$. The problem of designing a system so that its overall transfer function equals a given model with transfer function $G_m(s)$ is called the *model-matching* problem. Now if $G_m(s)$ is not implementable, no matter how hard we try, it is not possible to match $G_m(s)$ without violating the four constraints. On the other hand, if $G_m(s)$ is implementable, it is possible, as will be shown in the next chapter, to match $G_m(s)$. Therefore, the model-matching problem is the same as our implementability problem. In conclusion, in model matching, we can arbitrarily assign poles as well as minimum-phase zeros so long as they meet the pole-zero excess inequality.

To conclude this section, we mention that if G_o is implementable, it does not mean that it can be implemented using *any* configuration. For example, $G_o(s) = 1/(s + 1)^2$ is implementable for the plant $G(s) = 1/s(s - 1)$. This $G_o(s)$, however, cannot be implemented in the unity-feedback configuration shown in Figure 9.3; it can be implemented using some other configurations, as will be discussed in the next chapter. In conclusion, for any $G(s)$ and any implementable $G_o(s)$, there exists at least one configuration in which $G_s(s)$ can be implemented under the preceding four constraints.

9.2.1 Asymptotic Tracking and Permissible Pole-Zero Cancellation Region

A control system with overall transfer function

$$G_o(s) = \frac{\beta_0 + \beta_1 s + \beta_2 s^2 + \cdots + \beta_m s^m}{\alpha_0 + \alpha_1 s + \alpha_2 s^2 + \cdots + \alpha_n s^n} \tag{9.3}$$

with $\alpha_n > 0$ and $n \geq m$, is said to achieve *asymptotic tracking* if the plant output $y(t)$ tracks eventually the reference input $r(t)$ without an error, that is,

$$\lim_{t \to \infty} |y(t) - r(t)| = 0$$

Clearly if $G_o(s)$ is not stable, it cannot track any reference signal. Therefore, we require $G_o(s)$ to be stable, which in turn requires $\alpha_i > 0$ for all i[1]. Thus, the denominator of $G_o(s)$ cannot have any missing term or a term with a negative coefficient. Now the condition for $G_o(s)$ to achieve asymptotic tracking depends on the type of $r(t)$ to be tracked. The more complicated $r(t)$, the more complicated $G_o(s)$. From Section 6.3.1, we conclude that if $r(t)$ is a step function, the conditions for $G_o(s)$ to achieve tracking are $G_o(s)$ stable and $\alpha_0 = \beta_0$. If $r(t)$ is a ramp function, the conditions are $G_o(s)$ stable, $\alpha_0 = \beta_0$, and $\alpha_1 = \beta_1$. If $r(t) = at^2$, an acceleration function, then the conditions are $G_o(s)$ stable, $\alpha_0 = \beta_0$, $\alpha_1 = \beta_1$, and $\alpha_2 = \beta_2$. If $r(t) = 0$, the only condition for $y(t)$ to track $r(t)$ is $G_o(s)$ stable. In this case, the output may be excited by nonzero initial conditions, which in turn may be excited by noise or disturbance. To bring $y(t)$ to zero is called the *regulating problem*. In conclusion, the conditions for $G_o(s)$ to achieve asymptotic tracking are simple and can be easily met in the design.

Asymptotic tracking is a property of $G_o(s)$ as $t \to \infty$ or a steady-state property of $G_o(s)$. It is not concerned with the manner or the speed at which $y(t)$ approaches $r(t)$. This is the transient performance of $G_o(s)$. The transient performance depends on the location of the poles and zeros of $G_o(s)$. How to choose poles and zeros to meet the specification on transient performance, however, is not a simple problem.

---

[1]Also, they can all be negative. For convenience, we consider only the positive case.

In choosing an implementable overall transfer function, if a zero of $G(s)$ is not retained in $G_o(s)$, we must introduce a pole to cancel it in implementation. If the zero is a non-minimum-phase zero, the pole that is introduced to cancel it is not stable and the resulting system will not be totally stable. If the zero is minimum phase but has a large imaginary part or is very close to the imaginary axis, then, as was discussed in Section 6.6.2, the pole may excite a response that is very oscillatory or takes a very long time to vanish. Therefore, in practice, not only the non-minimum-phase zeros of $G(s)$ but also those minimum-phase zeros that are close to the imaginary axis should be retained in $G_o(s)$, or the zeros of $G(s)$ lying outside the region C shown in Figures 6.13 or 7.4 should be retained in $G_o(s)$. How to determine such a region, however, is not necessarily simple. See the discussion in Chapter 7.

Exercise 9.2.3

What types of reference signals can the following systems track without an error?

a. $\dfrac{s + 5}{s^3 + 2s^2 + 8s + 5}$

b. $\dfrac{8s + 5}{s^3 + 2s^2 + 8s + 5}$

c. $\dfrac{2s^2 + 9s + 68}{s^3 + 2s^2 + 9s + 68}$

[**Answers:** (a) Step functions. (b) Ramp functions. (c) None, because it is not stable.]

9.3   VARIOUS DESIGN CRITERIA

The performance of a control system is generally specified in terms of the rise time, settling time, overshoot, and steady-state error. Suppose we have designed two systems, one with a better transient performance but a poorer steady-state performance, the other with a poorer transient performance but a better steady-stage performance. The question is: Which system should we use? This difficulty arises from the fact that the criteria consist of more than one factor. In order to make comparisons, the criteria may be modified as

$$J := k_1 \times \text{(Rise time)} + k_2 \times \text{(Settling time)}$$
$$+ k_3 \times \text{(Overshoot)} + k_4 \times \text{(Steady-state error)} \tag{9.4}$$

where the k_i are weighting factors and are chosen according to the relative importance of the rise time, settling time, and so forth. The system that has the smallest J is called the *optimal* system with respect to the criterion J. Although the criterion is reasonable, it is not easy to track analytically. Therefore more trackable criteria are used in engineering.

We define

$$e(t) := r(t) - y(t)$$

It is the error between the reference input and the plant output at time t as shown in Figure 9.4. Because an error exists at every t, we must consider the *total* error in $[0, \infty)$. One way to define the total error is

$$J_1 := \int_0^\infty e(t)\,dt \tag{9.5}$$

This is not a useful criterion, however, because of possible cancellations between positive and negative errors. Thus a small J_1 may not imply a small $e(t)$ for all t. A better definition of the total error is

$$J_2 := \int_0^\infty |e(t)|\,dt \tag{9.6}$$

This is called the *integral of absolute error* (IAE). In this case, a small J_2 will imply a small $e(t)$. Other possible definitions are

$$J_3 := \int_0^\infty |e(t)|^2\,dt \tag{9.7}$$

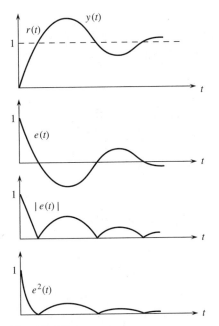

Figure 9.4 Errors.

and

$$J_4 := \int_0^\infty t|e(t)| \, dt \tag{9.8}$$

The former is called the *integral of square error* (ISE) or *quadratic error*, and the latter the *integral of time multiplied by absolute error* (ITAE). The ISE penalizes large errors more heavily than small errors, as is shown in Figure 9.4. Because of the unavoidable large errors at small t due to transient responses, it is reasonable not to put too much weight on those errors. This is achieved by multiplying t with $|e(t)|$. Thus the ITAE puts less weight on $e(t)$ for t small and more weight on $e(t)$ for t large. The total errors defined in J_2, J_3, and J_4 are all reasonable and can be used in design.

Although these criteria are reasonable, they should not be used without considering physical constraints. To illustrate this point, we consider a plant with transfer function $G(s) = (s + 2)/s(s + 3)$. Because $G(s)$ has no non-minimum-phase zero and has a pole-zero excess of 1, $G_o(s) = a/(s + a)$ is implementable for any positive a. We plot in Figure 9.5(a) the responses of $G_o(s)$ due to a unit-step reference input for $a = 1$ (solid line), $a = 10$ (dashed line), and $a = 100$ (dotted line). We see that the larger a is, the smaller J_2, J_3, and J_4 are. In fact, as a approaches infinity, J_2, J_3, and J_4 all approach zero. Therefore an optimal implementable $G_o(s)$ is $a/(s + a)$ with $a = \infty$.

As discussed in Section 6.7, the actuating signal of the plant is usually limited by

$$|u(t)| \leq M \qquad \text{for all } t \geq 0 \tag{9.9}$$

This arises from limited operational ranges of linear models or the physical constraints of devices such as the opening of valves or the rotation of rudders. Clearly, the larger the reference input, the larger the actuating signal. For convenience, the $u(t)$ in (9.9) will be assumed to be excited by a unit-step reference input and the constant M is proportionally scaled. Now we shall check whether this constraint will be met for all a. No matter how $G_o(s)$ is implemented, if there is no plant leakage, the closed-loop transfer function from the reference input r to the actuating signal u is given by

$$T(s) = \frac{G_o(s)}{G(s)} \tag{9.10}$$

If r is a step function, then the actuating signal u equals

$$U(s) = T(s)R(s) = \frac{G_o(s)}{G(s)} \cdot \frac{1}{s} = \frac{a(s + 3)}{(s + 2)(s + a)} \tag{9.11}$$

This response is plotted in Figure 9.5(b) for $a = 1, 10$, and 100. This can be obtained by analysis or by digital computer simulations. For this example, it happens that $|u(t)|_{\max} = u(0) = a$. For $a = 100$, $u(0)$ is outside the range of the plot. We see

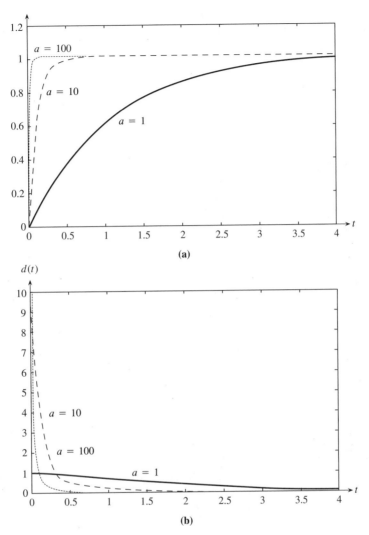

Figure 9.5 (a) Step responses. (b) Actuating signals.

that the larger a is, the larger the magnitude of the actuating signal. Therefore if a is very large, the constraint in (9.9) will be violated.

In conclusion, in using the performance indices in (9.6) to (9.8), we must include the constraint in (9.9). Otherwise we can make these indices as small as desired and the system will always be saturated. Another possible constraint is to limit the bandwidth of resulting overall systems. The reason for limiting the bandwidth is to avoid amplification of high-frequency noise. It is believed that both constraints will lead to comparable results. In this chapter we discuss only the constraint on actuating signals.

9.4   QUADRATIC PERFORMANCE INDICES

In this section we discuss the design of an overall system to minimize the quadratic performance index

$$\int_0^\infty [y(t) - r(t)]^2 \, dt \tag{9.12a}$$

subject to the constraint

$$|u(t)| \le M \tag{9.12b}$$

for all $t \ge 0$, and for some constant M. Unfortunately, no simple analytical method is available to design such a system. Furthermore, the resulting optimal system may not be linear and time-invariant. If we limit our design to linear time-invariant systems, then (9.12) must be replaced by the following quadratic performance index

$$J = \int_0^\infty [q(y(t) - r(t))^2 + u^2(t)] \, dt \tag{9.13}$$

where q is a weighting factor and is required to be positive. If q is a large positive number, more weight is placed on the error. As q approaches infinity, the contribution of u in (9.13) becomes less significant, and at the extreme, (9.13) reduces to (9.7). In this case, since no penalty is imposed on the actuating signal, the magnitude of the actuating signal may become very large or infinity; hence, the constraint in (9.12b) will be violated. If $q = 0$, then (9.13) reduces to

$$\int_0^\infty u^2(t) \, dt$$

and the optimal system that minimizes the criterion is the one with $u \equiv 0$. From these two extreme cases, we conclude that if q in (9.13) is adequately chosen, then the constraint in (9.12b) will be satisfied. Hence, although we are forced to use the quadratic performance index in (9.13) for mathematical convenience, if q is properly chosen, (9.13) is an acceptable substitution for (9.12).

9.4.1   Quadratic Optimal Systems

Consider a plant with transfer function

$$G(s) = \frac{N(s)}{D(s)} \tag{9.14}$$

It is assumed that $N(s)$ and $D(s)$ have no common factors and deg $N(s) \le$ deg $D(s)$ $= n$. The design problem is to find an overall transfer function to minimize the quadratic performance index

$$J = \int_0^\infty [q(y(t) - r(t))^2 + u^2(t)] \, dt \tag{9.15}$$

where q is a positive constant, r is the reference signal, y is the output, and u is the actuating signal. Before proceeding, we first discuss the spectral factorization.

Consider the polynomial

$$Q(s) := D(s)D(-s) + qN(s)N(-s) \tag{9.16}$$

It is formed from the denominator and numerator of the plant transfer function and the weighting factor q. It is clear that $Q(s) = Q(-s)$. Hence, if s_1 is a root of $Q(s)$, so is $-s_1$. Since all the coefficients of $Q(s)$ are real by assumption, if s_1 is a root of $Q(s)$, so is its complex conjugate s_1^*. Consequently all the roots of $Q(s)$ are symmetric with respect to the real axis, the imaginary axis, and the origin of the s-plane, as shown in Figure 9.6. We now show that $Q(s)$ has no root on the imaginary axis. Consider

$$
\begin{aligned}
Q(j\omega) &= D(j\omega)D(-j\omega) + qN(j\omega)N(-j\omega) \\
&= |D(j\omega)|^2 + q|N(j\omega)|^2
\end{aligned}
\tag{9.17}
$$

The assumption that $D(s)$ and $N(s)$ have no common factors implies that there exists no ω_0 such that $D(j\omega_0) = 0$ and $N(j\omega_0) = 0$. Otherwise $s^2 + \omega_0^2$ would be a common factor of $D(s)$ and $N(s)$. Thus if $q \neq 0$, $Q(j\omega)$ in (9.17) cannot be zero for any ω. Consequently, $Q(s)$ has no root on the imaginary axis. Now we shall divide the roots of $Q(s)$ into two groups, those in the open left half plane and those in the open right half plane. If all the open-left-half-plane roots are denoted by $D_o(s)$, then, because of the symmetry property, all the open right-half-plane roots can be denoted by $D_o(-s)$. Thus, we can always factor $Q(s)$ as

$$Q(s) = D(s)D(-s) + qN(s)N(-s) = D_o(s)D_o(-s) \tag{9.18}$$

where $D_o(s)$ is a Hurwitz polynomial. The factorization in (9.18) is called the *spectral factorization*.

With the spectral factorization, we are ready to discuss the optimal overall transfer function. The optimal overall transfer function depends on the reference signal $r(t)$. The more complicated $r(t)$, the more complicated the optimal overall transfer function. We discuss in the following only the case where $r(t)$ is a step function.

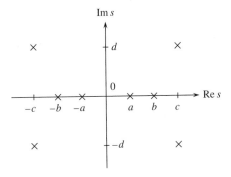

Figure 9.6 Distribution of the roots of $Q(s)$ in (9.16).

Problem   Consider a plant with transfer function $G(s) = N(s)/D(s)$, as shown in Figure 9.1, where $N(s)$ and $D(s)$ have no common factors and deg $N(s) \leq$ deg $D(s)$ $= n$. Find an *implementable* overall transfer function $G_o(s)$ to minimize the quadratic performance index

$$J = \int_0^\infty [q(y(t) - r(t))^2 + u^2(t)]\,dt$$

where $q > 0$, and $r(t) = 1$ for $t \geq 0$, that is, $r(t)$ is a step-reference signal.

Solution   First we compute the spectral factorization:

$$Q(s) := D(s)D(-s) + qN(s)N(-s) = D_o(s)D_o(-s)$$

where $D_o(s)$ is a Hurwitz polynomial. Then the optimal overall transfer function is given by

$$G_o(s) = \frac{qN(0)}{D_o(0)} \cdot \frac{N(s)}{D_o(s)} \qquad\qquad (9.19)$$

The proof of (9.19) is beyond the scope of this text; its employment, however, is very simple. This is illustrated by the following example.

Example 9.4.1

Consider a plant with transfer function

$$G(s) = \frac{N(s)}{D(s)} = \frac{1}{s(s + 2)} \qquad\qquad (9.20)$$

Find $G_o(s)$ to minimize

$$J = \int_0^\infty [9(y(t) - 1)^2 + u^2(t)]\,dt \qquad\qquad (9.21)$$

Clearly we have $q = 9$,

$$D(s) = s(s + 2) \qquad D(-s) = -s(-s + 2)$$

and

$$N(s) = 1 \qquad N(-s) = 1$$

We compute

$$
\begin{aligned}
Q(s) &:= D(s)D(-s) + qN(s)N(-s) \\
&= s(s + 2)(-s)(-s + 2) + 9 \cdot 1 \cdot 1 \\
&= -s^2(-s^2 + 4) + 9 = s^4 - 4s^2 + 9
\end{aligned}
\qquad\qquad (9.22)
$$

It is an even function of s. If terms with odd powers of s appear in $Q(s)$, an error must have been committed in the computation. Using the formula for computing the roots of quadratic equations, we have

$$s^2 = \frac{4 \pm \sqrt{16 - 4 \cdot 9}}{2} = \frac{4 \pm j\sqrt{20}}{2} = 2 \pm j\sqrt{5} = 3e^{\pm j\theta}$$

with

$$\theta = \tan^{-1}\left(\frac{\sqrt{5}}{2}\right) = 48°$$

Thus the four roots of $Q(s)$ are

$$\sqrt{3}e^{j\theta/2} = \sqrt{3}e^{j24°} \qquad -\sqrt{3}e^{j24°} = \sqrt{3}e^{j(180°+24°)} = \sqrt{3}e^{j204°}$$
$$\sqrt{3}e^{-j24°} \qquad\qquad -\sqrt{3}e^{-j24°} = \sqrt{3}e^{j(180°-24°)} = \sqrt{3}e^{j156°}$$

as shown in Figure 9.7. The two roots in the left column are in the open right half s-plane; the two roots in the right column are in the open left half s-plane. Using the two left-half-plane roots, we form

$$
\begin{aligned}
D_o(s) &= (s + \sqrt{3}e^{j24°})(s + \sqrt{3}e^{-j24°}) \\
&= s^2 + \sqrt{3}(e^{j24°} + e^{-j24°})s + 3 \\
&= s^2 + 2 \cdot \sqrt{3}(\cos 24°)s + 3 = s^2 + 3.2s + 3
\end{aligned}
\tag{9.23}
$$

This completes the spectral factorization. Because $q = 9$, $N(0) = 1$, and $D_o(0) = 3$, the optimal system is, using (9.19),

$$G_o(s) = \frac{9 \cdot 1}{3} \cdot \frac{1}{s^2 + 3.2s + 3} = \frac{3}{s^2 + 3.2s + 3} \tag{9.24}$$

This $G_o(s)$ is clearly implementable. Because $G_o(0) = 1$, the optimal system has a zero position error. The implementation of this optimal system will be discussed in the next chapter.

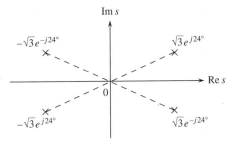

Figure 9.7 Roots of (9.22).

Exercise 9.4.1

Given $G(s) = (s - 1)/s(s + 1)$, find an implementable overall transfer function to minimize

$$J = \int_0^\infty [9(y(t) - 1)^2 + u^2(t)]\,dt \tag{9.25}$$

[**Answer:** $G_o(s) = -3(s - 1)/(s + 3)(s + 1)$.]

9.4.2   Computation of Spectral Factorizations

The design of quadratic optimal systems requires the computation of spectral factorizations. One way to carry out the factorization is to compute all the roots of $Q(s)$ and then group all the left-half-plane roots, as we did in (9.23). This method can be easily carried out if software for solving roots of polynomials is available. For example, if we use PC-MATLAB to carry out the spectral factorization of $Q(s)$ in (9.22), then the commands

```
q=[1 0 −4 0 9];
r=roots(q)
```

yield the following four roots:

```
r= − 1.5811 + 0.7071i
    − 1.5811 − 0.7071i
      1.5811 + 0.7071i
      1.5811 − 0.7071i
```

The first and second roots are in the open left half plane and will be used to form $D_o(s)$. The command

```
poly([r(1) r(2)])
```

yields a polynomial of degree 2 with coefficients

```
1.0000    3.1623    3.0000
```

This is $D_o(s)$. Thus the use of a digital computer to carry out spectral factorizations is very simple.

We now introduce a method of carrying out spectral factorizations without solving for roots. Consider the $Q(s)$ in (9.22). It is a polynomial of degree 4. In the spectral factorization of

$$Q(s) = s^4 - 4s^2 + 9 = D_o(s)D_o(-s) \tag{9.26}$$

the degrees of polynomials $D_o(s)$ and $D_o(-s)$ are the same. Therefore, the degree of $D_o(s)$ is half of that of $Q(s)$, or two for this example. Let

$$D_o(s) = b_0 + b_1 s + b_2 s^2 \qquad (9.27)$$

where b_i are required to be all positive.[2] If any one of them is zero or negative, then $D_o(s)$ is not Hurwitz. Clearly, we have

$$D_o(-s) = b_0 + b_1(-s) + b_2(-s)^2 = b_0 - b_1 s + b_2 s^2 \qquad (9.28)$$

The multiplication of $D_o(s)$ and $D_o(-s)$ yields

$$\begin{aligned} D_o(s)D_o(-s) &= (b_0 + b_1 s + b_2 s^2)(b_0 - b_1 s + b_2 s^2) \\ &= b_0^2 + (2b_0 b_2 - b_1^2)s^2 + b_2^2 s^4 \end{aligned}$$

It is an even function of s. In order to meet (9.26), we equate

$$b_0^2 = 9$$

$$2b_0 b_2 - b_1^2 = -4$$

and

$$b_2^2 = 1$$

Thus we have $b_0 = 3$, $b_2 = 1$ and

$$b_1^2 = 2b_0 b_2 + 4 = 2 \cdot 3 \cdot 1 + 4 = 10$$

which implies $b_1 = \sqrt{10}$. Note that we require all b_i to be positive; therefore, we have taken only the positive part of the square roots. Thus the spectral factorization of (9.26) is

$$D_o(s) = 3 + \sqrt{10}\, s + s^2 = 3 + 3.2s + s^2$$

We see that this procedure is quite simple and can be used if a digital computer and the required software are not available. The preceding result can be stated more generally as follows: If

$$Q(s) = a_0 + a_2 s^2 + a_4 s^4 \qquad (9.29)$$

and if

$$D_o(s) = b_0 + b_1 s + b_2 s^2 \qquad (9.30)$$

then

$$b_0 = \sqrt{a_0} \qquad b_2 = \sqrt{a_4} \qquad b_1 = \sqrt{(-a_2 + 2b_0 b_2)} \qquad (9.31)$$

Note that before computing b_1, we must compute first b_0 and b_2.

Now we shall extend the preceding procedure to a more general case. Consider

$$Q(s) = a_0 + a_2 s^2 + a_4 s^4 + a_6 s^6 \qquad (9.32)$$

It is an even polynomial of degree 6. Let

$$D_o(s) = b_0 + b_1 s + b_2 s^2 + b_3 s^3 \qquad (9.33)$$

---

[2]Also, they can all be negative. For convenience, we consider only the positive case.

Then

$$D_o(-s) = b_0 - b_1 s + b_2 s^2 - b_3 s^3$$

and

$$D_o(s)D_o(-s) = b_0^2 + (2b_0 b_2 - b_1^2)s^2 + (b_2^2 - 2b_1 b_3)s^4 - b_3^2 s^6 \quad (9.34)$$

Equating (9.32) and (9.34) yields

$$b_0^2 = a_0 \quad (9.35a)$$

$$2b_0 b_2 - b_1^2 = a_2 \quad (9.35b)$$

$$b_2^2 - 2b_1 b_3 = a_4 \quad (9.35c)$$

and

$$b_3^2 = -a_6 \quad (9.35d)$$

From (9.35a) and (9.35d), we can readily compute $b_0 = \sqrt{a_0}$ and $b_3 = \sqrt{-a_6}$. In other words, the leading and constant coefficients of $D_o(s)$ are simply the square roots of the magnitudes of the leading and constant coefficients of $Q(s)$. Once b_0 and b_3 are computed, there are only two unknowns, b_1 and b_2, in the two equations in (9.35b) and (9.35c). These two equations are not linear and can be solved iteratively as follows. We rewrite them as

$$b_1 = \sqrt{2b_0 b_2 - a_2} \quad (9.36a)$$

$$b_2 = \sqrt{a_4 + 2b_1 b_3} \quad (9.36b)$$

First we choose an arbitrary b_2—say, $b_2^{(0)}$—and use this $b_2^{(0)}$ to compute b_1 as

$$b_1^{(1)} = \sqrt{2b_0 b_2^{(0)} - a_2}$$

We then use this $b_1^{(1)}$ to compute b_2 as

$$b_2^{(1)} = \sqrt{a_4 + 2b_1^{(1)} b_3}$$

If $b_2^{(1)}$ happens to equal $b_2^{(0)}$, then the chosen $b_2^{(0)}$ is the solution of (9.36). Of course, the possibility of having $b_2^{(1)} = b_2^{(0)}$ is extremely small. We then use $b_2^{(1)}$ to compute a new b_1 as

$$b_1^{(2)} = \sqrt{2b_0 b_2^{(1)} - a_2}$$

and then a new b_2 as

$$b_2^{(2)} = \sqrt{a_4 + 2b_1^{(2)} b_3}$$

If $b_2^{(2)}$ is still quite different from $b_2^{(1)}$, we repeat the process. It can be shown that the process will converge to the true solutions.[3] This is an iterative method of carrying out the spectral factorization. In application, we may stop the iteration when the difference between two subsequent $b_2^{(i)}$ and $b_2^{(i+1)}$ is smaller than, say, 5%. This is illustrated by an example.

---

[3]If we compute $b_2 = (a_2 + b_1^2)/2b_0$ and $b_1 = (b_2^2 - a_4)/2b_3$ iteratively, the process will diverge.

Example 9.4.2

Compute the spectral factorization of

$$Q(s) = 25 - 41s^2 + 20s^4 - 4s^6 \qquad (9.37)$$

Let

$$D_o(s) = b_0 + b_1 s + b_2 s^2 + b_3 s^3$$

Its constant term and leading coefficient are simply the square roots of the corresponding coefficients of $Q(s)$:

$$b_0 = \sqrt{25} = 5 \qquad b_3 = \sqrt{|-4|} = \sqrt{4} = 2$$

The substitution of these into (9.36) yields

$$b_1 = \sqrt{10b_2 + 41}$$

$$b_2 = \sqrt{20 + 4b_1}$$

Now we shall solve these equations iteratively. Arbitrarily, we choose b_2 as $b_2^{(0)} = 0$ and compute

	(0)	(1)	(2)	(3)	(4)	(5)
b_1		6.4	10.42	10.93	10.99	10.999
b_2	0	6.75	7.85	7.98	7.998	7.9998

We see that they converge rapidly to the solutions $b_1 = 11$ and $b_2 = 8$. To verify the convergence, we now choose b_2 as $b_2^{(0)} = 100$ and compute

	(0)	(1)	(2)	(3)	(4)	(5)
b_1		32.26	12.77	11.19	11.02	11.002
b_2	100	12.21	8.43	8.05	8.005	8.0006

They also converge rapidly to the solutions $b_1 = 11$ and $b_2 = 8$.

The preceding iterative procedure can be extended to the general case. The basic idea is the same and will not be repeated.

Exercise 9.4.2

Carry out spectral factorizations for

a. $4s^4 - 9s^2 + 16$

b. $-4s^6 + 10s^4 - 20s^2 + 16$

[**Answers:** $2s^2 + 5s + 4$, $2s^3 + 6.65s^2 + 8.56s + 4$.]

9.4.3 Selection of Weighting Factors

In this subsection we discuss the problem of selecting a weighting factor in the quadratic performance index to meet the constraint $|u(t)| \leq M$ for all $t \geq 0$. It is generally true that a larger q yields a larger actuating signal and a faster response. Conversely, a smaller q yields a smaller actuating signal and a slower response. Therefore, by choosing q properly, the constraint on the actuating signal can be met. We use the example in (9.20) to illustrate the procedure.

Consider a plant with transfer function $G(s) = 1/s(s + 2)$. Design an overall system to minimize

$$J = \int_0^\infty [q(y(t) - 1)^2 + u^2(t)]dt$$

It is also required that the actuating signal due to a unit-step reference input meet the constraint $|u(t)| \leq 3$, for all $t \geq 0$. Arbitrarily, we choose $q = 100$ and compute

$$Q(s) = s(s + 2)(-s)(-s + 2) + 100 \cdot 1 \cdot 1 = s^4 - 4s^2 + 100$$

Its spectral factorization can be computed as, using (9.31),

$$D_o(s) = s^2 + \sqrt{24}s + 10 = s^2 + 4.9s + 10$$

Thus the quadratic optimal transfer function is

$$G_o(s) = \frac{Y(s)}{R(s)} = \frac{qN(0)}{D_o(0)} \cdot \frac{N(s)}{D_o(s)} = \frac{100 \cdot 1}{10} \cdot \frac{1}{s^2 + 4.9s + 10}$$

$$= \frac{10}{s^2 + 4.9s + 10}$$

The unit-step response of this system is simulated and plotted in Figure 9.8(a). Its rise time, settling time, and overshoot are 0.92 s, 1.70 s, and 2.13%, respectively. Although the response is quite good, we must check whether or not its actuating signal meets the constraint. No matter what configuration is used to implement $G_o(s)$, if there is no plant leakage, the transfer function from the reference signal r to the actuating signal u is

$$T(s) = \frac{G_o(s)}{G(s)} = \frac{10}{s^2 + 4.9s + 10} \cdot \frac{s(s + 2)}{1} = \frac{10s(s + 2)}{s^2 + 4.9s + 10}$$

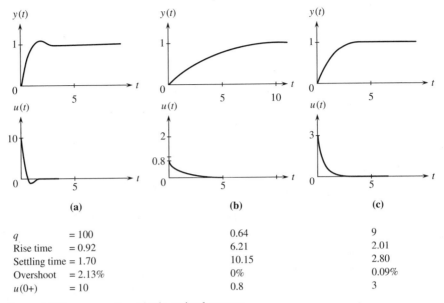

Figure 9.8 Responses of quadratic optimal systems.

The unit-step response of $T(s)$ is simulated and also plotted in Figure 9.8(a). We see that $u(0^+) = 10$ and the constraint $|u(t)| \leq 3$ is violated. Because the largest magnitude of $u(t)$ occurs at $t = 0^+$, it can also be computed by using the initial-value theorem (see Appendix A). The response $u(t)$ due to $r(t) = 1$ is

$$U(s) = T(s)R(s) = \frac{10s(s + 2)}{s^2 + 4.9s + 10} \cdot \frac{1}{s}$$

The application of the initial-value theorem yields

$$u(0^+) = \lim_{s \to \infty} sU(s) = \lim_{s \to \infty} sT(s)R(s) = \lim_{s \to \infty} T(s) = 10$$

Thus the constraint $|u(t)| \leq 3$ is not met and the selection of $q = 100$ is not acceptable.[4]

Next we choose $q = 0.64$ and repeat the design. The optimal transfer function is found as

$$G_o(s) = \frac{0.64 \times 1}{0.8} \cdot \frac{1}{s^2 + \sqrt{5.6}\,s + 0.8} = \frac{0.8}{s^2 + 2.4s + 0.8}$$

---

[4]It is shown by B. Seo [51] that if a plant transfer function is of the form $(b_1 s + b_0)/s(s + a)$, with $b_0 \neq 0$, then the maximum magnitude of the actuating signal of quadratic optimal systems occurs at $t = 0^+$ and $|u(t)| \leq u(0^+) = \sqrt{q}$.

Its unit-step response and the actuating signal are plotted in Figure 9.8(b). The response is fairly slow. Because $|u(t)| \leq u(0^+) = 0.8$ is much smaller than 3, the system can be designed to respond faster. Next we try $q = 9$, and compute

$$Q(s) = s(s + 2)(-s)(-s + 2) + 9 \cdot 1 \cdot 1$$

Its spectral factorization, using (9.31), is

$$D_o(s) = s^2 + \sqrt{10}\, s + 3 = s^2 + 3.2s + 3$$

Thus the optimal transfer function is

$$G_o(s) = \frac{qN(0)}{D_o(0)} \cdot \frac{N(s)}{D_o(s)} = \frac{9 \cdot 1}{3} \cdot \frac{1}{s^2 + 3.2s + 3} = \frac{3}{s^2 + 3.2s + 3} \tag{9.38}$$

and the transfer function from r to u is

$$T(s) = \frac{G_o(s)}{G(s)} = \frac{3s(s + 2)}{s^2 + 3.2s + 3}$$

Their unit-step responses are plotted in Figure 9.8(c). The rise time of $y(t)$ is 2.01 seconds, the settling time is 2.80 seconds, and the overshoot is 0.09%. We also have $|u(t)| \leq u(0^+) = T(\infty) = 3$, for all t. Thus this overall system has the fastest response under the constraint $|u(t)| \leq 3$.

From this example, we see that the weighting factor q is to be chosen by trial and error. We choose an arbitrary q, say $q = q_0$, and carry out the design. After the completion of the design, we then simulate the resulting overall system. If the response is slow or sluggish, we may increase q and repeat the design. In this case, the response will become faster. However, the actuating signal may also become larger and the plant may be saturated. Thus the choice of q is generally reached by a compromise between the speed of response and the constraint on the actuating signal.

Optimality is a fancy word because it means "the best." However, without introducing a performance index, it is meaningless to talk about optimality. Even if a performance index is introduced, if it is not properly chosen, the resulting system may not be satisfactory in practice. For example, the second system in Figure 9.8 is optimal with $q = 0.64$, but it is very slow. Therefore, the choice of a suitable performance index is not necessarily simple.

Exercise 9.4.3

Given a plant with transfer function $G(s) = (s + 2)/s(s - 2)$, find a quadratic optimal system under the constraint that the magnitude of the actuating signal due to a unit step reference input is less than 5.

[**Answer:** $G_o(s) = 5(s + 2)/(s^2 + 7s + 10)$.]

9.5   THREE MORE EXAMPLES

In this section we shall discuss three more examples. Every one of them will be redesigned in latter sections and be compared with quadratic optimal design.

Example 9.5.1

Consider a plant with transfer function [19, 34]

$$G(s) = \frac{2}{s(s^2 + 0.25s + 6.25)} \tag{9.39}$$

Design an overall system to minimize

$$J = \int_0^\infty [q(y(t) - 1)^2 + u^2(t)]dt \tag{9.40}$$

The weighting factor q is to be chosen so that the actuating signal $u(t)$ due to a unit-step reference input meets $|u(t)| \leq 10$ for $t \geq 0$. First we choose $q = 9$ and compute

$$
\begin{aligned}
Q(s) &= D(s)D(-s) + qN(s)N(-s) \\
&= s(s^2 + 0.25s + 6.25) \cdot (-s)(s^2 - 0.25s + 6.25) + 9 \cdot 2 \cdot 2 \quad \text{(9.41)} \\
&= -s^6 - 12.4375s^4 - 39.0625s^2 + 36
\end{aligned}
$$

The spectral factorization of (9.41) can be carried out iteratively as discussed in Section 9.4.2 or by solving its roots. As a review, we use both methods in this example. We first use the former method. Let

$$D_o(s) = b_0 + b_1s + b_2s^2 + b_3s^3$$

Its constant term and leading coefficient are simply the square roots of the corresponding coefficients of $Q(s)$:

$$b_0 = \sqrt{36} = 6 \qquad b_3 = \sqrt{|-1|} = \sqrt{1} = 1$$

The substitution of these into (9.36) yields

$$b_1 = \sqrt{12b_2 + 39.0625}$$

$$b_2 = \sqrt{2b_1 - 12.4375}$$

Now we shall solve these equations iteratively. Arbitrarily, we choose b_2 as $b_2^{(0)} = 0$ and compute

b_1		6.25	6.49	6.91	7.30	7.53	7.65	7.70	7.73	7.75	7.75
b_2	0	0.25	0.73	1.18	1.47	1.62	1.69	1.72	1.74	1.74	1.75

We see that they converge to the solution $b_1 = 7.75$ and $b_2 = 1.75$. Thus we have $Q(s) = D_o(s)D_o(-s)$ with

$$D_o(s) = s^3 + 1.75s^2 + 7.75s + 6$$

Thus the optimal overall transfer function that minimizes (9.40) with $q = 9$ is

$$G_o(s) = \frac{qN(0)}{D_o(0)} \cdot \frac{N(s)}{D_o(s)} = \frac{9 \cdot 2}{6} \cdot \frac{2}{s^3 + 1.75s^2 + 7.75s + 6}$$

$$= \frac{6}{s^3 + 1.75s^2 + 7.75s + 6} \qquad (9.42)$$

For this overall transfer function, it is found by computer simulation that $|u(t)| \leq 3$, for $t \geq 0$. Thus we may choose a larger q. We choose $q = 100$ and compute

$$Q(s) = D(s)D(-s) + 100N(s)N(-s)$$
$$= -s^6 - 12.4375s^4 - 39.0625s^2 + 400$$

Now we use the second method to carry out the spectral factorization. We use PC-MATLAB to compute its roots. The command

$$r = roots([-1\ 0\ -12.4375\ 0\ -39.0625\ 0\ 400])$$

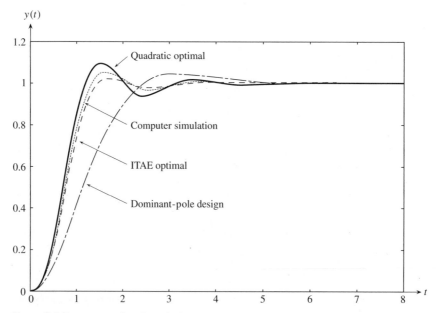

Figure 9.9 Responses of various designs of (9.39).

yields

$$r = -0.9917 + 3.0249i$$
$$-0.9917 - 3.0249i$$
$$0.9917 + 3.0249i$$
$$0.9917 - 3.0249i$$
$$1.9737$$
$$-1.9737$$

The first, second, and last roots are in the open left half plane. The command

poly([r(1) r(2) r(6)])

yields [1.000 3.9571 14.0480 20.0000]. Thus we have $D_o(s) = s^3 + 3.957s^2 + 14.048s + 20$ and the quadratic optimal overall transfer function is

$$G_o(s) = \frac{20}{s^3 + 3.957s^2 + 14.048s + 20} \tag{9.43}$$

For this transfer function, the maximum amplitude of the actuating signal due to a unit-step reference input is 10. Thus we cannot choose a larger q. The unit-step response of $G_o(s)$ in (9.43) is plotted in Figure 9.9 with the solid line. The response appears to be quite satisfactory.

Example 9.5.2

Consider a plant with transfer function

$$G(s) = \frac{s + 3}{s(s - 1)} \tag{9.44}$$

Find an overall transfer function to minimize the quadratic performance index

$$J = \int_0^\infty [100(y(t) - 1)^2 + u^2(t)]dt \tag{9.45}$$

where the weighting factor has been chosen as 100. We first compute

$$
\begin{aligned}
Q(s) &= D(s)D(-s) + qN(s)N(-s) \\
&= s(s - 1)(-s)(-s - 1) + 100(s + 3)(-s + 3) \\
&= s^4 - 101s^2 + 900 \\
&= (s + 9.5459)(s - 9.5459)(s + 3.1427)(s - 3.1427)
\end{aligned}
$$

where we have used PC-MATLAB to compute the roots of $Q(s)$. Thus we have $Q(s) = D_o(s)D_o(-s)$ with

$$D_o(s) = (s + 9.5459)(s + 3.1427) = s^2 + 12.7s + 30$$

and the quadratic optimal system is given by

$$G_o(s) = \frac{qN(0)}{D_o(0)} \cdot \frac{N(s)}{D_o(s)} = \frac{10(s + 3)}{s^2 + 12.7s + 30} \tag{9.46}$$

Its response due to a unit-step reference input is shown in Figure 9.10(a) with the solid line. The actuating signal due to a unit-step reference input is shown in Figure 9.10(b) with the solid line; it has the property $|u(t)| \le 10$ for $t \ge 0$.

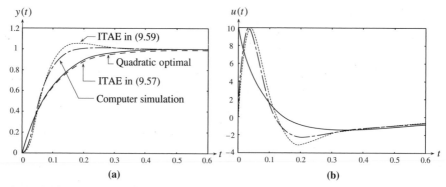

Figure 9.10 Responses of various designs of (9.44).

Example 9.5.3

Consider a plant with transfer function

$$G(s) = \frac{s - 1}{s(s - 2)} \tag{9.47}$$

It has a non-minimum-phase zero. To find the optimal system to minimize the quadratic performance index in (9.45), we compute

$$Q(s) = s(s - 2)(-s)(-s - 2) + 100(s - 1)(-s - 1) = s^4 - 104s^2 + 100$$
$$= (s + 10.1503)(s - 10.1503)(s + 0.9852)(s - 0.9852)$$

Thus we have $D_o(s) = s^2 + 11.14s + 10$ and

$$G_o(s) = \frac{-10(s - 1)}{s^2 + 11.14s + 10} \tag{9.48}$$

Its unit-step response is shown in Figure 9.11 with the solid line. By computer simulation we also find $|u(t)| \leq 10$ for $t \geq 0$ if the reference input is a unit-step function.

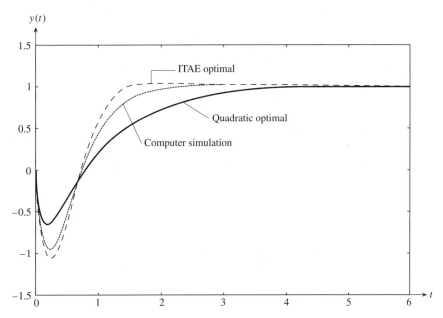

Figure 9.11 Responses of various designs of (9.47).

9.5.1 Symmetric Root Loci[5]

We discuss in this subsection the poles of $G_o(s)$ as a function of q by using the root-locus method. Consider the polynomial

$$D(s)D(-s) + qN(s)N(-s) \qquad (9.49)$$

The roots of (9.49) are the zeros of the rational function

$$1 + qG(s)G(-s)$$

or the solution of the equation

$$\frac{-1}{q} = G(s)G(-s) = \frac{N(s)N(-s)}{D(s)D(-s)} \qquad (9.50)$$

These equations are similar to (7.11) through (7.13), thus the root-locus method can be directly applied. The root loci of (9.50) for $G(s) = 1/s(s + 2)$ are plotted in

---

[5]This section may be skipped without loss of continuity.

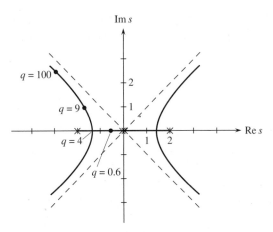

Figure 9.12 Root loci of (9.50).

Figure 9.12. The roots for $q = 0.64, 4, 9,$ and 100 are indicated as shown. We see that the root loci are symmetric with respect to the imaginary axis as well as the real axis. Furthermore the root loci will not cross the imaginary axis for $q > 0$. Although the root loci reveal the migration of the poles of the quadratic optimal system, they do not tell us how to pick a specific set of poles to meet the constraint on the actuating signal.

We discuss now the poles of $G_o(s)$ as $q \to \infty$. It is assumed that $G(s)$ has n poles and m zeros and has no non-minimum-phase zeros. Then, as $q \to \infty$, $2m$ root loci of $G(s)G(-s)$ will approach the $2m$ roots of $N(s)N(-s)$ and the remaining $(2n - 2m)$ root loci will approach the $(2n - 2m)$ asymptotes with angles

$$\frac{(2k + 1)\pi}{2n - 2m} \qquad k = 0, 1, 2, \ldots, 2n - 2m - 1$$

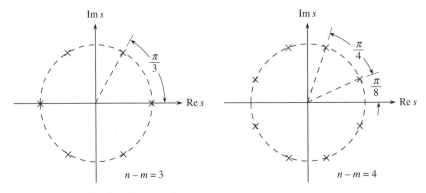

Figure 9.13 Distribution of optimal poles as $q \to \infty$.

(see Section 7.4, in particular, (7.27)). Thus as q approaches infinity, m poles of $G_o(s)$ will cancel m zeros of $G(s)$ and the remaining $(n - m)$ poles of $G_o(s)$ will distribute as shown in Figure 9.13, where we have assumed that the centroid defined in (7.27a) is at the origin. The pole pattern is identical to that of the Butterworth filter [13].

9.6   ITAE OPTIMAL SYSTEMS [33]

In this section we discuss the design of control systems to minimize the integral of time multiplied by absolute error (ITAE) in (9.8). For the quadratic overall system

$$G_o(s) = \frac{1}{s^2 + 2\zeta s + 1}$$

the ITAE, the integral of absolute error (IAE) in (9.6), and the integral of square error (ISE) in (9.7) as a function of the damping ratio ζ are plotted in Figure 9.14. The ITAE has largest changes as ζ varies, and therefore has the best selectivity. The ITAE also yields a system with a faster response than other criteria, therefore Graham and Lathrop [33] chose it as their design criterion. The system that has the smallest ITAE is called the *optimal system in the sense of ITAE* or the *ITAE optimal system*.
Consider the overall transfer function

$$G_o(s) = \frac{\alpha_0}{s^n + \alpha_{n-1}s^{n-1} + \cdots + \alpha_2 s^2 + \alpha_1 s + \alpha_0} \tag{9.51}$$

This transfer function contains no zeros. Because $G_o(0) = 1$, if $G_o(s)$ is stable, then the position error is zero, or the plant output will track asymptotically any step-reference input. By analog computer simulation, the denominators of ITAE optimal systems were found to assume the forms listed in Table 9.1. Their poles and unit-step responses, for $\omega_0 = 1$, are plotted in Figures 9.15 and 9.16. We see that the optimal poles are distributed evenly around the neighborhood of the unit circle. We also see that the overshoots of the unit-step responses are fairly large for large n. These systems are called the *ITAE zero-position-error optimal systems*.

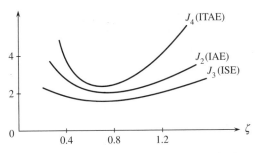

Figure 9.14 Comparison of various design criteria.

Table 9.1   ITAE Zero-Position-Error Optimal Systems

$$s + \omega_0$$
$$s^2 + 1.4\omega_0 s + \omega_0^2$$
$$s^3 + 1.75\omega_0 s^2 + 2.15\omega_0^2 s + \omega_0^3$$
$$s^4 + 2.1\omega_0 s^3 + 3.4\omega_0^2 s^2 + 2.7\omega_0^3 s + \omega_0^4$$
$$s^5 + 2.8\omega_0 s^4 + 5.0\omega_0^2 s^3 + 5.5\omega_0^3 s^2 + 3.4\omega_0^4 s + \omega_0^5$$
$$s^6 + 3.25\omega_0 s^5 + 6.60\omega_0^2 s^4 + 8.60\omega_0^3 s^3 + 7.45\omega_0^4 s^2 + 3.95\omega_0^5 s + \omega_0^6$$
$$s^7 + 4.475\omega_0 s^6 + 10.42\omega_0^2 s^5 + 15.08\omega_0^3 s^4 + 15.54\omega_0^4 s^3 + 10.64\omega_0^5 s^2 + 4.58\omega_0^6 s + \omega_0^7$$
$$s^8 + 5.20\omega_0 s^7 + 12.80\omega_0^2 s^6 + 21.60\omega_0^3 s^5 + 25.75\omega_0^4 s^4 + 22.20\omega_0^5 s^3 + 13.30\omega_0^6 s^2 + 5.15\omega_0^7 s + \omega_0^8$$

We now discuss the optimization of

$$G_o(s) = \frac{\alpha_1 s + \alpha_0}{s^n + \alpha_{n-1}s^{n-1} + \cdots + \alpha_2 s^2 + \alpha_1 s + \alpha_0} \tag{9.52}$$

with respect to the ITAE criterion. The transfer function has one zero; their coefficients, however, are constrained so that $G_o(s)$ has zero position error and zero velocity error. This system will track asymptotically any ramp-reference input. By analog computer simulation, the optimal step responses of $G_o(s)$ in (9.52) are found as shown in Figure 9.17. The optimal denominators of $G_o(s)$ in (9.52) are listed in Table 9.2. The systems are called the *ITAE zero-velocity-error optimal systems*.

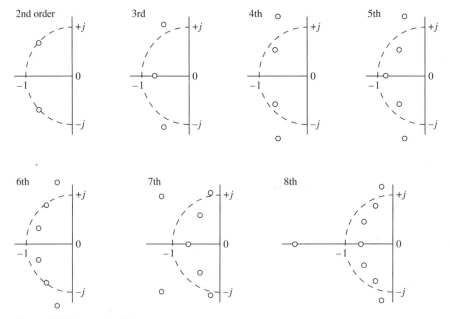

Figure 9.15 Optimal pole locations.

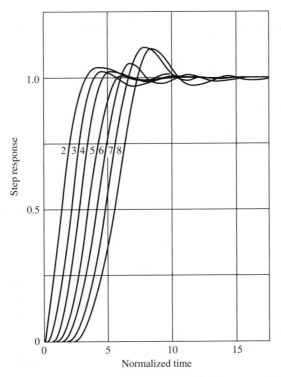

Figure 9.16 Step responses of ITAE optimal systems with zero position error.

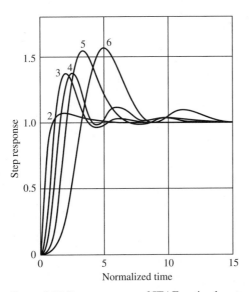

Figure 9.17 Step responses of ITAE optimal systems with zero velocity error.

Table 9.2 ITAE Zero-Velocity-Error Optimal Systems

$$s^2 + 3.2\omega_0 s + \omega_0^2$$
$$s^3 + 1.75\omega_0 s^2 + 3.25\omega_0^2 s + \omega_0^3$$
$$s^4 + 2.41\omega_0 s^3 + 4.93\omega_0^2 s^2 + 5.14\omega_0^3 s + \omega_0^4$$
$$s^5 + 2.19\omega_0 s^4 + 6.50\omega_0^2 s^3 + 6.30\omega_0^3 s^2 + 5.24\omega_0^4 s + \omega_0^5$$
$$s^6 + 6.12\omega_0 s^5 + 13.42\omega_0^2 s^4 + 17.16\omega_0^3 s^3 + 14.14\omega_0^4 s^2 + 6.76\omega_0^5 s + \omega_0^6$$

Similarly, for the following overall transfer function

$$G_o(s) = \frac{\alpha_2 s^2 + \alpha_1 s + \alpha_0}{s^n + \alpha_{n-1} s^{n-1} + \cdots + \alpha_2 s^2 + \alpha_1 s + \alpha_0} \tag{9.53}$$

the optimal step responses are shown in Figure 9.18 and the optimal denominators are listed in Table 9.3. They are called the *ITAE zero-acceleration-error optimal systems*. We see from Figures 9.16, 9.17, and 9.18 that the optimal step responses for $G_o(s)$ with and without zeros are quite different. It appears that if a system is required to track a more complicated reference input, then the transient performance will be poorer. For example, the system in Figure 9.18 tracks acceleration reference inputs, but its transient response is much worse than the one for the system in Figure 9.16, which can track only step-reference inputs. Therefore a price must be paid if we design a more complex system.

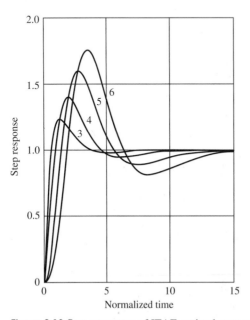

Figure 9.18 Step responses of ITAE optimal systems with zero acceleration error.

Table 9.3 ITAE Zero-Acceleration-Error Optimal Systems

$$s^3 + 2.97\omega_0 s^2 + 4.94\omega_0^2 s + \omega_0^3$$
$$s^4 + 3.71\omega_0 s^3 + 7.88\omega_0^2 s^2 + 5.93\omega_0^3 s + \omega_0^4$$
$$s^5 + 3.81\omega_0 s^4 + 9.94\omega_0^2 s^3 + 13.44\omega_0^3 s^2 + 7.36\omega_0^4 s + \omega_0^5$$
$$s^6 + 3.93\omega_0 s^5 + 11.68\omega_0^2 s^4 + 18.56\omega_0^3 s^3 + 19.3\omega_0^4 s^2 + 8.06\omega_0^5 s + \omega_0^6$$

9.6.1 Applications

In this subsection we discuss how to use Tables 9.1 through 9.3 to design ITAE optimal systems. These tables were developed without considering plant transfer functions. For example, for two different plant transfer functions such as $1/s(s + 2)$ and $1/s(s - 10)$, the optimal transfer function $G_o(s)$ can be chosen as

$$\frac{\omega_0^2}{s^2 + 1.4\omega_0 s + \omega_0^2}$$

The actuating signals for both systems, however, will be different. Therefore ω_0 in both systems should be different. We shall use the constraint on the actuating signal as a criterion in choosing ω_0. This will be illustrated in the following examples.

Example 9.6.1

Consider the plant transfer function in (9.20) or

$$G(s) = \frac{1}{s(s + 2)}$$

Find a zero-position-error system to minimize ITAE. It is also required that the actuating signal due to a unit-step reference input satisfy the constraint

$$|u(t)| \le 3$$

for all t.

The ITAE optimal overall transfer function is chosen from Table 9.1 as

$$G_o(s) = \frac{\omega_0^2}{s^2 + 1.4\omega_0 s + \omega_0^2}$$

It is implementable. Clearly the larger the ω_0, the faster the response. However, the actuating signal will also be larger. Now we shall choose ω_0 to meet $|u(t)| \le 3$. The transfer function from r to u is

$$T(s) := \frac{U(s)}{R(s)} = \frac{G_o(s)}{G(s)} = \frac{\omega_0^2 s(s + 2)}{s^2 + 1.4\omega_0 s + \omega_0^2}$$

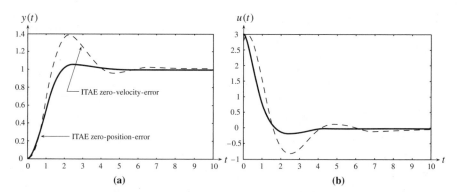

Figure 9.19 Step responses of (9.54) (with solid lines) and (9.55) (with dashed lines).

Consequently, we have

$$U(s) = T(s)R(s) = \frac{\omega_0^2 s(s + 2)}{s^2 + 1.4\omega_0 s + \omega_0^2} \cdot \frac{1}{s}$$

By computer simulation, we find that the largest magnitude of $u(t)$ occurs at $t = 0^+$.[6] Thus the largest magnitude of $u(t)$ can be computed by using the initial-value theorem as

$$u_{max} = u(0^+) = \lim_{s \to \infty} sU(s) = \lim_{s \to \infty} \frac{\omega_0^2 s(s + 2)}{s^2 + 1.4\omega_0 s + \omega_0^2} = \omega_0^2$$

In order to meet the constraint $|u(t)| \leq 3$, we set $\omega_0^2 = 3$. Thus the ITAE optimal system is

$$G_o(s) = \frac{3}{s^2 + 1.4 \times \sqrt{3}s + 3} = \frac{3}{s^2 + 2.4s + 3} \tag{9.54}$$

This differs from the quadratic optimal system in (9.38) only in one coefficient. Because they minimize different criteria, there is no reason that they have the same overall transfer function. The unit-step responses of $G_o(s)$ and $T(s)$ are shown in Figure 9.19 with solid lines. They appear to be satisfactory.

Exercise 9.6.1

Consider a plant with transfer function $2/s^2$. Find an optimal system to minimize the ITAE criterion under the constraint $|u(t)| \leq 3$.

[**Answer:** $6/(s^2 + 3.4s + 6)$.]

---

[6]If the largest magnitude of $u(t)$ does not occur at $t = 0$, then its analytical computation will be complicated. It is easier to find it by computer simulations.

Example 9.6.2

Consider the problem in Example 9.6.1 with the additional requirement that the velocity error be zero. A possible overall transfer function is, from Table 9.2,

$$G_o(s) = \frac{3.2\omega_0 s + \omega_0^2}{s^2 + 3.2\omega_0 s + \omega_0^2}$$

However, this is not implementable because it violates the pole-zero excess inequality. Now we choose from Table 9.2 the transfer function of degree 3:

$$G_o(s) = \frac{3.25\omega_0^2 s + \omega_0^3}{s^3 + 1.75\omega_0 s^2 + 3.25\omega_0^2 s + \omega_0^3}$$

This is implementable and has zero velocity error. Now we choose ω_0 so that the actuating signal due to a unit-step reference input meets $|u(t)| \leq 3$. The transfer function from r to u is

$$T(s) = \frac{G_o(s)}{G(s)} = \frac{(3.25\omega_0^2 s + \omega_0^3)s(s + 2)}{s^3 + 1.75\omega_0 s^2 + 3.25\omega_0^2 s + \omega_0^3}$$

Its unit-step response is shown in Figure 9.19(b) with the dashed line. We see that the largest magnitude of $u(t)$ does not occur at $t = 0^+$. Therefore, the procedure in Example 9.6.1 cannot be used to choose ω_0 for this problem. By computer simulation, we find that if $\omega_0 = 0.928$, then $|u(t)| \leq 3$. For this ω_0, $G_o(s)$ becomes

$$G_o(s) = \frac{2.799s + 0.799}{s^3 + 1.624s^2 + 2.799s + 0.799} \tag{9.55}$$

This is the ITAE zero-velocity-error optimal system. Its unit-step response is plotted in Figure 9.19(a) with the dashed line. It is much more oscillatory than that of the ITAE zero-position-error optimal system. The corresponding actuating signal is plotted in Figure 9.19(b).

Example 9.6.3

Consider the plant transfer function in (9.39), that is,

$$G(s) = \frac{2}{s(s^2 + 0.25s + 6.25)}$$

Find an ITAE zero-position-error optimal system. It is also required that the actuating signal $u(t)$ due to a unit-step reference input meet the constraint $|u(t)| \leq 10$, for $t \geq 0$. We choose from Table 9.1

$$G_o(s) = \frac{\omega_0^3}{s^3 + 1.75\omega_0 s^2 + 2.15\omega_0^2 s + \omega_0^3}$$

By computer simulation, we find that if $\omega_0 = 2.7144$, then $|u(t)| \leq u(0) = 10$ for all $t \geq 0$. Thus the ITAE optimal system is

$$G_o(s) = \frac{20}{s^3 + 4.75s^2 + 15.84s + 20} \tag{9.56}$$

Its unit-step response is plotted in Figure 9.9 with the dashed line. Compared with the quadratic optimal design, the ITAE design has a faster response and a smaller overshoot. Thus for this problem, the ITAE optimal system is more desirable.

Example 9.6.4

Consider the plant transfer function in (9.44) or

$$G(s) = \frac{s + 3}{s(s - 1)}$$

Find an ITAE zero-position-error optimal system. It is also required that the actuating signal $u(t)$ due to a unit-step reference input meet the constraint $|u(t)| \leq 10$, for $t \geq 0$. The pole-zero excess of $G(s)$ is 1 and $G(s)$ has no non-minimim-phase zero; therefore, the ITAE optimal transfer function

$$G_o(s) = \frac{\omega_0}{s + \omega_0} \tag{9.57}$$

is implementable. We find by computer simulation that if $\omega_0 = 10$, then $G_o(s)$ meets the design specifications. Its step response and actuating signal are plotted in Figure 9.10 with the dashed line. They are almost indistinguishable from those of the quadratic optimal system. Because $G_o(s)$ does not contain plant zero $(s + 3)$, its implementation will involve the pole-zero cancellation of $(s + 3)$, as will be discussed in the next chapter. Because it is a stable pole and has a fairly small time constant, its cancellation will not affect seriously the behavior of the overall system, as will be demonstrated in the next chapter.

In addition to (9.57), we may also choose the following ITAE optimal transfer function

$$G_o(s) = \frac{\omega_0^2}{s^2 + 1.4\omega_0 s + \omega_0^2} \tag{9.58}$$

It has pole-zero excess larger than that of $G(s)$ and is implementable. We find by computer simulation that if $\omega_0 = 24.5$, then the $G_o(s)$ in (9.58) or

$$G_o(s) = \frac{600.25}{s^2 + 34.3s + 600.25} \tag{9.59}$$

meets the design specifications. Its step response and actuating signal are plotted in Figure 9.10 with the dotted lines. The step response is much faster than the ones of (9.57) and the quadratic optimal system. However, it has an overshoot of about 4.6%.

Example 9.6.5

Consider the plant transfer function in (9.47) or

$$G(s) = \frac{s - 1}{s(s - 2)}$$

Find an ITAE zero-position-error optimal system. It is also required that the actuating signal $u(t)$ due to a unit-step reference input meet the constraint $|u(t)| \leq 10$, for $t \geq 0$. This plant transfer function has a non-minimum-phase zero and no ITAE standard form is available to carry out the design. However, we can employ the idea in [34] and use computer simulation to find its ITAE optimal transfer function as [54]

$$G_o(s) = \frac{-10(s - 1)}{s^2 + 5.1s + 10} \tag{9.60}$$

under the constraint $|u(t)| \leq 10$. We mention that the non-minimum-phase zero $(s - 1)$ of $G(s)$ must be retained in $G_o(s)$, otherwise $G_o(s)$ is not implementable. Its step response is plotted in Figure 9.11 with the dashed line. It has a faster response than the one of the quadratic optimal system in (9.48); however, it has a larger undershoot and a larger overshoot. Therefore it is difficult to say which system is better.

9.7 SELECTION BASED ON ENGINEERING JUDGMENT

In the preceding sections, we introduced two criteria for choosing overall transfer functions. The first criterion is the minimization of the quadratic performance index. The main reason for choosing this criterion is that it renders a simple and straight-forward procedure to compute the overall transfer function. The second criterion is the minimization of the integral of time multiplied by absolute error (ITAE). It was chosen in [33] because it has the best selectivity. This criterion, however, does not render an analytical method to find the overall transfer function; it is obtained by trial and error and by computer simulation. In this section, we forego the concept of minimization or optimization and select overall transfer functions based on engineering judgment. We require the system to have a zero position error and a *good* transient performance. By a good transient performance, we mean that the rise and settling times are small and the overshoot is also small. Without comparisons, it is not possible to say what is small. Fortunately, we have quadratic and ITAE optimal systems for comparisons. Therefore, we shall try to find an overall system that has a comparable or better transient performance than the quadratic or ITAE optimal system. Whether the transient performance is comparable or better is based on engineering judgment; no mathematical criterion will be used. Consequently, the selection will be subjective and the procedure of selection is purely trial and error.

Example 9.7.1

Consider the plant transfer function in (9.39), or

$$G(s) = \frac{2}{s(s^2 + 0.25s + 6.25)}$$

We use computer simulation to select the following two overall transfer functions

$$G_{o1}(s) = \frac{20}{(s + 2)(s^2 + 2.5s + 10)} = \frac{20}{s^3 + 4.5s^2 + 15s + 20} \tag{9.61}$$

$$G_{o2}(s) = \frac{20}{(s + 10)(s^2 + 2s + 2)} = \frac{20}{s^3 + 12s^2 + 22s + 20} \tag{9.62}$$

The actuating signals of both systems due to a unit-step reference input meet the constraint $|u(t)| \leq 10$ for $t \geq 0$. Their step responses are plotted in Figure 9.9 with, respectively, the dotted line and the dashed-and-dotted line. The step response of $G_{o1}(s)$ lies somewhere between those of the quadratic optimal system and the ITAE optimal system. Therefore, $G_{o1}(s)$ is a viable alternative of the quadratic or ITAE optimal system.

The concept of dominant poles can also be used to select $G_o(s)$. Consider the overall transfer function in $G_{o2}(s)$. It has a pair of complex-conjugate poles at $-1 \pm j1$ and a real pole at -10. Because the response due to the real pole dies out much faster than does the response due to the complex-conjugate poles, the response of $G_{o2}(s)$ is essentially determined or dominated by the complex-conjugate poles. The complex-conjugate poles have the damping ratio 0.707, and consequently the step response has an overshoot of about 5% (see Section 7.2.1), as can be seen from Figure 9.9. However, because the product of the three poles must equal 20 in order to meet the constraint on the actuating signal, if we choose the nondominant pole far away from the imaginary axis, then the complex-conjugate poles cannot be too far away from the origin of the s-plane. Consequently, the time constant of $G_{o2}(s)$ is larger than that of $G_{o1}(s)$ and its step response is slower, as is shown in Figure 9.9. Therefore for this problem, the use of dominant poles does not yield a satisfactory system. We mention that if the complex-conjugate poles are chosen at $-2 \pm j2$, then the system will respond faster. However, the real pole must be chosen as $20/8 = 2.5$ in order to meet the constraint on the actuating signal. In this case, the complex-conjugate poles no longer dominate over the real pole, and the concept of dominant poles cannot be used.

Example 9.7.2

Consider the plant transfer function in (9.44), or

$$G(s) = \frac{s + 3}{s(s - 1)} \qquad (9.63)$$

We have designed a quadratic optimal system in (9.46) and two ITAE optimal systems in (9.57) with $\omega_0 = 10$ and (9.59). Their step responses are shown in Figure 9.10. Now using computer simulation, we find that the following

$$G_o(s) = \frac{784}{s^2 + 50.4s + 784} \qquad (9.64)$$

has the response shown with the dashed-and-dotted line in Figure 9.10. It is comparable with that of the ITAE optimal system in (9.59) under the same constraint on the actuating signal. Thus, the overall transfer function in (9.64) can also be used, although it is not optimal in any sense.

Example 9.7.3

Consider the plant transfer function in (9.47) or

$$G(s) = \frac{s - 1}{s(s - 2)}$$

We have designed a quadratic optimal system in (9.48) and an ITAE optimal system in (9.60). Their step responses are shown in Figure 9.11. Now we find, by using computer simulation, that the response of

$$G_o(s) = \frac{-10(s - 1)}{(s + \sqrt{10})^2} \qquad (9.65)$$

lies somewhere between those of (9.48) and (9.60) under the same constraint on the actuating signal. Therefore, (9.65) can also be chosen as an overall transfer function.

In this section, we have shown by examples that it is possible to use computer simulation to select an overall transfer function whose performance is comparable to that of the quadratic or ITAE optimal system. The method, however, is a trial-and-error method. In the search, we vary the coefficients of the quadratic or ITAE system and see whether or not the performance could be improved. If we do not have the quadratic or ITAE optimal system as a starting point, it would be difficult to find a good system. Therefore, the computer simulation method cannot replace the quadratic design method, nor the standard forms of the ITAE optimal systems. It can be used to complement the two optimal methods.

9.8 SUMMARY AND CONCLUDING REMARKS

This chapter introduced the inward approach to design control systems. In this approach, we first find an overall transfer function to meet design specifications and then implement it. In this chapter, we discussed only the problem of choosing an overall transfer function. The implementation problem is discussed in the next chapter.

The choice of an overall transfer function is not entirely arbitrary; otherwise we may simply choose the overall transfer function as 1. Given a plant transfer function $G(s) = N(s)/D(s)$, an overall transfer function $G_o(s) = N_o(s)/D_o(s)$ is said to be *implementable* if there exists a configuration with no plant leakage such that $G_o(s)$ can be built using only proper compensators. Furthermore, the resulting system is required to be well posed and totally stable—that is, the closed-loop transfer function of every possible input-output pair of the system is proper and stable. The necessary and sufficient conditions for $G_o(s)$ to be implementable are that (1) $G_o(s)$ is stable, (2) $G_o(s)$ contains the non-minimum-phase zeros of $G(s)$, and (3) the pole-zero excess of $G_o(s)$ is equal to or larger than that of $G(s)$. These constraints are not stringent; poles of $G_o(s)$ can be arbitrarily assigned so long as they all lie in the open left half s-plane; other than retaining all zeros outside the region C in Figures 6.13 or 7.4, all other zeros of $G_o(s)$ can be arbitrarily assigned in the entire s-plane.

In this chapter, we discussed how to choose an implementable overall system to minimize the quadratic and ITAE performance indices. In using these performance indices, a constraint on the actuating signal or on the bandwidth of resulting systems must be imposed; otherwise, it is possible to design an overall system to have a performance index as small as desirable and the corresponding actuating signal will approach infinity. The procedure of finding quadratic optimal systems is simple and straightforward; after computing a spectral factorization, the optimal system can be readily obtained from (9.19). Spectral factorizations can be carried out by iteration without computing any roots, or computing all the roots of (9.16) and then grouping the open left half s-plane roots. ITAE optimal systems are obtainable from Tables 9.1 through 9.3. Because the tables are not exhaustive, for some plant transfer functions (for example, those with non-minimum-phase zeros), no standard forms are available to find ITAE optimal systems. In this case, we may resort to computer simulation to find an ITAE optimal system.

In this chapter, we also showed by examples that overall transfer functions that have comparable performance as quadratic or ITAE optimal systems can be obtained by computer simulation without minimizing any mathematical performance index. It is therefore suggested that after obtaining quadratic or ITAE optimal systems, we may change the parameters of the optimal systems to see whether a more desirable system can be obtained. In conclusion, we should make full use of computers to carry out the design.

We give some remarks concerning the quadratic optimal design to conclude this chapter.

1. The quadratic optimal system in (9.19) is reduced from a general formula in Reference [10]. The requirement of implementability is included in (9.19). If no

such requirement is included, the optimal transfer function that minimizes (9.15) with $r(t) = 1$ is

$$\overline{G}_o(s) = \frac{qN(0)}{D_o(0)} \cdot \frac{N_+(s)}{D_o(s)} \tag{9.66}$$

where $N_+(s)$ is $N(s)$ with all its right-half-plane roots reflected into the left half plane. In this case, the resulting overall transfer function may not be implementable. For example, if $G(s) = (s - 1)/s(s + 1)$, then the optimal system that minimizes

$$J = \int_0^\infty [q(y(t) - 1)^2 + u^2(t)]dt \tag{9.67}$$

with $q = 9$ is

$$\overline{G}_o(s) = \frac{3(s + 1)}{s^2 + 4s + 3}$$

which does not retain the non-minimum-phase zero and is not implementable. For this optimal system, J can be computed as $J = 3$. See Chapter 11 of Reference [12] for a discussion of computing J. The *implementable* optimal system that minimizes J in (9.67) is

$$G_o(s) = \frac{-3(s - 1)}{s^2 + 4s + 3}$$

For this implementable $G_o(s)$, J can be computed as $J = 21$. It is considerably larger than the J for $\overline{G}_o(s)$. Although $\overline{G}_o(s)$ has a smaller performance index, it cannot be implemented.

2. If $r(t)$ in (9.15) is a ramp function, that is $r(t) = at$, $t \geq 0$, then the optimal system that minimizes (9.15) is

$$G_o(s) = q(k_1 + k_2 s)\frac{N(s)}{D_o(s)} = \left(1 + \frac{k_2}{k_1}s\right)\frac{qN(0)}{D_o(0)} \cdot \frac{N(s)}{D_o(s)} \tag{9.68}$$

where

$$k_1 = \frac{N(0)}{D_o(0)} \quad \text{and} \quad k_2 = \frac{d}{ds}\left[\frac{N(-s)}{D_o(-s)}\right]_{s=0}$$

or, if $N(s) = N_0 + N_1 s + \cdots + N_m s^m$ and $D_o(s) = D_0 + D_1 s + \cdots + D_n s^n$,

$$k_1 = \frac{N_0}{D_0} \quad \text{and} \quad k_2 = \frac{N_0 D_1 - D_0 N_1}{D_0^2}$$

The optimal system in (9.68) is not implementable because it violates the pole-zero excess inequality. However, if we modify (9.68) as

$$\overline{G}_o(s) = \left(1 + \frac{k_2}{k_1}s\right)\frac{qN(0)}{D_o(0)}\frac{N(s)}{D_o(s) + \epsilon s^{n+1}} \tag{9.69}$$

where $n := \deg D_o(s)$ and ϵ is a very small positive number, then $\overline{G}_o(s)$ will be implementable. Furthermore, for a sufficiently small ϵ, $D_o(s) + \epsilon s^{n+1}$ is Hurwitz, and the frequency response of $\overline{G}_o(s)$ is very close to that of $G_o(s)$ in (9.68). Thus (9.69) is a simple and reasonable modification of (9.68).[7]

3. The quadratic optimal design can be carried out using transfer functions or using state-variable equations. In using state-variable equations, the concepts of controllability and observability are needed. The optimal design requires solving an algebraic Riccati equation and designing a state estimator (see Chapter 11). For the single-variable systems studied in this text, the transfer function approach is simpler and intuitively more transparent. The state-variable approach, however, can be more easily extended to multivariable systems.

PROBLEMS

9.1. Given $G(s) = (s + 2)/(s - 1)$, is $G_o(s) = 1$ implementable? Given $G(s) = (s - 1)/(s + 2)$, is $G_o(s) = 1$ implementable?

9.2. Given $G(s) = (s + 3)(s - 2)/s(s + 2)(s - 3)$, which of the following $G_o(s)$ are implementable?

$$\frac{s - 2}{s(s + 2)} \qquad \frac{s + 3}{(s + 2)(s - 3)} \qquad \frac{s - 2}{(s + 2)^2}$$

$$\frac{(s + 4)(s - 2)}{s^4 + 4s^2 + 3s + 6} \qquad \frac{s - 2}{s^3 + 4s + 2}$$

9.3. Consider a plant with transfer function $G(s) = (s + 3)/s(s - 2)$.

a. Find an implementable overall transfer function that has all poles at -2 and has a zero position error.

b. Find an implementable overall transfer function that has all poles at -2 and has a zero velocity error. Is the choice unique? Do you have to retain $s + 3$ in $G_o(s)$? Find two sets of solutions: One retains $s + 3$ and the other does not.

9.4. Consider a plant with transfer function $G(s) = (s - 3)/s(s - 2)$.

a. Find an implementable overall transfer function that has all poles at -2 and has a zero position error.

b. Find an implementable overall transfer function that has all poles at -2 and has a zero velocity error. Is the choice unique if we require the degree of $G_o(s)$ to be as small as possible?

---

[7]This modification was suggested by Professor Jong-Lick Lin of Cheng Kung University, Taiwan.

9.5. What types of reference signals will the following $G_o(s)$ track without an error?

a. $G_o(s) = \dfrac{-5s - 2}{-s^2 - 5s - 2}$

b. $G_o(s) = \dfrac{4s^2 + s + 3}{s^5 + 3s^4 + 4s^2 + s + 3}$

c. $G_o(s) = \dfrac{-2s^2 + 154s + 120}{s^4 + 14s^3 + 71s^2 + 154s + 120}$

9.6. Consider two systems. One has a settling time of 10 seconds and an overshoot of 5%, the other has a settling time of 7 seconds and an overshoot of 10%. Is it possible to state which system is better? Now we introduce a performance index as

$$J = k_1 \cdot (\text{Settling time}) + k_2 \cdot (\text{Percentage overshoot})$$

If $k_1 = k_2 = 0.5$, which system is better? If $k_1 = 0.8$ and $k_2 = 0.2$, which system is better?

9.7. Is the function

$$J = \int_0^\infty [q(y(t) - r(t)) + u(t)]dt$$

with $q > 0$ a good performance criterion?

9.8. Consider the design problem in Problem 7.15 or a plant with transfer function $G(s) = -2/s^2$. Design an overall system to minimize the quadratic performance index in (9.15) with $q = 4$. What are its position error and velocity error?

9.9. In Problem 9.8, design a quadratic optimal system that is as fast as possible under the constraint that the actuating signal due to a step-reference input must have a magnitude less than 5.

9.10. Plot the poles of $G_o(s)$ as a function of q in Problem 9.9.

9.11. Consider the design problem in Problem 7.14 or a plant with transfer function

$$G(s) = \dfrac{0.015}{s^2 + 0.11s + 0.3}$$

Design an overall system to minimize the quadratic performance index in (9.15) with $q = 9$. Is the position error of the optimal system zero? Is the index of the optimal system finite?

9.12. Consider the design problem in Problem 7.12 or a plant with transfer function

$$G(s) = \dfrac{4(s + 0.05)}{s(s + 2)(s - 1.2)}$$

Find a quadratic optimal system with $q = 100$. Carry out the spectral factorization by using the iterative method discussed in Section 9.4.2.

9.13. Let $Q(s) = D_o(s)D_o(-s)$ with

$$Q(s) = a_0 + a_2 s^2 + a_4 s^4 + \cdots + a_{2n} s^{2n}$$

and

$$D_o(s) = b_0 + b_1 s + b_2 s^2 + \cdots + b_n s^n$$

Show

$$a_0 = b_0^2$$
$$a_2 = 2b_0 b_2 - b_1^2$$
$$a_4 = 2b_0 b_4 - 2b_1 b_3 + b_2^2$$

$$\vdots$$

$$a_{2n} = 2b_0 b_{2n} - 2b_1 b_{2n-1} + 2b_2 b_{2n-2} - \cdots + (-1)^n b_n^2$$

where $b_i = 0$, for $i > n$.

9.14. The depth of a submarine can be maintained automatically by a control system, as discussed in Problem 7.8. The transfer function of the submarine from the stern angle θ to the actual depth y can be approximated as

$$G(s) = \frac{10(s + 2)^2}{(s + 10)(s^2 + 0.1)}$$

Find an overall system to minimize the performance index

$$J = \int_0^\infty [(y(t) - 1)^2 + \theta^2] dt$$

9.15. Consider a plant with transfer function $s/(s^2 - 1)$. Design an overall system to minimize the quadratic performance index in (9.15) with $q = 1$. Does the optimal system have zero position error? If not, modify the overall system to yield a zero position error.

9.16. Consider a plant with transfer function $G(s) = 1/s(s + 1)$. Find an implementable transfer function to minimize the ITAE criterion and to have zero position error. It is also required that the actuating signal due to a unit-step reference input have a magnitude less than 10.

9.17. Repeat Problem 9.16 with the exception that the overall system is required to have a zero velocity error.

9.18. Repeat Problem 9.16 for $G(s) = 1/s(s - 1)$.

9.19. Repeat Problem 9.17 for $G(s) = 1/s(s - 1)$.

9.20. Find an ITAE zero-position-error optimal system for the plant given in Problem 9.8. The magnitude of the actuating signal is required to be no larger than the one in Problem 9.8.

9.21. Find an ITAE zero-position-error optimal system for the plant in Problem 9.11. The real part of the poles of the optimal system is required to equal that in Problem 9.11.

9.22. Is it possible to obtain an ITAE optimal system for the plant in Problem 9.12 from Table 9.1 or 9.2? If yes, what will happen to the plant zero?

9.23. Repeat Problem 9.22 for the plant in Problem 9.14.

9.24. **a.** Consider a plant with transfer function $G(s) = (s + 4)/s(s + 1)$. Design an ITAE zero-position-error optimal system of degree 1. It is required that the actuating signal due to a unit-step reference input have a magnitude less than 10.

b. Consider a plant with transfer function $G(s) = (s + 4)/s(s + 1)$. Design an ITAE zero-position-error optimal system of degree 2. It is required that the actuating signal due to a unit-step reference input have a magnitude less than 10.

c. Compare their unit-step responses.

9.25. Consider the generator-motor set in Figure 6.1. Its transfer function is assumed to be

$$G(s) = \frac{300}{s^4 + 184s^3 + 760s^2 + 162s}$$

It is a type 1 transfer function. Design a quadratic optimal system with $q = 25$. Design an ITAE optimal system with $u(0^+) = 5$. Plot their poles. Are there many differences?

9.26. Consider a plant with transfer function $1/s^2$. Find an optimal system with zero velocity error to minimize the ITAE criterion under the constraint $|u(t)| \leq 6$. [Answer: $(6s + 2.5)/(s^3 + 2.38s^2 + 6s + 2.5)$.]

9.27. If software for computing step responses is available, adjust the coefficients of the quadratic optimal system in Problem 9.8, 9.11, 9.12, 9.14, or 9.15 to see whether a comparable or better transient performance can be obtained.

10 Implementation—Linear Algebraic Method

10.1 INTRODUCTION

The first step in the design of control systems using the inward approach is to find an overall transfer function to meet design specifications. This step was discussed in Chapter 9. Now we discuss the second step—namely, implementation of the chosen overall transfer function. In other words, given a plant transfer function $G(s)$ and an implementable $G_o(s)$, we shall find a feedback configuration without plant leakage and compute compensators so that the transfer function of the resulting system equals $G_o(s)$. The compensators used must be proper and the resulting system must be well posed and totally stable.

The preceding problem can also be stated as follows: Given a plant $G(s)$ and given a model $G_o(s)$, design an overall system so that the overall transfer function equals or matches $G_o(s)$. Thus the problem can also be called the *model-matching problem*. In the model-matching problem, we match not only poles but also zeros; therefore, it can also be called the *pole-and-zero placement problem*. There is a closely related problem, called the *pole-placement problem*. In the pole-placement problem, we match or control only poles of resulting overall systems; zeros are not specified. In this chapter, we study both the model-matching and pole-placement problems.

This chapter introduces three control configurations. They are the unity-feedback, two-parameter, and plant input/output feedback configurations. The unity-feedback configuration can be used to achieve any pole placement but not any model matching. The other two configurations, however, can be used to achieve any model matching. In addition to model matching and pole placement, this chapter also studies robust tracking and disturbance rejection.

The idea used in this chapter is very simple. The design is carried out by matching coefficients of compensators with desired polynomials. If the denominator $D(s)$ and numerator $N(s)$ of a plant transfer function have common factors, then it is not possible to achieve any pole placement or any model matching. Therefore, we require $D(s)$ and $N(s)$ to have no common factors or to be coprime. Under this assumption, the conditions of achieving matching depend on the degree of compensators. The larger the degree, the more parameters we have for matching. If the degree of compensators is large enough, matching is always possible. The design procedures in this chapter are essentially developed from these concepts and conditions.

10.2 UNITY-FEEDBACK CONFIGURATION—MODEL MATCHING

We discuss in this section the implementation of an implementable $G_o(s)$ by using the unity-feedback configuration shown in Figure 10.1. Let $G(s)$ and $C(s)$ be respectively the transfer function of the plant and compensator. If the overall transfer function from r to y is $G_o(s)$, then we have

$$G_o(s) = \frac{C(s)G(s)}{1 + C(s)G(s)} \tag{10.1}$$

which implies

$$G_o(s) + C(s)G(s)G_o(s) = C(s)G(s)$$

and

$$C(s)G(s)(1 - G_o(s)) = G_o(s)$$

Thus, the compensator can be computed from

$$C(s) = \frac{G_o(s)}{G(s)(1 - G_o(s))} \tag{10.2}$$

We use examples to illustrate its computation and discuss the issues that arise in its implementation.

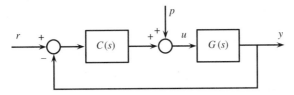

Figure 10.1 Unity-feedback system.

Example 10.2.1

Consider a plant with transfer function $1/s(s + 2)$. The optimal system that minimizes the ITAE criterion and meets the constraint $|u(t)| \leq 3$ was computed in (9.54) as $3/(s^2 + 2.4s + 3)$. If we implement this $G_o(s)$ in Figure 10.1, then the compensator is

$$C(s) = \frac{\dfrac{3}{s^2 + 2.4s + 3}}{\dfrac{1}{s(s + 2)} \cdot \left(1 - \dfrac{3}{s^2 + 2.4s + 3}\right)}$$

(10.3)

$$= \frac{\dfrac{3}{s^2 + 2.4s + 3}}{\dfrac{1}{s(s + 2)} \cdot \dfrac{s^2 + 2.4s}{s^2 + 2.4s + 3}} = \frac{3(s + 2)}{s + 2.4}$$

It is a proper compensator. This implementation has a pole-zero cancellation. Because the cancelled pole $(s + 2)$ is a stable pole, the system is totally stable. The condition for the unity-feedback configuration in Figure 10.1 to be well posed is $1 + G(\infty)C(\infty) \neq 0$. This is the case for this implementation; therefore, the system is well posed. In fact, if $G(s)$ is strictly proper and if $C(s)$ is proper, then the unity-feedback configuration is always well posed.

This implementation has the cancellation of the stable pole $(s + 2)$. This canceled pole is dictated by the plant, and the designer has no control over it. If this cancellation is acceptable, then the design is completed. In a latter section, we shall discuss a different implementation where the designer has freedom in choosing canceled poles.

Example 10.2.2

Consider a plant with transfer function $2/(s + 1)(s - 1)$. If $G_o(s) = 2/(s^2 + 2s + 2)$, then

$$C(s) = \frac{\dfrac{2}{s^2 + 2s + 2}}{\dfrac{2}{(s + 1)(s - 1)} \cdot \left(1 - \dfrac{2}{s^2 + 2s + 2}\right)}$$

$$= \frac{\dfrac{2}{s^2 + 2s + 2}}{\dfrac{2}{(s + 1)(s - 1)} \cdot \dfrac{s^2 + 2s}{s^2 + 2s + 2}} = \frac{(s + 1)(s - 1)}{s(s + 2)}$$

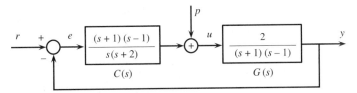

Figure 10.2 Unity-feedback system.

The compensator is proper. The implementation is plotted in Figure 10.2. The implementation has a stable pole-zero cancellation and an unstable pole-zero cancellation between $C(s)$ and $G(s)$.

As discussed in Chapter 6, noise and/or disturbance may enter a control system at every terminal. Therefore we require every control system to be totally stable. We compute the closed-loop transfer function from p to y shown in Figure 10.2:

$$G_{yp}(s) := \frac{Y(s)}{P(s)} = \frac{\dfrac{2}{(s+1)(s-1)}}{1 + \dfrac{(s+1)(s-1)}{s(s+2)} \cdot \dfrac{2}{(s+1)(s-1)}}$$

$$= \frac{\dfrac{2}{(s+1)(s-1)}}{\dfrac{s(s+2)+2}{s(s+2)}} = \frac{2s(s+2)}{(s-1)(s+1)(s^2+2s+2)}$$

It is unstable. Thus the output will approach infinity if there is any nonzero, no matter how small, disturbance. Consequently, the system is not totally stable, and the implementation is not acceptable.

Exercise 10.2.1

Consider a plant with transfer function $1/s^2$. Implement $G_o(s) = 6/(s^2 + 3.4s + 6)$ in the unity-feedback configuration. Is the implementation acceptable?

[**Answer:** $C(s) = 6s/(s + 3.4)$, unacceptable.]

These examples show that the implementation of $G_o(s)$ in the unity-feedback configuration will generally involve pole-zero cancellations. The canceled poles are determined by the plant transfer function, and we have no control over them. In general, if $G(s)$ has open right-half-plane poles or two or more poles at $s = 0$, then the unity-feedback configuration cannot be used to implement any $G_o(s)$. Thus, the unity-feedback configuration cannot, in general, be used in model matching.

10.3 UNITY-FEEDBACK CONFIGURATION—POLE PLACEMENT BY MATCHING COEFFICIENTS

Although the unity-feedback configuration cannot be used in *every* model matching, it can be used to achieve arbitrary pole placement. In pole placement, we assign only poles and leave zeros unspecified. We first use an example to illustrate the basic idea and to discuss the issues involved.

Example 10.3.1

Consider a plant with transfer function

$$G(s) = \frac{1}{s(s + 2)}$$

and consider the unity-feedback configuration shown in Figure 10.1. If the compensator $C(s)$ is a gain of k (a transfer function of degree 0), then the overall transfer function can be computed as

$$G_o(s) = \frac{kG(s)}{1 + kG(s)} = \frac{k}{s^2 + 2s + k}$$

This $G_o(s)$ has two poles. These two poles cannot be arbitrarily assigned by choosing a value for k. For example, if we assign the two poles at -2 and -3, then the denominator of $G_o(s)$ must equal

$$s^2 + 2s + k = (s + 2)(s + 3) = s^2 + 5s + 6$$

Clearly, there is no k to meet the equation. Therefore, if the compensator is of degree 0, it is not possible to achieve arbitrary pole placement.[1]

Next let the compensator be proper and of degree 1 or

$$C(s) = \frac{B_0 + B_1 s}{A_0 + A_1 s}$$

with $A_1 \neq 0$. Then the overall transfer function can be computed as

$$G_o(s) = \frac{C(s)G(s)}{1 + C(s)G(s)} = \frac{B_0 + B_1 s}{s(s + 2)(A_1 s + A_0) + B_1 s + B_0}$$

$$= \frac{B_1 s + B_0}{A_1 s^3 + (2A_1 + A_0)s^2 + (2A_0 + B_1)s + B_0}$$

This $G_o(s)$ has three poles. We show that all these three poles can be arbitrarily assigned by choosing a suitable $C(s)$. Let the denominator of $G_o(s)$ be

$$D_o(s) = s^3 + F_2 s^2 + F_1 s + F_0$$

---

[1] The root loci of this problem are plotted in Figure 7.5. If $C(s) = k$, we can assign the two poles only along the root loci.

where F_i are entirely arbitrary. Now we equate the denominator of $G_o(s)$ with $D_o(s)$ or

$$A_1 s^3 + (2A_1 + A_0)s^2 + (2A_0 + B_1)s + B_0 = s^3 + F_2 s^2 + F_1 s + F_0$$

Matching the coefficients of like power of s yields

$$A_1 = 1 \quad 2A_1 + A_0 = F_2 \quad 2A_0 + B_1 = F_1 \quad B_0 = F_0$$

which imply

$$A_1 = 1 \quad A_0 = F_2 - 2A_1 \quad B_1 = F_1 - 2F_2 + 4A_1 \quad B_0 = F_0$$

For example, if we assign the three poles $G_o(s)$ as -2 and $-2 \pm 2j$, then $D_o(s)$ becomes

$$D_o(s) = s^3 + F_2 s^2 + F_1 s + F_0 = (s + 2)(s + 2 + 2j)(s + 2 - 2j)$$
$$= s^3 + 6s^2 + 16s + 16$$

We mention that if a complex pole is assigned in $D_o(s)$, its complex conjugate must also be assigned. Otherwise, $D_o(s)$ will have complex coefficients. For this set of poles, we have

$$A_1 = 1 \quad A_0 = 6 - 2 = 4 \quad B_1 = 16 - 2 \cdot 6 + 4 = 8 \quad B_0 = 16$$

and the compensator is

$$C(s) = \frac{8s + 16}{s + 4}$$

This compensator will place the poles of $G_o(s)$ at -2 and $-2 \pm 2j$. To verify this, we compute

$$G_o(s) = \frac{C(s)G(s)}{1 + C(s)G(s)} = \frac{\dfrac{8s + 16}{s + 4} \cdot \dfrac{1}{s(s + 2)}}{1 + \dfrac{8s + 16}{s + 4} \cdot \dfrac{1}{s(s + 2)}}$$

$$= \frac{8s + 16}{s(s + 2)(s + 4) + 8s + 16} = \frac{8s + 16}{s^3 + 6s^2 + 16s + 16}$$

Indeed $G_o(s)$ has poles at -2 and $-2 \pm 2j$. Note that the compensator also introduces zero $(8s + 16)$ into $G_o(s)$. The zero is obtained from solving a set of equations, and we have no control over it. Thus, pole placement is different from pole-and-zero placement or model matching.

This example shows the basic idea of pole placement in the unity-feedback configuration. It is achieved by matching coefficients. In the following, we shall extend the procedure to the general case and also establish the condition for achieving pole placement. Consider the unity-feedback system shown in Figure 10.1. Let

$$G(s) = \frac{N(s)}{D(s)} \qquad C(s) = \frac{B(s)}{A(s)} \qquad G_o(s) = \frac{N_o(s)}{D_o(s)}$$

and deg $N(s) \leq$ deg $D(s) = n$. The substitution of these into (10.1) yields

$$G_o(s) = \frac{C(s)G(s)}{1 + C(s)G(s)} = \frac{\dfrac{B(s)}{A(s)} \cdot \dfrac{N(s)}{D(s)}}{1 + \dfrac{B(s)}{A(s)} \cdot \dfrac{N(s)}{D(s)}}$$

which becomes

$$G_o(s) = \frac{N_o(s)}{D_o(s)} = \frac{B(s)N(s)}{A(s)D(s) + B(s)N(s)} \tag{10.4}$$

Given $G(s)$, if there exists a proper compensator $C(s) = B(s)/A(s)$ so that all poles of $G_o(s)$ can be arbitrarily assigned, the design is said to achieve arbitrary pole placement. In the placement, if a complex number is assigned as a pole, its complex conjugate must also be assigned. From (10.4), we see that the pole-placement problem is equivalent to solving

$$A(s)D(s) + B(s)N(s) = D_o(s) \tag{10.5}$$

This polynomial equation is called a *Diophantine equation*. In the equation, $D(s)$ and $N(s)$ are given, the roots of $D_o(s)$ are the poles of the overall system to be assigned, and $A(s)$ and $B(s)$ are unknown polynomials to be solved. Note that $B(s)$ also appears in the numerator $N_o(s)$ of $G_o(s)$. Because $B(s)$ is obtained from solving (10.5), we have no direct control over it and, consequently, no control over the zeros of $G_o(s)$. Note that before solving (10.5), we don't know what $G_o(s)$ will be; therefore $C(s) = B(s)/A(s)$ cannot be computed from (10.2).

10.3.1 Diophantine Equations

The crux of pole placement is solving the Diophantine equation in (10.5) or

$$A(s)D(s) + B(s)N(s) = D_o(s)$$

In this equation, $D(s)$ and $N(s)$ are given, $D_o(s)$ is to be chosen by the designer. The questions are: Under what conditions will solutions $A(s)$ and $B(s)$ exist? and will the compensator $C(s) = B(s)/A(s)$ be proper? First we show that if $D(s)$ and $N(s)$ have common factors, then $D_o(s)$ cannot be arbitrarily chosen or, equivalently, arbitrary pole placement is not possible. For example, if $D(s)$ and $N(s)$ both contain the factor $(s - 2)$ or $D(s) = (s - 2)\overline{D}(s)$ and $N(s) = (s - 2)\overline{N}(s)$, then (10.5) becomes

$$A(s)D(s) + B(s)N(s) = (s - 2)[A(s)\overline{D}(s) + B(s)\overline{N}(s)] = D_o(s)$$

This implies that $D_o(s)$ must contain the same common factor $(s - 2)$. Thus, if $N(s)$ and $D(s)$ have common factors, then not every root of $D_o(s)$ can be arbitrarily assigned. Therefore we assume from now on that $D(s)$ and $N(s)$ have no common factors.

Because $G(s)$ and $C(s)$ are proper, we have deg $N(s) \leq$ deg $D(s) = n$ and deg $B(s) \leq$ deg $A(s) = m$, where deg stands for "the degree of." Thus, $D_o(s)$ in (10.5) has degree $n + m$ or, equivalently, the unity-feedback system in Figure 10.1 has $(n + m)$ number of poles. We develop in the following the conditions under which all $(n + m)$ number of poles can be arbitrarily assigned.

If deg $C(s) = 0$ (that is, $C(s) = k$, where k is a real number), then from the root-locus method, we see immediately that it is not possible to achieve *arbitrary* pole placement. We can assign poles only along the root loci. If the degree of $C(s)$ is 1, that is

$$C(s) = \frac{B(s)}{A(s)} = \frac{B_0 + B_1 s}{A_0 + A_1 s}$$

then we have four adjustable parameters for pole placement. Thus, the larger the degree of the compensator, the more parameters we have for pole placement. Therefore, if the degree of the compensator is sufficiently large, it is possible to achieve arbitrary pole placement.

Conventionally, the Diophantine equation is solved directly by using polynomials and the solution is expressed as a general solution. The general solution, however, is not convenient for our application. See Problem 10.19 and Reference [41]. In our application, we require deg $B(s) \leq$ deg $A(s)$ to insure properness of compensators. We also require the degree of compensators to be as small as possible. Instead of solving (10.5) directly, we shall transform it into a set of linear algebraic equations. We write

$$D(s) := D_0 + D_1 s + D_2 s^2 + \cdots + D_n s^n \qquad D_n \neq 0 \qquad (10.6a)$$

$$N(s) := N_0 + N_1 s + N_2 s^2 + \cdots + N_n s^n \qquad\qquad (10.6b)$$

$$A(s) := A_0 + A_1 s + A_2 s^2 + \cdots + A_m s^m \qquad\qquad (10.7a)$$

and

$$B(s) := B_0 + B_1 s + B_2 s^2 + \cdots + B_m s^m \qquad\qquad (10.7b)$$

where D_i, N_i, A_i, B_i are all real numbers, not necessarily nonzero. Because deg $D_o(s) = n + m$, we can express $D_o(s)$ as

$$D_o(s) = F_0 + F_1 s + F_2 s^2 + \cdots + F_{n+m} s^{n+m} \qquad\qquad (10.8)$$

The substitution of these into (10.5) yields

$$(A_0 + A_1 s + \cdots + A_m s^m)(D_0 + D_1 s + \cdots + D_n s^n)$$
$$+ (B_0 + B_1 s + \cdots + B_m s^m)(N_0 + N_1 s + \cdots + N_n s^n)$$
$$= F_0 + F_1 s + F_2 s^2 + \cdots + F_{n+m} s^{n+m}$$

which becomes, after grouping the coefficients associated with the same powers of s,

$$(A_0 D_0 + B_0 N_0) + (A_0 D_1 + B_0 N_1 + A_1 D_0 + B_1 N_0)s + \cdots$$
$$+ (A_m D_n + B_m N_n)s^{n+m} = F_0 + F_1 s + F_2 s^2 + \cdots + F_{n+m} s^{n+m}$$

Matching the coefficients of like powers of s yields

$$A_0 D_0 + B_0 N_0 = F_0$$
$$A_0 D_1 + B_0 N_1 + A_1 D_0 + B_1 N_0 = F_1$$
$$\vdots$$
$$A_m D_n + B_m N_n = F_{n+m}$$

There are a total of $(n + m + 1)$ equations. These equations can be arranged in matrix form as

$$\mathbf{S_m c_m} := \begin{bmatrix} D_0 & N_0 & 0 & 0 & & 0 & 0 \\ D_1 & N_1 & D_0 & N_0 & & \vdots & \vdots \\ \vdots & \vdots & \vdots & \vdots & \cdots & 0 & 0 \\ D_n & N_n & D_{n-1} & N_{n-1} & & D_0 & N_0 \\ 0 & 0 & D_n & N_n & & D_1 & N_1 \\ \vdots & \vdots & \vdots & \vdots & & \vdots & \vdots \\ 0 & 0 & 0 & 0 & & D_n & N_n \end{bmatrix} \begin{bmatrix} A_0 \\ B_0 \\ \hline A_1 \\ B_1 \\ \vdots \\ \hline A_m \\ B_m \end{bmatrix} = \begin{bmatrix} F_0 \\ F_1 \\ F_2 \\ \vdots \\ F_{n+m} \end{bmatrix} \quad (10.9)$$

This is a set of $(n + m + 1)$ linear algebraic equations. The matrix $\mathbf{S_m}$ has $(n + m + 1)$ rows and $2(m + 1)$ columns and is formed from the coefficients of $D(s)$ and $N(s)$. The first two columns are simply the coefficients of $D(s)$ and $N(s)$ arranged in ascending order. The next two columns are the first two columns shifted down by one position. We repeat the process until we have $(m + 1)$ sets of coefficients. We see that solving the Diophantine equation in (10.5) has now been transformed into solving the linear algebraic equation in (10.9).

As discussed in Appendix B, the equation in (10.9) has a solution for any F_i or, equivalently, for any $D_o(s)$ if and only if the matrix $\mathbf{S_m}$ has a full row rank. A necessary condition for $\mathbf{S_m}$ to have a full row rank is that $\mathbf{S_m}$ has more columns than rows or an equal number of columns and rows:

$$n + m + 1 \le 2(m + 1) \quad \text{or} \quad n - 1 \le m \quad (10.10)$$

Thus in order to achieve arbitrary pole placement, the degree of compensators in the unity-feedback configuration must be $n - 1$ or higher. If the degree is less than $n - 1$, it may be possible to assign *some* set of poles but not *every* set of poles. Therefore, we assume from now on that $m \ge n - 1$.

With $m \ge n - 1$, it is shown in Reference [15] that the matrix $\mathbf{S_m}$ has a full row rank if and only if $D(s)$ and $N(s)$ are coprime or have no common factors. We mention that if $m = n - 1$, the matrix $\mathbf{S_m}$ is a square matrix of order $2n$. In this case, for every $D_o(s)$, the solution of (10.9) is unique. If $m \geq n$, then (10.9) has more unknowns than equations and solutions of (10.9) are not unique.

Next we discuss the condition for the compensator to be proper or deg $B(s) \leq$ deg $A(s)$. We consider first $G(s)$ strictly proper and then $G(s)$ biproper. If $G(s)$ is strictly proper, then $N_n = 0$ and the last equation of (10.9) becomes

$$A_m D_n + B_m N_n = A_m D_n = F_{n+m}$$

which implies

$$A_m = \frac{F_{n+m}}{D_n} \tag{10.11}$$

Thus if $F_{n+m} \neq 0$, then $A_m \neq 0$ and the compensator $C(s) = B(s)/A(s)$ is proper. Note that if $m = n - 1$, the solution of (10.9) is unique and for any desired poles, there is a unique proper compensator to achieve the design. If $m \geq n$, then the solution of (10.9) is not unique, and some parameters of the compensator can be used, in addition to arbitrary pole placement, to achieve other design objective, as will be discussed later.

If $G(s)$ is biproper and if $m = n - 1$, then $\mathbf{S_m}$ in (10.9) is a square matrix and the solution of (10.9) is unique. In this case, there is no guarantee that $A_{n-1} \neq 0$ and the compensator may become improper. See Reference [15, p. 463.]. If $m \geq n$, then solutions of (10.9) are not unique and we can always find a strictly proper compensator to achieve pole placement. The preceding discussion is summarized as theorems.

THEOREM 10.1

Consider the unity-feedback system shown in Figure 10.1 with a strictly proper plant transfer function $G(s) = N(s)/D(s)$ with deg $N(s) <$ deg $D(s) = n$. It is assumed that $N(s)$ and $D(s)$ are coprime. If $m \geq n - 1$, then for *any* polynomial $D_o(s)$ of degree $(n + m)$, a proper compensator $C(s) = B(s)/A(s)$ of degree m exists to achieve the design. If $m = n - 1$, the compensator is unique. If $m \geq n$, the compensators are not unique and some of the coefficients of the compensators can be used to achieve other design objectives. Furthermore, the compensator can be determined from the linear algebraic equation in (10.9). ■

THEOREM 10.2

Consider the unity-feedback system shown in Figure 10.1 with a biproper plant transfer function $G(s) = N(s)/D(s)$ with deg $N(s) =$ deg $D(s) = n$. It is assumed that $N(s)$ and $D(s)$ are coprime. If $m \geq n$, then for *any* polynomial $D_o(s)$ of degree $(n + m)$, a proper compensator $C(s) = B(s)/A(s)$ of degree m exists to achieve the design. If $m = n$, and if the compensator is chosen to be strictly proper, then the compensator is unique. If $m \geq n + 1$, compensators are not unique and some of the coefficients of the compensators can be used to achieve other design objectives. Furthermore, the compensator can be determined from the linear algebraic equation in (10.9). ■

Example 10.3.2

Consider a plant with transfer function

$$G(s) = \frac{N(s)}{D(s)} = \frac{s - 2}{(s + 1)(s - 1)} = \frac{-2 + s + 0 \cdot s^2}{-1 + 0 \cdot s + s^2} \qquad (10.12)$$

Clearly $D(s)$ and $N(s)$ have no common factor and $n = 2$. Let

$$C(s) = \frac{B_0 + B_1 s}{A_0 + A_1 s}$$

It is a compensator of degree $m = n - 1 = 1$. Arbitrarily, we choose the three poles of the overall system as -3, $-2 + j1$ and $-2 - j1$. Then we have

$$D_o(s) := (s + 3)(s + 2 - j1)(s + 2 + j1) = 15 + 17s + 7s^2 + s^3$$

We form the linear algebraic equation in (10.9) as

$$\begin{bmatrix} -1 & -2 & 0 & 0 \\ 0 & 1 & -1 & -2 \\ 1 & 0 & 0 & 1 \\ 0 & 0 & 1 & 0 \end{bmatrix} \begin{bmatrix} A_0 \\ B_0 \\ A_1 \\ B_1 \end{bmatrix} = \begin{bmatrix} 15 \\ 17 \\ 7 \\ 1 \end{bmatrix} \qquad (10.13)$$

Its solution can easily be obtained as

$$A_0 = \frac{79}{3} \qquad A_1 = 1 \qquad B_0 = -\frac{62}{3} \qquad B_1 = -\frac{58}{3}$$

Thus the compensator is

$$C(s) = \frac{-\dfrac{62}{3} - \dfrac{58}{3} s}{\dfrac{79}{3} + s} = \frac{-(58s + 62)}{3s + 79} = \frac{-(19.3s + 20.7)}{s + 26.3} \qquad (10.14)$$

and the resulting overall system is, from (10.4),

$$G_o(s) = \frac{B(s)N(s)}{D_o(s)} = \frac{\left(-\dfrac{62}{3} - \dfrac{58}{3} s\right)(s - 2)}{s^3 + 7s^2 + 17s + 15} = \frac{-(58s + 62)(s - 2)}{3(s^3 + 7s^2 + 17s + 15)}$$

Note that the zero $58s + 62$ is solved from the Diophantine equation and we have no control over it. Because $G_o(0) = 124/45$, if we apply a unit-step reference input, the output will approach $124/45 = 2.76$. See (4.25). Thus the output of this overall system will not track asymptotically step-reference inputs.

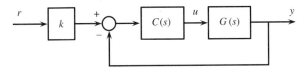

Figure 10.3 Unity-feedback system with a precompensator.

The design in Example 10.3.2 achieves pole placement but not tracking of step-reference inputs. This problem, however, can be easily corrected by introducing a constant gain k as shown in Figure 10.3. If we choose k so that $kG_o(0) = k \times 124/45 = 1$ or $k = 45/124$, then the plant output in Figure 10.3 will track any step-reference input. We call the constant gain in Figure 10.3 a *precompensator*. In practice, the precompensator may be incorporated into the reference input r by calibration or by resetting. For example, in temperature control, if r_0, which corresponds to 70°, yields a steady-state temperature of 67° and r_1 yields a steady-state temperature of 70°. We can simply change the scale so that r_0 corresponds to 67° and r_1 corresponds to 70°. By so doing, no steady-state error will be introduced in tracking step-reference inputs.

Example 10.3.3

Consider a plant with transfer function $1/s^2$. Find a compensator in Figure 10.1 so that the resulting system has all poles at $s = -2$. This plant transfer function has degree 2. If we choose a compensator of degree $m = n - 1 = 2 - 1 = 1$, then we can achieve arbitrary pole placement. Clearly we have

$$D_o(s) = (s + 2)^3 = s^3 + 6s^2 + 12s + 8$$

From the coefficients of

$$\frac{1}{s^2} = \frac{1 + 0 \cdot s + 0 \cdot s^2}{0 + 0 \cdot s + 1 \cdot s^2}$$

and $D_o(s)$, we form

$$\begin{bmatrix} 0 & 1 & 0 & 0 \\ 0 & 0 & 0 & 1 \\ 1 & 0 & 0 & 0 \\ 0 & 0 & 1 & 0 \end{bmatrix} \begin{bmatrix} A_0 \\ B_0 \\ A_1 \\ B_1 \end{bmatrix} = \begin{bmatrix} 8 \\ 12 \\ 6 \\ 1 \end{bmatrix}$$

Its solution is

$$A_0 = 6 \qquad A_1 = 1 \qquad B_0 = 8 \qquad B_1 = 12$$

Thus the compensator is

$$C(s) = \frac{12s + 8}{s + 6}$$

and the resulting overall system is

$$G_o(s) = \frac{B(s)N(s)}{D_o(s)} = \frac{12s + 8}{s^3 + 6s^2 + 12s + 8} \tag{10.15}$$

Note that the zero $12s + 8$ is solved from the Diophantine equation and we have no control over it. Because the constant and s terms of the numerator and denominator of $G_o(s)$ are the same, the overall system has zero position error and zero velocity error. See Section 6.3.1. Thus, the system will track asymptotically any ramp-reference input. For this problem, there is no need to introduce a precompensator as in Figure 10.3. The reason is that the plant transfer function is of type 2 or has double poles at $s = 0$. In this case, the unity-feedback system, if it is stable, will automatically have zero velocity error.

From the preceding two examples, we see that arbitrary pole placement in the unity-feedback configuration can be used to achieve asymptotic tracking. If a plant transfer function is of type 0, we need to introduce a precompensator to achieve tracking of step-reference inputs. In practice, we can simply reset the reference input rather than introduce the precompensator. If $G(s)$ is of type 1, after pole placement, the unity-feedback system will automatically track step-reference inputs. If $G(s)$ is of type 2, the unity-feedback system will track ramp-reference inputs. If $G(s)$ is of type 3, the unity-feedback system will track acceleration-reference inputs. In pole placement, some zeros will be introduced. These zeros will affect the transient response of the system. Therefore, it is important to check the response of the resulting system before the system is actually built in practice.

To conclude this section, we mention that the algebraic equation in (10.9) can be solved by using MATLAB. For example, to solve (10.13), we type

```
a=[-1 -2 0 0;0 1 -1 -2;1 0 0 1;0 0 1 0];b=[15;17;7;1];
a\b
```

Then MATLAB will yield

    26.3333

    -20.6667

    1.0000

    -19.3333

This yields the compensator in (10.14).

Exercise 10.3.1

Redesign the problem in Example 10.3.1 by solving (10.9).

Exercise 10.3.2

Consider a plant with transfer function

$$G(s) = \frac{1}{s^2 - 1}$$

Design a proper compensator $C(s)$ and a gain k such that the overall system in Figure 10.3 has all poles located at -2 and will track asymptotically step-reference inputs.

[**Answers:** $C(s) = (13s + 14)/(s + 6)$, $k = 4/7$.]

10.3.2   Pole Placement with Robust Tracking

Consider the design problem in Example 10.3.2, that is, given $G(s) = (s - 2)/(s + 1)(s - 1)$ in (10.12), if we use the compensator in (10.14) and a precompensator $k = 45/124$, then the resulting overall system in Figure 10.3 will track any step-reference input. As discussed in Section 6.4, the transfer function of a plant may change due to changes of load, aging, wearing or external perturbation such as wind gust on an antenna. Now suppose, after implementation, the plant transfer function changes to $\overline{G}(s) = (s - 2.1)/(s + 1)(s - 0.9)$, the question is: Will the system in Figure 10.3 with the perturbed transfer function still track any step-reference input without an error?

In order to answer this question, we compute the overall transfer function with $G(s)$ replaced by $\overline{G}(s)$:

$$\overline{G}_o(s) = \frac{45}{124} \cdot \frac{\dfrac{s - 2.1}{(s + 1)(s - 0.9)} \cdot \dfrac{-(58s + 62)}{3s + 79}}{1 + \dfrac{s - 2.1}{(s + 1)(s - 0.9)} \cdot \dfrac{-(58s + 62)}{3s + 79}}$$

$$= \frac{45}{124} \cdot \frac{-58s^2 + 59.8s + 130.2}{3s^3 + 21.3s^2 + 65s + 59.1} \tag{10.16}$$

Because $\overline{G}_o(0) = (45 \times 130.2)/(124 \times 59.1) = 0.799 \neq 1$, the system no longer tracks asymptotically step-reference inputs. If the reference input is 1, the output approaches 0.799 and the tracking error is about 20%. This type of design is called *nonrobust tracking* because plant parameter perturbations destroy the tracking property.

Now we shall redesign the system so that the tracking property will not be destroyed by plant parameter perturbations. This is achieved by increasing the degree of the compensator by one and then using the extra parameters to design a type 1 compensator. The original compensator in (10.14) has degree 1. We increase its degree and consider

$$C(s) = \frac{B_0 + B_1 s + B_2 s^2}{A_0 + A_1 s + A_2 s^2} \tag{10.17}$$

Both the plant and compensator have degree 2, therefore the unity-feedback system in Figure 10.1 has four poles. We assign the four poles arbitrarily as -3, -3, $-2 \pm j$, then we have

$$D_o(s) = (s + 3)^2(s + 2 - j1)(s + 2 + j1)$$
$$= s^4 + 10s^3 + 38s^2 + 66s + 45 \tag{10.18}$$

The compensator that achieves this pole placement can be solved from the following linear algebraic equation

$$\begin{bmatrix} -1 & -2 & 0 & 0 & 0 & 0 \\ 0 & 1 & -1 & -2 & 0 & 0 \\ 1 & 0 & 0 & 1 & -1 & -2 \\ 0 & 0 & 1 & 0 & 0 & 1 \\ 0 & 0 & 0 & 0 & 1 & 0 \end{bmatrix} \begin{bmatrix} A_0 \\ B_0 \\ A_1 \\ B_1 \\ A_2 \\ B_2 \end{bmatrix} = \begin{bmatrix} 45 \\ 66 \\ 38 \\ 10 \\ 1 \end{bmatrix} \tag{10.19}$$

This equation has six unknowns and five equations. After deleting the first column, the remaining square matrix of order 5 in (10.19) still has a full row rank. Therefore A_0 can be arbitrarily assigned. (See Appendix B). If we assign it as zero, then the compensator in (10.17) has a pole at $s = 0$ or becomes type 1. With $A_0 = 0$, the solution of (10.19) can be computed as $B_0 = -22.5$, $A_1 = 68.83$, $B_1 = -78.67$, $A_2 = 1$, and $B_2 = -58.83$. Therefore, the compensator in (10.17) becomes

$$C(s) = \frac{-(58.83s^2 + 78.67s + 22.5)}{s(s + 68.83)} \tag{10.20}$$

and the overall transfer function is

$$G_o(s) = \frac{\dfrac{-(58.83s^2 + 78.67s + 22.5)}{s(s + 68.83)} \cdot \dfrac{s - 2}{(s + 1)(s - 1)}}{1 + \dfrac{-(58.83s^2 + 78.67s + 22.5)}{s(s + 68.83)} \cdot \dfrac{s - 2}{(s + 1)(s - 1)}}$$
$$= \frac{-58.83s^3 + 38.99s^2 + 134.84s + 45}{s^4 + 10s^3 + 37.99s^2 + 66.01s + 45} \tag{10.21}$$

We see that, other than truncation errors, the denominator of $G_o(s)$ practically equals $D_o(s)$ in (10.18). Thus the compensator in (10.20) achieves the pole placement. Because $G_o(0) = 45/45 = 1$, the unity-feedback system achieves asymptotic tracking of step-reference inputs. Note that there is no need to introduce a precompensator as in Figure 10.3 in this design, because the compensator is designed to be of type 1.

Now we show that even if the plant transfer function changes to $\overline{G}(s) = (s - 2.1)/(s + 1)(s - 0.9)$, the overall system will still achieve tracking. With the perturbed $\overline{G}(s)$, the overall transfer function becomes

$$\overline{G}_o(s) = \frac{\dfrac{-(58.83s^2 + 78.67s + 22.5)}{s(s + 68.83)} \cdot \dfrac{s - 2.1}{(s + 1)(s - 0.9)}}{1 + \dfrac{-(58.83s^2 + 78.67s + 22.5)}{s(s + 68.83)} \cdot \dfrac{s - 2.1}{(s + 1)(s - 0.9)}} \quad (10.22)$$

$$= \frac{-58.83s^3 + 44.873s^2 + 142.707s + 47.25}{s^4 + 10.1s^3 + 50.856s^2 + 80.76s + 47.25}$$

We first check the stability of $\overline{G}_o(s)$. If $\overline{G}_o(s)$ is not stable, it cannot track any signal. The application of the Routh test to the denominator of $\overline{G}_o(s)$ yields

	s^4	1	50.856	47.25				
$1/10.1$	s^3	10.1	80.76		[0	42.86	47.25]	
$10.1/42.86$	s^2	42.86	47.25		[0	69.63]		
	s	69.63						
	1	47.25						

All entries in the Routh table are positive, therefore $\overline{G}_o(s)$ is stable. Because $\overline{G}_o(0) = 47.25/47.25 = 1$, the overall system with the perturbed plant transfer function still tracks asymptotically any step-reference input. In fact, because the compensator is of type 1, no matter how large the changes in the coefficients of the plant transfer function, so long as the unity-feedback system remains stable, the system will track asymptotically any step-reference input. Therefore the tracking property of this design is *robust*.

Exercise 10.3.3

Given a plant with transfer function $1/(s - 1)$, design a compensator of degree 0 so that the pole of the unity-feedback system in Figure 10.3 is -2. Find also a precompensator so that the overall system will track asymptotically any step-reference input. If the plant transfer function changes to $1/(s - 1.1)$, is the unity-feedback system still stable? Will it still track any step-reference input without an error?

[**Answers:**   3, 2/3, yes, no.]

Exercise 10.3.4

Given a plant with transfer function $1/(s - 1)$, design a compensator of degree 1 so that the poles of the unity-feedback system in Figure 10.3 are -2 and -2. In this design, do you have freedom in choosing some of the coefficients of the compensator? Can you choose compensator coefficients so that the system in Figure 10.3 will track asymptotically step-reference inputs with $k = 1$? In this design, will the system remain stable and track any step-reference input after the plant transfer function changes to $1/(s - 1.1)$?

[**Answers:** $[(5 - \alpha)s + (4 + \alpha)]/(s + \alpha)$; yes; yes by choosing $\alpha = 0$; yes.]

To conclude this subsection, we remark on the choice of poles in pole placement. Clearly, the poles chosen must be stable. Furthermore, they should not be very close to the imaginary axis. As a guide, we may place them evenly inside the region C shown in Figure 6.13. Pole placement, however, will introduce some zeros into the resulting system. These zeros will also affect the response of the system. (See Figure 2.16.) There is no way to predict where these zeros will be located; therefore, it would be difficult to predict from the poles chosen what the final response of the resulting system will be. It is therefore advisable to simulate the resulting system before it is actually built.

10.3.3 Pole Placement and Model Matching

In this subsection, we give some remarks regarding the design based on pole placement and the design based on model matching. We use an example to illustrate the issues. Consider a plant with transfer function $G(s) = (s + 3)/s(s - 1)$. Its quadratic optimal system was computed in (9.46) as $G_o(s) = 10(s + 3)/(s^2 + 12.7s + 30)$ under the constraint that the actuating signal due to a unit-step reference input meets $|u(t)| \leq 10$ for all $t \geq 0$. Now we shall redesign the problem by using the method of pole placement. In other words, given $G(s) = (s + 3)/s(s - 1)$, we shall find a compensator of degree 1 in the unity-feedback configuration in Figure 10.1 so that the resulting system has a set of three desired poles. Because the poles of the quadratic optimal system or, equivalently, the roots of $(s^2 + 12.7s + 30)$ are -9.56 and -3.14, we choose the three poles as -9.56, -3.14 and -10. Thus we have

$$D_o(s) = (s^2 + 12.7s + 30)(s + 10) = s^3 + 22.7s^2 + 157s + 300$$

and the compensator $C(s) = (B_0 + B_1s)/(A_0 + A_1s)$ to achieve this set of pole placement can be computed from

$$\begin{bmatrix} 0 & 3 & 0 & 0 \\ -1 & 1 & 0 & 3 \\ 1 & 0 & -1 & 1 \\ 0 & 0 & 1 & 0 \end{bmatrix} \begin{bmatrix} A_0 \\ B_0 \\ A_1 \\ B_1 \end{bmatrix} = \begin{bmatrix} 300 \\ 157 \\ 22.7 \\ 1 \end{bmatrix}$$

as

$$C(s) = \frac{20.175s + 100}{s + 3.525}$$

Thus, from (10.4), the overall transfer function from r to y is

$$G_o(s) = \frac{(20.175s + 100)(s + 3)}{s^3 + 22.7s^2 + 157s + 300}$$

and the transfer function from r to u is

$$T(s) := \frac{U(s)}{R(s)} = \frac{G_o(s)}{G(s)} = \frac{(20.175s + 100)s(s - 1)}{s^3 + 22.7s^2 + 157s + 300}$$

The unit-step responses of $G_o(s)$ and $T(s)$ are plotted in Figure 10.4 with solid lines. For comparison, the corresponding responses of the quadratic optimal system are also plotted with the dashed lines. We see that for the set of poles chosen, the resulting system has a larger overshoot than that of the quadratic optimal system. Furthermore, the actuating signal does not meet the constraint $|u(t)| \leq 10$.

It is conceivable that a set of poles can be chosen in pole placement so that the resulting system has a comparable response as one obtained by model matching. The problem is that, in pole placement, there is no way to predict the response of the resulting system from the set of poles chosen, because the response also depends on the zeros which are yet to be solved from the Diophantine equation. Therefore, pole placement design should consist of the following steps: (1) choose a set of poles, (2) compute the required compensator, (3) compute the resulting overall transfer function, and (4) check the response of the resulting system and check whether the actuating signal meets the constraint. If the design is not satisfactory, we go back to the first step and repeat the design. In model matching, we can choose an overall system to meet design specifications. Only after a satisfactory overall system is chosen do we compute the required compensator. Therefore, it is easier and simpler to obtain a good control system by model matching than by pole placement.

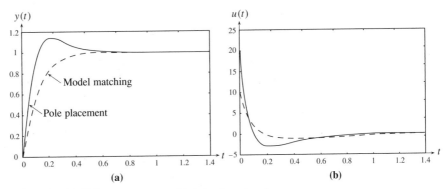

Figure 10.4 (a) Unit-step response. (b) Actuating signal.

10.4   TWO-PARAMETER COMPENSATORS

Although the unity-feedback configuration can be used to achieve arbitrary pole placement, it generally cannot be used to achieve model matching. In this section, we introduce a configuration, called the *two-parameter* configuration, that can be used to implement *any* implementable overall transfer function.

The actuating signal in Figure 10.1 is of the form, in the Laplace transform domain,

$$U(s) = C(s)(R(s) - Y(s)) = C(s)R(s) - C(s)Y(s) \tag{10.23}$$

That is, the *same* compensator is applied to the reference input and plant output to generate the actuating signal. Now we shall generalize it to

$$U(s) = C_1(s)R(s) - C_2(s)Y(s) \tag{10.24}$$

as shown in Figure 10.5(a). This is the most general form of compensators. We call $C_1(s)$ *feedforward* compensator and $C_2(s)$ *feedback* compensator. Let $C_1(s)$ and $C_2(s)$ be

$$C_1(s) = \frac{L(s)}{A_1(s)} \qquad C_2(s) = \frac{M(s)}{A_2(s)}$$

where $L(s)$, $M(s)$, $A_1(s)$, and $A_2(s)$ are polynomials. In general, $A_1(s)$ and $A_2(s)$ need not be the same. It turns out that even if they are chosen to be the same, the two compensators can be used to achieve any model matching. Furthermore, simple and straightforward design procedures can be developed. Therefore we assume $A_1(s) = A_2(s) = A(s)$ and the compensators become

$$C_1(s) = \frac{L(s)}{A(s)} \qquad C_2(s) = \frac{M(s)}{A(s)} \tag{10.25}$$

and the configuration in Figure 10.5(a) becomes the one in Figure 10.5(b). If $A(s)$, which is yet to be designed, contains unstable roots, the signal at the output of $L(s)/A(s)$ will grow to infinity and the system cannot be totally stable. Therefore the configuration in Figure 10.5(b) cannot be used in actual implementation. If we move $L(s)/A(s)$ into the feedback loop, then the configuration becomes the one shown in Figure 10.5(c). This configuration is also not satisfactory for two reasons: First, if $L(s)$ contains unstable roots, the design will involve unstable pole-zero cancellations and the system cannot be totally stable. Second, because the two compensators $L(s)/A(s)$ and $M(s)/L(s)$ have different denominators, if they are implemented using operational amplifier circuits, they will use twice as many integrators as the one to be discussed immediately. See Section 5.6.1. Therefore the configuration in Figure 10.5(c) should not be used. If we move $M(s)/L(s)$ outside the loop, then the resulting system is as shown in Figure 10.5(d). This configuration should not be used for the same reasons as for the configuration in Figure 10.5(c). Therefore, the three configurations in Figures 10.5(b), (c), and (d) will not be used in actual implementation.

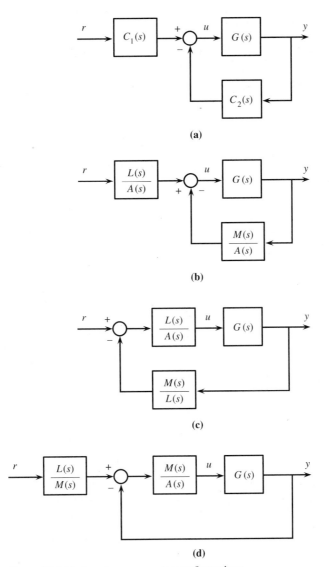

Figure 10.5 Various two-parameter configurations.

We substitute (10.25) into (10.24) and rewrite it into matrix form as

$$U(s) = \frac{L(s)}{A(s)} R(s) - \frac{M(s)}{A(s)} Y(s) = \begin{bmatrix} \dfrac{L(s)}{A(s)} & -\dfrac{M(s)}{A(s)} \end{bmatrix} \begin{bmatrix} R(s) \\ Y(s) \end{bmatrix} \quad (10.26)$$

Thus the compensator

$$\mathbf{C}(s) := [C_1(s) \quad -C_2(s)] = \begin{bmatrix} \dfrac{L(s)}{A(s)} & -\dfrac{M(s)}{A(s)} \end{bmatrix} = A^{-1}(s)[L(s) \quad -M(s)] \quad (10.27)$$

is a 1×2 rational matrix; it has two inputs r and y and one output u, and can be plotted as shown in Figure 10.6. The minus sign in (10.27) is introduced to take care of the negative feedback in Figure 10.6. Mathematically, this configuration is no different from the ones in Figure 10.5. However, if we implement $C(s)$ in (10.27) as a unit, then the problem of possible unstable pole-zero cancellation will not arise. Furthermore, its implementation will use the minimum number of integrators. Therefore, the configuration in Figure 10.6 will be used exclusively in implementation. We call the compensator in Figure 10.6 a two-parameter compensator [63]. The configurations in Figure 10.5 are called two-degree-of-freedom structures in [36].

We can use the procedure in Section 5.6.1 to implement a two-parameter compensator as a unit. For example, consider

$$[C_1(s) \quad -C_2(s)] = \left[\frac{10(s + 30)^2}{s(s - 15.2)} \quad -\frac{88.9s^2 + 140s + 9000}{s(s - 15.2)} \right]$$

We first expand it as

$$[C_1(s) \quad -C_2(s)] = [10 \quad -88.9] + \left[\frac{752s + 9000}{s^2 - 15.2s + 0} \quad \frac{-1491.28s - 9000}{s^2 - 15.2s + 0} \right]$$

From its coefficients and (5.44), we can obtain the following two-dimensional state-variable equation realization

$$\begin{bmatrix} \dot{x}_1(t) \\ \dot{x}_2(t) \end{bmatrix} = \begin{bmatrix} 15.2 & 1 \\ 0 & 0 \end{bmatrix} \begin{bmatrix} x_1(t) \\ x_2(t) \end{bmatrix} + \begin{bmatrix} 752 & -1491.28 \\ 9000 & -9000 \end{bmatrix} \begin{bmatrix} r(t) \\ y(t) \end{bmatrix}$$

$$u(t) = [1 \quad 0] \begin{bmatrix} x_1(t) \\ x_2(t) \end{bmatrix} + [10 \quad -88.9] \begin{bmatrix} r(t) \\ y(t) \end{bmatrix}$$

From this equation, we can easily plot a basic block diagram as shown in Figure 10.7. It consists of two integrators. The block diagram can be built using operational amplifier circuits. We note that the adder in Figure 10.6 does not correspond to any adder in Figure 10.7.

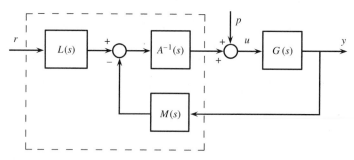

Figure 10.6 Two-parameter feedback configuration.

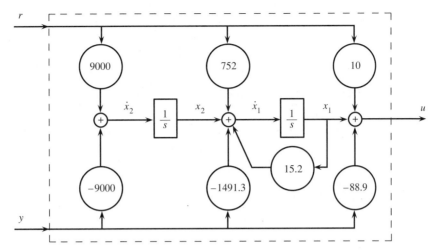

Figure 10.7 Basic block diagram.

Exercise 10.4.1

Find a minimal realization of

$$[C_1(s) \quad -C_2(s)] = \left[\frac{10(s + 30)}{s + 5.025} \quad -\frac{38.7s + 300}{s + 5.025} \right]$$

and then draw a basic block diagram for it.

[**Answer:** $\dot{x}(t) = -5.025x(t) + [249.75 \quad -105.5325] \begin{bmatrix} r(t) \\ y(t) \end{bmatrix}$

$u(t) = x(t) + [10 \quad -38.7] \begin{bmatrix} r(t) \\ y(t) \end{bmatrix}.]$

10.4.1 Two-Parameter Configuration—Model Matching

In this subsection, we present a procedure to implement any implementable $G_o(s)$ in the two-parameter configuration. From Mason's formula, the transfer function from r to y in Figure 10.6 is

$$\frac{Y(s)}{R(s)} = \frac{L(s)A^{-1}(s)G(s)}{1 + A^{-1}(s)M(s)G(s)}$$

which becomes, after substituting $G(s) = N(s)/D(s)$ and multiplying by $A(s)D(s)/A(s)D(s)$

$$\frac{Y(s)}{R(s)} = \frac{L(s)N(s)}{A(s)D(s) + M(s)N(s)} \tag{10.28}$$

Now we show that this can be used to achieve any model matching. For convenience, we discuss only the case where $G(s)$ is strictly proper.

Problem   Given $G(s) = N(s)/D(s)$, where $N(s)$ and $D(s)$ are coprime, deg $N(s) <$ deg $D(s) = n$, and given an implementable $G_o(s) = N_o(s)/D_o(s)$, find proper compensators $L(s)/A(s)$ and $M(s)/A(s)$ such that

$$G_o(s) = \frac{N_o(s)}{D_o(s)} = \frac{L(s)N(s)}{A(s)D(s) + M(s)N(s)} \qquad (10.29)$$

Procedure:

Step 1:   Compute

$$\frac{G_o(s)}{N(s)} = \frac{N_o(s)}{D_o(s)N(s)} =: \frac{N_p(s)}{D_p(s)} \qquad (10.30)$$

where $N_p(s)$ and $D_p(s)$ are coprime. Since $N_o(s)$ and $D_o(s)$ are coprime by assumption, common factors may exist only between $N_o(s)$ and $N(s)$. Cancel all common factors between them and denote the rest $N_p(s)$ and $D_p(s)$. Note that if $N_o(s) = N(s)$, then $D_p(s) = D_o(s)$ and $N_p(s) = 1$. Using (10.30), we rewrite (10.29) as

$$G_o(s) = \frac{N_p(s)N(s)}{D_p(s)} = \frac{L(s)N(s)}{A(s)D(s) + M(s)N(s)} \qquad (10.31)$$

From this equation, one might be tempted to set $L(s) = N_p(s)$ and to solve for $A(s)$ and $M(s)$ from $D_p(s) = A(s)D(s) + M(s)N(s)$. Unfortunately, the resulting compensators are generally not proper. Therefore, some more manipulation is needed.

Step 2:   Introduce an arbitrary Hurwitz polynomial $\overline{D}_p(s)$ so that the degree of $D_p(s)\overline{D}_p(s)$ is at least $2n - 1$. In other words, if deg $D_p(s) := p$, then the degree of $\overline{D}_p(s)$ must be at least $2n - 1 - p$. Because the polynomial $\overline{D}_p(s)$ will be canceled in the design, its roots should be chosen to lie inside an acceptable pole-zero cancellation region.

Step 3:   Rewrite (10.31) as

$$G_o(s) = \frac{N(s)N_p(s)}{D_p(s)} = \frac{N(s)[N_p(s)\overline{D}_p(s)]}{D_p(s)\overline{D}_p(s)} = \frac{N(s)L(s)}{A(s)D(s) + M(s)N(s)} \qquad (10.32)$$

Now we set

$$L(s) = N_p(s)\overline{D}_p(s) \qquad (10.33)$$

and solve $A(s)$ and $M(s)$ from

$$A(s)D(s) + M(s)N(s) = D_p(s)\overline{D}_p(s) =: F(s) \qquad (10.34)$$

If we write

$$A(s) = A_0 + A_1 s + \cdots + A_m s^m \qquad (10.35a)$$

$$M(s) = M_0 + M_1 s + \cdots + M_m s^m \qquad (10.35b)$$

and

$$F(s) := D_p(s)\overline{D}_p(s) = F_0 + F_1 s + F_2 s^2 + \cdots + F_{n+m} s^{n+m} \qquad (10.36)$$

with $m \geq n - 1$, then $A(s)$ and $M(s)$ in (10.34) can be solved from the following linear algebraic equation:

$$
\begin{bmatrix}
D_0 & N_0 & 0 & 0 & & 0 & 0 \\
D_1 & N_1 & D_0 & N_0 & & \vdots & \vdots \\
\vdots & \vdots & \vdots & \vdots & \cdots & 0 & 0 \\
D_n & N_n & D_{n-1} & N_{n-1} & & D_0 & N_0 \\
0 & 0 & D_n & N_n & & D_1 & N_1 \\
\vdots & \vdots & \vdots & \vdots & & \vdots & \vdots \\
0 & 0 & 0 & 0 & & D_n & N_n
\end{bmatrix}
\begin{bmatrix}
A_0 \\
M_0 \\
\hline
A_1 \\
M_1 \\
\vdots \\
\hline
A_m \\
M_m
\end{bmatrix}
=
\begin{bmatrix}
F_0 \\
F_1 \\
F_2 \\
\vdots \\
F_{n+m}
\end{bmatrix}
\qquad (10.37)
$$

The solution and (10.33) yield the compensators $L(s)/A(s)$ and $M(s)/A(s)$. This completes the design.

Now we show that the compensators are proper. Equation (10.37) becomes (10.9) if M_i is replaced by B_i. Thus, Theorem 10.1 is directly applicable to (10.37). Because deg $N(s) <$ deg $D(s)$ and deg $F(s) \geq 2n - 1$, Theorem 10.1 implies the existence of $M(s)$ and $A(s)$ in (10.37) or (10.34) with deg $M(s) \leq$ deg $A(s)$. Thus, the compensator $M(s)/A(s)$ is proper. Furthermore, (10.34) implies

$$\text{deg } A(s) = \text{deg } [D_p(s)\overline{D}_p(s)] - \text{deg } D(s) = \text{deg } F(s) - n$$

Now we show deg $L(s) \leq$ deg $A(s)$. Applying the pole-zero excess inequality of $G_o(s)$ to (10.32) and using (10.33), we have

$$\text{deg } [D_p(s)\overline{D}_p(s)] - (\text{deg } N(s) + \text{deg } L(s)) \geq \text{deg } D(s) - \text{deg } N(s)$$

which implies

$$\text{deg } L(s) \leq \text{deg } [D_p(s)\overline{D}_p(s)] - \text{deg } D(s) = \text{deg } A(s)$$

Thus the compensator $L(s)/A(s)$ is also proper.

The design always involves the pole-zero cancellation of $\overline{D}_p(s)$. The polynomial $\overline{D}_p(s)$, however, is chosen by the designer. Thus if $\overline{D}_p(s)$ is chosen to be Hurwitz or to have its roots lying inside the region C shown in Figure 6.13, then the two-parameter system in Figure 10.6 is totally stable. The condition for the two-parameter configuration to be well posed is $1 + G(\infty)C_2(\infty) \neq 0$ where $C_2(s) = M(s)/A(s)$. This condition is always met if $G(s)$ is strictly proper and $C_2(s)$ is proper. Thus the system is well posed. The configuration clearly has no plant leakage. Thus this design meets all the constraints discussed in Chapter 9. In conclusion, the two-parameter configuration in Figure 10.6 can be used to implement any implementable overall transfer function.

Example 10.4.1

Consider the plant with transfer function

$$G(s) = \frac{N(s)}{D(s)} = \frac{s + 3}{s(s - 1)}$$

studied in (9.44). Its ITAE optimal system was found in (9.59) as

$$G_o(s) = \frac{600.25}{s^2 + 34.3s + 600.25} \qquad (10.38)$$

Note that the zero $(s + 3)$ of $G(s)$ does not appear in $G_o(s)$, thus the design will involve the pole-zero cancellation of $s + 3$. Now we implement $G_o(s)$ by using the two-parameter configuration in Figure 10.6. We compute

$$\frac{G_o(s)}{N(s)} = \frac{600.25}{(s^2 + 34.3s + 600.25)(s + 3)} =: \frac{N_p(s)}{D_p(s)}$$

Next we choose $\overline{D}_p(s)$ so that the degree of $D_p(s)\overline{D}_p(s)$ is at least $2n - 1 = 3$. Because the degree of $D_p(s)$ is 3, the degree of $\overline{D}_p(s)$ can be chosen as 0. We choose $\overline{D}_p(s) = 1$. Thus we have

$$L(s) = N_p(s)\overline{D}_p(s) = 600.25 \times 1 = 600.25$$

The polynomials $A(s) = A_0 + A_1 s$ and $M(s) = M_0 + M_1 s$ can be solved from $A(s)D(s) + M(s)N(s) = D_p(s)\overline{D}_p(s)$ with

$$F(s) := D_p(s)\overline{D}_p(s) = (s^2 + 34.3s + 600.25)(s + 3)$$
$$= s^3 + 37.3s^2 + 703.15s + 1800.75$$

or from the following linear algebraic equation:

$$\begin{bmatrix} D_0 & N_0 & 0 & 0 \\ D_1 & N_1 & D_0 & N_0 \\ D_2 & N_2 & D_1 & N_1 \\ 0 & 0 & D_2 & N_2 \end{bmatrix}\begin{bmatrix} A_0 \\ M_0 \\ A_1 \\ M_1 \end{bmatrix} = \begin{bmatrix} 0 & 3 & 0 & 0 \\ -1 & 1 & 0 & 3 \\ 1 & 0 & -1 & 1 \\ 0 & 0 & 1 & 0 \end{bmatrix}\begin{bmatrix} A_0 \\ M_0 \\ A_1 \\ M_1 \end{bmatrix} = \begin{bmatrix} 1800.75 \\ 703.15 \\ 37.3 \\ 1 \end{bmatrix}$$

The solution is $A(s) = A_0 + A_1 s = 3 + s$ and $M(s) = M_0 + M_1 s = 600.25 + 35.3s$. Thus the compensator is

$$[C_1(s) \quad -C_2(s)] = \begin{bmatrix} \dfrac{600.25}{s + 3} & -\dfrac{35.3s + 600.25}{s + 3} \end{bmatrix} \qquad (10.39)$$

This completes the design.

Example 10.4.2

Consider the same plant transfer function in the preceding example. Now we shall implement its quadratic optimal transfer function $G_o(s) = 10(s + 3)/(s^2 + 12.7s + 30)$ developed in (9.46). First we compute

$$\frac{G_o(s)}{N(s)} = \frac{10(s + 3)}{(s^2 + 12.7s + 30)(s + 3)} = \frac{10}{s^2 + 12.7s + 30} =: \frac{N_p(s)}{D_p(s)}$$

Because the degree of $D_p(s)$ is 2, we must introduce $\overline{D}_p(s)$ of degree at least 1 so that the degree of $D_p(s)\overline{D}_p(s)$ is at least $2n - 1 = 3$. Arbitrarily, we choose

$$\overline{D}_p(s) = s + 3 \tag{10.40}$$

(This issue will be discussed further in the next section.) Thus we have

$$L(s) = N_p(s)\overline{D}_p(s) = 10(s + 3)$$

and

$$F(s) = D_p(s)\overline{D}_p(s) = (s^2 + 12.7s + 30)(s + 3)$$
$$= s^3 + 15.7s^2 + 68.1s + 90$$

The polynomials $A(s)$ and $M(s)$ can be solved from

$$\begin{bmatrix} 0 & 3 & 0 & 0 \\ -1 & 1 & 0 & 3 \\ 1 & 0 & -1 & 1 \\ 0 & 0 & 1 & 0 \end{bmatrix} \begin{bmatrix} A_0 \\ M_0 \\ A_1 \\ M_1 \end{bmatrix} = \begin{bmatrix} 90 \\ 68.1 \\ 15.7 \\ 1 \end{bmatrix} \tag{10.41}$$

as $A(s) = A_1 s + A_0 = s + 3$ and $M(s) = M_1 s + M_0 = 13.7s + 30$. Thus, the compensator is

$$[C_1(s) \quad -C_2(s)] = \begin{bmatrix} \dfrac{10(s + 3)}{s + 3} & -\dfrac{13.7s + 30}{s + 3} \end{bmatrix} \tag{10.42}$$
$$= \begin{bmatrix} 10 & -\dfrac{13.7s + 30}{s + 3} \end{bmatrix}$$

This completes the design. Note that $C_1(s)$ reduces to 10 because $\overline{D}_p(s)$ was chosen as $s + 3$. For different $\overline{D}_p(s)$, $C_1(s)$ is not a constant, as is seen in the next section.

Example 10.4.3

Consider a plant with transfer function $G(s) = 1/s(s + 2)$. Implement its ITAE optimal system $G_o(s) = 3/(s^2 + 2.4s + 3)$. This $G_o(s)$ was implemented by using the unity-feedback system in Example 10.2.1. The design had the pole-zero cancellation of $s + 2$, which was dictated by the given plant transfer function. Now we implement $G_o(s)$ in the two-parameter configuration and show that the designer has the freedom of choosing canceled poles. First we compute

$$\frac{G_o(s)}{N(s)} = \frac{3}{(s^2 + 2.4s + 3) \cdot 1} =: \frac{N_p(s)}{D_p(s)}$$

In this example, we have $N_p(s) = N_o(s) = 3$ and $D_p(s) = D_o(s) = s^2 + 2.4s + 3$. We choose a polynomial $\bar{D}_p(s)$ of degree 1 so that the degree of $D_p(s)\bar{D}_p(s)$ is $2n - 1 = 3$. Arbitrarily, we choose $\bar{D}_p(s) = s + 10$. Then we have $L(s) = N_p(s)\bar{D}_p(s) = 3(s + 10)$ and

$$F(s) = D_p(s)\bar{D}_p(s) = (s^2 + 2.4s + 3)(s + 10) = s^3 + 12.4s^2 + 27s + 30$$

The polynomials $A(s)$ and $M(s)$ can be solved from

$$\begin{bmatrix} 0 & 1 & 0 & 0 \\ 2 & 0 & 0 & 1 \\ 1 & 0 & 2 & 0 \\ 0 & 0 & 1 & 0 \end{bmatrix} \begin{bmatrix} A_0 \\ M_0 \\ \hline A_1 \\ M_1 \end{bmatrix} = \begin{bmatrix} 30 \\ 27 \\ 12.4 \\ 1 \end{bmatrix}$$

as $A(s) = A_1 s + A_0 = s + 10.4$ and $M(s) = M_1 s + M_0 = 6.2s + 30$. This completes the design.

Although this problem can be implemented in the unity-feedback and two-parameter configurations, the former involves the pole-zero cancellation of $(s + 2)$, which is dictated by the plant; the latter involves the cancellation of $(s + 10)$, which is chosen by the designer. Therefore, the two-parameter configuration is more flexible and may be preferable to the unity-feedback configuration in achieving model matching.

Exercise 10.4.2

Given $G(s) = 1/s(s - 1)$, implement

a. $G_o(s) = 4/(s^2 + 2.8s + 4)$

b. $G_o(s) = (13s + 8)/(s^3 + 3.5s^2 + 13s + 8)$

in the two-parameter configuration. All canceled poles are to be chosen at $s = -4$.

[**Answers:** (a) $L(s)/A(s) = 4(s + 4)/(s + 7.8)$, $M(s)/A(s) = (23s + 16)/(s + 7.8)$. (b) $L(s)/A(s) = (13s + 8)/(s + 4.5)$, $M(s)/A(s) = (17.5s + 8)/(s + 4.5)$.]

10.5   EFFECT OF $\overline{D}_p(s)$ ON DISTURBANCE REJECTION AND ROBUSTNESS[2]

In the two-parameter configuration, we must introduce a Hurwitz polynomial $\overline{D}_p(s)$ in (10.32) or

$$G_o(s) = \frac{N(s)N_p(s)\overline{D}_p(s)}{D_p(s)\overline{D}_p(s)} = \frac{N(s)L(s)}{A(s)D(s) + M(s)N(s)}$$

to insure that the resulting compensators are proper. Because $\overline{D}_p(s)$ is completely canceled in $G_o(s)$, the tracking of $r(t)$ by the plant output $y(t)$ is not affected by the choice of $\overline{D}_p(s)$. Neither is the actuating signal affected by $\overline{D}_p(s)$, because the transfer function from r to u is

$$T(s) := \frac{U(s)}{R(s)} = \frac{G_o(s)}{G(s)} = \frac{N(s)N_p(s)\overline{D}_p(s)D(s)}{N(s)D_p(s)\overline{D}_p(s)} = \frac{N_p(s)D(s)}{D_p(s)} \tag{10.43}$$

where $\overline{D}_p(s)$ does not appear directly or indirectly. Therefore the choice of $\overline{D}_p(s)$ does not affect the tracking property of the overall system and the magnitude of the actuating signal.

Although $\overline{D}_p(s)$ does not appear in $G_o(s)$, it will appear in the closed-loop transfer functions of some input/output pairs. We compute the transfer function from the disturbance input p to the plant output y in Figure 10.6:

$$\begin{aligned} H(s) &:= \frac{Y(s)}{P(s)} = \frac{G(s)}{1 + G(s)M(s)A^{-1}(s)} \\ &= \frac{N(s)A(s)}{A(s)D(s) + M(s)N(s)} = \frac{N(s)A(s)}{D_p(s)\overline{D}_p(s)} \end{aligned} \tag{10.44}$$

We see that $\overline{D}_p(s)$ appears directly in $H(s)$; it also affects $A(s)$ through the Diophantine equation in (10.34). Therefore the choice of $\overline{D}_p(s)$ will affect the disturbance rejection property of the system. This problem will be studied in this section by using examples.

In this section, we also study the effect of $\overline{D}_p(s)$ on the stability range of the overall system. As discussed in Section 6.4, the plant transfer function $G(s)$ may change due to changes of the load, power supplies, or other reasons. Therefore it is of practical interest to see how much the coefficients of $G(s)$ may change before the overall system becomes unstable. The larger the region in which the coefficients of $G(s)$ are permitted to change, the more *robust* the overall system is. In the following examples, we also study this problem.

---

[2]May be skipped without loss of continuity.

Example 10.5.1

Consider a plant with transfer function $G(s) = (s + 3)/s(s - 1)$. Implement its quadratic optimal system $G_o(s) = 10(s + 3)/(s^2 + 12.7s + 30)$ in the two-parameter configuration. As shown in Example 10.4.2, we must choose $\overline{D}_p(s)$ of degree 1 to achieve the design. If $\overline{D}_p(s)$ is chosen as $(s + 3)$, then the compensator was computed in (10.42). For this $\overline{D}_p(s)$ and compensator, the transfer function from p to y can be computed as

$$H(s) = \frac{N(s)A(s)}{D_p(s)\overline{D}_p(s)} = \frac{(s + 3)(s + 3)}{(s^2 + 12.7s + 30)(s + 3)} = \frac{s + 3}{s^2 + 12.7s + 30} \quad (10.45)$$

If $\overline{D}_p(s)$ is chosen as $(s + 30)$, using the same procedure as in Example 10.4.2, we can compute the compensator as

$$[C_1(s) \quad -C_2(s)] = \left[\frac{10(s + 30)}{s + 5.025} \quad -\frac{38.675s + 300}{s + 5.025} \right] \quad (10.46)$$

For this $\overline{D}_p(s)$ and compensator, we have

$$H(s) = \frac{(s + 3)(s + 5.025)}{(s^2 + 12.7s + 30)(s + 30)} \quad (10.47)$$

Next we choose $\overline{D}_p(s) = s + 300$, and compute the compensator as

$$[C_1(s) \quad -C_2(s)] = \left[\frac{10(s + 300)}{s + 25.275} \quad -\frac{288.425s + 3000}{s + 25.275} \right] \quad (10.48)$$

For this $\overline{D}_p(s)$ and compensator, we have

$$H(s) = \frac{(s + 3)(s + 25.275)}{(s^2 + 12.7s + 30)(s + 300)} \quad (10.49)$$

Now we assume the disturbance p to be a unit-step function and compute the plant outputs for the three cases in (10.45), (10.47), and (10.49). The results are plotted in Figure 10.8 with the solid line for $\overline{D}_p(s) = s + 3$, the dashed line for $\overline{D}_p(s) = s + 30$, and the dotted line for $\overline{D}_p(s) = s + 300$. We see that the system with $\overline{D}_p(s) = s + 300$ attenuates the disturbance most. We plot in Figure 10.9 the amplitude characteristics of the three $H(s)$. The one corresponding to $\overline{D}_p(s) = s + 300$ again has the best attenuation property for all ω. Therefore, we conclude that, for this example, the faster the root of $\overline{D}_p(s)$, the better the disturbance rejection property.

Now we study the robustness property of the system. First we consider the case $\overline{D}_p(s) = s + 300$ with the compensator in (10.48). Suppose that after the imple-

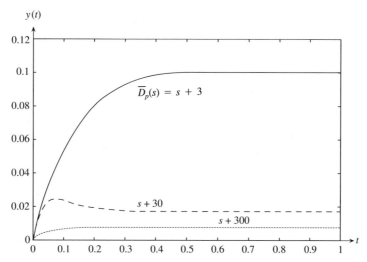

Figure 10.8 Effect of canceled poles on disturbance rejection (time domain).

mentation, the plant transfer function $G(s)$ changes to

$$\overline{G}(s) = \frac{\overline{N}(s)}{\overline{D}(s)} = \frac{s + 3 + \epsilon_2}{s(s - 1 + \epsilon_1)} \tag{10.50}$$

With this perturbed transfer function, the transfer function from r to y becomes

$$\overline{G}_o(s) = \frac{L(s)\overline{N}(s)}{A(s)\overline{D}(s) + M(s)\overline{N}(s)}$$

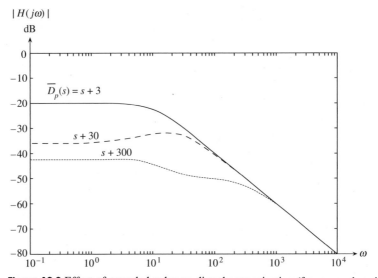

Figure 10.9 Effect of canceled poles on disturbance rejection (frequency domain).

The perturbed overall system is stable if

$$
\begin{aligned}
A(s)\overline{D}(s) + M(s)\overline{N}(s) &= (s + 25.275) \cdot s(s - 1 + \epsilon_1) \\
&\quad + (288.425s + 3000)(s + 3 + \epsilon_2) \\
&= s^3 + (312.7 + \epsilon_1)s^2 \\
&\quad + (3840 + 25.275\epsilon_1 + 288.45\epsilon_2)s + 3000(3 + \epsilon_2)
\end{aligned}
$$

(10.51)

is a Hurwitz polynomial. The application of the Routh test to (10.51) yields the following stability conditions:

$$
312.7 + \epsilon_1 > 0 \qquad 3 + \epsilon_2 > 0 \tag{10.52a}
$$

and

$$
3840 + 25.275\epsilon_1 + 288.425\epsilon_2 - \frac{3000(3 + \epsilon_2)}{312.7 + \epsilon_1} > 0 \tag{10.52b}
$$

(See Exercise 4.6.3.) These conditions can be simplified to

$$
\epsilon_2 > -3 \qquad \text{if } \epsilon_1 > -117.7
$$

and

$$
\epsilon_2 > -\frac{1191768 + 11743.4925\epsilon_1 + 25.275\epsilon_1^2}{87190.4975 + 288.425\epsilon_1} \qquad \text{if } -302.29 < \epsilon_1 < -117.7
$$

It is plotted in Figure 10.10 with the dotted line.

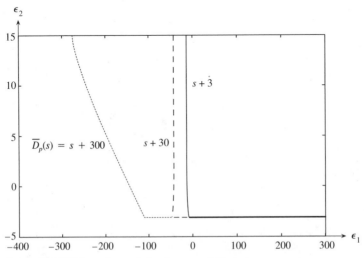

Figure 10.10 Effect of canceled poles on stability range.

In order to make comparisons, we repeat the computation for the cases $\overline{D}_p(s) = s + 30$ and $\overline{D}_p(s) = s + 3$. Their stability regions are also plotted in Figure 10.10 respectively with the dashed line and solid line. We see that the region corresponding to $\overline{D}_p(s) = s + 300$ is the largest and the one corresponding to $\overline{D}_p(s) = s + 3$ is the smallest. Thus we conclude that for this problem, the faster the root of $\overline{D}_p(s)$, the more robust the resulting system is.

Example 10.5.2

Consider a plant with transfer function $G(s) = N(s)/D(s) = (s - 1)/s(s - 2)$. Implement its quadratic optimal system $G_o(s) = -10(s - 1)/(s^2 + 11.14s + 10)$ in the two-parameter configuration. First, we compute

$$\frac{G_o(s)}{N(s)} = \frac{-10(s - 1)}{(s^2 + 11.14s + 10)(s - 1)} = \frac{-10}{s^2 + 11.14s + 10} =: \frac{N_p(s)}{D_p(s)} \tag{10.53}$$

Because the degree of $D_p(s)$ is 2, which is smaller than $2n - 1 = 3$, we must choose a Hurwitz polynomial $\overline{D}_p(s)$ of degree at least 1. We choose

$$\overline{D}_p(s) = s + \beta$$

Then we have

$$\begin{aligned}
D_p(s)\overline{D}_p(s) &= (s^2 + 11.14s + 10)(s + \beta) \\
&= s^3 + (11.14 + \beta)s^2 + (10 + 11.14\beta)s + 10\beta
\end{aligned}$$

and

$$L(s) = N_p(s)\overline{D}_p(s) = -10(s + \beta)$$

The polynomials $A(s)$ and $M(s)$ can be solved from

$$\begin{bmatrix} 0 & -1 & 0 & 0 \\ -2 & 1 & 0 & -1 \\ 1 & 0 & -2 & 1 \\ 0 & 0 & 1 & 0 \end{bmatrix} \begin{bmatrix} A_0 \\ M_0 \\ \hline A_1 \\ M_1 \end{bmatrix} = \begin{bmatrix} 10\beta \\ 10 + 11.14\beta \\ 11.14 + \beta \\ 1 \end{bmatrix}$$

This can be solved directly. It can also be solved by computing

$$\begin{bmatrix} 0 & -1 & 0 & 0 \\ -2 & 1 & 0 & -1 \\ 1 & 0 & -2 & 1 \\ 0 & 0 & 1 & 0 \end{bmatrix}^{-1} = \begin{bmatrix} -1 & -1 & -1 & -2 \\ -1 & 0 & 0 & 0 \\ 0 & 0 & 0 & 1 \\ 1 & 1 & 2 & 4 \end{bmatrix}$$

Thus we have

$$
\begin{bmatrix} A_0 \\ M_0 \\ A_1 \\ M_1 \end{bmatrix} = \begin{bmatrix} -1 & -1 & -1 & -2 \\ -1 & 0 & 0 & 0 \\ 0 & 0 & 0 & 1 \\ 1 & 1 & 2 & 4 \end{bmatrix} \begin{bmatrix} 10\beta \\ 10 + 11.14\beta \\ 11.14 + \beta \\ 1 \end{bmatrix} = \begin{bmatrix} -23.14 - 22.14\beta \\ -10\beta \\ 1 \\ 36.28 + 23.14\beta \end{bmatrix}
$$

and the compensator is

$$
\begin{aligned}
[C_1(s) \quad -C_2(s)] &= \begin{bmatrix} \dfrac{L(s)}{A(s)} & -\dfrac{M(s)}{A(s)} \end{bmatrix} \\[2mm]
&= \begin{bmatrix} \dfrac{-10(s + \beta)}{s - 23.14 - 22.14\beta} & -\dfrac{(36.28 + 23.14\beta)s - 10\beta}{s - 23.14 - 22.14\beta} \end{bmatrix}
\end{aligned}
$$

$$(10.54)$$

This completes the implementation of the quadratic optimal system. Note that no matter what value β assumes, as long as it is positive, $\bar{D}_p(s) = s + \beta$ will not affect the tracking property of the overall system. Neither will it affect the magnitude of the actuating signal.

Now we study the effect of $\bar{D}_p(s)$ on disturbance rejection. The transfer function from the disturbance p to the plant output y is

$$
H(s) := \frac{Y(s)}{P(s)} = \frac{N(s)A(s)}{D_p(s)\bar{D}_p(s)} = \frac{(s - 1)(s - 23.14 - 22.14\beta)}{(s^2 + 11.14s + 10)(s + \beta)} \qquad (10.55)
$$

Let the disturbance be a unit-step function. We compute the unit-step responses of (10.55) on a personal computer and plot the results in Figure 10.11 for $\bar{D}_p(s) =$

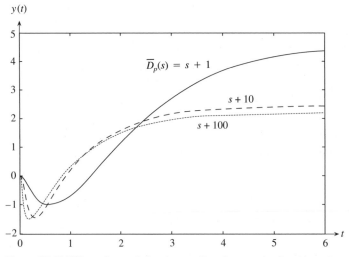

Figure 10.11 Effect of canceled poles on disturbance rejection (time domain).

$s + 1$ (solid line), $\overline{D}_p(s) = s + 10$ (dashed line), and $\overline{D}_p(s) = s + 100$ (dotted line). We see that the choice of $\overline{D}_p(s)$ does affect the disturbance rejection property of the system. Although the one corresponding to $\overline{D}_p(s) = s + 100$ has the smallest steady-state value, its undershoot is the largest. We plot in Figure 10.12 the amplitude characteristics of $H(s)$ for $\beta = 1$, 10, and 100 respectively with the solid line, dashed line, and dotted line. The one corresponding to $\beta = 100$ has the largest attenuation for small ω, but it has less attenuation for $\omega \geq 2$. Therefore, for this example, the choice of $\overline{D}_p(s)$ is not as clear-cut as in the preceding example. To have a small steady-state effect, we should choose a large β. If the frequency spectrum of disturbance lies mainly between 2 and 1000 radians per second, then we should choose a small β.

Now we study the effect of $\overline{D}_p(s)$ on the robustness of the overall system. Suppose after the implementation of $G_o(s)$, the plant transfer function $G(s)$ changes to

$$\overline{G}(s) = \frac{\overline{N}(s)}{\overline{D}(s)} = \frac{s - 1 + \epsilon_2}{s(s - 2 + \epsilon_1)} \tag{10.56}$$

With this plant transfer function and the compensators in (10.54), the overall transfer function becomes

$$\overline{G}_o(s) = \frac{L(s)\overline{N}(s)}{A(s)\overline{D}(s) + M(s)\overline{N}(s)}$$

with

$$\begin{aligned} A(s)\overline{D}(s) &+ M(s)\overline{N}(s) \\ &= (s - 23.14 - 22.14\beta) \cdot s(s - 2 + \epsilon_1) \\ &+ [(36.28 + 23.14\beta)s - 10\beta] \cdot (s - 1 + \epsilon_2) \end{aligned} \tag{10.57}$$

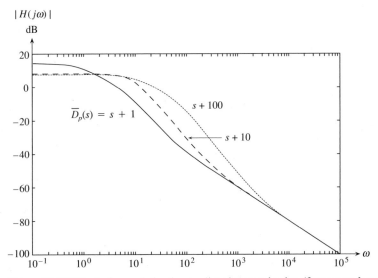

Figure 10.12 Effect of canceled poles on disturbance rejection (frequency domain).

We compute its stability ranges for the three cases with $\overline{D}_p(s) = s + 1$, $\overline{D}_p(s) = s + 10$, and $\overline{D}_p(s) = s + 100$. If $\overline{D}_p(s) = s + 1$, (10.57) becomes

$$
\begin{aligned}
A(s)\overline{D}(s) + M(s)\overline{N}(s) &= (s - 45.28) \cdot s(s - 2 + \epsilon_1) \\
&\quad + (59.42s - 10)(s - 1 + \epsilon_2) \\
&= s^3 + (12.14 + \epsilon_1)s^2 \\
&\quad + (21.14 - 45.28\epsilon_1 + 59.42\epsilon_2)s + 10(1 - \epsilon_2)
\end{aligned}
$$

It is Hurwitz under the following three conditions

$$
12.14 + \epsilon_1 > 0 \qquad 1 - \epsilon_2 > 0
$$

and

$$
(21.14 - 45.28\epsilon_1 + 59.42\epsilon_2) - \frac{10(1 - \epsilon_2)}{(12.14 + \epsilon_1)} > 0
$$

(See Exercise 4.6.3.) These conditions can be simplified as

$$
\frac{-246.6396 + 528.5592\epsilon_1 + 45.28\epsilon_1^2}{731.3588 + 59.42\epsilon_1} < \epsilon_2 < 1 \qquad \text{for } -12.14 < \epsilon_1 < 1.78
$$

From these inequalities, the stability range of ϵ_1 and ϵ_2 can be plotted as shown in Figure 10.13 with the solid line. We repeat the computation for $\overline{D}_p(s) = s + 10$ and $\overline{D}_p(s) = s + 100$ and plot the results in Figure 10.13, respectively, with the

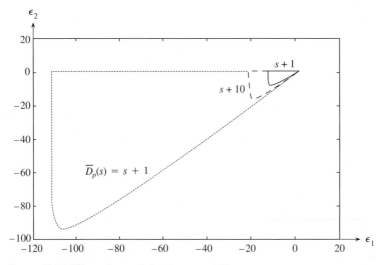

Figure 10.13 Effect of canceled poles on stability range.

dashed line and dotted line. We see that, roughly speaking, the larger β, the larger the stability range. (Note that in the neighborhood of $\epsilon_1 = 0$ and $\epsilon_2 = 0$, the stability region corresponding to $\overline{D}_p(s) = s + 100$ does not include completely the regions corresponding to $\overline{D}_p(s) = s + 10$ and $\overline{D}_p(s) = s + 1$.) Therefore we conclude that the faster the root of $\overline{D}_p(s)$, the more robust the overall system.

From the preceding two examples, we conclude that the choice of $\overline{D}_p(s)$ in the two-parameter configuration does affect the disturbance rejection and robustness properties of the resulting system. For the system in Example 10.5.1, which has no non-minimum-phase zeros, the faster the root of $\overline{D}_p(s)$, the better the step disturbance rejection and the more robust the resulting system. For the system in Example 10.5.2, which has a non-minimum-phase zero, the choice of $\overline{D}_p(s)$ is no longer clear-cut. In conclusion, in using the two-parameter configuration to achieve model matching, although the choice of $\overline{D}_p(s)$ does not affect the tracking property of the system and the magnitude of the actuating signal, it does affect the disturbance rejection and robustness properties of the system. Therefore, we should utilize this freedom in the design. No general rule of choosing $\overline{D}_p(s)$ seems available at present. However, we may always choose it by trial and error.

10.5.1   Model Matching and Disturbance Rejection

A system is said to achieve step *disturbance rejection* if the plant output excited by any step disturbance eventually vanishes, that is,

$$\lim_{t\to\infty} y_p(t) = 0 \tag{10.58}$$

where $y_p(t)$ is the plant output excited by the disturbance $p(t)$ shown in Figure 10.6. In the examples in the preceding section, we showed that by choosing the root of $\overline{D}_p(s)$ appropriately, the effect of step disturbances can be reduced. However, no matter where the root is chosen, the steady-state effect of step disturbances can never be completely eliminated. Now we shall show that by increasing the degree of $\overline{D}_p(s)$, step disturbances can be completely eliminated as $t\to\infty$.

The transfer function from p to y in Figure 10.6 is

$$H(s) = \frac{N(s)A(s)}{D_p(s)\overline{D}_p(s)} \tag{10.59}$$

If p is a step function with magnitude a, then

$$Y_p(s) = H(s)P(s) = \frac{N(s)A(s)}{D_p(s)\overline{D}_p(s)} \cdot \frac{a}{s} \tag{10.60}$$

The application of the final-value theorem to (10.60) yields

$$\lim_{t\to\infty} y_p(t) = \lim_{s\to 0} s \cdot \frac{N(s)A(s)}{D_p(s)\overline{D}_p(s)} \cdot \frac{a}{s} = \frac{aN(0)A(0)}{D_o(0)\overline{D}_p(0)} \tag{10.61}$$

This becomes zero for any a if and only if $N(0) = 0$ or $A(0) = 0$. Note that $D_p(0) \neq 0$ and $\overline{D}_p(0) \neq 0$ because $D_p(s)$ and $\overline{D}_p(s)$ are Hurwitz. The constant $N(0)$ is given and is often nonzero. Therefore, the only way to achieve disturbance rejection is to design $A(s)$ with $A(0) = 0$. Recall that $A(s)$ is to be solved from the Diophantine equation in (10.34) or the linear algebraic equation in (10.37). If the degree of $\overline{D}_p(s)$ is chosen so that the degree of $D_p(s)\overline{D}_p(s)$ is $2n - 1$, where $n =$ deg $D(s)$, then the solution $A(s)$ is unique and we have no control over $A(0)$. However, if we increase the degree of $\overline{D}_p(s)$, then solutions $A(s)$ are no longer unique and we may have the freedom of choosing $A(0)$. This will be illustrated by an example.

Example 10.5.3

Consider a plant with transfer function $G(s) = (s + 3)/s(s - 1)$. Implement its quadratic optimal system $G_o(s) = 10(s + 3)/(s^2 + 12.7s + 30)$. This was implemented in Example 10.5.1 by choosing the degree of $\overline{D}_p(s)$ as 1. Now we shall increase the degree of $\overline{D}_p(s)$ to 2 and repeat the design. First we compute

$$\frac{G_o(s)}{N(s)} = \frac{10}{s^2 + 12.7s + 30} =: \frac{N_p(s)}{D_p(s)}$$

Arbitrarily, we choose

$$\overline{D}_p(s) = (s + 30)^2 \tag{10.62}$$

Then we have

$$D_p(s)\overline{D}_p(s) = (s^2 + 12.7s + 30)(s + 30)^2$$
$$= s^4 + 72.7s^3 + 1692s^2 + 13{,}230s + 27{,}000$$

and

$$L(s) = N_p(s)\overline{D}_p(s) = 10(s + 30)^2$$

The polynomials $A(s) = A_0 + A_1 s + A_2 s^2$ and $M(s) = M_0 + M_1 s + M_2 s^2$ can be solved from

$$\begin{bmatrix} 0 & 3 & 0 & 0 & 0 & 0 \\ -1 & 1 & 0 & 3 & 0 & 0 \\ 1 & 0 & -1 & 1 & 0 & 3 \\ 0 & 0 & 1 & 0 & -1 & 1 \\ 0 & 0 & 0 & 0 & 1 & 0 \end{bmatrix} \begin{bmatrix} A_0 \\ M_0 \\ \hline A_1 \\ M_1 \\ \hline A_2 \\ M_2 \end{bmatrix} = \begin{bmatrix} 27{,}000 \\ 13{,}230 \\ 1692 \\ 72.7 \\ 1 \end{bmatrix} \tag{10.63}$$

This has 5 equations and 6 unknowns. Because the first column of the 5×6 matrix is linearly dependent on the remaining columns, A_0 can be arbitrarily assigned, in particular, assigned as 0. With $A_0 = 0$, the solution of (10.63) can be computed as

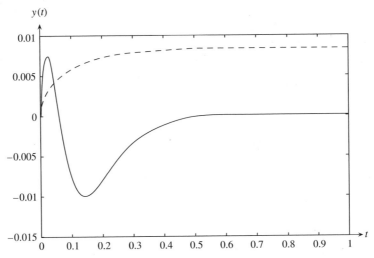

Figure 10.14 Disturbance rejection.

$A_1 = -15.2$, $A_2 = 1$, $M_0 = 9000$, $M_1 = 1410$, and $M_2 = 88.9$. Thus the compensator is

$$[C_1(s) \quad -C_2(s)] = \left[\frac{10(s + 30)^2}{s(s - 15.2)} \quad -\frac{88.9s^2 + 1410s + 9000}{s(s - 15.2)} \right] \quad \text{(10.64)}$$

This completes the design. With this compensator, the transfer function from r to y is still $G_o(s) = 10(s + 3)/(s^2 + 12.7s + 30)$. Therefore, the tracking property of the system remains unchanged. Now we compute the transfer function from the disturbance p to the plant output y:

$$H(s) := \frac{Y(s)}{P(s)} = \frac{N(s)A(s)}{D_p(s)\bar{D}_p(s)} = \frac{(s + 3)s(s - 15.2)}{(s^2 + 12.7s + 30)(s + 30)^2} \quad \text{(10.65)}$$

Because $H(0) = 0$, if the disturbance is a step function, the excited plant output will approach zero as $t \to \infty$, as shown in Figure 10.14 with the solid line. As a comparison, we also show in Figure 10.14 with the dashed line the plant output due to a step disturbance for the design using $\bar{D}_p(s) = s + 300$. We see that by increasing the degree of $\bar{D}_p(s)$, it is possible to achieve step disturbance rejection. In actual design, we may try several $\bar{D}_p(s)$ of degree 2 and then choose one which suppresses most disturbances. In conclusion, by increasing the degree of $\bar{D}_p(s)$, we may achieve disturbance rejection.

Exercise 10.5.1

Repeat the preceding example by choosing $\bar{D}_p(s) = (s + 300)^2$.

[**Answer:** $H(s) = s(s + 3)(s - 2200)/(s^2 + 12.7s + 30)(s + 300)^2$.]

10.6   PLANT INPUT/OUTPUT FEEDBACK CONFIGURATION

Consider the configuration shown in Figure 10.15(a) in which $G(s)$ is the plant transfer function and $C_1(s)$, $C_2(s)$, and $C_0(s)$ are proper compensators. This configuration introduces feedback from the plant input and output; therefore, it is called the *plant input/output feedback configuration* or *plant I/O feedback configuration* for short. This configuration can be used to implement *any* implementable $G_o(s)$. Instead of discussing the general case, we discuss only the case where

$$\deg D_o(s) \; - \; \deg N_o(s) \; = \; \deg D(s) \; - \; \deg N(s) \qquad (10.66)$$

In other words, the pole-zero excess of $G_o(s)$ equals that of $G(s)$. In this case, we can always assume $C_0(s) = 1$ and

$$C_1(s) \; = \; \frac{L(s)}{A(s)} \qquad C_2(s) \; = \; \frac{M(s)}{A(s)} \qquad (10.67)$$

and the plant I/O feedback configuration can be simplified as shown in Figure 10.15(b). Note that $A(s)$, $L(s)$, and $M(s)$ in Figure 10.15(b) are different from those in the two-parameter configuration in Figure 10.6. The two compensators enclosed by the dashed line can be considered as a two-input, one-output compensator and must be implemented as a unit as discussed in Section 10.4. The configuration has two loops, one with loop gain $-L(s)/A(s)$, the other $-G(s)M(s)/A(s)$. Thus its

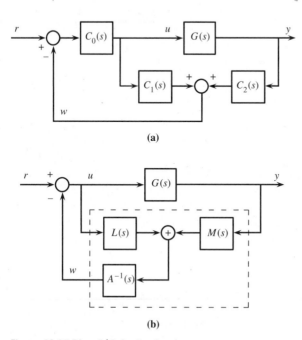

(a)

(b)

Figure 10.15 Plant I/O feedback system.

characteristic function equals

$$\Delta(s) = 1 - \left(-\frac{L(s)}{A(s)} - G(s)\frac{M(s)}{A(s)} \right) = 1 + \frac{L(s)}{A(s)} + \frac{N(s)M(s)}{D(s)A(s)} \quad \text{(10.68)}$$

and the transfer function from r to y equals, using Mason's formula,

$$G_o(s) = \frac{G(s)}{\Delta(s)} = \frac{N(s)A(s)}{A(s)D(s) + L(s)D(s) + M(s)N(s)} \quad \text{(10.69)}$$

Now we develop a procedure to implement $G_o(s)$.

Problem   Given $G(s) = N(s)/D(s)$, where $N(s)$ and $D(s)$ are coprime and deg $N(s)$ $\leq$ deg $D(s)$. Implement an implementable $G_o(s) = N_o(s)/D_o(s)$ with deg $D_o(s) -$ deg $N_o(s) = $ deg $D(s) - $ deg $N(s)$.

Procedure:
Step 1:   Compute

$$\frac{G_o(s)}{N(s)} = \frac{N_o(s)}{D_o(s)N(s)} =: \frac{N_p(s)}{D_p(s)} \quad \text{(10.70)}$$

where $N_p(s)$ and $D_p(s)$ have no common factors.
Step 2:   If deg $N_p(s) =: \overline{m} < n - 1$, introduce an arbitrary Hurwitz polynomial $\overline{A}(s)$ of degree $n - 1 - \overline{m}$. If $\overline{m} \geq n - 1$, set $\overline{A}(s) = 1$
Step 3:   Rewrite (10.70) and equate it with (10.69):

$$G_o(s) = \frac{N(s)N_p(s)}{D_p(s)} = \frac{N(s)N_p(s)\overline{A}(s)}{D_p(s)\overline{A}(s)} \quad \text{(10.71)}$$

$$= \frac{N(s)A(s)}{A(s)D(s) + L(s)D(s) + M(s)N(s)}$$

From this equation, we have

$$A(s) = N_p(s)\overline{A}(s)$$
$$A(s)D(s) + L(s)D(s) + M(s)N(s) = D_p(s)\overline{A}(s) \quad \text{(10.72)}$$

which becomes, after substituting (10.72)

$$L(s)D(s) + M(s)N(s) = D_p(s)\overline{A}(s) - A(s)D(s)$$
$$= D_p(s)\overline{A}(s) - N_p(s)\overline{A}(s)D(s) \quad \text{(10.73)}$$
$$= \overline{A}(s)(D_p(s) - N_p(s)D(s)) =: F(s)$$

This Diophantine equation can be used to solve $L(s)$ and $M(s)$. Because of the introduction of $\overline{A}(s)$, $A(s)$ in (10.72) has a degree at least $n - 1$ and the degrees of $L(s)$ and $M(s)$ are insured to be at most equal to that of $A(s)$. Therefore, the resulting compensators $L(s)/A(s)$ and $M(s)/A(s)$ are proper. Equation (10.73) is similar to (10.34), therefore it can also be solved using a linear algebraic equation. For example, if deg $A(s) = n - 1$, then deg

$F(s) = 2n - 1$. Let

$$L(s) = L_0 + L_1s + \cdots + L_{n-1}s^{n-1} \tag{10.74a}$$

$$M(s) = M_0 + M_1s + \cdots + M_{n-1}s^{n-1} \tag{10.74b}$$

and

$$F(s) = F_0 + F_1s + F_2s^2 + \cdots + F_{2n-1}s^{2n-1} \tag{10.74c}$$

Then $L(s)$ and $M(s)$ can be solved from the following linear algebraic equation:

$$
\begin{bmatrix}
D_0 & N_0 & 0 & 0 & & 0 & 0 \\
D_1 & N_1 & D_0 & N_0 & & \vdots & \vdots \\
\vdots & \vdots & \vdots & \vdots & \cdots & 0 & 0 \\
D_n & N_n & D_{n-1} & N_{n-1} & & D_0 & N_0 \\
0 & 0 & D_n & N_n & & D_1 & N_1 \\
\vdots & \vdots & \vdots & \vdots & & \vdots & \vdots \\
0 & 0 & 0 & 0 & & D_n & N_n
\end{bmatrix}
\begin{bmatrix}
L_0 \\
M_0 \\
\hline
L_1 \\
M_1 \\
\vdots \\
\hline
L_{n-1} \\
M_{n-1}
\end{bmatrix}
=
\begin{bmatrix}
F_0 \\
F_1 \\
F_2 \\
\vdots \\
F_{2n-1}
\end{bmatrix}
\tag{10.75}
$$

This is illustrated by an example.

Example 10.6.1

Consider $G(s) = (s + 3)/s(s - 1)$. Implement its quadratic optimal system $G_o(s) = 10(s + 3)/(s^2 + 12.7s + 30)$. This problem was implemented in the two-parameter configuration in Example 10.4.2. Now we shall implement it in the plant I/O feedback configuration shown in Figure 10.15(b). First we compute

$$\frac{G_o(s)}{N(s)} = \frac{10(s + 3)}{(s^2 + 12.7s + 30)(s + 3)} = \frac{10}{s^2 + 12.7s + 30} =: \frac{N_p(s)}{D_p(s)}$$

Because the degree of $N_p(s)$ is 0, we must introduce a Hurwitz polynomial $\overline{A}(s)$ of degree at least $n - 1 - \deg N_p(s) = 1$. Arbitrarily, we choose

$$\overline{A}(s) = s + 3 \tag{10.76}$$

Then we have

$$A(s) = N_p(s)\overline{A}(s) = 10(s + 3)$$

and, from (10.73),

$$F(s) = \overline{A}(s)(D_p(s) - N_p(s)D(s)) = (s + 3)[s^2 + 12.7s + 30 - 10s(s - 1)]$$
$$= -9s^3 - 4.3s^2 + 98.1s + 90$$

The polynomials $L(s)$ and $M(s)$ can be solved from

$$
\begin{bmatrix}
0 & 3 & 0 & 0 \\
-1 & 1 & 0 & 3 \\
1 & 0 & -1 & 1 \\
0 & 0 & 1 & 0
\end{bmatrix}
\begin{bmatrix}
L_0 \\
M_0 \\
L_1 \\
M_1
\end{bmatrix}
=
\begin{bmatrix}
90 \\
98.1 \\
-4.3 \\
-9
\end{bmatrix}
\tag{10.77}
$$

as $L(s) = L_1 s + L_0 = -9s - 27$ and $M(s) = M_1 s + M_0 = 13.7s + 30$. Thus the compensator is

$$
C_1(s) = \frac{L(s)}{A(s)} = \frac{-9s - 27}{10(s + 3)} = \frac{-9}{10}
\tag{10.78a}
$$

$$
C_2(s) = \frac{M(s)}{A(s)} = \frac{13.7s + 30}{10(s + 3)}
\tag{10.78b}
$$

This completes the design. Note that $C_1(s)$ reduces to a constant because $\overline{A}(s)$ was chosen as $s + 3$. Different $\overline{A}(s)$ will yield nonconstant $C_1(s)$.

We see that the design using the plant I/O feedback configuration is quite similar to that of the two-parameter configuration. Because $\overline{A}(s)$ is completely canceled in $G_o(s)$, the choice of $\overline{A}(s)$ will not affect the tracking property and actuating signal of the system. However, its choice may affect disturbance rejection and stability robustness of the resulting system. The idea is similar to the two-parameter case and will not be repeated.

Exercise 10.6.1

Consider a plant with transfer function $G(s) = 1/s(s - 1)$. Find compensators in the plant I/O feedback configuration to yield (a) $G_o(s) = 4/(s^2 + 2.8s + 4)$ and (b) $G_o(s) = (13s + 8)/(s^3 + 3.5s^2 + 13s + 8)$. All canceled poles are to be chosen at $s = -4$.

[**Answers:** (a) $L(s)/A(s) = (-3s - 8.2)/4(s + 4)$, $M(s)/A(s) = (23s + 16)/4(s + 4)$. (b) $L(s)/A(s) = (-12s - 3.5)/(13s + 8)$, $M(s)/A(s) = (17.5s + 8)/(13s + 8)$.]

10.7 SUMMARY AND CONCLUDING REMARKS

This chapter discusses the problem of implementing overall transfer functions. We discuss first the unity-feedback configuration. Generally, the unity-feedback configuration cannot be used to achieve model matching. However, it can always be used to achieve pole placement. If the degree of $G(s)$ is n, the minimum degree of compensators to achieve arbitrary pole placement is $n - 1$ if $G(s)$ is strictly proper, or n if $G(s)$ is biproper. If we increase the degree of compensators, then the unity-feedback configuration can be used to achieve pole placement and robust tracking.

The two-parameter configuration can be used to achieve any model matching. In this configuration, generally we have freedom in choosing canceled poles. The choice of these poles will not affect the tracking property of the system and the magnitude of the actuating signal. Therefore, these canceled poles can be chosen to suppress the effect of disturbance and to increase the stability robustness of the system. If we increase the degree of compensators, then it is possible to achieve model matching and disturbance rejection.

Finally we introduced the plant input/output feedback configuration. This configuration is developed from state estimator (or observer) and state feedback (or controller) in state-variable equations. See Chapter 11. The configuration can also be used to achieve any model matching. For a comparison of the two-parameter configuration and the plant I/O feedback configuration, see References [16, 44].

In this chapter all compensators for pole placement and model matching are obtained by solving sets of linear algebraic equations. Thus the method is referred to as the *linear algebraic method*.

We now compare the inward approach and the outward approach. We introduced the root-locus method and the frequency-domain method in the outward approach. In the root-locus method, we try to shift the poles of overall systems to the desired pole region. The region is developed from a quadratic transfer function with a constant numerator. Therefore, if an overall transfer function is not of the form, even if the poles are shifted into the region, there is no guarantee that the resulting system has the desired performance. In the frequency-domain method, because the relationship among the phase margin, gain margin, and time response is not exact, even if the design meets the requirement on the phase and gain margins, there is no guarantee that the time response of the resulting system will be satisfactory. Furthermore, if a plant has open right-half-plane poles, the frequency-domain method is rarely used. The constraint on actuating signals is not considered in the root-locus method, nor in the frequency-domain method.

In the inward approach, we first choose an overall transfer function; it can be chosen to minimize the quadratic or ITAE performance index or simply by engineering judgment. The constraint on actuating signals can be included in the choice. Once an overall transfer function is chosen, we may implement it in the unity-feedback configuration. If it cannot be so implemented, we can definitely implement it in the two-parameter or plant input/output feedback configuration. In the implementation, we may also choose canceled poles to improve disturbance rejection property and to increase stability robustness property of the resulting system. Thus, the inward approach appears to be more general and more versatile than the outward approach. Therefore, the inward approach should be a viable alternative in the design of control systems.

We give a brief history about various design methods to conclude this chapter. The earliest systematic method to design feedback systems was developed by Bode in 1945 [7]. It is carried out by using frequency plots, which can be obtained by measurement. Thus, the method is very useful to systems whose mathematical equations are difficult to develop. The method, however, is difficult to employ if a system, such as an aircraft, has unstable poles. In order to overcome the unstable poles of aircrafts, Evans proposed the root-locus method in 1950 [27]. The method has since been widely used in practice. The inward approach was first discussed by Truxal in 1955 [57]. He called the method synthesis through pole-zero configuration. The conditions in Corollary 9.1 were developed for the unity-feedback configuration. In spite of its importance, the method was mentioned only in a small number of control texts. ITAE optimal systems were developed by Graham and Lathrop [33] in 1953. Newton and colleagues [48] and Chang [10] were among the earliest to develop quadratic optimal systems by using transfer functions.

The development of implementable transfer functions for *any* control configuration was attempted in [12]. The conditions of proper compensators and no plant leakage, which was coined by Horowitz [36], were employed. Total stability was not considered. Although the condition of well-posedness was implicitly used, the concept was not fully understood and the proof was incomplete. It was found in [14] that, without imposing well-posedness, the plant input/output feedback configuration can be used to implement $G_o(s) = 1$ using exclusively proper compensators. It was a clear violation of physical constraints. Thus the well-posedness condition was explicitly used in [15, 16] to design control systems. By requiring proper compensators, total stability, well-posedness and no plant leakage, the necessity of the implementability conditions follows immediately for *any* control configuration. Although these constraints were intuitively apparent, it took many years to be fully understood and be stated without any ambiguity. A similar problem was studied by Youla, Bongiorno, and Lu [68], where the conditions for $G_o(s)$ to be implementable in the single-loop configuration in Figure P10.3 by using stable proper compensators are established. The conditions involve the interlacing of real poles of $G_o(s)$ and $G(s)$, and the problem is referred to as *strong stabilization*.

The computation of compensators by solving linear algebraic equations was first suggested by Shipley [55] in 1963. Conditions for the existence of compensators were not discussed. The coprimeness condition was used in 1969 [11] to establish the existence of compensators in the plant input/output feedback system and to establish its relationship with state feedback [66] and state estimators [45]. The two-parameter configuration seems to be first introduced by Aström [2] in 1980 and was employed in [3]. The design procedure discussed in this text follows [21].

Most results in Chapter 10 are available for multivariable systems, systems with more than one input or more than one output. The results were developed from the polynomial fractional approach, which was developed mainly by Rosenbrock [53], Wolovich [65], Kučera [41], and Callier and Desoer [9]. The polynomial fractional method was translated into linear algebraic equations in [14, 15]. Using elementary results in linear algebra, it was possible to develop most results in the polynomial fractional approach in [15]. More importantly, minimum-degree compensators can now be easily obtained by solving linear algebraic equations. This method of computing compensators is therefore called the linear algebraic method.

PROBLEMS

10.1. Consider a plant with transfer function $1/s(s + 3)$. Implement the overall transfer function $G_o(s) = 4/(s^2 + 4s + 4)$ in the unity-feedback configuration in Figure 10.1. Does the implementation involve pole-zero cancellations? Do you have the freedom of choosing canceled poles?

10.2. Consider a plant with transfer function $2/s(s - 3)$. Can you implement the overall transfer function $G_o(s) = 4/(s^2 + 4s + 4)$ in the unity-feedback configuration in Figure 10.1? Will the resulting system be well posed? totally stable?

10.3. Given $G(s) = 1/(s - 1)$. Can $G_o(s) = 1/(s + 1)$ be implemented in the unity-feedback configuration in Figure 10.1? Can $G_o(s)$ be implemented in the single-loop feedback system shown in Figure P10.3 with $C_1(s) = 1$?

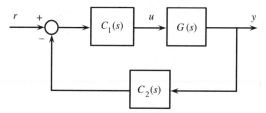

Figure P10.3

10.4. Consider a plant with transfer function $1/s(s + 2)$. Find a proper compensator of degree 1 in the unity-feedback configuration such that the overall system has poles at $-1 + j$, $-1 - j$, and -3. What is the compensator? Use the root-locus method to give an explanation of the result.

10.5. Consider a plant with transfer function $1/s(s - 2)$. Find a proper compensator of degree 1 in the unity-feedback configuration such that the overall system has poles at $-1 + j$, $-1 - j$, and -3. Will the resulting system track asymptotically every step-reference input? Is this tracking robust?

10.6. **a.** Consider a plant with transfer function $1/s(s + 2)$. Design a compensator in the unity-feedback configuration such that the overall system has the dominant poles $-1.4 \pm j1.43$ and a pole a with $a = -4$. What is $u(0^+)$?

b. Repeat (a) for $a = -5$. Can you conclude that, for the same dominant poles, the farther away the pole a, the larger the actuating signal?

10.7. **a.** Consider a plant with transfer function $G(s) = 1/s(s + 2)$ and its quadratic optimal system $G_o(s) = 3/(s^2 + 3.2s + 3)$. Can $G_o(s)$ be implemented in Figure P10.7 by adjusting k_1 and k_2?

b. Consider a plant with transfer function $G(s) = (s - 1)/(s^2 - 4)$ and its quadratic optimal system $G_o(s) = -1.8(s - 1)/(s^2 + 5.2s + 5)$. Can $G_o(s)$ be implemented in Figure P10.7 by adjusting k_1 and k_2? Will this optimal system track asymptotically step-reference inputs? Note that if

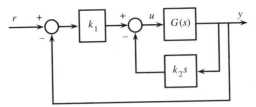

Figure P10.7

$G(s)$ is of type 0 as in this example, then the quadratic performance index equals infinity.

10.8. Consider the single-loop configuration with $C_1(s) = 1$ shown in Figure P10.3. Is it possible to achieve arbitrary pole placement by choosing a suitable proper compensator $C(s)$? If yes, develop a design procedure. Do you have control over the zeros of the resulting overall system?

10.9. **a.** Consider a plant with transfer function $G(s) = (s - 1)/(s^2 - 4)$. Find a compensator $C(s)$ and a precompensator k so that the unity-feedback system in Figure 10.3 has all poles at -3 and tracks asymptotically step-reference inputs.

   b. If the plant transfer function in (a) becomes $\overline{G}(s) = (s - 0.9)/(s^2 - 4)$ after implementation, is the system still stable? Will the system still track asymptotically every step-reference input? Is this design robust in tracking?

10.10. In Problem 10.9, is it possible to design a system which is robust in tracking by increasing the degree of the compensator?

10.11. Implement $G_o(s)$ in Problem 10.7(a) in the unity-feedback configuration. Also implement it in the two-parameter configuration. Do you have control over the canceled pole in both cases?

10.12. Can you implement $G_o(s)$ in Problem 10.7(b) in the unity-feedback configuration? Can you implement it in the two-parameter configuration?

10.13. Consider $G(s) = 2/s(s^2 + 0.25s + 6.25)$. Implement

$$G_o(s) = \frac{20}{s^3 + 3.957s^2 + 14.048s + 20}$$

in the unity-feedback and two-parameter configurations. Which implementation is more desirable? Give your reasons.

10.14. **a.** Consider $G(s) = (s + 3)/s(s - 1)$ and its overall system $G_o(s) = 784/(s^2 + 50.4s + 784)$. Implement the system in the two-parameter configuration by choosing $\overline{D}_p(s) = 1$. Will the plant output of the system reject completely, as $t \to \infty$, any step disturbance function entering at the plant input terminal?

b. Repeat the design in (a) by choosing $\overline{D}_p(s) = s + \beta$ with $\beta = 3$. Implement $G_o(s)$ and at the same time achieve disturbance rejection.

c. Repeat (b) for $\beta = 30$ and $\beta = 300$. If computer software is available, simulate and compare the results in (a), (b), and (c).

10.15. Consider the system in Problem 10.14. Suppose, after implementation, the plant transfer function changes to $\overline{G}(s) = (s + 3 + \epsilon_2)/s(s - 1 + \epsilon_1)$. Find the ranges of ϵ_1 and ϵ_2 in which the system remains stable for the three cases.

10.16. Implement the systems in Problems 10.7 and 10.13 in the plant input/output feedback configuration.

10.17. Implement the system in Problem 10.14(a) in the plant input/output feedback configuration. Are the compensators proper? Can we apply the procedure in Section 10.6 to design this problem?

10.18. a. Consider the unity-feedback system shown in Figure 10.1. Show that if the plant transfer function is of type 1 and if the system is stable, then the plant output will track asymptotically any step-reference input. The tracking is robust with plant and compensator variations as long as the system remains stable.

b. Consider the two-parameter system shown in Figure 10.6. Show that if the plant transfer function is of type 1 and if the overall system is stable, then the plant output will track asymptotically any step-reference input if and only if $M(0) = L(0)$. The tracking is robust with plant and compensator variations as long as the system remains stable and the dc gains of $M(s)$ and $L(s)$ remain equal.

c. Consider the plant input/output feedback system shown in Figure 10.15(b). Show that if the plant transfer function is of type 1 and if the overall system is stable, then the plant output will track asymptotically any step-reference input if and only if $A(0) = M(0)$. The tracking is robust with plant and compensator variations as long as the system remains stable and the dc gains of $A(s)$ and $M(s)$ remain equal.

10.19. a. It is known that two polynomials $D(s)$ and $N(s)$ are coprime if and only if there exist two polynomials $\overline{A}(s)$ and $\overline{B}(s)$ such that

$$\overline{A}(s)D(s) + \overline{B}(s)N(s) = 1$$

Show that if $D(s) = s^2 - 1$ and $N(s) = s - 2$, then $\overline{A}(s) = 1/3$ and $\overline{B}(s) = -(s + 2)/3$ meet the condition.

b. For any two polynomials $D(s)$ and $N(s)$, there exist two polynomials $\hat{A}(s)$ and $\hat{B}(s)$ such that

$$\hat{A}(s)D(s) + \hat{B}(s)N(s) = 0$$

Find two such polynomials for $D(s)$ and $N(s)$ in (a).

c. Consider the Diophantine equation

$$A(s)D(s) + B(s)N(s) = D_o(s)$$

where $D(s)$ and $N(s)$ are coprime. Show that, for any polynomial $Q(s)$,

$$A(s) = \bar{A}(s)D_o(s) + Q(s)\hat{A}(s) \qquad B(s) = \bar{B}(s)D_o(s) + Q(s)\hat{B}(s)$$

is a general solution of the Diophantine equation. For $D(s)$ and $N(s)$ in (a) and $D_o(s) = s^3 + 7s^2 + 17s + 15$, show that the following set of two polynomials

$$A(s) = \frac{1}{3}(s^3 + 7s^2 + 17s + 15) + Q(s)(-s + 2)$$

$$B(s) = -\frac{1}{3}(s + 2)(s^3 + 7s^2 + 17s + 15) + Q(s)(s^2 - 1)$$

is a general solution.

d. Show that if $Q(s) = (s^2 + 9s + 32)/3$, then the degrees of $A(s)$ and $B(s)$ are smallest possible. Compare the result with the one in Example 10.3.2. Which procedure is simpler? (It is true that many engineering problems can be solved by applying existing mathematical results. However, emphasis in mathematics is often different from that in engineering. Mathematicians are interested in existence conditions of solutions and general forms of solutions. Engineers are interested in particular solutions and in efficient methods of solving them. This problem illustrates well their differences in emphases.)

11 State Space Design

11.1 INTRODUCTION

In this chapter we discuss the design of control systems using state-variable equations. We use simple networks to introduce the concepts of controllability and observability, and we develop their conditions intuitively. We then discuss equivalence transformations. Using an equivalence transformation, we show how to design pole placement using state feedback under the assumption of controllability. Because the concept and condition of controllability are dual to those of observability, design of full-dimensional state estimators can then be established. We also discuss the design of reduced-dimensional state estimators by solving Lyapunov equations. The connection of state feedback to the output of state estimators is justified by establishing the separation property. This design is also compared with the transfer function approach discussed in Section 10.6. Finally, we introduce the Lyapunov stability theorem and then apply it to establish the Routh stability test.

The state in state-variable equations forms a linear space, called *state space*; therefore, the design using state-variable equations is also called *state space design*.

11.2 CONTROLLABILITY AND OBSERVABILITY

Consider the n-dimensional state-variable equation

$$\dot{\mathbf{x}} = \mathbf{A}\mathbf{x} + \mathbf{b}u \qquad (11.1a)$$

$$y = \mathbf{c}\mathbf{x} + du \qquad (11.1b)$$

where **A**, **b**, and **c** are respectively $n \times n$, $n \times 1$, and $1 \times n$ matrices. The constant d is scalar. Equation (11.1) is said to be *controllable* if we can transfer any state to any other state in a finite time by applying an input. The equation is said to be *observable* if we can determine the initial state from the knowledge of the input and output over a finite time interval. We first use examples to illustrate these concepts.

Example 11.2.1

Consider the network shown in Figure 11.1(a). The input is a current source, and the output is the voltage across the 2-Ω resistor shown. The voltage x across the capacitor is the only state variable of the network. If $x(0) = 0$, no matter what input is applied, because of the symmetry of the four resistors, the voltage across the capacitor is always zero. Therefore, it is not possible to transfer $x(0) = 0$ to any nonzero x. Thus, the state-variable equation describing the network is not controllable. If $x(0)$ is different from zero, then it will generate a response across the output y. Thus, it is possible that the equation is observable, as will be established later.

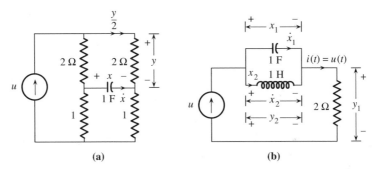

(a)   (b)

Figure 11.1 (a) An uncontrollable network. (b) An unobservable network.

Example 11.2.2

Consider the network shown in Figure 11.1(b). The input is a current source and the output y_1 is the voltage across the resistor. The network has two state variables: the voltage x_1 across the capacitor, and the current x_2 through the inductor. Nonzero $x_1(0)$ and/or $x_2(0)$ will excite a response inside the LC loop. However, the current $i(t)$ always equals $u(t)$ and the output always equals $2u(t)$ no matter what $x_1(0)$ and $x_2(0)$ are. Therefore, there is no way to determine the initial state from $u(t)$ and $y_1(t)$. Thus, the state-variable equation describing the network will not be observable.

Because the LC loop is connected directly to the input, it is possible that the equation is controllable, as will be shown later. We mention that controllability and observability depend on what are considered as input and output. If y_2 in Figure 11.1(b) is considered as the output, then the equation will be observable.

We develop in the following a condition for (11.1) to be controllable. The purpose is to show how the condition arises; therefore, technical details are skipped. The solution of (11.1a) due to any $x(0)$ and $u(t)$ is, as derived in (2.72),

$$x(t) = e^{At}x(0) + \int_0^t e^{A(t-\tau)}bu(\tau)d\tau$$

which can be written as, using (2.67),

$$x(t) - e^{At}x(0) = \int_0^t [I + A(t - \tau) + A^2\frac{(t - \tau)^2}{2 \cdot} + \cdots]bu(\tau)d\tau$$

$$= b\int_0^t u(\tau)d\tau + Ab\int_0^t (t - \tau)u(\tau)d\tau$$

$$+ A^2b\int_0^t 0.5(t - \tau)^2u(\tau)d\tau + \cdots$$

$$= [b \quad Ab \quad A^2b \quad \cdots] \begin{bmatrix} \int_0^t u(\tau)d\tau \\ \int_0^t (t - \tau)u(\tau)d\tau \\ \int_0^t 0.5(t - \tau)^2u(\tau)d\tau \\ \vdots \end{bmatrix}$$

Using Theorem B.1, we conclude that for any $x(0)$ and $x(t)$, a solution $u(t)$ exists if and only if the matrix

$$[b \quad Ab \quad A^2b \quad \cdots \quad A^{n-1}b \quad \cdots] \tag{11.2}$$

has rank n. The matrix has n rows but infinitely many columns. Fortunately, using the Cayley–Hamilton theorem (see Problem 2.37), we can show that (11.2) has rank n if and only if the $n \times n$ matrix

$$U = [b \quad Ab \quad A^2b \quad \cdots \quad A^{n-1}b] \tag{11.3}$$

has rank n. Thus we conclude that (11.1) is controllable if and only if the matrix in (11.3) has rank n or, equivalently, its determinant is different from zero. The matrix in (11.3) is called the *controllability matrix*. The first column is b, the second column is Ab, and the last column is $A^{n-1}b$. Similarly, we define the $n \times n$ matrix

$$V = \begin{bmatrix} c \\ cA \\ \vdots \\ cA^{n-1} \end{bmatrix} \tag{11.4}$$

Its first row is c, second row is cA, and the last row is cA^{n-1}. Then the n-dimensional equation in (11.1) is observable if and only if the matrix V has rank n or, equivalently, its determinant is nonzero. We call $\mathbf{V}$ the *observability matrix*. Controllability is a property of $(\mathbf{A}, \mathbf{b})$; observability is a property of $(\mathbf{A}, \mathbf{c})$. They are independent of the direct transmission part d. We use examples to illustrate how to use these conditions.

Example 11.2.3

Consider

$$G(s) = \frac{N(s)}{D(s)} = \frac{2s - 1}{s^2 - 1.5s - 1} = \frac{2s - 1}{(s - 2)(s + 0.5)} \tag{11.5}$$

Its controllable-form realization, as developed in (5.17), is

$$\dot{\mathbf{x}} = \begin{bmatrix} 1.5 & 1 \\ 1 & 0 \end{bmatrix} \mathbf{x} + \begin{bmatrix} 1 \\ 0 \end{bmatrix} u \tag{11.6a}$$

$$y = [2 \quad -1]\mathbf{x} \tag{11.6b}$$

Because the dimension of (11.6) equals the number of poles of $G(s)$ in (11.5), (11.6) is a minimal state-variable equation (see Section 2.8). Every minimal equation is controllable and observable. We demonstrate this fact for the equation in (11.6).

To check controllability, we first compute

$$\mathbf{Ab} = \begin{bmatrix} 1.5 & 1 \\ 1 & 0 \end{bmatrix} \begin{bmatrix} 1 \\ 0 \end{bmatrix} = \begin{bmatrix} 1.5 \\ 1 \end{bmatrix} \tag{11.7}$$

Thus the controllability matrix of (11.6) is

$$\mathbf{U} = [\mathbf{b} \quad \mathbf{Ab}] = \begin{bmatrix} 1 & 1.5 \\ 0 & 1 \end{bmatrix} \tag{11.8}$$

Its determinant is 1 (see (B.2)). Thus (11.6) is controllable.

To check observability, we first compute

$$\mathbf{cA} = [2 \quad -1]\begin{bmatrix} 1.5 & 1 \\ 1 & 0 \end{bmatrix} = [2 \quad 2] \tag{11.9}$$

Thus the observability matrix of (11.6) is

$$\mathbf{V} = \begin{bmatrix} \mathbf{c} \\ \mathbf{cA} \end{bmatrix} = \begin{bmatrix} 2 & -1 \\ 2 & 2 \end{bmatrix} \tag{11.10}$$

Its determinant is $4 + 2 = 6$. Thus the equation is observable.

Example 11.2.4

Consider the transfer function

$$G(s) = \frac{2s + 1}{s^2 - 1.5s - 1} = \frac{2s + 1}{(s - 2)(s + 0.5)} \tag{11.11}$$

Its controllable-form realization is

$$\dot{\mathbf{x}} = \begin{bmatrix} 1.5 & 1 \\ 1 & 0 \end{bmatrix} \mathbf{x} + \begin{bmatrix} 1 \\ 0 \end{bmatrix} u \tag{11.12a}$$

$$y = [2 \quad 1]\mathbf{x} \tag{11.12b}$$

This equation differs from (11.6) only in the output equation; therefore, its controllability matrix is the same as (11.8), and (11.12) is controllable. To check observability, we compute

$$\mathbf{cA} = [2 \quad 1]\begin{bmatrix} 1.5 & 1 \\ 1 & 0 \end{bmatrix} = [4 \quad 2]$$

Thus the observability matrix of (11.12) is

$$\mathbf{V} = \begin{bmatrix} \mathbf{c} \\ \mathbf{cA} \end{bmatrix} = \begin{bmatrix} 2 & 1 \\ 4 & 2 \end{bmatrix} \tag{11.13}$$

Its determinant is $4 - 4 = 0$. Thus, (11.12) is not observable. Note that the number of poles of $G(s)$ in (11.11) is 1 (why?), whereas the dimension of (11.12) is 2. Thus (11.12) is not a minimal equation. The equation is controllable but not observable.

The observable-form realization of $G(s)$ in (11.11) is

$$\dot{\mathbf{x}} = \begin{bmatrix} 1.5 & 1 \\ 1 & 0 \end{bmatrix} \mathbf{x} + \begin{bmatrix} 2 \\ 1 \end{bmatrix} u \tag{11.14a}$$

$$y = [1 \quad 0]\mathbf{x} \tag{11.14b}$$

Its controllability and observability matrices are

$$\mathbf{U} = \begin{bmatrix} 2 & 4 \\ 1 & 2 \end{bmatrix} \qquad \mathbf{V} = \begin{bmatrix} 1 & 0 \\ 1.5 & 1 \end{bmatrix}$$

Because det $\mathbf{U} = 0$ and det $\mathbf{V} = 1$, where det stands for the determinant, the equation is observable but not controllable.

The preceding discussion can be extended to the general case. Consider

$$G(s) = \frac{N(s)}{D(s)} = \frac{b_1 s^3 + b_2 s^2 + b_3 s + b_4}{s^4 + a_1 s^3 + a_2 s^2 + a_3 s + a_4} \tag{11.15}$$

Note that the assignment of coefficients in (11.15) is different from the one in Section

5.5.1. This is done for convenience in developing the design procedure in Section 11.4. The controllable-form realization of (11.15) is, as shown in (5.17),

$$\dot{\mathbf{x}} = \begin{bmatrix} -a_1 & -a_2 & -a_3 & -a_4 \\ 1 & 0 & 0 & 0 \\ 0 & 1 & 0 & 0 \\ 0 & 0 & 1 & 0 \end{bmatrix} \mathbf{x} + \begin{bmatrix} 1 \\ 0 \\ 0 \\ 0 \end{bmatrix} u \qquad (11.16a)$$

$$y = [b_1 \quad b_2 \quad b_3 \quad b_4]\mathbf{x} \qquad (11.16b)$$

Its controllability matrix can be computed as

$$\mathbf{U} = \begin{bmatrix} 1 & -a_1 & e_2 & e_3 \\ 0 & 1 & -a_1 & e_2 \\ 0 & 0 & 1 & -a_1 \\ 0 & 0 & 0 & 1 \end{bmatrix}$$

with $e_2 = -a_2 + a_1^2$ and $e_3 = -a_3 + 2a_1a_2 - a_1^3$. It is a triangular matrix; its determinant is always 1 no matter what the a_i are (see (B.2)). Therefore, it is always controllable, and this is the reason for calling it *controllable-form*. To check whether (11.16) is observable, we must first compute the observability matrix $\mathbf{V}$ in (11.4) and then check its rank. An alternative method is to check whether $N(s)$ and $D(s)$ have common factors or not. If they have common factors, then the equation cannot be observable. This is the case in Example 11.2.4. If $N(s)$ and $D(s)$ have no common factors, then (11.16) will be observable as well. Similarly, the observable-form realization of (11.15) is, as discussed in (5.18),

$$\dot{x} = \begin{bmatrix} -a_1 & 1 & 0 & 0 \\ -a_2 & 0 & 1 & 0 \\ -a_3 & 0 & 0 & 1 \\ -a_4 & 0 & 0 & 0 \end{bmatrix} \mathbf{x} + \begin{bmatrix} b_1 \\ b_2 \\ b_3 \\ b_4 \end{bmatrix} u \qquad (11.17a)$$

$$y = [1 \quad 0 \quad 0 \quad 0]\mathbf{x} \qquad (11.17b)$$

This equation is always observable. It is controllable if $N(s)$ and $D(s)$ have no common factors; otherwise, it is not controllable.

Exercise 11.2.1

Show that the network in Figure 11.1(a) can be described by the following state-variable equation

$$\dot{x} = -0.75x + 0 \cdot u \qquad y = 0.5x + u$$

Is the equation controllable? observable?

[**Answers:** No, yes.]

Exercise 11.2.2

Show that the network in Figure 11.1(b) with y_1 as the output can be described by

$$\dot{\mathbf{x}} = \begin{bmatrix} 0 & -1 \\ 1 & 0 \end{bmatrix} \mathbf{x} + \begin{bmatrix} 1 \\ 0 \end{bmatrix} u$$

$$y_1 = [0 \quad 0]\mathbf{x} + 2u$$

Is the equation controllable? observable?

[**Answers:**   Yes, no.]

Exercise 11.2.3

Show that the network in Figure 11.1(b) with y_2 as the output can be described by

$$\dot{\mathbf{x}} = \begin{bmatrix} 0 & -1 \\ 1 & 0 \end{bmatrix} \mathbf{x} + \begin{bmatrix} 1 \\ 0 \end{bmatrix} u$$

$$y_2 = [1 \quad 0]\mathbf{x}$$

Is the equation controllable? observable?

[**Answers:**   Yes, yes.]

In addition to (11.3) and (11.4), there are many other controllability and observability conditions. Although (11.3) and (11.4) are most often cited conditions in control texts, they are very sensitive to parameter variations, and consequently not suitable for computer computation. For a discussion of this problem, see Reference [15, p. 217]. If state-variable equations are in diagonal or, more generally, Jordan form, then controllability can be determined from **b** alone and observability, from **c** alone. See Problems 11.2, 11.3, and Reference [15, p. 209].

11.2.1   Pole-Zero Cancellations

Controllability and observability are closely related to pole-zero cancellations in block diagrams. This is illustrated by an example.

Example 11.2.5

Consider the block diagram in Figure 11.2(a) with input u and output y. It consists of two blocks with transfer functions $1/(s-1)$ and $(s-1)/(s+1)$. If we assign the output of the first block as x_1, then we have

$$\dot{x}_1 = x_1 + u$$

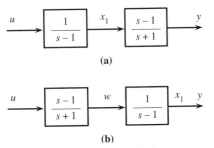

Figure 11.2 Pole-zero cancellations.

See Section 5.5.2. The second block has input x_1 and output y, thus we have

$$\frac{Y(s)}{X_1(s)} = \frac{s - 1}{s + 1} = 1 + \frac{-2}{s + 1}$$

Therefore, it can be realized as

$$\dot{x}_2 = -x_2 - 2x_1 \qquad y = x_2 + x_1$$

Thus, the tandem connection of the two blocks can be described by

$$\begin{bmatrix} \dot{x}_1 \\ \dot{x}_2 \end{bmatrix} = \begin{bmatrix} 1 & 0 \\ -2 & -1 \end{bmatrix} \begin{bmatrix} x_1 \\ x_2 \end{bmatrix} + \begin{bmatrix} 1 \\ 0 \end{bmatrix} u$$

$$y_2 = \begin{bmatrix} 1 & 1 \end{bmatrix} \mathbf{x}$$

Because its controllability matrix

$$\mathbf{U} = \begin{bmatrix} \mathbf{b} & \mathbf{Ab} \end{bmatrix} = \begin{bmatrix} 1 & 1 \\ 0 & -2 \end{bmatrix}$$

is nonsingular, the equation is controllable. The observability matrix is

$$\mathbf{V} = \begin{bmatrix} \mathbf{c} \\ \mathbf{cA} \end{bmatrix} = \begin{bmatrix} 1 & 1 \\ -1 & -1 \end{bmatrix}$$

which is singular. Thus, the equation is not observable. The equation has two state variables, but its overall transfer function

$$G_o(s) = \frac{1}{s - 1} \cdot \frac{s - 1}{s + 1} = \frac{1}{s + 1}$$

has only one pole. Therefore, the equation is not minimal and cannot be both controllable and observable. Similarly, we can show that the equation describing Figure 11.2(b) is observable but not controllable. In general, if there are pole-zero cancellations in tandem connection, the state-variable equation describing the connection cannot be both controllable and observable. If poles of the input block are canceled by zeros of the output block, the equation cannot be observable; if poles of the output block are canceled by zeros of the input block, the equation cannot be controllable.

11.3 EQUIVALENT STATE-VARIABLE EQUATIONS

We first use an example to illustrate the concept of equivalent state-variable equations. Consider the network shown in Figure 11.3. First we choose the inductor current x_1 and the capacitor voltage x_2 as state variables, as shown in Figure 11.3(a). Then the voltage across the inductor is $2\dot{x}_1$ and the current through the capacitor is $\dot{x}_2$. Because the voltage across the resistor is x_2, the current through it is $x_2/2 = 0.5x_2$. From the outer loop, we have

$$2\dot{x}_1 + x_2 - u = 0 \qquad \text{or} \qquad \dot{x}_1 = -0.5x_2 + 0.5u$$

From the node denoted by A, we have

$$\dot{x}_2 = x_1 - 0.5x_2$$

Thus, the network can be described by the following state-variable equation

$$\begin{bmatrix} \dot{x}_1 \\ \dot{x}_2 \end{bmatrix} = \begin{bmatrix} 0 & -0.5 \\ 1 & -0.5 \end{bmatrix} \mathbf{x} + \begin{bmatrix} 0.5 \\ 0 \end{bmatrix} u \tag{11.18a}$$

$$y = [0 \quad 1]\mathbf{x} \tag{11.18b}$$

For the same network, we now choose the two loop currents $\bar{x}_1$ and $\bar{x}_2$ shown in Figure 11.3(b) as state variables. Then the voltages across the inductor, resistor, and capacitor are respectively $2\dot{\bar{x}}_1$, $2(\bar{x}_1 - \bar{x}_2)$ and $\int_0^t \bar{x}_2(\tau)d\tau$. From the left-hand side loop, we have

$$u = 2\dot{\bar{x}}_1 + 2(\bar{x}_1 - \bar{x}_2)$$

which implies

$$\dot{\bar{x}}_1 = -\bar{x}_1 + \bar{x}_2 + 0.5u$$

From the right-hand side loop, we have

$$\int_0^t \bar{x}_2(\tau)d\tau = 2(\bar{x}_1 - \bar{x}_2) = y$$

which implies

$$\bar{x}_2 = 2(\dot{\bar{x}}_1 - \dot{\bar{x}}_2)$$

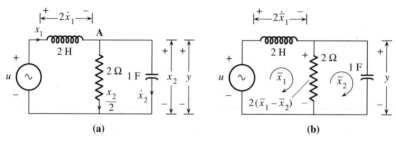

(a)   (b)

Figure 11.3 Network with different choices of state variables.

This becomes, after the substitution for $\dot{\bar{x}}_1$,

$$\dot{\bar{x}}_2 = -0.5\bar{x}_2 + (-\bar{x}_1 + \bar{x}_2 + 0.5u) = -\bar{x}_1 + 0.5\bar{x}_2 + 0.5u$$

Thus, the network in Figure 11.3 can also be described by the following state-variable equation

$$\begin{bmatrix} \dot{\bar{x}}_1 \\ \dot{\bar{x}}_2 \end{bmatrix} = \begin{bmatrix} -1 & 1 \\ -1 & 0.5 \end{bmatrix} \bar{\mathbf{x}} + \begin{bmatrix} 0.5 \\ 0.5 \end{bmatrix} u \tag{11.19a}$$

$$y = [2 \quad -2]\mathbf{x} \tag{11.19b}$$

Both (11.18) and (11.19) describe the same network; therefore, they must be related in some way. Indeed, from the currents through the inductor and resistor in Figure 11.3(a) and (b), we have $x_1 = \bar{x}_1$ and $x_2/2 = \bar{x}_1 - \bar{x}_2$. Thus we have

$$\mathbf{x} := \begin{bmatrix} x_1 \\ x_2 \end{bmatrix} = \begin{bmatrix} 1 & 0 \\ 2 & -2 \end{bmatrix} \begin{bmatrix} \bar{x}_1 \\ \bar{x}_2 \end{bmatrix} =: \begin{bmatrix} 1 & 0 \\ 2 & -2 \end{bmatrix} \bar{\mathbf{x}} \tag{11.20}$$

The square matrix in (11.20) is nonsingular because its determinant is -2. Two state-variable equations are said to be *equivalent* if their states can be related by a nonsingular matrix such as in (11.20). Thus (11.18) and (11.19) are equivalent.

Consider the n-dimensional state-variable equation

$$\dot{\mathbf{x}} = \mathbf{A}\mathbf{x} + \mathbf{b}u \tag{11.21a}$$

$$y = \mathbf{c}\mathbf{x} + du \tag{11.21b}$$

Let $\mathbf{P}$ be an arbitrary nonsingular matrix. Define $\bar{\mathbf{x}} = \mathbf{P}\mathbf{x}$. The substitution of $\mathbf{x} = \mathbf{P}^{-1}\bar{\mathbf{x}}$ and $\dot{\mathbf{x}} = \mathbf{P}^{-1}\dot{\bar{\mathbf{x}}}$ into (11.21) yields

$$\mathbf{P}^{-1}\dot{\bar{\mathbf{x}}} = \mathbf{A}\mathbf{P}^{-1}\bar{\mathbf{x}} + \mathbf{b}u$$

$$y = \mathbf{c}\mathbf{P}^{-1}\bar{\mathbf{x}} + du$$

which become

$$\dot{\bar{\mathbf{x}}} = \bar{\mathbf{A}}\bar{\mathbf{x}} + \bar{\mathbf{b}}u \tag{11.22a}$$

$$y = \bar{\mathbf{c}}\bar{\mathbf{x}} + \bar{d}u \tag{11.22b}$$

with

$$\bar{\mathbf{A}} := \mathbf{P}\mathbf{A}\mathbf{P}^{-1} \quad \bar{\mathbf{b}} := \mathbf{P}\mathbf{b} \quad \bar{\mathbf{c}} := \mathbf{c}\mathbf{P}^{-1} \quad \bar{d} := d \tag{11.22c}$$

Equations (11.21) and (11.22) are said to be *equivalent*, and $\mathbf{P}$ is called an *equivalence transformation*. For (11.18) and (11.19), we have, from (11.20),

$$\mathbf{P}^{-1} = \begin{bmatrix} 1 & 0 \\ 2 & -2 \end{bmatrix} \quad \text{and} \quad \mathbf{P} = \begin{bmatrix} 1 & 0 \\ 2 & -2 \end{bmatrix}^{-1} = \begin{bmatrix} 1 & 0 \\ 1 & -0.5 \end{bmatrix}$$

and

$$\mathbf{PAP}^{-1} = \begin{bmatrix} 1 & 0 \\ 1 & -0.5 \end{bmatrix} \begin{bmatrix} 0 & -0.5 \\ 1 & -0.5 \end{bmatrix} \begin{bmatrix} 1 & 0 \\ 2 & -2 \end{bmatrix} = \begin{bmatrix} -1 & 1 \\ -1 & 0.5 \end{bmatrix} = \overline{\mathbf{A}}$$

$$\mathbf{Pb} = \begin{bmatrix} 1 & 0 \\ 1 & -0.5 \end{bmatrix} \begin{bmatrix} 0.5 \\ 0 \end{bmatrix} = \begin{bmatrix} 0.5 \\ 0.5 \end{bmatrix} = \overline{\mathbf{b}}$$

$$\mathbf{cP}^{-1} = [0 \quad 1] \begin{bmatrix} 1 & 0 \\ 2 & -2 \end{bmatrix} = [2 \quad -2] = \overline{\mathbf{c}}$$

Thus (11.18) and (11.19) are indeed related by (11.22c).

The transformation in $\overline{\mathbf{A}} = \mathbf{PAP}^{-1}$ is called a *similarity* transformation. Any similarity transformation will not change the eigenvalues of a matrix. Indeed, using (B.3), we have

$$\det \mathbf{P} \det \mathbf{P}^{-1} = \det (\mathbf{PP}^{-1}) = \det \mathbf{I} = 1$$

and

$$\begin{aligned}
\det (s\mathbf{I} - \overline{\mathbf{A}}) &= \det (s\mathbf{PP}^{-1} - \mathbf{PAP}^{-1}) = \det [\mathbf{P}(s\mathbf{I} - \mathbf{A})\mathbf{P}^{-1}] \\
&= \det \mathbf{P} \det (s\mathbf{I} - \mathbf{A}) \det \mathbf{P}^{-1} = \det (s\mathbf{I} - \mathbf{A})
\end{aligned} \tag{11.23}$$

Thus $\mathbf{A}$ and $\overline{\mathbf{A}}$ have the same characteristic polynomial and, consequently, the same eigenvalues. Equivalent state-variable equations also have the same transfer function. Indeed, the transfer function of (11.22) is

$$\begin{aligned}
\overline{G}(s) &= \overline{c}(s\mathbf{I} - \overline{\mathbf{A}})^{-1}\overline{\mathbf{b}} = \mathbf{cP}^{-1}[\mathbf{P}(s\mathbf{I} - \mathbf{A})\mathbf{P}^{-1}]^{-1}\mathbf{Pb} \\
&= \mathbf{cP}^{-1}\mathbf{P}(s\mathbf{I} - \mathbf{A})^{-1}\mathbf{P}^{-1}\mathbf{Pb} = \mathbf{c}(s\mathbf{I} - \mathbf{A})^{-1}\mathbf{b} = G(s)
\end{aligned}$$

Thus any equivalence transformation will not change the transfer function. The properties of controllability and observability are also preserved. Indeed, using

$$\overline{\mathbf{A}}^2 = \overline{\mathbf{A}}\,\overline{\mathbf{A}} = \mathbf{PAP}^{-1}\mathbf{PAP}^{-1} = \mathbf{PA}^2\mathbf{P}^{-1}$$

and, in general, $\overline{\mathbf{A}}^n = \mathbf{PA}^n\mathbf{P}^{-1}$, we have

$$\begin{aligned}
\overline{\mathbf{U}} &:= [\overline{\mathbf{b}} \quad \overline{\mathbf{A}}\overline{\mathbf{b}} \quad \overline{\mathbf{A}}^2\overline{\mathbf{b}} \cdots \overline{\mathbf{A}}^{n-1}\overline{\mathbf{b}}] \\
&= [\mathbf{Pb} \quad \mathbf{PAP}^{-1}\mathbf{Pb} \quad \mathbf{PA}^2\mathbf{P}^{-1}\mathbf{Pb} \cdots \mathbf{PA}^{n-1}\mathbf{P}^{-1}\mathbf{Pb}] \\
&= \mathbf{P}[\mathbf{A} \quad \mathbf{Ab} \quad \mathbf{A}^2\mathbf{b} \cdots \mathbf{A}^{n-1}\mathbf{b}] = \mathbf{PU}
\end{aligned} \tag{11.24}$$

Because $\mathbf{P}$ is nonsingular, the rank of $\overline{\mathbf{U}}$ equals the rank of $\mathbf{U}$. Thus, (11.22) is controllable if and only if (11.21) is controllable. This shows that the property of controllability is invariant under any equivalence transformation. The observability part can be similarly established.

11.4   POLE PLACEMENT

In this section, we discuss the design of pole placement by using state feedback. We first use an example to illustrate the basic idea.

Example 11.4.1

Consider a plant with transfer function

$$G(s) = \frac{10}{s(s + 1)}$$

The plant could be an armature-controlled dc motor driving an antenna and can be represented by the block diagram enclosed by the dotted line in Figure 11.4(a). If we assign the output of each block as a state variable as shown, then we can readily obtain

$$\dot{x}_1 = x_2$$

and

$$\dot{x}_2 = -x_2 + 10u$$

(see Section 5.5.2). Thus, the plant can be described by the following state-variable equation

$$\dot{\mathbf{x}} = \begin{bmatrix} \dot{x}_1 \\ \dot{x}_2 \end{bmatrix} = \begin{bmatrix} 0 & 1 \\ 0 & -1 \end{bmatrix} \mathbf{x} + \begin{bmatrix} 0 \\ 10 \end{bmatrix} u \qquad (11.25a)$$

$$y = [1 \quad 0]\mathbf{x} \qquad (11.25b)$$

Now we introduce feedback from x_1 and x_2 as shown in Figure 11.4(a) with real constant gains k_1 and k_2. This is called *state feedback*. With the feedback, the transfer function from r to y becomes, using Mason's formula,

$$G_o(s) = \frac{\dfrac{10}{s(s + 1)}}{1 + \dfrac{10k_2}{s + 1} + \dfrac{10k_1}{s(s + 1)}} = \frac{10}{s(s + 1) + 10k_2 s + 10k_1}$$

$$= \frac{10}{s^2 + (1 + 10k_2)s + 10k_1} \qquad (11.26)$$

We see that by choosing k_1 and k_2, the poles of $G_o(s)$ can be arbitrarily assigned

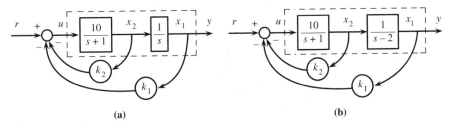

(a)   (b)

Figure 11.4 Block diagrams.

provided complex-conjugate poles are assigned in pairs. For example, if we assign the poles of $G_o(s)$ as $-2 \pm j2$, then k_1 and k_2 can be determined from

$$(s + 2 + j2)(s + 2 - j2) = s^2 + 4s + 8$$
$$= s^2 + (1 + 10k_2)s + 10k_1 \qquad (11.27)$$

or

$$1 + 10k_2 = 4 \qquad 10k_1 = 8$$

as $k_1 = 0.8$ and $k_2 = 0.3$. This shows that by state feedback, the poles of the resulting system can be arbitrarily assigned.

Because the feedback paths consist of only gains (transfer functions with degree 0), the number of the poles of the resulting system remains the same as the original plant. Thus, the introduction of constant-gain state feedback does not increase the number of poles of the system, it merely *shifts* the poles of the plant to new positions. Note that the numerator of $G_o(s)$ equals that of $G(s)$. Thus, the state feedback does not affect the numerator or the zeros of $G(s)$. This is a general property and will be established later.

Exercise 11.4.1

Consider a plant with transfer function $G(s) = 10/(s + 1)(s - 2)$ as shown in Figure 11.4(b). Find two feedback gains k_1 and k_2 such that the poles of the resulting system are located at $-2 \pm j2$.

[**Answer:** $k_1 = 2$, $k_2 = 0.5$.]

Although Example 11.4.1 illustrates the basic idea of pole placement by state feedback, the procedure cannot easily be extended to general $G(s) = N(s)/D(s)$. In the following, we use state-variable equations to discuss the design. Consider the n-dimensional state-variable equation

$$\dot{\mathbf{x}} = \mathbf{A}\mathbf{x} + \mathbf{b}u \qquad (11.28a)$$

$$y = \mathbf{c}\mathbf{x} \qquad (11.28b)$$

with transfer function

$$G(s) = \mathbf{c}(s\mathbf{I} - \mathbf{A})^{-1}\mathbf{b} \qquad (11.28c)$$

To simplify discussion, we have assumed the direct transmission part to be zero. The equation is plotted in Figure 11.5, where single lines denote single signals and double lines denote two or more signals. Let

$$u(t) = r(t) - \mathbf{k}\mathbf{x}(t) \qquad (11.29)$$

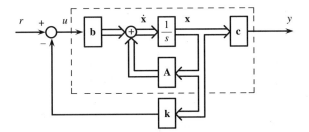

Figure 11.5 State feedback.

where $\mathbf{k} = [k_1 \quad k_2 \quad \cdots \quad k_n]$ is a $1 \times n$ real constant vector and $r(t)$ is the scalar reference signal. Equation (11.29) is called *constant-gain state feedback*, or simply *state feedback*, and is plotted in Figure 11.5. The substitution of (11.29) into (11.28) yields

$$\dot{\mathbf{x}} = \mathbf{Ax} - \mathbf{bkx} + \mathbf{b}r = (\mathbf{A} - \mathbf{bk})\mathbf{x} + \mathbf{b}r \qquad (11.30a)$$

$$y = \mathbf{cx} \qquad (11.30b)$$

with transfer function

$$G_o(s) = \mathbf{c}(s\mathbf{I} - \mathbf{A} + \mathbf{bk})^{-1}\mathbf{b} \qquad (11.30c)$$

We show that if $(\mathbf{A}, \mathbf{b})$ is controllable, then the eigenvalues of $(\mathbf{A} - \mathbf{bk})$ can be arbitrarily assigned by choosing a real constant feedback gain $\mathbf{k}$. Because $\mathbf{A}$ and $\mathbf{b}$ are implicitly assumed to be real matrices and $\mathbf{k}$ is required to be real, if a complex number is assigned as an eigenvalue of $(\mathbf{A} - \mathbf{bk})$, its complex conjugate must also be assigned. To simplify discussion, we assume that (11.28) is of dimension 4 and its characteristic polynomial and transfer function from u to y are

$$\Delta(s) = s^4 + a_1 s^3 + a_2 s^2 + a_3 s + a_4 \qquad (11.31)$$

and

$$G(s) = \frac{N(s)}{D(s)} = \frac{b_1 s^3 + b_2 s^2 + b_3 s + b_4}{s^4 + a_1 s^3 + a_2 s^2 + a_3 s + a_4} \qquad (11.32)$$

We show in the following that if (11.28) is controllable, then it can be transformed by an equivalence transformation into

$$\dot{\bar{\mathbf{x}}} = \bar{\mathbf{A}}\bar{\mathbf{x}} + \bar{\mathbf{b}}u = \begin{bmatrix} -a_1 & -a_2 & -a_3 & -a_4 \\ 1 & 0 & 0 & 0 \\ 0 & 1 & 0 & 0 \\ 0 & 0 & 1 & 0 \end{bmatrix} \bar{\mathbf{x}} + \begin{bmatrix} 1 \\ 0 \\ 0 \\ 0 \end{bmatrix} u \qquad (11.33a)$$

$$y = \bar{\mathbf{c}}\bar{\mathbf{x}} = [b_1 \quad b_2 \quad b_3 \quad b_4]\bar{\mathbf{x}} \qquad (11.33b)$$

The controllability matrix of (11.33) can be computed as

$$
\overline{\mathbf{U}} = \begin{bmatrix} 1 & -a_1 & e_2 & e_3 \\ 0 & 1 & -a_1 & e_2 \\ 0 & 0 & 1 & -a_1 \\ 0 & 0 & 0 & 1 \end{bmatrix}
\tag{11.34a}
$$

with $e_2 = -a_2 + a_1^2$ and $e_3 = -a_3 + 2a_1a_2 - a_1^3$. It is triangular and its inverse is also triangular and equals

$$
\overline{\mathbf{U}}^{-1} = \begin{bmatrix} 1 & a_1 & a_2 & a_3 \\ 0 & 1 & a_1 & a_2 \\ 0 & 0 & 1 & a_1 \\ 0 & 0 & 0 & 1 \end{bmatrix}
\tag{11.34b}
$$

The easiest way to verify this is to show $\overline{\mathbf{U}}^{-1}\overline{\mathbf{U}} = \mathbf{I}$. Now, as developed in (11.24), the controllability matrices of (11.28) and (11.33) are related by

$$
\overline{\mathbf{U}} = \mathbf{P}\mathbf{U}
\tag{11.35}
$$

which implies, for $n = 4$,

$$
\mathbf{S} := \mathbf{P}^{-1} = \mathbf{U}\overline{\mathbf{U}}^{-1} = [\mathbf{b} \quad \mathbf{Ab} \quad \mathbf{A}^2\mathbf{b} \quad \mathbf{A}^3\mathbf{b}] \begin{bmatrix} 1 & a_1 & a_2 & a_3 \\ 0 & 1 & a_1 & a_2 \\ 0 & 0 & 1 & a_1 \\ 0 & 0 & 0 & 1 \end{bmatrix}
\tag{11.36}
$$

Now if (11.28) is controllable, then $\mathbf{U}$ is nonsingular and, consequently, $\mathbf{S} = \mathbf{P}^{-1}$ is nonsingular. Thus, by the equivalence transformation $\overline{\mathbf{x}} = \mathbf{P}\mathbf{x}$ or $\mathbf{x} = \mathbf{P}^{-1}\overline{\mathbf{x}} = \mathbf{S}\overline{\mathbf{x}}$, (11.28) can be transformed into (11.33).

Consider the controllable-form equation in (11.33). If we introduce the state feedback

$$
u = r - \overline{\mathbf{k}}\overline{\mathbf{x}} = r - [\overline{k}_1 \quad \overline{k}_2 \quad \overline{k}_3 \quad \overline{k}_4]\overline{\mathbf{x}}
\tag{11.37}
$$

then the equation becomes

$$
\dot{\overline{\mathbf{x}}} = (\overline{\mathbf{A}} - \overline{\mathbf{b}}\overline{\mathbf{k}}) + \overline{\mathbf{b}}r
$$

$$
= \begin{bmatrix} -a_1 - \overline{k}_1 & -a_2 - \overline{k}_2 & -a_3 - \overline{k}_3 & -a_4 - \overline{k}_4 \\ 1 & 0 & 0 & 0 \\ 0 & 1 & 0 & 0 \\ 0 & 0 & 1 & 0 \end{bmatrix} \overline{\mathbf{x}} + \begin{bmatrix} 1 \\ 0 \\ 0 \\ 0 \end{bmatrix} r
\tag{11.38a}
$$

$$
y = [b_1 \quad b_2 \quad b_3 \quad b_4]\overline{\mathbf{x}}
\tag{11.38b}
$$

We see that (11.38) is still of the controllable form, so its transfer function from r to y is

$$G_o(s) = \frac{b_1 s^3 + b_2 s^2 + b_3 s + b_4}{s^4 + (a_1 + \bar{k}_1)s^3 + (a_2 + \bar{k}_2)s^2 + (a_3 + \bar{k}_3)s + (a_4 + \bar{k}_4)} \qquad (11.39)$$

Now by choosing $\bar{k}_i$, the denominator of $G_o(s)$, and consequently the poles of $G_o(s)$, can be arbitrarily assigned. We also see that the numerator of $G_o(s)$ equals that of $G(s)$. Thus the state feedback does not affect the zeros of $G(s)$.

To relate $\bar{\mathbf{k}}$ in (11.37) with $\mathbf{k}$ in (11.29), we substitute $\bar{\mathbf{x}} = \mathbf{P}\mathbf{x}$ into (11.37) to yield

$$u = r - \bar{\mathbf{k}}\bar{x} = r - \bar{\mathbf{k}}\mathbf{P}\mathbf{x}$$

Thus we have $\mathbf{k} = \bar{\mathbf{k}}\mathbf{P}$. The preceding procedure is summarized in the following.

Procedure for Designing Pole Placement

Given a 4-dimensional controllable $(\mathbf{A}, \mathbf{b})$ and a set of eigenvalues λ_i, $i = 1, 2, 3, 4$. Find a real 1×4 vector $\mathbf{k}$ so that the matrix $(\mathbf{A} - \mathbf{b}\mathbf{k})$ has the set as its eigenvalues.

1. Compute the characteristic polynomial of $\mathbf{A}$: $\Delta(s) = \det(s\mathbf{I} - \mathbf{A}) = s^4 + a_1 s^3 + a_2 s^2 + a_3 s + a_4$.
2. Compute the desired characteristic polynomial

$$\bar{\Delta}(s) = (s - \lambda_1)(s - \lambda_2)(s - \lambda_3)(s - \lambda_4)$$
$$= s^4 + \bar{a}_1 s^3 + \bar{a}_2 s^2 + \bar{a}_3 s + \bar{a}_4$$

3. Compute the feedback gain for the equivalent controllable-form equation

$$\bar{\mathbf{k}} = [\bar{a}_1 - a_1 \quad \bar{a}_2 - a_2 \quad \bar{a}_3 - a_3 \quad \bar{a}_4 - a_4]$$

4. Compute the equivalence transformation

$$\mathbf{S} := \mathbf{P}^{-1} = [\mathbf{b} \quad \mathbf{A}\mathbf{b} \quad \mathbf{A}^2\mathbf{b} \quad \mathbf{A}^3\mathbf{b}] \begin{bmatrix} 1 & a_1 & a_2 & a_3 \\ 0 & 1 & a_1 & a_2 \\ 0 & 0 & 1 & a_1 \\ 0 & 0 & 0 & 1 \end{bmatrix} \qquad (11.40)$$

5. Compute the feedback gain $\mathbf{k} = \bar{\mathbf{k}}\mathbf{P} = \bar{\mathbf{k}}\mathbf{S}^{-1}$.

It is clear that the preceding procedure can be easily extended to the general case. There are other design procedures. For example, the following formula

$$\mathbf{k} = [0 \quad 0 \quad 0 \quad 1][\mathbf{b} \quad \mathbf{A}\mathbf{b} \quad \mathbf{A}^2\mathbf{b} \quad \mathbf{A}^3\mathbf{b}]^{-1}\bar{\Delta}(\mathbf{A})$$
$$= [0 \quad 0 \quad 0 \quad 1][\mathbf{b} \quad \mathbf{A}\mathbf{b} \quad \mathbf{A}^2\mathbf{b} \quad \mathbf{A}^3\mathbf{b}]^{-1}[\mathbf{A}^4 + \bar{a}_1\mathbf{A}^3 + \bar{a}_2\mathbf{A}^2 + \bar{a}_3\mathbf{A} + \bar{a}_4\mathbf{I}]$$

called the *Ackermann formula*, is widely quoted in control texts. Note that $\bar{\Delta}(s)$ is not the characteristic polynomial of $\mathbf{A}$ (it is the characteristic polynomial of $(\mathbf{A} - \mathbf{b}\mathbf{k})$), thus $\bar{\Delta}(\mathbf{A}) \neq 0$. The derivation of the Ackermann formula is more complex; its computation is no simpler. This is why we introduced the preceding procedure. The procedure is probably the easiest to introduce but is not necessarily the best for computer computation. See Reference [15].

Example 11.4.2

Consider the equation in (11.25) or

$$\dot{\mathbf{x}} = \begin{bmatrix} \dot{x}_1 \\ \dot{x}_2 \end{bmatrix} = \begin{bmatrix} 0 & 1 \\ 0 & -1 \end{bmatrix} \mathbf{x} + \begin{bmatrix} 0 \\ 10 \end{bmatrix} u \qquad (11.41a)$$

$$y = [1 \quad 0]\mathbf{x} \qquad (11.41b)$$

with transfer function

$$G(s) = \frac{10}{s^2 + s} \qquad (11.41c)$$

Find the feedback gain $\mathbf{k}$ in $u = r - \mathbf{kx}$ such that the resulting equation has $-2 \pm j2$ as its eigenvalues.

We compute the characteristic polynomial of $\mathbf{A}$

$$\det(s\mathbf{I} - \mathbf{A}) = \det \begin{bmatrix} s & -1 \\ 0 & s+1 \end{bmatrix} = s(s+1) = s^2 + s + 0$$

and the desired characteristic polynomial

$$(s + 2 + j2)(s + 2 - j2) = s^2 + 4s + 8$$

Thus, the feedback gain for the equivalent controllable-form equation is

$$\mathbf{k} = [\bar{a}_1 - a_1 \quad \bar{a}_2 - a_2] = [4 - 1 \quad 8 - 0] = [3 \quad 8]$$

Next we compute the equivalence transformation

$$\mathbf{S} = \mathbf{P}^{-1} = [\mathbf{b} \quad \mathbf{Ab}] \begin{bmatrix} 1 & a_1 \\ 0 & 1 \end{bmatrix} = \begin{bmatrix} 0 & 10 \\ 10 & -10 \end{bmatrix} \begin{bmatrix} 1 & 1 \\ 0 & 1 \end{bmatrix} = \begin{bmatrix} 0 & 10 \\ 10 & 0 \end{bmatrix}$$

Thus we have

$$\mathbf{k} = \bar{\mathbf{k}}\mathbf{P} = [3 \quad 8] \begin{bmatrix} 0 & 10 \\ 10 & 0 \end{bmatrix}^{-1} = [3 \quad 8] \begin{bmatrix} 0 & 0.1 \\ 0.1 & 0 \end{bmatrix} = [0.8 \quad 0.3]$$

As we expected, this result is the same as the one in Example 11.4.1. To verify the result, we compute

$$s\mathbf{I} - \mathbf{A} + \mathbf{bk} = \begin{bmatrix} s & 0 \\ 0 & s \end{bmatrix} - \begin{bmatrix} 0 & 1 \\ 0 & -1 \end{bmatrix} + \begin{bmatrix} 0 \\ 10 \end{bmatrix}[0.8 \quad 0.3] = \begin{bmatrix} s & -1 \\ 8 & s+4 \end{bmatrix}$$

Thus, the transfer function of the resulting system is

$$G_o(s) = [1 \quad 0] \begin{bmatrix} s & -1 \\ 8 & s+4 \end{bmatrix}^{-1} \begin{bmatrix} 0 \\ 10 \end{bmatrix}$$

$$= [1 \quad 0] \frac{1}{s^2 + 4s + 8} \begin{bmatrix} s+4 & 1 \\ -8 & s \end{bmatrix} \begin{bmatrix} 0 \\ 10 \end{bmatrix} = \frac{10}{s^2 + 4s + 8}$$

as expected. We see that the state feedback shifts the poles of $G(s)$ from 0 and -1 to $-2 \pm j2$. However, it has no effect on the numerator of $G(s)$.

If $(\mathbf{A}, \mathbf{b})$ is not controllable, then the matrix in (11.36) is not nonsingular and (11.28) cannot be transformed into the controllable-form equation in (11.33). In this case, it is not possible to assign all eigenvalues of $(\mathbf{A} - \mathbf{bk})$; however, it is possible to assign some of them. See Problem 11.8 and Reference [15].

To conclude this section, we mention that state feedback gain can be obtained by using MATLAB. For the problem in Example 11.4.2, we type

```
a=[0 1;0 −1];b=[0;10];
i=sqrt(−1);
p=[−2+2*i;−2−2*i];
k=place(a,b,p)
```

Then k=[0.8 0.3] will appear on the screen. The command acker(a,b,p), which uses the Ackermann formula, will also yield the same result. But the MATLAB manual states that acker(a,b,p) is not numerically reliable and starts to break down rapidly for equations with dimension larger than 10.

11.5 QUADRATIC OPTIMAL REGULATOR

Consider

$$\dot{\mathbf{x}} = \mathbf{A}\mathbf{x} + \mathbf{b}u \tag{11.42a}$$

$$y = \mathbf{c}\mathbf{x} \tag{11.42b}$$

In the preceding section, the input u was expressed as $r - \mathbf{kx}$, where r is the reference input and $\mathbf{k}$ is the feedback gain, also called the *control law*. Now we assume that the reference input r is zero and that the response of the system is excited by nonzero initial state $\mathbf{x}(0)$, which in turn may be excited by external disturbances. The problem is then to find a feedback gain to force the response to zero as quickly as possible. This is called the *regulator problem*. If $r = 0$, then (11.29) reduces to $u = -\mathbf{kx}$, and (11.42a) becomes

$$\dot{\mathbf{x}} = (\mathbf{A} - \mathbf{bk})\mathbf{x} \tag{11.43}$$

Clearly, the response of (11.43) depends on the eigenvalues of $(\mathbf{A} - \mathbf{bk})$. If all eigenvalues have negative real parts or lie inside the open left half s-plane, then for any nonzero $\mathbf{x}(0)$, the response of (11.43) approaches zero as $t \to \infty$. The larger the negative real parts, the faster the response approaches zero. Now if $(\mathbf{A}, \mathbf{b})$ is controllable, we can find a $\mathbf{k}$ to place all eigenvalues of (11.43) in any desired positions. This is one way of designing the regulator problem.

How to choose the eigenvalues of (11.43) is not simple, however. One possibility is to use the pole patterns in Figure 9.15. The radius of the semicircles in Figure 9.15 is to be determined from the constraint on the magnitude of $u(t)$. Generally, the larger the radius, the larger the feedback gain **k** and the larger the amplitude of $u(t)$. Therefore, the constraint on $u(t)$ will limit the rate for the response to approach zero. The most systematic and popular method is to find **k** to minimize the quadratic performance index

$$J = \int_0^\infty [\mathbf{x}'(t)\mathbf{Q}\mathbf{x}(t) + Ru^2(t)]dt \tag{11.44}$$

where the prime denotes the transpose, **Q** is a symmetric positive semidefinite matrix, and R is a positive constant. Before proceeding, we digress to discuss positive definite and semidefinite matrices.

A matrix **Q** is *symmetric* if its transpose equals itself, that is, $\mathbf{Q}' = \mathbf{Q}$. All eigenvalues of symmetric matrices are real, that is, no symmetric matrices can have complex eigenvalues. See Reference [15, p. 566.]. A symmetric matrix is *positive definite* if $\mathbf{x}'\mathbf{Q}\mathbf{x} > 0$ for all nonzero **x**; it is *positive semidefinite* if $\mathbf{x}'\mathbf{Q}\mathbf{x} \geq 0$ for all **x** and the equality holds for some nonzero **x**. Then we have the following theorem. See Reference [15, p. 413.].

THEOREM 11.1

A symmetric matrix **Q** of order n is positive definite (positive semidefinite) if and only if any one of the following conditions holds:

1. All n eigenvalues of **Q** are positive (zero or positive).
2. It is possible to decompose **Q** as $\mathbf{Q} = \mathbf{N}'\mathbf{N}$, where **N** is a nonsingular square matrix (where **N** is an $m \times n$ matrix with $0 < m < n$).
3. All the *leading* principal minors of **Q** are positive (all the principal minors of **Q** are zero or positive). ∎

If **Q** is symmetric and of order 3, or

$$\mathbf{Q} = \begin{bmatrix} q_{11} & q_{21} & q_{31} \\ q_{21} & q_{22} & q_{32} \\ q_{31} & q_{32} & q_{33} \end{bmatrix}$$

then the *leading* principal minors are

$$q_{11} \qquad \det \begin{bmatrix} q_{11} & q_{21} \\ q_{21} & q_{22} \end{bmatrix} \qquad \det \mathbf{Q}$$

that is, the determinants of the submatrices by deleting the last k rows and the last k columns for $k = 2, 1, 0$. The principal minors of **Q** are

$$q_{11} \qquad q_{22} \qquad q_{33} \qquad \det \begin{bmatrix} q_{11} & q_{21} \\ q_{21} & q_{22} \end{bmatrix}$$

$$\det \begin{bmatrix} q_{11} & q_{31} \\ q_{31} & q_{33} \end{bmatrix} \qquad \det \begin{bmatrix} q_{22} & q_{32} \\ q_{32} & q_{33} \end{bmatrix} \qquad \det \mathbf{Q}$$

that is, the determinants of all submatrices whose diagonal elements are also diagonal elements of **Q**. Principal minors include all leading principal minors but not conversely. To check positive definiteness, we check only the leading principal minors. To check positive semidefiniteness, however, it is not enough to check only the leading principal minors. We must check all principal minors. For example, the leading principal minors of

$$\begin{bmatrix} 1 & 0 & 1 \\ 0 & 0 & 0 \\ 1 & 0 & 0 \end{bmatrix}$$

are 1, 0, and 0, which are zero or positive, but the matrix is not positive semidefinite because one principal minor is -1 (which one?).

If **Q** is positive semidefinite and R is positive, then the two integrands in (11.44) will not cancel each other and J is a good performance criterion. The reasons for choosing the quadratic index in (11.44) are similar to those in (9.13). It yields a simple analytical solution, and if **Q** and R are chosen properly, the solution is acceptable in practice.

If **Q** is chosen as $\mathbf{c'c}$, then (11.44) becomes

$$J = \int_0^\infty [\mathbf{x}'(t)\mathbf{c'c}\mathbf{x}(t) + Ru^2(t)]dt = \int_0^\infty [y^2(t) + Ru^2(t)]dt \qquad (11.45)$$

This performance index is the same as (9.13) with $r(t) = 0$ and $R = 1/q$. Now, if the state-variable equation in (11.42) is controllable and observable, then the feedback gain that minimizes (4.45) is given by

$$\mathbf{k} = R^{-1}\mathbf{b'K} \qquad (11.46)$$

where **K** is the symmetric and positive definite matrix meeting

$$-\mathbf{KA} - \mathbf{A'K} + \mathbf{K}bR^{-1}\mathbf{b'K} - \mathbf{c'c} = 0 \qquad (11.47)$$

This is called the *algebraic Riccati equation*. This equation may have one or more solutions, but only one solution is symmetric and positive definite. The derivation of (4.46) and (4.47) is beyond the scope of this text and can be found in References [1, 5]. We show in the following its application.

Example 11.5.1

Consider the plant with transfer function $1/s(s + 2)$ studied in Example 9.4.1. Its controllable form realization is

$$\dot{\mathbf{x}} = \begin{bmatrix} -2 & 0 \\ 1 & 0 \end{bmatrix}\mathbf{x} + \begin{bmatrix} 1 \\ 0 \end{bmatrix}u \qquad (11.48a)$$

$$y = [0 \quad 1]\mathbf{x} \qquad (11.48b)$$

Find the feedback gain to minimize the performance index

$$J = \int_0^\infty \left[y^2(t) + \frac{1}{9} u^2(t) \right] dt \qquad (11.49)$$

The comparison of (11.45) and (11.49) yields $\mathbf{Q} = \mathbf{c'c}$ and $R = 1/9$. Let

$$\mathbf{K} = \begin{bmatrix} k_{11} & k_{21} \\ k_{21} & k_{22} \end{bmatrix}$$

It is a symmetric matrix. For this problem, (11.47) becomes

$$- \begin{bmatrix} k_{11} & k_{21} \\ k_{21} & k_{22} \end{bmatrix} \begin{bmatrix} -2 & 0 \\ 1 & 0 \end{bmatrix} - \begin{bmatrix} -2 & 1 \\ 0 & 0 \end{bmatrix} \begin{bmatrix} k_{11} & k_{21} \\ k_{21} & k_{22} \end{bmatrix}$$

$$+ 9 \begin{bmatrix} k_{11} & k_{21} \\ k_{21} & k_{22} \end{bmatrix} \begin{bmatrix} 1 \\ 0 \end{bmatrix} [1 \quad 0] \begin{bmatrix} k_{11} & k_{21} \\ k_{21} & k_{22} \end{bmatrix} - \begin{bmatrix} 0 \\ 1 \end{bmatrix} [0 \quad 1] = \begin{bmatrix} 0 & 0 \\ 0 & 0 \end{bmatrix}$$

Equating the corresponding entries yields

$$4k_{11} - 2k_{21} + 9k_{11}^2 = 0 \qquad (11.50a)$$

$$2k_{21} - k_{22} + 9k_{11}k_{21} = 0 \qquad (11.50b)$$

and

$$9k_{21}^2 - 1 = 0 \qquad (11.50c)$$

From (11.50c), we have $k_{21} = \pm 1/3$. If $k_{21} = -1/3$, then the resulting $\mathbf{K}$ will not be positive definite. Thus we choose $k_{21} = 1/3$. The substitution of $k_{21} = 1/3$ into (11.50a) yields

$$9k_{11}^2 + 4k_{11} - \frac{2}{3} = 0$$

whose solutions are 0.129 and -0.68. If $k_{11} = -0.68$, then the resulting $\mathbf{K}$ will not be positive definite. Thus we choose $k_{11} = 0.129$. From (11.50b), we can solve k_{22} as 1.05. Therefore, we have

$$\mathbf{K} = \begin{bmatrix} 0.129 & 0.333 \\ 0.333 & 1.05 \end{bmatrix}$$

which can be easily verified as positive definite. Thus the feedback gain is given by

$$\mathbf{k} = R^{-1}\mathbf{b'K} = 9[1 \quad 0] \begin{bmatrix} 0.129 & 0.333 \\ 0.333 & 1.05 \end{bmatrix} = [1.2 \quad 3] \qquad (11.51)$$

and (11.43) becomes

$$\dot{\mathbf{x}} = (\mathbf{A} - \mathbf{bk})\mathbf{x} = \left(\begin{bmatrix} -2 & 0 \\ 1 & 0 \end{bmatrix} - \begin{bmatrix} 1 \\ 0 \end{bmatrix} [1.2 \quad 3] \right) \mathbf{x}$$

$$= \begin{bmatrix} -3.2 & -3 \\ 1 & 0 \end{bmatrix} \mathbf{x} \qquad (11.52)$$

The characteristic polynomial of the matrix in (11.52) is

$$\det \begin{bmatrix} s + 3.2 & 3 \\ -1 & s \end{bmatrix} = s^2 + 3.2s + 3$$

which equals the denominator of the optimal transfer function in (9.24). This is not surprising, because the performance index in (11.49) is essentially the same as (9.21) with zero reference input. Therefore the quadratic optimal regulator problem using state-variable equations is closely related to the quadratic optimal transfer function in Chapter 9. In fact, it can be shown that $D_o(s)$ obtained by spectral factorization in Chapter 9 equals the characteristic polynomial of $(\mathbf{A} - \mathbf{bk})$. See Reference [1]. We also mention that the conditions of controllability and observability are essential here. These conditions are equivalent to the requirement in Chapter 9 that $N(s)$ and $D(s)$ in $G(s) = N(s)/D(s)$ have no common factors.

The optimal gain in quadratic regulators, also called *linear quadratic regulator* or *lqr*, can be obtained by using MATLAB. For the example, we type

```
a=[-2 0;1 0];b=[1;0];
q=[0 0;0 1];r=1/9;
k=lqr(a,b,q,r)
```

then k=[1.1623 3.000] will appear on the screen. Thus the use of MATLAB is very simple.

11.6   STATE ESTIMATORS

The state feedback in the preceding sections is introduced under the assumption that all state variables are available for connection to a gain. This assumption may or may not hold in practice. For example, for the dc motor discussed in Figure 11.4, the two state variables can be generated by using a potentiometer and a tachometer. However, if no tachometer is available or if it is available but is very expensive and we have decided to use only a potentiometer in the design, then the state feedback cannot be directly applied. In this case, we must design a state estimator. This and the following sections will discuss this problem.

Consider

$$\dot{\mathbf{x}} = \mathbf{A}\mathbf{x} + \mathbf{b}u \qquad (11.53a)$$

$$y = \mathbf{c}\mathbf{x} \qquad (11.53b)$$

with known $\mathbf{A}$, $\mathbf{b}$, and $\mathbf{c}$. The problem is to use the available input u and output y to drive a system, called a *state estimator*, whose output $\hat{\mathbf{x}}$ approaches the actual state $\mathbf{x}$. The easiest way of building such an estimator is to simulate the system, as shown in Figure 11.6. Note that the original system could be an electromechanical one, and the estimator in Figure 11.6 may be built using operational amplifier circuits. Be-

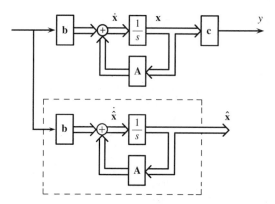

Figure 11.6 Open-loop state estimator.

cause the original system and the estimator are driven by the same input, their states $\mathbf{x}(t)$ and $\hat{\mathbf{x}}(t)$ should be equal for all t if their initial states are the same. Now if (11.53) is observable, its initial state can be computed and then applied to the estimator. Therefore, in theory, the estimator in Figure 11.6 can be used, especially if both systems start with $\mathbf{x}(0) = \hat{\mathbf{x}}(0) = \mathbf{0}$. We call the estimator in Figure 11.6 the *open-loop state estimator*.

Let the output of the estimator in Figure 11.6 be denoted by $\hat{\mathbf{x}}$. Then it is described by

$$\dot{\hat{\mathbf{x}}} = \mathbf{A}\hat{\mathbf{x}} + \mathbf{b}u \tag{11.54}$$

Subtracting this equation from (11.53a) yields

$$\dot{\mathbf{x}} - \dot{\hat{\mathbf{x}}} = \mathbf{A}(\mathbf{x} - \hat{\mathbf{x}})$$

Define $\mathbf{e}(t) := \mathbf{x}(t) - \hat{\mathbf{x}}(t)$. It is the error between the actual state and the estimated state at time t. Then it is governed by

$$\dot{\mathbf{e}} = \mathbf{A}\mathbf{e} \tag{11.55}$$

and its solution is

$$\mathbf{e}(t) = e^{\mathbf{A}t}\mathbf{e}(0) = \mathbf{e}^{\mathbf{A}t}(\mathbf{x}(0) - \hat{\mathbf{x}}(0))$$

Although it is possible to estimate $\mathbf{x}(0)$ and then set $\hat{\mathbf{x}}(0) = \mathbf{x}(0)$, in practice $\mathbf{e}(0)$ is often nonzero due to estimation error or disturbance. Now if $\mathbf{A}$ has eigenvalues in the open right half plane, then the error $\mathbf{e}(t)$ will grow with time. Even if all eigenvalues of $\mathbf{A}$ have negative real parts, we have no control over the rate at which $\mathbf{e}(t)$ approaches zero. Thus, the open-loop state estimator in Figure 11.6 is not desirable in practice.

Although the output y is available, it is not utilized in the open-loop estimator in Figure 11.6. Now we shall compare it with $\mathbf{c}\hat{\mathbf{x}}$ and use the difference to drive an estimator through a constant vector $\mathbf{l}$ as shown in Figure 11.7(a). Then the output $\hat{\mathbf{x}}$ of the estimator is governed by

$$\dot{\hat{\mathbf{x}}} = \mathbf{A}\hat{\mathbf{x}} + \mathbf{b}u + \mathbf{l}(y - \mathbf{c}\hat{\mathbf{x}})$$

or

$$\dot{\hat{\mathbf{x}}} = (\mathbf{A} - \mathbf{lc})\hat{\mathbf{x}} + \mathbf{b}u + \mathbf{l}y \qquad (11.56)$$

and is replotted in Figure 11.7(b). We see that the estimator is now driven by u as well as y. We show in the following that if $(\mathbf{A}, \mathbf{c})$ is observable, then (11.56) can be designed so that the estimated state $\hat{\mathbf{x}}$ will approach the actual state $\mathbf{x}$ as quickly as desired.

The subtraction of (11.56) from (11.53a) yields, using $y = \mathbf{cx}$,

$$\dot{\mathbf{x}} - \dot{\hat{\mathbf{x}}} = \mathbf{Ax} + \mathbf{b}u - (\mathbf{A} - \mathbf{lc})\hat{\mathbf{x}} - \mathbf{b}u - \mathbf{lcx}$$
$$= (\mathbf{A} - \mathbf{lc})\mathbf{x} - (\mathbf{A} - \mathbf{lc})\hat{\mathbf{x}} = (\mathbf{A} - \mathbf{lc})(\mathbf{x} - \hat{\mathbf{x}})$$

which becomes, after the substitution of $\mathbf{e} = \mathbf{x} - \hat{\mathbf{x}}$,

$$\dot{\mathbf{e}} = (\mathbf{A} - \mathbf{lc})\dot{\mathbf{e}} \qquad (11.57)$$

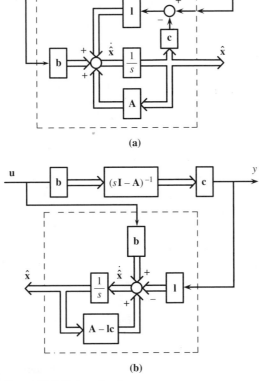

(a)

(b)

Figure 11.7 State estimator.

Now we show that if $(\mathbf{A}, \mathbf{c})$ is observable, then the eigenvalues of $(\mathbf{A} - \mathbf{lc})$ can be arbitrarily assigned (provided complex-conjugate eigenvalues are assigned in pairs) by choosing a suitable vector $\mathbf{l}$. If $(\mathbf{A}, \mathbf{c})$ is observable, its observability matrix in (11.4) has rank n. The transpose of (11.4) is

$$\mathbf{V}' = [\mathbf{c}' \quad \mathbf{A}'\mathbf{c}' \quad (\mathbf{A}')^2\mathbf{c}' \quad \cdots \quad (\mathbf{A}')^{n-1}\mathbf{c}']$$

and is also of rank n. By comparing this with (11.3), we conclude that $(\mathbf{A}', \mathbf{c}')$ is controllable. Consequently, the eigenvalues of $(\mathbf{A}' - \mathbf{c}'\mathbf{l}')$ or its transpose $(\mathbf{A} - \mathbf{lc})$ can be arbitrarily assigned by choosing a suitable $\mathbf{l}$. This completes the argument. We list in the following a procedure of designing $\mathbf{l}$. It is a simple modification of the procedure for pole placement.

Procedure for Designing Estimators

Given a 4-dimensional observable $(\mathbf{A}, \mathbf{c})$ and a set of eigenvalues λ_i, $i = 1, 2, 3, 4$, find a real 4×1 vector $\mathbf{l}$ so that the matrix $(\mathbf{A} - \mathbf{lc})$ has the set as its eigenvalues.

1. Compute the characteristic polynomial of $\mathbf{A}$: $\det (s\mathbf{I} - \mathbf{A}) = s^4 + a_1 s^3 + a_2 s^2 + a_3 s + a_4$.
2. Compute the desired characteristic polynomial of the estimator in (11.56):

$$(s - \lambda_1)(s - \lambda_2)(s - \lambda_3)(s - \lambda_4) = s^4 + \bar{a}_1 s^3 + \bar{a}_2 s^2 + \bar{a}_3 s + \bar{a}_4$$

3. Compute $\bar{\mathbf{l}}' = [\bar{a}_1 - a_1 \quad \bar{a}_2 - a_2 \quad \bar{a}_3 - a_3 \quad \bar{a}_4 - a_4]$.
4. Compute the equivalence transformation

$$\mathbf{S} := \mathbf{P}^{-1} = \begin{bmatrix} 1 & 0 & 0 & 0 \\ a_1 & 1 & 0 & 0 \\ a_2 & a_1 & 1 & 0 \\ a_3 & a_2 & a_1 & 1 \end{bmatrix} \begin{bmatrix} \mathbf{c} \\ \mathbf{cA} \\ \mathbf{cA}^2 \\ \mathbf{cA}^3 \end{bmatrix} \tag{11.58}$$

5. Compute $\mathbf{l} = \mathbf{P}\bar{\mathbf{l}} = \mathbf{S}^{-1}\bar{\mathbf{l}}$.

Now if the eigenvalues of (11.57) are designed to have large negative real parts by choosing a suitable $\mathbf{l}$, then no matter what $\mathbf{e}(0)$ is, $\mathbf{e}(t)$ will approach zero rapidly. Therefore, in using the estimator in Figure 11.7, there is no need to estimate $\mathbf{x}(0)$. The state estimator in (11.56) has the same dimension as (11.53) and is called a *full-dimensional estimator*.

11.6.1   Reduced-Dimensional Estimators

Consider the n-dimensional equation

$$\dot{\mathbf{x}} = \mathbf{Ax} + \mathbf{b}u \tag{11.59a}$$

$$y = \mathbf{cx} \tag{11.59b}$$

The estimator in (11.56) has dimension n and is thus a full-dimensional estimator. In this section, we discuss the design of $(n- 1)$-dimensional estimators. Such an estimator can be designed using similarity transformations as in the preceding section. See Reference [15]. In this section, we discuss a different approach that does not require any transformation.

Consider the $(n - 1)$-dimensional state-variable equation

$$\dot{\mathbf{z}} = \mathbf{Fz} + \mathbf{g}y + \mathbf{h}u \qquad (11.60)$$

where $\mathbf{F}$, $\mathbf{g}$, and $\mathbf{h}$ are respectively $(n - 1) \times (n - 1)$, $(n - 1) \times 1$ and $(n - 1) \times 1$ constant matrices. The equation is driven by y and u. The state vector $\mathbf{z}(t)$ is called an estimate of $\mathbf{Tx}(t)$, where $\mathbf{T}$ is an $(n - 1) \times n$ constant matrix, if

$$\lim_{t \to \infty} |\mathbf{z}(t) - \mathbf{Tx}(t)| = 0$$

for any $\mathbf{x}(0)$, $\mathbf{z}(0)$, and $u(t)$. The conditions for (11.60) to be an estimate of $\mathbf{Tx}$ are

1. $\mathbf{TA} - \mathbf{FT} = \mathbf{gc}$
2. $\mathbf{h} = \mathbf{Tb}$
3. All eigenvalues of $\mathbf{F}$ have negative real parts.

Indeed, if we define $\mathbf{e} := \mathbf{z} - \mathbf{Tx}$, then

$$\dot{\mathbf{e}} = \dot{\mathbf{z}} - \mathbf{T}\dot{\mathbf{x}} = \mathbf{Fz} + \mathbf{g}y + \mathbf{h}u - \mathbf{T}(\mathbf{Ax} + \mathbf{b}u)$$

which becomes, after the substitution of $\mathbf{z} = \mathbf{e} + \mathbf{Tx}$ and $y = \mathbf{cx}$,

$$\dot{\mathbf{e}} = \mathbf{Fe} + (\mathbf{FT} - \mathbf{TA} + \mathbf{gc})\mathbf{x} + (\mathbf{h} - \mathbf{Tb})u$$

This equation reduces to, after using the conditions in 1 and 2,

$$\dot{\mathbf{e}} = \mathbf{Fe}$$

If all eigenvalues of $\mathbf{F}$ have negative parts, then $\mathbf{e}(t) = e^{\mathbf{F}t}\mathbf{e}(0)$ approaches zero for any $\mathbf{e}(0)$. This shows that under the three conditions, (11.60) is an estimate of $\mathbf{Tx}$. The matrix equation $\mathbf{TA} - \mathbf{FT} = \mathbf{gc}$ is called a *Lyapunov equation*. Now we list the design procedure in the following.

Procedure for Designing Reduced-Dimensional Estimators

1. Choose an $(n - 1) \times (n - 1)$ matrix $\mathbf{F}$ so that all its eigenvalues have negative real parts and are different from those of $\mathbf{A}$.
2. Choose an $(n - 1) \times 1$ vector $\mathbf{g}$ so that $(\mathbf{F}, \mathbf{g})$ is controllable.
3. Solve the $(n - 1) \times n$ matrix $\mathbf{T}$ from the Lyapunov equation $\mathbf{TA} - \mathbf{FT} = \mathbf{gc}$.
4. If the square matrix of order n

$$\mathbf{P} := \begin{bmatrix} \mathbf{c} \\ \mathbf{T} \end{bmatrix}$$

is singular, go back to step 1 and/or step 2 and repeat the process. If $\mathbf{P}$ is nonsingular, compute $\mathbf{h} = \mathbf{Tb}$. Then

$$\dot{\mathbf{z}} = \mathbf{Fz} + \mathbf{g}y + \mathbf{h}u \qquad (11.61a)$$

$$\hat{\mathbf{x}} = \begin{bmatrix} \mathbf{c} \\ \mathbf{T} \end{bmatrix}^{-1} \begin{bmatrix} y \\ z \end{bmatrix}$$ (11.61b)

is an estimator of $\mathbf{x}$ in (11.59).

We give some remarks about the procedure. The conditions that the eigenvalues of $\mathbf{F}$ differ from those of $\mathbf{A}$ and that $(\mathbf{F}, \mathbf{g})$ be controllable are introduced to insure that a solution $\mathbf{T}$ in $\mathbf{TA} - \mathbf{FT} = \mathbf{gc}$ exists and has full rank. The procedure can also be used to design a full-dimensional estimator if $\mathbf{F}$ is chosen to be of order n. For a more detailed discussion, see Reference [15]. In this design, we have freedom in choosing the form of $\mathbf{F}$. It can be chosen as one of the companion forms shown in (2.77); it can also be chosen as a diagonal matrix. In this case, all eigenvalues must be distinct, otherwise no $\mathbf{g}$ exists so that $(\mathbf{F}, \mathbf{g})$ is controllable. See Problem 11.2. If $\mathbf{F}$ is chosen as a Jordan-form matrix, then its eigenvalues can be repeated. If $\mathbf{F}$ is of Jordan form, all solutions of $\mathbf{TA} - \mathbf{FT} = \mathbf{gc}$ can be parameterized. See Reference [59]. We mention that Lyapunov equations can be solved in MATLAB by calling the command lyap.

Example 11.6.1

Consider the equation in (11.41) or

$$\dot{\mathbf{x}} = \begin{bmatrix} \dot{x}_1 \\ \dot{x}_2 \end{bmatrix} = \begin{bmatrix} 0 & 1 \\ 0 & -1 \end{bmatrix} \mathbf{x} + \begin{bmatrix} 0 \\ 10 \end{bmatrix} u$$ (11.62a)

$$y = [1 \quad 0]\mathbf{x}$$ (11.62b)

Design a reduced-dimensional state estimator with eigenvalue -4. Equation (11.62) has dimension $n = 2$. Thus, its reduced-dimensional estimator has dimension 1 and $\mathbf{F}$ reduces to a scalar. We set $\mathbf{F} = -4$ and choose $\mathbf{g} = 1$. Clearly $(\mathbf{F}, \mathbf{g})$ is controllable. The matrix $\mathbf{T}$ is a 1×2 matrix. Let $\mathbf{T} = [t_1 \quad t_2]$. Then it can be solved from

$$[t_1 \quad t_2]\begin{bmatrix} 0 & 1 \\ 0 & -1 \end{bmatrix} - (-4)[t_1 \quad t_2] = 1 \times [1 \quad 0]$$

or

$$[0 \quad t_1 - t_2] + [4t_1 \quad 4t_2] = [1 \quad 0]$$

Thus, we have $4t_1 = 1$ and $t_1 - t_2 + 4t_2 = 0$, which imply $k_1 = 0.25$ and $t_2 = -t_1/3 = -1/12 = -0.083$. The matrix

$$\mathbf{P} := \begin{bmatrix} \mathbf{c} \\ \mathbf{T} \end{bmatrix} = \begin{bmatrix} 1 & 0 \\ 0.25 & -0.083 \end{bmatrix}$$

is clearly nonsingular and its inverse is

$$\begin{bmatrix} 1 & 0 \\ 0.25 & -0.083 \end{bmatrix}^{-1} = \begin{bmatrix} 1 & 0 \\ 3 & -12 \end{bmatrix}$$

We compute

$$\mathbf{h} = \mathbf{Tb} = [0.25 \quad -0.083]\begin{bmatrix} 0 \\ 10 \end{bmatrix} = -0.83$$

Thus the one-dimensional state estimator is

$$\dot{z} = -4z + y - 0.83u \qquad (11.63a)$$

$$\hat{\mathbf{x}} = \begin{bmatrix} 1 & 0 \\ 3 & -12 \end{bmatrix}\begin{bmatrix} y \\ z \end{bmatrix} = \begin{bmatrix} y \\ 3y - 12\mathbf{z} \end{bmatrix} \qquad (11.63b)$$

This completes the design.

Exercise 11.6.1

In Example 11.6.1, if $\mathbf{F}$ is chosen as -1 and $\mathbf{g} = 1$, can you find a $\mathbf{T}$ to meet the Lyapunov equation?

[**Answer:** No. In this case, the eigenvalue of $\mathbf{F}$ coincides with one of the eigenvalues of $\mathbf{A}$, and the first condition of the procedure is not met.]

11.7 CONNECTION OF STATE FEEDBACK AND STATE ESTIMATORS

Consider the n-dimensional equation

$$\dot{\mathbf{x}} = \mathbf{Ax} + \mathbf{b}u \qquad (11.64a)$$

$$y = \mathbf{cx} \qquad (11.64b)$$

If (11.64) is controllable, by introducing the state feedback

$$u = r - \mathbf{kx}$$

the eigenvalues of $(\mathbf{A} - \mathbf{bk})$ can be arbitrarily assigned. If the state is not available, we must design a state estimator such as

$$\dot{\hat{\mathbf{x}}} = (\mathbf{A} - \mathbf{lc})\hat{\mathbf{x}} + \mathbf{b}u + \mathbf{l}y \qquad (11.65)$$

to generate an estimate $\hat{\mathbf{x}}$ of $\mathbf{x}$. We then apply the state feedback to $\hat{\mathbf{x}}$ as shown in Figure 11.8, that is,

$$u = r - \mathbf{k}\hat{\mathbf{x}} \qquad (11.66)$$

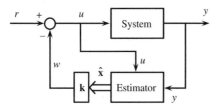

Figure 11.8 State feedback and state estimator.

The feedback gain is designed for the original state $\mathbf{x}$. Now it is connected to the estimated state $\hat{\mathbf{x}}$. Will the resulting system still have the desired eigenvalues? This will be answered in the following.

The substitution of (11.66) and $y = \mathbf{cx}$ into (11.64a) and (11.65) yields

$$\dot{\mathbf{x}} = \mathbf{Ax} + \mathbf{b}(r - \mathbf{k}\hat{\mathbf{x}})$$

and

$$\dot{\hat{\mathbf{x}}} = (\mathbf{A} - \mathbf{lc})\hat{\mathbf{x}} + \mathbf{lcx} + \mathbf{b}(r - \mathbf{k}\hat{\mathbf{x}})$$

They can be combined as

$$\begin{bmatrix} \dot{\mathbf{x}} \\ \dot{\hat{\mathbf{x}}} \end{bmatrix} = \begin{bmatrix} \mathbf{A} & -\mathbf{bk} \\ \mathbf{lc} & \mathbf{A} - \mathbf{lc} - \mathbf{bk} \end{bmatrix} \begin{bmatrix} \mathbf{x} \\ \hat{\mathbf{x}} \end{bmatrix} + \begin{bmatrix} \mathbf{b} \\ \mathbf{b} \end{bmatrix} r \qquad (11.67a)$$

$$y = \mathbf{cx} = \begin{bmatrix} \mathbf{c} & \mathbf{0} \end{bmatrix} \begin{bmatrix} \mathbf{x} \\ \hat{\mathbf{x}} \end{bmatrix} \qquad (11.67b)$$

This equation describes the overall system in Figure 11.8. Consider

$$\mathbf{P} = \begin{bmatrix} \mathbf{I} & \mathbf{0} \\ \mathbf{I} & -\mathbf{I} \end{bmatrix} = \mathbf{P}^{-1} \qquad (11.68)$$

The inverse of $\mathbf{P}$ happens to equal itself. By applying the equivalence transformation

$$\begin{bmatrix} \overline{\mathbf{x}} \\ \overline{\hat{\mathbf{x}}} \end{bmatrix} = \begin{bmatrix} \mathbf{I} & \mathbf{0} \\ \mathbf{I} & -\mathbf{I} \end{bmatrix} \begin{bmatrix} \mathbf{x} \\ \hat{\mathbf{x}} \end{bmatrix}$$

to (11.67), we will finally obtain, using (11.22c),

$$\begin{bmatrix} \dot{\overline{\mathbf{x}}} \\ \dot{\overline{\hat{\mathbf{x}}}} \end{bmatrix} = \begin{bmatrix} \mathbf{A} - \mathbf{bk} & \mathbf{bk} \\ \mathbf{0} & \mathbf{A} - \mathbf{lc} \end{bmatrix} \begin{bmatrix} \overline{\mathbf{x}} \\ \overline{\hat{\mathbf{x}}} \end{bmatrix} + \begin{bmatrix} \mathbf{b} \\ \mathbf{0} \end{bmatrix} r \qquad (11.69a)$$

$$y = \begin{bmatrix} \mathbf{c} & \mathbf{0} \end{bmatrix} \begin{bmatrix} \overline{\mathbf{x}} \\ \overline{\hat{\mathbf{x}}} \end{bmatrix} \qquad (11.69b)$$

Because any equivalence transformation will not change the characteristic polynomial and transfer function, the characteristic polynomial of (11.67) equals that of (11.69) and, using (B.3), is

$$\det \begin{bmatrix} s\mathbf{I} - \mathbf{A} + \mathbf{bk} & -\mathbf{bk} \\ \mathbf{0} & s\mathbf{I} - \mathbf{A} + \mathbf{lc} \end{bmatrix}$$

$$= \det(s\mathbf{I} - \mathbf{A} + \mathbf{bk}) \det(s\mathbf{I} - \mathbf{A} + \mathbf{lc})$$

(11.70)

Similarly, the transfer function of (11.67) equals that of (11.69) and, using (B.6), is

$$\begin{bmatrix} \mathbf{c} & \mathbf{0} \end{bmatrix} \begin{bmatrix} s\mathbf{I} - \mathbf{A} + \mathbf{bk} & -\mathbf{bk} \\ \mathbf{0} & s\mathbf{I} - \mathbf{A} + \mathbf{lc} \end{bmatrix}^{-1} \begin{bmatrix} \mathbf{b} \\ \mathbf{0} \end{bmatrix}$$

$$= \begin{bmatrix} \mathbf{c} & \mathbf{0} \end{bmatrix} \begin{bmatrix} (s\mathbf{I} - \mathbf{A} + \mathbf{bk})^{-1} & \alpha \\ \mathbf{0} & (s\mathbf{I} - \mathbf{A} + \mathbf{lc})^{-1} \end{bmatrix} \begin{bmatrix} \mathbf{b} \\ \mathbf{0} \end{bmatrix}$$

(11.71)

$$= \begin{bmatrix} \mathbf{c}(s\mathbf{I} - \mathbf{A} + \mathbf{bk})^{-1} & \mathbf{c}\alpha \end{bmatrix} \begin{bmatrix} \mathbf{b} \\ \mathbf{0} \end{bmatrix} = \mathbf{c}(s\mathbf{I} - \mathbf{A} + \mathbf{bk})^{-1}\mathbf{b}$$

where $\alpha = (s\mathbf{I} - \mathbf{A} + \mathbf{bk})^{-1}\mathbf{bk}(s\mathbf{I} - \mathbf{A} + \mathbf{lc})^{-1}$. From (11.70), we see that the eigenvalues of the overall system in Figure 11.8 consist of the eigenvalues of the state feedback and the eigenvalues of the state estimator. Thus the connection of the feedback gain to the output of the estimator does not change the original designs. Thus the state feedback and the state estimator can be designed separately. This is often referred to as the *separation property*.

The transfer function of the overall system in Figure 11.8 is computed in (11.71). It equals (11.30c) and has only n poles. The overall system, however, has $2n$ eigenvalues. Therefore, the state-variable equations in (11.67) and (11.69) are not minimal equations. In fact, they can be shown to be uncontrollable and unobservable. The transfer function of the state feedback system with a state estimator in Figure 11.8 equals the transfer function of the state feedback system without a state estimator in Figure 11.5; thus, the state estimator is hidden from the input r and output y and its transfer function is canceled in the design. This can be explained physically as follows. In computing transfer functions, all initial states are assumed to be zero, therefore, we have $\mathbf{x}(0) = \hat{\mathbf{x}}(0)$ and, consequently, $\mathbf{x}(t) = \hat{\mathbf{x}}(t)$ for all t and for any $u(t)$. Thus the estimator does not appear in (11.71). This situation is similar to the pole-zero cancellation design in Chapter 10.

We have shown the separation property by using the full-dimensional estimator. The property still holds if we use reduced-dimensional estimators. The proof, however, is slightly more complicated. See Reference [15].

11.7.1 Comparison with Linear Algebraic Method

Consider a minimal state-variable equation with transfer function

$$G(s) = \frac{N(s)}{D(s)}$$

After introducing state feedback and the state estimator, the transfer function of the resulting overall system will be of the form

$$G_o(s) = \frac{N(s)}{D_o(s)}$$

where $D_o(s)$ has the same degree and the same leading coefficient as $D(s)$. Note that the numerator of $G_o(s)$ is the same as that of $G(s)$. Clearly $G_o(s)$ is implementable for the given $G(s)$ (see Section 9.2) and the linear algebraic methods discussed in Chapter 10 can be used to implement such $G_o(s)$. In this subsection, we use an example to compare the design using the state-variable method with that using the linear algebraic method.

Consider the minimal equation in Example 11.4.2 or

$$\dot{\mathbf{x}} = \begin{bmatrix} \dot{x}_1 \\ \dot{x}_2 \end{bmatrix} = \begin{bmatrix} 0 & 1 \\ 0 & -1 \end{bmatrix} \mathbf{x} + \begin{bmatrix} 0 \\ 10 \end{bmatrix} u$$

$$y = [1 \quad 0]\mathbf{x}$$

with transfer function

$$G(s) = \frac{N(s)}{D(s)} = \frac{10}{s^2 + s} \tag{11.72}$$

It is computed in Example 11.4.2 that the feedback gain $\mathbf{k} = [0.8 \quad 0.3]$ in $u = r - \mathbf{k}\mathbf{x}$ will shift the poles of $G(s)$ to $-2 \pm j2$ and that the resulting transfer function from r to y is

$$G_o(s) = \frac{N_o(s)}{D_o(s)} = \frac{10}{s^2 + 4s + 8} \tag{11.73}$$

Now if the state is not available for feedback, we must design a state estimator. A reduced-dimensional state estimator with eigenvalue -4 is designed in Example 11.6.1 as

$$\dot{z} = -4z + y - 0.83u$$

$$\hat{\mathbf{x}} = \begin{bmatrix} y \\ 3y - 12z \end{bmatrix}$$

Its basic block diagram is plotted in Figure 11.9(a). Now we apply the feedback gain $\mathbf{k}$ to $\hat{\mathbf{x}}$ as shown in Figure 11.9(b). This completes the state-variable design.

In order to compare with the linear algebraic method, we compute the transfer functions from u to w and y to w of the block bounded by the dashed line in Figure 11.9(b). There is no loop inside the block; therefore, using Mason's formula, we have

$$C_1(s) = \frac{W(s)}{U(s)} = (-0.83) \cdot \frac{1}{s + 4} \cdot (-12) \cdot (0.3) = \frac{3}{s + 4}$$

and

$$C_2(s) = \frac{W(s)}{Y(s)} = \frac{(-12) \cdot (0.3)}{s + 4} + 0.3 \cdot 3 + 0.8 = \frac{1.7s + 3.2}{s + 4}$$

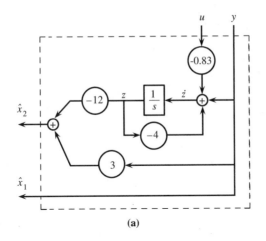

(a)

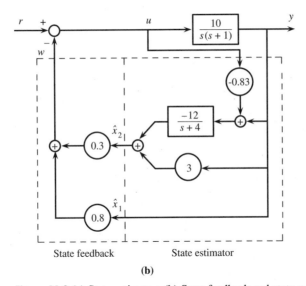

State feedback   State estimator

(b)

Figure 11.9 (a) State estimator. (b) State feedback and state estimator.

These two transfer functions are plotted in Figure 11.10(a) and then rearranged in Figure 11.10(b). It is the plant input/output feedback configuration studied in Figure 10.15.

Now we shall redesign the problem using the method in Section 10.6, namely, given $G(s)$ in (11.72) and implementable $G_o(s)$ in (11.73), find two compensators of the form

$$C_1(s) = \frac{L(s)}{A(s)} \qquad C_2(s) = \frac{M(s)}{A(s)}$$

so that the resulting system in Figure 10.15(b) or Figure 11.10 has $G_o(s)$ as its transfer

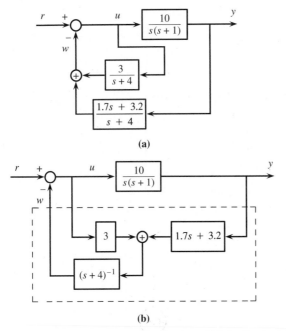

Figure 11.10 (a) Compensators. (b) Plant I/O feedback configuration.

function. Using the procedure in Section 10.6, we first compute

$$\frac{G_o(s)}{N(s)} = \frac{10}{(s^2 + 4s + 8) \cdot 10} = \frac{1}{s^2 + 4s + 8} =: \frac{N_p(s)}{D_p(s)}$$

with $N_p(s) = 1$ and $D_p(s) = s^2 + 4s + 8$. Because deg $N_p(s) = 0$, we must introduce a polynomial $\bar{A}(s)$ of degree $2 - 1 = 1$. We choose it as $\bar{A}(s) = s + 4$ for the eigenvalue of the estimator is chosen as -4 in the state-variable approach. Thus we have

$$A(s) = N_p(s)\bar{A}(s) = 1 \cdot (s + 4) = s + 4$$

The polynomials $L(s) = L_1 s + L_0$ and $M(s) = M_1 s + M_0$ can be determined, using (10.73), from

$$L(s)D(s) + M(s)N(s) = \bar{A}(s)(D_p(s) - N_p(s)D(s))$$
$$= (s + 4)(s^2 + 4s + 8 - s^2 - s) = 3s^2 + 20s + 32$$

or from the following linear algebraic equation, using (10.75),

$$\begin{bmatrix} 0 & 10 & 0 & 0 \\ 1 & 0 & 0 & 10 \\ 1 & 0 & 1 & 0 \\ 0 & 0 & 1 & 0 \end{bmatrix} \begin{bmatrix} L_0 \\ M_0 \\ L_1 \\ M_1 \end{bmatrix} = \begin{bmatrix} 32 \\ 20 \\ 3 \\ 0 \end{bmatrix} \qquad (11.74)$$

The first equation of (11.74) yields $10M_0 = 32$ or $M_0 = 3.2$. The fourth equation of (11.74) yields $L_1 = 0$. The second and third equations of (11.74) are $L_0 + 10M_1 = 20$ and $L_0 + L_1 = 3$, which yield $L_0 = 3$ and $M_1 = 1.7$. Thus the compensators are

$$C_1(s) = \frac{L(s)}{A(s)} = \frac{3}{s + 4} \qquad C_2(s) = \frac{M(s)}{A(s)} = \frac{1.7s + 3.2}{s + 4}$$

They are the same as those computed using state-variable equations.

Now we compare the state-variable approach and the transfer function approach in designing this problem. The state-variable approach requires the concepts of controllability and observability. The design requires computing similarity transformations and solving Lyapunov matrix equations. In the transfer function approach, we require the concept of coprimeness (that is, two polynomials have no common factors). The design is completed by solving a set of linear algebraic equations. Therefore, the transfer function approach is simpler in concept and computation than the state-variable approach. The transfer function approach can be used to design any implementable transfer function. The design of any implementable transfer function by using state-variable equations would be more complicated and has not yet appeared in any control text.

11.8 LYAPUNOV STABILITY THEOREM

A square matrix $\mathbf{A}$ is said to be *stable* if all its eigenvalues have negative real parts. One way to check this is to compute its characteristic polynomial

$$\Delta(s) = \det(s\mathbf{I} - \mathbf{A})$$

We then apply the Routh test to check whether or not $\Delta(s)$ is a Hurwitz polynomial. If it is, $\mathbf{A}$ is stable; otherwise, it is not stable.

In addition to the preceding method, there is another popular method of checking the stability of $\mathbf{A}$. It is stated as a theorem.

THEOREM 11.2 (Lyapunov Theorem)

All eigenvalues of $\mathbf{A}$ have negative real parts if for any symmetric positive definite matrix $\mathbf{N}$, the Lyapunov equation

$$\mathbf{A}'\mathbf{M} + \mathbf{M}\mathbf{A} = -\mathbf{N}$$

has a symmetric positive definite solution $\mathbf{M}$. ■

COROLLARY 11.2

All eigenvalues of $\mathbf{A}$ have negative real parts if, for any symmetric positive semidefinite matrix $\mathbf{N} = \mathbf{n}'\mathbf{n}$ with the property that $(\mathbf{A}, \mathbf{n})$ is observable, the Lyapunov equation

$$\mathbf{A}'\mathbf{M} + \mathbf{M}\mathbf{A} = -\mathbf{N} \tag{11.75}$$

has a symmetric positive definite solution $\mathbf{M}$. ■

We give an intuitive argument to establish these results. Consider

$$\dot{\mathbf{x}}(t) = \mathbf{A}\mathbf{x}(t) \tag{11.76}$$

Its solution is $\mathbf{x}(t) = e^{\mathbf{A}t}\mathbf{x}(0)$. If the eigenvalues of $\mathbf{A}$ are c_1, c_2, and c_3, then every component of $\mathbf{x}(t)$ is a linear combination of e^{c_1t}, e^{c_2t}, and e^{c_3t}. These time functions will approach zero as $t\rightarrow\infty$ if and only if every c_i has a negative real part. Thus we conclude that the response of (11.76) due to any nonzero initial state will approach zero if and only if $\mathbf{A}$ is stable.

We define

$$V(\mathbf{x}) := \mathbf{x}'\mathbf{M}\mathbf{x} \tag{11.77}$$

If $\mathbf{M}$ is symmetric positive definite, $V(\mathbf{x})$ is positive for any nonzero $\mathbf{x}$ and is zero only at $\mathbf{x} = \mathbf{0}$. Thus the plot of $V(\mathbf{x})$ will be bowl-shaped, as is shown in Figure 11.11. Such a $V(\mathbf{x})$ is called a *Lyapunov function*.

Now we consider the time history of $V(\mathbf{x}(t))$ along the trajectory of (11.76). Using $\dot{\mathbf{x}}' = \mathbf{x}'\mathbf{A}'$, we compute,

$$\frac{d}{dt} V(\mathbf{x}(t)) = \frac{d}{dt}(\mathbf{x}'\mathbf{M}\mathbf{x}) = \dot{\mathbf{x}}'\mathbf{M}\mathbf{x} + \mathbf{x}'\mathbf{M}\dot{\mathbf{x}}$$

$$= \mathbf{x}'\mathbf{A}'\mathbf{M}\mathbf{x} + \mathbf{x}'\mathbf{M}\mathbf{A}\mathbf{x} = \mathbf{x}'(\mathbf{A}'\mathbf{M} + \mathbf{M}\mathbf{A})\mathbf{x}$$

which becomes, after the substitution of (11.75),

$$\frac{d}{dt} V(\mathbf{x}(t)) = -\mathbf{x}'\mathbf{N}\mathbf{x} \tag{11.78}$$

If $\mathbf{N}$ is positive definite, then $dV(\mathbf{x})/dt$ is strictly negative for all nonzero $\mathbf{x}$. Thus, for any initial state $\mathbf{x}(0)$, the Lyapunov function $V(\mathbf{x}(t))$ decreases *monotonically* until it reaches the origin as shown in Figure 11.11(a). Thus $\mathbf{x}(t)$ approaches zero as $t\rightarrow\infty$ and $\mathbf{A}$ is stable. This establishes the Lyapunov theorem.

If $\mathbf{N}$ is symmetric positive semidefinite, then $dV(\mathbf{x})/dt \leq 0$, and $V(\mathbf{x}(t))$ may not decrease monotonically to zero. It may stay constant along some part of a trajectory such as AB shown in Figure 11.11(b). The condition that $(\mathbf{A}, \mathbf{n})$ is observable, however, will prevent $dV(\mathbf{x}(t))/dt = 0$ for all t. Therefore, even if $dV(\mathbf{x}(t))/dt = 0$

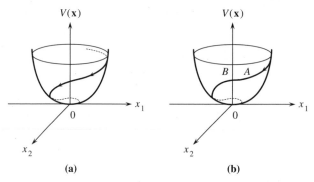

Figure 11.11 Lyapunov functions.

for some t, $V(\mathbf{x}(t))$ will eventually continue to decrease until it reaches zero as $t \to \infty$. This establishes the corollary. For a more detailed discussion of these results, see Reference [15].

11.8.1 Application—A Proof of the Routh Test

Consider the polynomial of degree 3

$$D(s) = s^3 + a_1 s^2 + a_2 s + a_3 \qquad (11.79)$$

We form the Routh table:

$$
\begin{array}{c|cc}
s^3 & 1 & a_2 \\
k_3 = \dfrac{1}{a_1} \quad s^2 & a_1 & a_3 \\
k_2 = \dfrac{a_1}{c} \quad s & c & \\
k_1 = \dfrac{c}{a_3} \quad 1 & a_3 &
\end{array}
\qquad
\left[0 \quad a_2 - \dfrac{a_3}{a_1} =: c \right]
$$

Then the Routh test states that the polynomial is Hurwitz if and only if

$$a_1 > 0 \qquad a_3 > 0 \qquad c := a_2 - \frac{a_3}{a_1} > 0 \qquad (11.80)$$

Now we use Corollary 11.2 to establish the conditions in (11.80). Consider the block diagram in Figure 11.12 with k_i defined in the Routh table. The block diagram has three loops with loop gains $-1/k_3 s$, $-1/k_1 k_2 s^2$, and $-1/k_2 k_3 s^2$, where we have assumed implicitly that all k_i are different from 0. The loop with loop gain $-1/k_3 s$ and the one with $-1/k_1 k_2 s^2$ do not touch each other. Therefore, the characteristic function in Mason's formula is

$$\Delta = 1 - \left(-\frac{1}{k_3 s} - \frac{1}{k_1 k_2 s^2} - \frac{1}{k_2 k_3 s^2} \right) + \left(\frac{-1}{k_3 s} \right)\left(\frac{-1}{k_1 k_2 s^2} \right)$$

Let us consider x_3 as the output, that is, $y = x_3$. Then the forward path gain from u to y is

$$P_1 = \frac{1}{k_3 s}$$

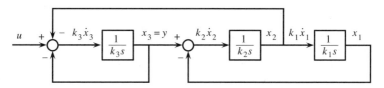

Figure 11.12 Block diagram.

and, because the path does not touch the loop with loop gain $-1/k_1k_2s^2$, the corresponding Δ_1 is,

$$\Delta_1 = 1 - \left(-\frac{1}{k_1k_2s^2}\right)$$

Thus the transfer function from u to y is

$$G(s) = \frac{\dfrac{1}{k_3s} \cdot \left(1 + \dfrac{1}{k_1k_2s^2}\right)}{1 + \dfrac{1}{k_3s} + \dfrac{1}{k_1k_2s^2} + \dfrac{1}{k_2k_3s^2} + \dfrac{1}{k_1k_2k_3s^3}}$$

which can be simplified as

$$G(s) = \frac{\dfrac{1}{k_3}s^2 + \dfrac{1}{k_1k_2k_3}}{s^3 + \dfrac{1}{k_3}s^2 + \dfrac{k_1 + k_3}{k_1k_2k_3}s + \dfrac{1}{k_1k_2k_3}}$$

From k_i in the Routh table, we can readily show $1/k_3 = a_1$, $1/k_1k_2k_3 = a_3$, and $(k_1 + k_3)/k_1k_2k_3 = a_2$. Thus, the transfer function from u to y of the block diagram is

$$G(s) = \frac{a_1s^2 + a_3}{s^3 + a_1s^2 + a_2s + a_3} =: \frac{N(s)}{D(s)} \tag{11.81}$$

Clearly, $N(s)$ and $D(s)$ have no common factor. Thus $G(s)$ has three poles. Now we develop a state-variable equation to describe the system in Figure 11.12. We assign the state variables as shown. Then we have

$$k_1\dot{x}_1 = x_2 \qquad\qquad k_2\dot{x}_2 = -x_1 + x_3$$
$$k_3\dot{x}_3 = -x_2 - x_3 + u \qquad y = x_3$$

These can be expressed in matrix form as

$$\dot{\mathbf{x}} = \begin{bmatrix} 0 & \dfrac{1}{k_1} & 0 \\[2mm] \dfrac{-1}{k_2} & 0 & \dfrac{1}{k_2} \\[2mm] 0 & \dfrac{-1}{k_3} & \dfrac{-1}{k_3} \end{bmatrix} \mathbf{x} + \begin{bmatrix} 0 \\[1mm] 0 \\[1mm] \dfrac{1}{k_3} \end{bmatrix} u \tag{11.82a}$$

$$y = \begin{bmatrix} 0 & 0 & 1 \end{bmatrix}\mathbf{x} \tag{11.82b}$$

Both (11.81) and (11.82) describe the same block diagram, thus (11.81) is the transfer function of the state-variable equation in (11.82). Because the dimension of (11.82) equals the number of poles of $G(s)$ in (11.81), (11.82) is a minimal realization of $G(s)$ and the characteristic polynomial of $\mathbf{A}$ in (11.82) equals $D(s) = s^3 + a_1s^2 + a_2s + a_3$.

Now we use Corollary 11.2 to develop a condition for $\mathbf{A}$ to be stable or, equivalently, for $D(s)$ to be Hurwitz. Define

$$
\mathbf{N} = \mathbf{n}'\mathbf{n} = \begin{bmatrix} 0 \\ 0 \\ \sqrt{2} \end{bmatrix} \begin{bmatrix} 0 & 0 & \sqrt{2} \end{bmatrix} = \begin{bmatrix} 0 & 0 & 0 \\ 0 & 0 & 0 \\ 0 & 0 & 2 \end{bmatrix}
$$

It is symmetric and positive semidefinite. Because the matrix

$$
\begin{bmatrix} \mathbf{n} \\ \mathbf{nA} \\ \mathbf{nA}^2 \end{bmatrix} = \begin{bmatrix} 0 & 0 & \sqrt{2} \\ 0 & \dfrac{-\sqrt{2}}{k_3} & \dfrac{-\sqrt{2}}{k_3} \\ \dfrac{\sqrt{2}}{k_2k_3} & \dfrac{\sqrt{2}}{k_2k_3} & d \end{bmatrix}
$$

with $d = -\sqrt{2}(k_3 - k_2)/k_2k_3^2$, is nonsingular, the pair $(\mathbf{A}, \mathbf{n})$ is observable. Therefore Corollary 11.2 can be used to establish a stability condition for $\mathbf{A}$. It is straightforward to verify the following

$$
\begin{bmatrix} 0 & \dfrac{-1}{k_2} & 0 \\ \dfrac{1}{k_1} & 0 & \dfrac{-1}{k_3} \\ 0 & \dfrac{1}{k_2} & \dfrac{-1}{k_3} \end{bmatrix} \begin{bmatrix} k_1 & 0 & 0 \\ 0 & k_2 & 0 \\ 0 & 0 & k_3 \end{bmatrix}
$$

$$ \tag{11.83} $$

$$
+ \begin{bmatrix} k_1 & 0 & 0 \\ 0 & k_2 & 0 \\ 0 & 0 & k_3 \end{bmatrix} \begin{bmatrix} 0 & \dfrac{1}{k_1} & 0 \\ \dfrac{-1}{k_2} & 0 & \dfrac{1}{k_2} \\ 0 & \dfrac{-1}{k_3} & \dfrac{-1}{k_3} \end{bmatrix} = - \begin{bmatrix} 0 & 0 & 0 \\ 0 & 0 & 0 \\ 0 & 0 & 2 \end{bmatrix}
$$

Therefore the symmetric matrix

$$
\mathbf{M} = \begin{bmatrix} k_1 & 0 & 0 \\ 0 & k_2 & 0 \\ 0 & 0 & k_3 \end{bmatrix}
$$

is a solution of the Lyapunov equation in (11.83). Consequently, the condition for $\mathbf{A}$ in (11.82) to be stable is $\mathbf{M}$ positive definite or, equivalently,

$$k_1 > 0 \qquad k_1 k_2 > 0 \qquad k_1 k_2 k_3 > 0$$

which implies

$$k_1 = \frac{c}{a_3} > 0 \qquad k_2 = \frac{a_1}{c} > 0 \qquad k_3 = \frac{1}{a_1} > 0$$

This set of conditions implies

$$a_1 > 0 \qquad c = a_2 - \frac{a_3}{a_1} > 0 \qquad a_3 > 0$$

which is the same as (11.80). This is one way to establish the Routh stability test. For a more general discussion, see Reference [15].

11.9 SUMMARY AND CONCLUDING REMARKS

This chapter introduced state space designs. We first introduced the concepts of controllability and observability. We showed by examples that a state-variable equation is minimal if and only if it is controllable and observable. We then introduced equivalent state-variable equations; they are obtained by using a nonsingular matrix as an equivalence transformation. Any equivalent transformation will not change the eigenvalues of the original equation or its transfer function. Neither are the properties of controllability and observability affected by any equivalence transformation.

If a state-variable equation is controllable, then it can be transformed, using an equivalence transformation, into the controllable-form equation. Using this form, we developed a procedure to achieve arbitrary pole placement by using constant-gain state feedback. Although state feedback will shift the eigenvalues of the original system, it does not affect the numerator of the system's transfer function.

If $(\mathbf{A}, \mathbf{c})$ is observable, then $(\mathbf{A}', \mathbf{c}')$, where the prime denotes the transpose, is controllable. Using this property, we showed that if $(\mathbf{A}, \mathbf{c})$ is observable, a state estimator with any eigenvalues can be designed. We also discussed a method of designing estimators by solving Lyapunov equations. The connection of state feedback gains to estimated states, rather than to the original state, was justified by establishing the separating property. We then compared the state space design with the linear algebraic method developed in Chapter 10. It was shown that the transfer function approach is simpler, in both concept and computation, than the state-variable approach. Finally, we introduced the Lyapunov stability theorem.

To conclude this chapter, we discuss constant gain output feedback. In constant-gain *state* feedback, we can assign all n poles arbitrarily. In constant-gain *output* feedback of single-variable systems, we can arbitrarily assign only one pole; the remaining poles cannot be assigned. For example, consider the constant-gain output feedback system shown in Figure 11.13(a). We can assign one pole in any place. For example, if we assign it at -3, then from the root loci shown in Figure 11.13(b), we can see that the other two poles will move into the unstable region. Therefore, the design is useless. For constant-gain output feedback, it is better to use the root-

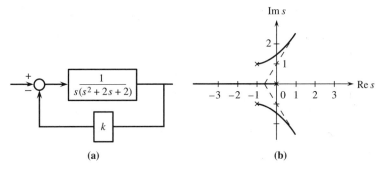

Figure 11.13 (a) Constant gain output feedback. (b) Root loci.

locus method in Chapter 7 to carry out the design. The design using a compensator of degree 1 or higher in the feedback path is called *dynamic output feedback*. The design of dynamic output feedback is essentially the same as the design of state estimators and the design in Chapter 10. Therefore, it will not be discussed.

PROBLEMS

11.1. Check the controllability and observability of the following state-variable equations:

a. $\dot{x} = -x + u \quad y = 2x$

b. $\dot{\mathbf{x}} = \begin{bmatrix} 0 & 1 \\ 1 & 1 \end{bmatrix} \mathbf{x} + \begin{bmatrix} 1 \\ 0 \end{bmatrix} u$

$y = [2 \quad -2]\mathbf{x}$

c. $\dot{\mathbf{x}} = \begin{bmatrix} -3 & 1 \\ -2 & 0 \end{bmatrix} \mathbf{x} + \begin{bmatrix} 2 \\ 4 \end{bmatrix} u$

$y = [1 \quad 0]\mathbf{x}$

d. $\dot{\mathbf{x}} = \begin{bmatrix} -3 & 1 & 0 \\ -2 & 0 & 0 \\ 1 & 0 & 0 \end{bmatrix} \mathbf{x} + \begin{bmatrix} 2 \\ 4 \\ 1 \end{bmatrix} u$

$y = [1 \quad 0 \quad 0]\mathbf{x}$

Which are minimal equations?

11.2. Show that the equation

$$\dot{\mathbf{x}} = \begin{bmatrix} \lambda_1 & 0 \\ 0 & \lambda_2 \end{bmatrix} \mathbf{x} + \begin{bmatrix} 1 \\ -2 \end{bmatrix} u$$

$$y = [2 \quad 0]\mathbf{x}$$

is controllable if and only if $\lambda_1 \neq \lambda_2$. Show that the equation is always not observable.

11.3. Show that the equation

$$\dot{\mathbf{x}} = \begin{bmatrix} \lambda_1 & 1 \\ 0 & \lambda_1 \end{bmatrix} \mathbf{x} + \begin{bmatrix} b_1 \\ b_2 \end{bmatrix} u$$

$$y = [2 \quad 0]\mathbf{x}$$

is controllable if and only if $b_2 \neq 0$. It is independent of b_1. Show that the equation is always observable.

11.4. Let $\bar{\mathbf{x}} = \mathbf{P}\mathbf{x}$ with

$$\mathbf{P} = \begin{bmatrix} 1 & 2 \\ -1 & 3 \end{bmatrix}$$

Find equivalent state-variable equations for the equations in Problem 11.1(b) and (c).

11.5. Check the controllability and observability of the equations in Problem 11.4. Also compute their transfer functions. Does the equivalence transformation change these properties and transfer functions?

11.6. Given a plant with transfer function

$$G(s) = \frac{10}{s(s - 1)(s + 2)}$$

use the procedure in Example 11.4.1 to find the feedback gain such that the resulting system has poles at -2, -3, and -4.

11.7. Redesign Problem 11.6 using the state-variable method. Is the feedback gain the same?

11.8. Consider the state-variable equation

$$\dot{\mathbf{x}} = \begin{bmatrix} \mathbf{A}_{11} & \mathbf{A}_{12} \\ 0 & \mathbf{A}_{22} \end{bmatrix} \mathbf{x} + \begin{bmatrix} \mathbf{b}_1 \\ 0 \end{bmatrix} u$$

Show that the equation is not controllable. Show also that the eigenvalues of $\mathbf{A}_{22}$ will not be affected by any state feedback. If all eigenvalues of $\mathbf{A}_{22}$ have negative real parts and if $(\mathbf{A}_{11}, \mathbf{b}_1)$ is controllable, then the equation is said to be *stabilizable*.

11.9. Consider

$$\dot{\mathbf{x}} = \begin{bmatrix} 1 & 1 \\ 1 & 1 \end{bmatrix} \mathbf{x} + \begin{bmatrix} 1 \\ 0 \end{bmatrix} u$$

$$y = [2 \quad -1]\mathbf{x}$$

Find the feedback gain **k** in $u = r - \mathbf{k}\mathbf{x}$ such that the resulting system has eigenvalues at -2 and -3.

11.10. Consider

$$\dot{\mathbf{x}} = \begin{bmatrix} -3 & 1 \\ -2 & 0 \end{bmatrix} \mathbf{x} + \begin{bmatrix} 1 \\ -2 \end{bmatrix} u$$

$$y = \begin{bmatrix} 1 & 0 \end{bmatrix} \mathbf{x}$$

Find the feedback gain **k** in $u = r - \mathbf{k}\mathbf{x}$ such that the resulting system has eigenvalues at $-2 \pm 2j$.

11.11. Design a full-dimensional state estimator with eigenvalues -3 and -4 for the state-variable equation in Problem 11.9.

11.12. Design a full-dimensional state estimator with eigenvalues -3 and -4 for the state-variable equation in Problem 11.10.

11.13. Design a reduced-dimensional state estimator with eigenvalue -3 for the state-variable equation in Problem 11.9.

11.14. Design a reduced-dimensional state estimator with eigenvalue -3 for the state-variable equation in Problem 11.10.

11.15. Consider a controllable n-dimensional $(\mathbf{A}, \mathbf{b})$. Let **F** be an arbitrary $n \times n$ matrix and let $\bar{\mathbf{k}}$ be an arbitrary $n \times 1$ vector. Show that if the solution **T** of $\mathbf{A}\mathbf{T} - \mathbf{T}\mathbf{F} = \mathbf{b}\bar{\mathbf{k}}$ is nonsingular, then $(\mathbf{A} - \mathbf{b}\bar{\mathbf{k}}\mathbf{T}^{-1})$ has the same eigenvalues as **F**.

11.16. Connect the state feedback in Problem 11.9 to the estimator designed in Problem 11.11. Compute the compensators from u to w and from y to w in Figure 11.8. Also compute the overall transfer function from r to y. Does the overall transfer function completely characterize the overall system? What are the missing poles?

11.17. Repeat Problem 11.16 by using the estimator in Problem 11.13. Does the overall transfer function equal the one in Problem 11.16?

11.18. Connect the state feedback in Problem 11.10 to the estimator designed in Problem 11.12. Compute the compensators from u to w and from y to w in Figure 11.8. Also compute the overall transfer function from r to y. Does the overall transfer function completely characterize the overall system? What are the missing poles?

11.19. Repeat Problem 11.18 by using the estimator in Problem 11.14. Does the overall transfer function equal the one in Problem 11.18?

11.20. Redesign Problem 11.17 using the linear algebraic method in Section 10.6. Which method is simpler?

11.21. Redesign Problem 11.19 using the linear algebraic method in Section 10.6. Which method is simpler?

11.22. Check whether the following matrices are positive definite or semidefinite.

$$\begin{bmatrix} 2 & 1 \\ 1 & 1 \end{bmatrix} \qquad \begin{bmatrix} 0 & 0 & -2 \\ 0 & 3 & 0 \\ -2 & 0 & 1 \end{bmatrix} \qquad \begin{bmatrix} 5 & 0 & -2 \\ 0 & 3 & 0 \\ -2 & 0 & 1 \end{bmatrix}$$

11.23. Compute the eigenvalues of the matrices in Problem 11.22. Are they all real? From the eigenvalues, check whether the matrices are positive definite or semidefinite.

11.24. Consider the system in Problem 11.9. Find the state feedback gain to minimize the quadratic performance index in (11.45) with $R = 1$.

11.25. Consider

$$\mathbf{A}_1 = \begin{bmatrix} 2 & -4 \\ 3 & 1 \end{bmatrix} \qquad \mathbf{A}_2 = \begin{bmatrix} 0 & 1 & 0 \\ 0 & 0 & 1 \\ -2 & -3 & -1 \end{bmatrix}$$

a. Use the Routh test to check their stability.

b. Use the Lyapunov theorem to check their stability.

12 *Discrete-Time System Analysis*

12.1 INTRODUCTION

Signals can be classified as continuous-time and discrete-time. A continuous-time signal is defined for all time, whereas a discrete-time signal is defined only at discrete instants of time. Similarly, systems are classified as analog and digital. Analog systems are excited by continuous-time signals and generate continuous-time signals as outputs. The input and output of digital systems are discrete-time signals, or sequences of numbers. A system with an analog input and a digital output or vice versa can be modeled as either an analog or a digital system, depending on convenience of analysis and design.

A control system generally consists of a plant, a transducer, and a compensator or controller, as shown in Figure 12.1(a). The plant may consist of an object, such as an antenna, to be controlled, a motor, and possibly a power amplifier. Most plants in control systems are analog systems. The transducers and the compensators discussed in the preceding chapters are analog devices. However, because of their reliability, flexibility, and accuracy, digital transducers and compensators are increasingly being used to control analog plants, as is shown in Figure 12.1(b). In this and the following chapters, we discuss how to design digital compensators to control analog plants.

The organization of this chapter is as follows. In Section 12.2 we discuss the reasons for using digital compensators. Section 12.3 discusses A/D and D/A converters; they are needed to connect digital compensators with analog plants and vice versa. We then introduce the z-transform; its role is similar to the Laplace transform in the continuous-time case. The z-transform is then used to solve difference equa-

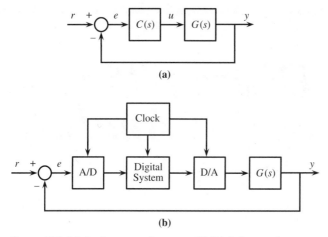

(a)

(b)

Figure 12.1 (a) Analog control system. (b) Digital control system.

tions and to develop transfer functions. Discrete-time state-variable equations together with the properties of controllability and observability are then discussed. We also discuss basic block diagrams and the realization problem. Finally we discuss stability, the Jury test, the Lyapunov Theorem, and frequency responses. Most concepts and results in the discrete-time case are quite similar to those in the continuous-time case; therefore, some results are stated without much discussion.

12.2 WHY DIGITAL COMPENSATORS?

A signal is called a *continuous-time* or *analog signal* if it is defined at every instant of time, as shown in Figure 12.2(a). Note that a continuous-time signal is not necessarily a continuous function of time; as is shown, it may be discontinuous.

The temperature in a room is a continuous-time signal. If it is *recorded* or *sampled* only at 2 P.M. and 2 A.M. of each day, then the data will appear as shown

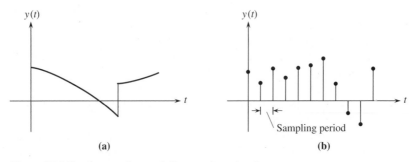

(a)   **(b)**

Figure 12.2 Continuous-time and discrete-time signals.

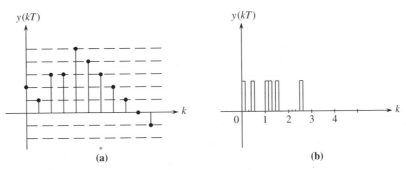

Figure 12.3 Digital signal and its binary representation.

in Figure 12.2(b). It is called a *discrete-time* signal, for it is defined only at discrete instants of time. The instants at which the data appear are called the *sampling instants*; the time intervals between two subsequent sampling instants are called the *sampling intervals*. In this text, we study only discrete-time signals with equal sampling intervals, in which case the sampling interval is called the *sampling period*.

The amplitude of a discrete-time signal can assume any value in a continuous range. If its amplitude can assume values only from a finite set, then it is called a *digital* signal, as shown in Figure 12.3(a). The gap between assumable values is called the *quantization step*. For example, if the temperature is recorded with a digital thermometer with integers read out, then the amplitude can assume only integers, and the quantization step is 1. Thus, a digital signal is *discretized* in time and *quantized* in amplitude, whereas a discrete-time signal is discretized in time but not quantized in amplitude. Signals processed on digital computers are all digital signals. Clearly, an error arises when a discrete-time signal is quantized. The error depends on the number of bits used to represent the digital signal. The error may be appreciable for a 4-bit representation. However, if 16 or more bits are used to represent digital signals, then the errors between discrete-time and digital signals are often negligible.

Consider digital signals that are limited up to the first decimal point—such as $10.1, 0.7$, and 2.9. The product of two such nonzero numbers may become zero. For example, $0.2 \times 0.1 = 0.0$, which is indistinguishable from $0.2 \times 0 = 0$. For this and other reasons, it is difficult to analyze digital signals. Therefore, digital signals are often considered to be discrete-time signals in analysis and design. The discrepancies between them are then studied separately as an error problem. Representation of discrete-time signals, such as $1/3$, may require an infinite number of bits, which is not possible in practice; therefore, in implementation, discrete-time signals must be quantized to become digital signals. In conclusion, in analysis, we consider only discrete-time signals; in implementation, we consider only digital signals. In this text, we make no distinction between these two types of signals, and *discrete-time* and *digital* are often used interchangeably.

In processing and transmission, digital signals are expressed in binary form, a string of zeros and ones or bits, as shown in Figure 12.3(b). Because strings of zeros and ones do not resemble the waveforms of original signals, we may call a digital signal a *nonanalog* signal. A continuous-time signal usually has the same waveform as the physical variable, thus it is also called an *analog* signal.

Systems that receive and generate analog signals are called *analog systems*. Systems that receive and generate digital signals are called *digital systems*. However, an analog system can be *modeled* as a digital system for convenience of analysis and design. For example, the system described by (2.90) is an analog system. However, if the input is stepwise, as shown in Figure 2.23 (which is still an analog signal), and if we consider the output only at sampling instants, then the system can be modeled as a digital system and described by the discrete-time equation in (2.92). This type of modeling is used widely in digital control systems. A system that has an analog input and generates a digital output, such as the transducer in Problem 3.11, can be modeled as either an analog system or a digital system.

We compare analog and digital techniques in the following:

1. Digital signals are coded in sequences of 0 and 1, which in terms are represented by ranges of voltages (for example, 0 from 0 to 1 volt and 1 from 2 to 4 volts). This representation is less susceptible to noise and drift of power supply.
2. The accuracy of analog systems is often limited. For example, if an analog system is to be built using a resistor with resistance 980.5 ohms and a capacitor with capacitance 81.33 microfarads, it would be difficult and expensive to obtain components with exactly these values. The accuracy of analog transducers is also limited. It is difficult to read an exact value if it is less than 0.1% of the full scale. In digital systems, there is no such problem, however. The accuracy of a digital device can be increased simply by increasing the number of bits. Thus, digital systems are generally more accurate and more reliable than analog systems.
3. Digital systems are more flexible than analog systems. Once an analog system is built, there is no way to alter it without replacing some components or the entire system. Except for special digital hardware, digital systems can often be changed by programming. If a digital computer is used, it can be used not only as a compensator but also to collect data, to carry out complex computation, and to monitor the status of the control system. Thus, a digital system is much more flexible and versatile than an analog system.
4. Because of the advance of very large scale integration (VLSI) technology, the price of digital systems has been constantly decreasing during the last decade. Now the use of a digital computer or microprocessor is cost effective even for small control systems.

For these reasons, it is often desirable to design digital compensators in control systems.

12.3 A/D AND D/A CONVERSIONS

Although compensators are becoming digital, most plants are still analog. In order to connect digital compensators and analog plants, analog signals must be converted into digital signals and vice versa. These conversions can be achieved by using

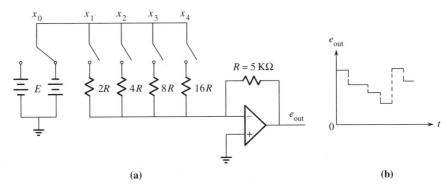

Figure 12.4 D/A converter.

analog-to-digital (A/D) and digital-to-analog (D/A) converters. We now discuss how these conversions are achieved.

Consider the operational amplifier circuit shown in Figure 12.4. It is essentially the operational amplifier circuit shown in Figure 3.15. As in (3.39), the output can be shown to be

$$v_o = -\left(x_1 \frac{R}{2R} + x_2 \frac{R}{4R} + x_3 \frac{R}{8R} + x_4 \frac{R}{16R}\right) E$$
$$= -(x_1 2^{-1} + x_2 2^{-2} + x_3 2^{-3} + x_4 2^{-4})E$$

where E is the supplied voltage, and x_i is either 1 or 0, closed or open. The bit x_0 is called the *sign bit*. If $x_0 = 0$, then $E > 0$; if $x_0 = 1$, then $E < 0$. If $x_0 x_1 x_2 x_3 x_4 = 11011$, and if $E = 10$ volts, then

$$v_o = -(1 \cdot 2^{-1} + 1 \cdot 2^{-3} + 1 \cdot 2^{-4}) \cdot (-10) = 0.6875$$

The circuit will hold this value until the next set of x_i is received. Thus the circuit changes a five-bit digital signal $x_0 x_1 x_2 x_3 x_4$ into an analog signal of magnitude 0.6875, as shown in Figure 12.4(b). Thus the circuit can convert a digital signal into an analog signal, and is called a D/A converter. The D/A converter in Figure 12.4 is used only to illustrate the basic idea of conversion; practical D/A converters usually use different circuit arrangements so that resistors have resistances closer to each other and are easier to fabricate. The output of a D/A converter is discontinuous, as is shown in Figure 12.4. It can be smoothed by passing through a low-pass filter. This may not be necessary if the converter is connected to a plant, because most plants are low-pass in nature and can act as low-pass filters.

The analog-to-digital conversion can be achieved by using the circuit shown in Figure 12.5(a). The circuit consists of a D/A converter, a counter, a comparator, and control logic. In the conversion, the counter starts to drive the D/A converter. The output of the converter is compared with the analog signal to be converted. The counter is stopped when the output of the D/A converter exceeds the value of the analog signal, as shown in Figure 12.5(b). The value of the counter is then transferred to the output register and is the digital representation of the analog signal.

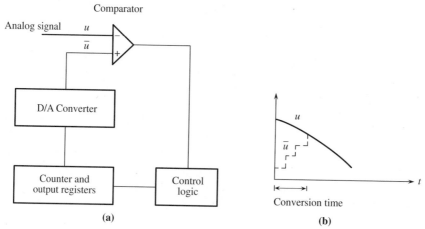

Figure 12.5 A/D converter and conversion time.

We see from Figure 12.5(b) that the A/D conversion cannot be achieved instantaneously; it takes a small amount of time to complete the conversion (for example, 2 microseconds for a 12-bit A/D converter). Because of this conversion time, if an analog signal changes rapidly, then the value converted may be different from the value intended for conversion. This problem can be resolved by connecting a sample-and-hold circuit in front of an A/D converter. Such a circuit is shown in Figure 12.6. The field-effect transistor (FET) is used as a switch; its on and off states are controlled by a control logic. The voltage followers [see Figure 3.14(a)] in front and in back of the switch are used to eliminate the loading problem or to shield the capacitor from other parts of circuits. When the switch is closed, the input voltage will rapidly charge the capacitor to the input voltage. When the switch is off, the capacitor voltage remains almost constant. Hence the output of the circuit is stepwise as shown. Using this device, the problem due to the conversion time can be eliminated. Therefore, a sample-and-hold circuit is often used together with an A/D converter.

With A/D and D/A converters, the analog control system in Figure 12.1(a) can be implemented as shown in Figure 12.1(b). We call the system in Figure 12.1(b) a *digital control system*. In the remainder of this chapter, we discuss digital system analysis; design will be discussed in the next chapter.

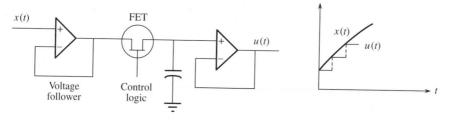

Figure 12.6 Sample-and-hold circuit.

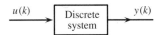

Figure 12.7 Discrete-time system.

12.4   THE z-TRANSFORM

Consider the discrete-time system shown in Figure 12.7. If we apply an input sequence $u(k) := u(kT)$, $k = 0, 1, 2, \ldots$, then the system will generate an output sequence $y(k) := y(kT)$. This text studies only the class of discrete-time systems whose inputs and outputs can be described by linear difference equations with constant real coefficients such as

$$3y(k + 2) + 2y(k + 1) - y(k) = 2u(k + 1) - 3u(k) \qquad (12.1)$$

or

$$3y(k) + 2y(k - 1) - y(k - 2) = 2u(k - 1) - 3u(k - 2) \qquad (12.2)$$

In order to be describable by such an equation, the system must be linear, time-invariant, and lumped (LTIL). Difference equations are considerably simpler than the differential equations discussed in Chapter 2. For example, if we write (12.2) as

$$y(k) = \frac{1}{3} [-2y(k - 1) + y(k - 2) + 2u(k - 1) - 3u(k - 2)]$$

then its response due to the initial conditions $y(-2) = 1$, $y(-1) = -2$ and the unit-step input sequence $u(k) = 1$, for $k = 0, 1, 2, \ldots$, and $u(k) = 0$ for $k < 0$, can be computed recursively as

$$y(0) = \frac{1}{3} [-2y(-1) + y(-2) + 2u(-1) - 3u(-2)]$$

$$= \frac{1}{3} [-2 \times (-2) + 1 + 2 \times 0 - 3 \times 0] = \frac{5}{3}$$

$$y(1) = \frac{1}{3} [-2y(0) + y(-1) + 2u(0) - 3u(-1)]$$

$$= \frac{1}{3} \left[-2 \times \frac{5}{3} - 2 + 2 \right] = \frac{-10}{9}$$

$$y(2) = \frac{1}{3} [-2y(1) + y(0) + 2u(1) - 3u(0)]$$

$$= \frac{1}{3} \left[-2 \times \frac{-10}{9} + \frac{5}{3} + 2 - 3 \right] = \frac{26}{27}$$

and so forth. Thus, the solution of difference equations can be obtained by direct substitution. The solution obtained by this process is generally not in closed form, and it is difficult to abstract from the solution general properties of the equation. For this and other reasons, we introduce the z-transform.

Consider a sequence $f(k)$. The z-transform of $f(k)$ is defined as

$$F(z) := \mathcal{Z}[f(k)] := \sum_0^\infty f(k)z^{-k} \tag{12.3}$$

where z is a complex variable. The z-transform is defined for $f(k)$ with $k \geq 0$; $f(k)$ with $k < 0$ does not appear in $F(z)$. If we write $F(z)$ explicitly as

$$F(z) = f(0) + f(1)z^{-1} + f(2)z^{-2} + f(3)z^{-3} + \cdots$$

then z^{-i} can be used to indicate the ith sampling instant. In other words, z^0 indicates the initial time $k = 0$; z^{-1} indicates the time instant at $k = 1$ and, in general, z^{-i} indicates the ith sampling instant. For this reason, z^{-1} is called the *unit-delay* element.

The infinite power series in (12.3) is not easy to manipulate. Fortunately, the z-transforms of sequences encountered in this text can be expressed in closed form. This will be obtained by using the following formula

$$1 + r + r^2 + r^3 + \cdots = \sum_0^\infty r^k = \frac{1}{1 - r} \tag{12.4}$$

where r is a real or complex constant with amplitude less than 1, or $|r| < 1$.

Example 12.4.1

Consider $f(k) = e^{-akT}$, $k = 0, 1, 2, \ldots$. Its z-transform is

$$F(z) = \sum_0^\infty e^{-akT}z^{-k} = \sum_0^\infty (e^{-aT}z^{-1})^k = \frac{1}{1 - e^{-aT}z^{-1}} = \frac{z}{z - e^{-aT}} \tag{12.5}$$

This holds only if $|e^{-aT}z^{-1}| < 1$ or $|e^{-aT}| < |z|$. This condition, called the *region of convergence*, will be disregarded, however, and (12.5) is considered to be defined for all z except at $z = e^{-aT}$. See Reference [18] for a justification.

If $a = 0$, e^{-akT} equals 1 for all positive k. This is called the *unit-step sequence*, as is shown in Figure 12.8(a), and will be denoted by $q(k)$. Thus we have

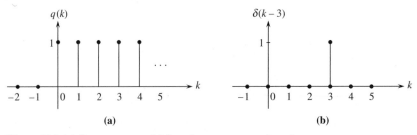

Figure 12.8 (a) Step sequence. (b) Impulse sequence at $k = 3$.

$$\mathscr{Z}[q(k)] = \frac{1}{1 - z^{-1}} = \frac{z}{z - 1}$$

If we define $b = e^{-aT}$, then (12.5) becomes

$$\mathscr{Z}[b^k] = \frac{1}{1 - bz^{-1}} = \frac{z}{z - b}$$

The z-transform has the following linearity property

$$\mathscr{Z}[a_1 f_1(k) + a_2 f_2(k)] = a_1 \mathscr{Z}[f_1(k)] + a_2 \mathscr{Z}[f_2(k)]$$

Using this property, we have

$$\mathscr{Z}[\sin \omega kT] = \mathscr{Z}\left[\frac{e^{j\omega kT} - e^{-j\omega kT}}{2j}\right] = \frac{1}{2j}\left[\frac{z}{z - e^{j\omega T}} - \frac{z}{z - e^{-j\omega T}}\right]$$

$$= \frac{z(z - e^{-j\omega T} - z + e^{j\omega T})}{2j(z - e^{j\omega T})(z - e^{-j\omega T})}$$

$$= \frac{z(e^{j\omega T} - e^{-j\omega T})}{2j(z^2 - (e^{j\omega T} + e^{-j\omega T})z + 1)} = \frac{z \sin \omega T}{z^2 - 2(\cos \omega T)z + 1}$$

An impulse sequence or a Kronecker sequence is defined as

$$\delta(k) = \begin{cases} 1 & \text{if } k = 0 \\ 0 & \text{if } k \neq 0 \end{cases} \qquad (12.6a)$$

or, more generally,

$$\delta(k - n) = \begin{cases} 1 & \text{if } k = n \\ 0 & \text{if } k \neq n \end{cases} \qquad (12.6b)$$

as shown in Figure 12.8(b). The impulse sequence $\delta(k - n)$ equals 1 at $k = n$ and zero elsewhere. The z-transforms of (12.6) are, from definition,

$$\mathscr{Z}[\delta(k)] = 1 \qquad \mathscr{Z}[\delta(k - n)] = z^{-n}$$

All sequences, except the impulse sequence, studied in this text will be obtained from sampling of analog signals. For example, the sequence $f(kT) = e^{-akT}$ in Example 12.4.1 is the sampled sequence of $f(t) = e^{-at}$ with sampling period T. Let $\overline{F}(s)$ be the Laplace transform of $f(t)$ and $F(z)$ be the z-transform of $f(kT)$. Note that we must use different notations to denote the Laplace transform and z-transform, or confusion will arise. If $f(kT)$ is the sample of $f(t)$, then we have

$$F(z) = \mathscr{Z}[f(kT)] = \mathscr{Z}[f(t)|_{t=kT}] = \mathscr{Z}[[\mathscr{L}^{-1}\overline{F}(s)]|_{t=kT}] \qquad (12.7a)$$

This is often written simply as

$$F(z) = \mathscr{Z}[\overline{F}(s)] \qquad (12.7b)$$

Example 12.4.2

Consider $f(t) = e^{-at}$. Then we have

$$\overline{F}(s) = \mathscr{L}[f(t)] = \mathscr{L}[e^{-at}] = \frac{1}{s + a}$$

and

$$F(z) = \mathscr{Z}[f(kT)] = \mathscr{Z}[e^{-akT}] = \frac{z}{z - e^{-aT}}$$

Thus we have

$$\frac{z}{z - e^{-aT}} = \mathscr{Z}\left[\frac{1}{s + a}\right]$$

Exercise 12.4.1

Let $f(t) = \sin \omega t$. Verify

$$\frac{z \sin \omega T}{(z - e^{j\omega T})(z - e^{-j\omega T})} = \mathscr{Z}\left[\frac{\omega}{(s - j\omega)(s + j\omega)}\right]$$

From the preceding example and exercise, we see that a pole α in $\overline{F}(s)$ is mapped into the pole $e^{\alpha T}$ in $F(z)$ by the analog-to-digital transformation in (12.7). This property will be further established in the next subsection. We list in Table 12.1 some pairs of $\overline{F}(s)$ and $F(z)$. In the table, we use $\overline{\delta}(t)$ to denote the impulse defined for the continuous-time case in Appendix A and $\delta(k)$ to denote the impulse sequence defined for the discrete-time case in (12.6). Because $\overline{\delta}(t)$ is not defined at $t = 0$, $\delta(k)$ is not the sample of $\overline{\delta}(t)$. The sixth and eighth pairs of the table are obtained by using

$$\mathscr{Z}[kf(k)] = -z\frac{dF(z)}{dz}$$

For example, because $\mathscr{Z}[b^k] = z/(z - b)$, we have

$$\mathscr{Z}[kb^k] = -z\frac{d}{dz}\left(\frac{z}{z - b}\right) = -z \cdot \frac{(z - b) - z}{(z - b)^2} = \frac{bz}{(z - b)^2}$$

12.4.1 The Laplace Transform and the z-Transform

The Laplace transform is defined for continuous-time signals and the z-transform is defined for discrete-time signals. If we apply the Laplace transform directly to a discrete-time signal $f(kT)$, then the result will be zero. Now we shall modify $f(kT)$

Table 12.1 z-Transform Pairs.

$\overline{F}(s)$	$f(t)$	$f(kT)$	$F(z)$
1	$\overline{\delta}(t)$	—	—
e^{-Ts}	$\overline{\delta}(t-T)$	—	—
—	—	$\delta(kT)$	1
—	—	$\delta((k-n)T)$	z^{-n}
$\dfrac{1}{s}$	1	1	$\dfrac{z}{z-1}$
$\dfrac{1}{s^2}$	t	kT	$\dfrac{Tz}{(z-1)^2}$
$\dfrac{1}{s+a}$	e^{-at}	e^{-akT}	$\dfrac{z}{z-e^{-aT}}$
$\dfrac{1}{(s+a)^2}$	te^{-at}	kTe^{-akT}	$\dfrac{Tze^{-aT}}{(z-e^{-aT})^2}$
$\dfrac{\omega}{s^2+\omega^2}$	$\sin \omega t$	$\sin \omega kT$	$\dfrac{z\sin \omega T}{z^2 - 2z(\cos \omega T)+1}$
$\dfrac{s}{s^2+\omega^2}$	$\cos \omega t$	$\cos \omega kT$	$\dfrac{z(z-\cos \omega T)}{z^2 - 2z(\cos \omega T)+1}$
$\dfrac{\omega}{(s+a)^2+\omega^2}$	$e^{-at}\sin \omega t$	$e^{-akT}\sin \omega kT$	$\dfrac{ze^{-aT}\sin \omega T}{z^2 - 2ze^{-aT}(\cos \omega T)+e^{-2aT}}$
$\dfrac{s+a}{(s+a)^2+\omega^2}$	$e^{-at}\cos \omega t$	$e^{-akT}\cos \omega kT$	$\dfrac{z^2 - ze^{-aT}(\cos \omega T)}{z^2 - 2ze^{-aT}(\cos \omega T)+e^{-2aT}}$

so that the Laplace transform can be applied. Consider $f(kT)$, for integer $k \geq 0$ and positive sampling period $T > 0$. We define

$$f^*(t) := \sum_{k=0}^{\infty} f(kT)\overline{\delta}(t-kT) \tag{12.8}$$

where $\overline{\delta}(t-kT)$ is the impulse defined in Appendix A. It is zero everywhere except at $t = kT$. Therefore, $f^*(t)$ is zero everywhere except at sampling instants kT, where it is an impulse with weight $f(kT)$. Thus, we may consider $f^*(t)$ to be a continuous-time representation of the discrete-time sequence $f(kT)$. The Laplace transform of $f^*(t)$ is, using (A.21),

$$F^*(s) = \mathcal{L}[f^*(t)] = \sum_{k=0}^{\infty} f(kT)\mathcal{L}[\overline{\delta}(t-kT)] = \sum_{k=0}^{\infty} f(kT)e^{-kTs} \tag{12.9}$$

If we define $z = e^{Ts}$, then (12.9) becomes

$$F^*(s)\big|_{z=e^{Ts}} = \sum_{k=0}^{\infty} f(kT)z^{-k}$$

Its right-hand side is the z-transform of $f(kT)$. Thus the z-transform of $f(kT)$ is the Laplace transform of $f^*(t)$ with the substitution of $z = e^{Ts}$ or

$$\mathcal{Z}[f(kT)] = \mathcal{L}[f^*(t)]\big|_{z=e^{Ts}} \qquad (12.10)$$

This is an important relationship between the Laplace transform and z-transform. We now discuss the implication of

$$z = e^{Ts} \qquad \text{or} \qquad s = \frac{1}{T}\ln z \qquad (12.11)$$

where ln stands for the natural logarithm. If $s = 0$, then $z = e^0 = 1$; that is, the origin of the s-plane is mapped into $z = 1$ in the z-plane as shown in Figure 12.9. In fact, because

$$e^{\left(\frac{jm2\pi}{T}\right)T} = e^{jm2\pi} = 1$$

for all positive and negative integer m, the points $s = 0, j2\pi/T, -j2\pi/T, j4\pi/T,$ ... are all mapped into $z = 1$. Thus, the mapping is not one-to-one. If $s = j\omega$, then

$$|z| = |e^{j\omega T}| = 1$$

for all ω. This implies that the imaginary axis of the s-plane is mapped into the unit circle on the z-plane. If $s = a + j\omega$, then

$$|z| = |e^{(a+j\omega)T}| = |e^{aT}|\,|e^{j\omega T}| = e^{aT}$$

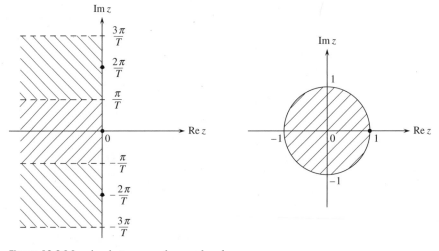

Figure 12.9 Mapping between s-plane and z-plane.

Thus, a vertical line in the s-plane is mapped into the circle in the z-plane with radius e^{aT}. If $a < 0$, the vertical line is in the left half s-plane and the radius of the circle is smaller than 1; if $a > 0$, the radius is larger than 1. Thus the entire open left half s-plane is mapped into the interior of the unit circle on the z-plane. To be more specific, the strip between $-\pi/T$ and π/T shown in Figure 12.9 is mapped into the unit circle. The upper and lower strips shown will also be mapped into the unit circle. We call the strip between $-\pi/T$ and π/T the *primary strip*.

12.4.2 Inverse z-Transform

Consider a z-transform $F(z) = N(z)/D(z)$, where $N(z)$ and $D(z)$ are polynomials of z. One way to obtain the inverse z-transform of $G(z)$ is to express $G(z)$ as a power series as in (12.3). This can be done by direct division. For example, consider

$$F(z) = \frac{2z - 3}{3z^2 + 2z - 1} \tag{12.12}$$

If we compute

$$
\begin{array}{r}
\frac{2}{3}z^{-1} - \frac{13}{9}z^{-2} + \frac{32}{27}z^{-3} + \cdots \\[4pt]
3z^2 + 2z - 1 \overline{\smash{\big)}\, 2z - 3 } \\
2z + \frac{4}{3} - \frac{2}{3}z^{-1} \\
\hline
\frac{-13}{3} + \frac{2}{3}z^{-1} \\
\frac{-13}{3} - \frac{26}{9}z^{-1} + \frac{13}{9}z^{-2} \\
\hline
\frac{32}{9}z^{-1} - \frac{13}{9}z^{-2} \\
\vdots
\end{array}
$$

then we have

$$F(z) = \frac{2z - 3}{3z^2 + 2z - 1} = 0 + \frac{2}{3}z^{-1} - \frac{13}{9}z^{-2} + \frac{32}{27}z^{-3} + \cdots$$

Thus, the inverse z-transform of $F(z)$ is

$$f(0) = 0, \ f(1) = \frac{2}{3}, \ f(2) = -\frac{13}{9}, \ f(3) = \frac{32}{27}, \ \ldots \tag{12.13}$$

Therefore, the inverse z-transform of $F(z)$ can be easily obtained by direct division.

The inverse z-transform can also be obtained by partial fraction expansion and table look-up. We use the z-transform pairs in Table 12.1. Although the procedure is similar to the Laplace transform case, we must make one modification. Instead of expanding $F(z)$, we expand $F(z)/z$. For example, for $F(z)$ in (12.12), we expand

$$\frac{F(z)}{z} = \frac{2z - 3}{z(3z^2 + 2z - 1)} = \frac{2z - 3}{z(3z - 1)(z + 1)} \tag{12.14}$$

$$= \frac{k_1}{z} + \frac{k_2}{3z - 1} + \frac{k_3}{z + 1}$$

with

$$k_1 = \left.\frac{2z - 3}{(3z - 1)(z + 1)}\right|_{z=0} = \frac{-3}{-1} = 3$$

$$k_2 = \left.\frac{2z - 3}{z(z + 1)}\right|_{z=\frac{1}{3}} = \frac{\frac{2}{3} - 3}{\frac{1}{3} \cdot \frac{4}{3}} = \frac{-21}{4}$$

and

$$k_3 = \left.\frac{2z - 3}{z(3z - 1)}\right|_{z=-1} = \frac{-5}{(-1)(-4)} = \frac{-5}{4}$$

We then multiply (12.14) by z to yield

$$F(z) = 3\frac{z}{z} - \frac{21}{4}\frac{z}{3z - 1} - \frac{5}{4}\frac{z}{z + 1} = 3 - \frac{7}{4}\frac{z}{z - \frac{1}{3}} - \frac{5}{4}\frac{z}{z + 1}$$

Therefore, the inverse z-transform of $F(z)$ is, using Table 12.1,

$$f(k) = 3\delta(k) - \frac{7}{4}\left(\frac{1}{3}\right)^k - \frac{5}{4}(-1)^k$$

for $k = 0, 1, 2, 3, \ldots$. For example, we have

$$f(0) = 3 - \frac{7}{4} - \frac{5}{4} = 0$$

$$f(1) = 0 - \frac{7}{4} \cdot \frac{1}{3} - \frac{5}{4} \cdot (-1) = \frac{-7}{12} + \frac{5}{4} = \frac{-7 + 15}{12} = \frac{2}{3}$$

$$f(2) = 0 - \frac{7}{4} \cdot \left(\frac{1}{9}\right) - \frac{5}{4} = \frac{-13}{9}$$

and so forth. The result is the same as (12.13).

From the preceding example, we see that the procedure here is quite similar to that in the inverse Laplace transform. The only difference is that we expand $F(z)/z$ in partial fraction expansion and then multiply the expansion by z. The reason for doing this is that the z-transform pairs in Table 12.1 are mostly of the form $z/(z - b)$.

12.4.3   Time Delay and Time Advance

Consider the time sequence $f(k)$ shown in Figure 12.10(a). It is not necessarily zero for $k \leq 0$. Its z-transform is

$$F(z) = \mathcal{Z}[f(k)] = \sum_0^\infty f(k)z^{-k} \tag{12.15}$$

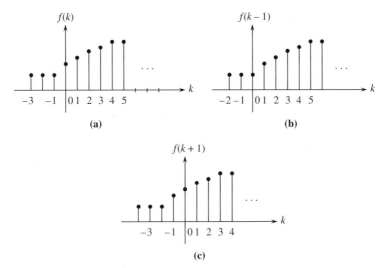

Figure 12.10 (a) Sequence. (b) Time delay. (c) Time advance.

It is defined only for $f(k)$ with $k \geq 0$, and $f(-1)$, $f(-2)$, ... do not appear in $F(z)$. Consider $f(k-1)$. It is $f(k)$ shifted to the right or delayed by one sampling period as shown in Figure 12.10(b). Its z-transform is

$$\mathcal{Z}[f(k-1)] = \sum_{0}^{\infty} f(k-1)z^{-k} = z^{-1}\sum_{k=0}^{\infty} f(k-1)z^{-(k-1)}$$

which becomes, after the substitution of $\bar{k} = k - 1$,

$$\mathcal{Z}[f(k-1)] = z^{-1}\sum_{\bar{k}=-1}^{\infty} f(\bar{k})z^{-\bar{k}} = z^{-1}\left[f(-1)z + \sum_{\bar{k}=0}^{\infty} f(\bar{k})z^{-\bar{k}} \right]$$

$$= z^{-1}[f(-1)z + F(z)]$$

(12.16a)

This has a simple physical interpretation. $F(z)$ consists of $f(k)$ with $k \geq 0$. If $f(k)$ is delayed by one sampling period, $f(-1)$ will move into $k = 0$ and must be included in the z-transform of $f(k-1)$. Thus we add $f(-1)z$ to $F(z)$ and then delay it (multiplying by z^{-1}) to yield $\mathcal{Z}[f(k-1)]$. Using the same argument, we have

$$\mathcal{Z}[f(k-2)] = z^{-2}[f(-2)z^2 + f(-1)z + F(z)]$$

(12.16b)

$$\mathcal{Z}[f(k-3)] = z^{-3}[f(-3)z^3 + f(-2)z^2 + f(-1)z + F(z)]$$

(12.16c)

and so forth.

Now we consider $f(k+1)$. It is the shifting of $f(k)$ to the left (or advancing by one sampling period) as shown in Figure 12.10(c). Because $f(0)$ is moved to $k = -1$, it will not be included in the z-transform of $f(k+1)$, so it must be excluded from $F(z)$. Thus, we subtract $f(0)$ from $F(z)$ and then advance it (multiplying by z), to yield $\mathscr{Z}[f(k + 1)]$ or

$$\mathscr{Z}[f(k + 1)] = z[F(z) - f(0)] \qquad (12.17a)$$

Similarly, we have

$$\mathscr{Z}[f(k + 2)] = z^2[F(z) - f(0) - f(1)z^{-1}] \qquad (12.17b)$$

$$\mathscr{Z}[f(k + 3)] = z^3[F(z) - f(0) - f(1)z^{-1} - f(2)z^{-2}] \qquad (12.17c)$$

and so forth.

12.5   SOLVING LTIL DIFFERENCE EQUATIONS

Consider the LTIL difference equation in (12.2) or

$$3y(k) + 2y(k - 1) - y(k - 2) = 2u(k - 1) - 3u(k - 2) \qquad (12.18)$$

As was discussed earlier, its solution can be obtained by direct substitution. The solution, however, will not be in closed form, and it will be difficult to develop from the solution general properties of the equation. Now we apply the z-transform to study the equation. The equation is of second order; therefore, the response $y(k)$ depends on the input $u(k)$ and two initial conditions. To simplify discussion, we assume that $u(k) = 0$ for $k \leq 0$ and that the two initial conditions are $y(-1)$ and $y(-2)$. The application of the z-transform to (12.18) yields, using (12.16),

$$3Y(z) + 2z^{-1}[Y(z) + y(-1)z] - z^{-2}[Y(z) + y(-1)z + y(-2)z^2]$$
$$= 2z^{-1}U(z) - 3z^{-2}U(z) \qquad (12.19)$$

which can be grouped as

$$(3 + 2z^{-1} - z^{-2})Y(z) = [-2y(-1) + y(-1)z^{-1} + y(-2)]$$
$$+ (2z^{-1} - 3z^{-2})U(z)$$

Thus we have

$$Y(z) = \frac{[-2y(-1) + y(-1)z^{-1} + y(-2)]}{3 + 2z^{-1} - z^{-2}} + \frac{(2z^{-1} - 3z^{-2})}{3 + 2z^{-1} - z^{-2}} U(z) \qquad (12.20a)$$

$$= \underbrace{\frac{[-2y(-1)z^2 + y(-1)z + y(-2)z^2]}{3z^2 + 2z - 1}}_{\text{Zero-Input Response}} + \underbrace{\frac{(2z - 3)}{3z^2 + 2z - 1} U(z)}_{\text{Zero-State Response}} \qquad (12.20b)$$

Now if $y(-2) = 1$, $y(-1) = -2$, and if $u(k)$ is a unit-step sequence, then $U(z) = z/(z - 1)$ and

$$Y(z) = \frac{5 - 2z^{-1}}{3 + 2z^{-1} - z^{-2}} + \frac{(2z^{-1} - 3z^{-2})}{3 + 2z^{-1} - z^{-2}} \cdot \frac{z}{z - 1}$$

$$= \frac{5z^2 - z}{3z^2 + 2z - 1} + \frac{z(2z - 3)}{(3z - 1)(z + 1)(z - 1)} = \frac{z(5z^2 - 4z - 2)}{(3z - 1)(z + 1)(z - 1)}$$

To find its time response, we expand $Y(z)/z$ as

$$\frac{Y(z)}{z} = \frac{5z^2 - 4z - 2}{(3z - 1)(z + 1)(z - 1)} = = \frac{19}{8} \cdot \frac{1}{3z - 1} + \frac{9}{8} \cdot \frac{1}{z + 1} - \frac{1}{4} \cdot \frac{1}{z - 1}$$

Thus we have

$$Y(z) = \frac{19}{8} \cdot \frac{z}{3z - 1} + \frac{9}{8} \cdot \frac{z}{z + 1} - \frac{1}{4} \cdot \frac{z}{z - 1}$$

and its inverse z-transform is, using Table 12.1,

$$y(k) = \frac{19}{24} \left(\frac{1}{3}\right)^k + \frac{9}{8} (-1)^k - \frac{1}{4} (1)^k \qquad (12.21)$$

for $k = 0, 1, 2, \ldots$. We see that using the z-transform, we can obtain closed-form solutions of LTIL difference equations.

12.5.1 Characteristic Polynomials and Transfer Functions

The response of any LTIL difference equations can be decomposed into the zero-state response and zero-input response as shown in (12.20). Consider the nth order LTIL difference equation

$$a_n y(k + n) + a_{n-1} y(k + n - 1) + \cdots$$
$$+ a_1 y(k + 1) + a_0 y(k)$$
$$= b_m u(k + m) + b_{m-1} u(k + m - 1) + \cdots \qquad (12.22)$$
$$+ b_1 u(k + 1) + b_0 u(k)$$

We define

$$D(p) := a_n p^n + a_{n-1} p^{n-1} + \cdots + a_1 p + a_0 \qquad (12.23a)$$

and

$$N(p) := b_m p^m + b_{m-1} p^{m-1} + \cdots + b_1 p + b_0 \qquad (12.23b)$$

where the variable p is the unit-time advance operator, defined as

$$py(k) := y(k + 1) \qquad p^2 y(k) := y(k + 2) \qquad p^3 y(k) := y(k + 3) \quad (12.24)$$

and so forth. Using this notation, (12.22) can be written as

$$D(p)y(k) = N(p)u(k) \qquad (12.25)$$

In the study of the zero-input response, we assume $u(k) \equiv 0$. Then (12.25) becomes

$$D(p)y(k) = 0 \qquad (12.26)$$

This is the homogeneous equation. Its solution is excited exclusively by initial conditions. The application of the z-transform to (12.26) yields, as in (12.20),

$$Y(z) = \frac{I(z)}{D(z)}$$

where $D(z)$ is defined in (12.23a) with p replaced by z and $I(z)$ is a polynomial of z depending on initial conditions $y(-k)$, $k = 1, 2, \ldots, n$. As in the continuous-time case, we call $D(z)$ the characteristic polynomial of (12.25) because it governs the *free, unforced,* or *natural* response of (12.25). The roots of the polynomial $D(s)$ are called the *modes*. For example, if

$$D(z) = (z - 2)(z + 1)^2(z + 2 - j3)(z + 2 + j3)$$

then the modes are 2, -1, -1, $-2 + j3$ and $-2 - j3$. Root 2 and complex roots $-2 \pm j3$ are simple modes and root -1 is a repeated mode with multiplicity 2. Thus, for any initial conditions, the zero-input response due to any initial conditions will be of the form

$$y(k) = k_1(2)^k + k_2(-2 + j3)^k + k_3(-2 - j3)^k + c_1(-1)^k + c_2k(-1)^k$$

for $k = 0, 1, 2, \ldots$. Thus the zero-input response is essentially determined by the modes of the system.

Consider the difference equation in (12.18). If all initial conditions are zero, then the response is excited exclusively by the input and is called the *zero-state response*. In the z-transform domain, the zero-state response of (12.18) is governed by, as computed in (12.20),

$$Y(z) = \frac{2z^{-1} - 3z^{-2}}{3 + 2z^{-1} - z^{-2}} U(z) = \frac{2z - 3}{3z^2 + 2z - 1} U(z) =: G(z)U(z) \quad \text{(12.27)}$$

where the rational function $G(z)$ is called the *discrete, digital, pulse,* or *sampled transfer function* or simply the *transfer function*. It is the ratio of the z-transforms of the output and input when all initial conditions are zero or

$$G(z) = \left.\frac{Y(z)}{U(z)}\right|_{Initial\ conditions = 0} = \left.\frac{\mathscr{Z}[\text{Output}]}{\mathscr{Z}[\text{Input}]}\right|_{Initial\ conditions = 0} \quad \text{(12.28)}$$

The transfer function describes only the zero-state responses of LTIL systems.

The transfer function can easily be obtained from difference equations. For example, if a system is described by the difference equation

$$D(p)y(k) = N(p)u(k)$$

where $D(p)$ and $N(p)$ are defined as in (12.23), then its transfer function is

$$G(z) = \frac{N(z)}{D(z)} \quad \text{(12.29)}$$

Poles and zeros of $G(z)$ are defined exactly as in the continuous-time case. For example, given

$$G(z) = \frac{N(z)}{D(z)} = \frac{2(z + 3)(z - 1)(z + 1)}{(z - 1)(z + 2)(z + 1)^3} = \frac{2(z + 3)}{(z + 2)(z + 1)^2} \quad \text{(12.30)}$$

Its poles are -2, -1 and -1; its zero is -3. Thus, if $N(z)$ and $D(z)$ in $G(z) = N(z)/D(z)$ have no common factors, then the roots of $N(z)$ are the zeros and the roots of $D(z)$ are the poles of $G(z)$.

Consider a discrete-time system described by the difference equation

$$D(p)y(k) = N(p)u(k) \tag{12.31}$$

with $D(p)$ and $N(p)$ defined in (12.23). The zero-input response of the system is governed by the modes, the roots of the characteristic polynomial $D(z)$. If $N(p)$ and $D(p)$ have no common factors, then the set of the poles of the transfer function in (12.29) equals the set of the modes. In this case, the system is said to be completely characterized by its transfer function and there is no loss of essential information in using the transfer function to study the system. On the other hand, if $D(z)$ and $N(z)$ have common factors—say, $R(s)$—then

$$G(z) = \frac{N(z)}{D(z)} = \frac{\overline{N}(z)R(z)}{\overline{D}(z)R(z)} = \frac{\overline{N}(z)}{\overline{D}(z)}$$

In this case, the poles of $G(z)$ consist of only the roots of $\overline{D}(z)$. The roots of $R(z)$ are not poles of $G(z)$, even though they are modes of the system. Therefore, if $D(z)$ and $N(z)$ have common factors, not every mode will be a pole of $G(z)$ ($G(z)$ is said to have *missing poles*), and the system is not completely characterized by the transfer function. In this case, we cannot disregard the zero-input response, and care must be exercised in using the transfer function.

To conclude this section, we plot in Figure 12.11 the time responses of some poles. If a simple or repeated pole lies inside the unit circle of the z-plane, its time response will approach zero as $k \to \infty$. If a simple or repeated pole lies outside the unit circle, its time response will approach infinity. The time response of a simple pole at $z = 1$ is a constant; the time response of a simple pole on the unit circle other than $z = 1$ is a sustained oscillation. The time response of a repeated pole on the unit circle will approach infinity as $k \to \infty$. In conclusion, the time response of a simple or repeated pole approaches zero if and only if the pole lies inside the unit circle. The time response of a pole approaches a nonzero constant if and only if the pole is simple and is located at $z = 1$.

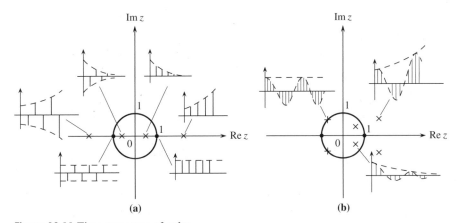

(a)   (b)

Figure 12.11 Time responses of poles.

12.5.2   Causality and Time Delay

Consider the digital transfer function

$$G(z) = \frac{N(z)}{D(z)}$$

with deg $N(z) = m$ and deg $D(z) = n$. The transfer function is improper if $m > n$ and proper if $n \geq m$. A system with an improper transfer function is called a *noncausal* or an *anticipatory system* because the output of the system may appear before the application of an input. For example, if $G(z) = z^2/(z - 0.5)$, then its unit-step response is

$$Y(z) = G(z)U(z) = \frac{z^2}{z - 0.5} \cdot \frac{z}{z - 1} = z + 1.5 + 1.75z^{-1} + 1.875z^{-2} + \cdots$$

and is plotted in Figure 12.12(a). We see that the output appears at $k = -1$, before the application of the input at $k = 0$. Thus the system can *predict* what will be applied in the future. No physical system has such capability. Therefore no physical discrete-time system can have an improper digital transfer function.

The output of a noncausal system depends on future input. For example, if $G(z) = (z^3 + 1)/(z - 0.1)$, then $y(k)$ depends on past input $u(m)$ with $m \leq k$ and future input $u(k + 1)$ and $u(k + 2)$. Therefore, a noncausal system cannot operate on real time. If we store the input on a tape and start to compute $y(k)$ after receiving $u(k + 2)$, then the transfer function can be used. However, in this case, we are not using $G(z)$, but rather $G(z)/z^2 = (z^3 + 1)/z^2(z - 0.1)$, which is no longer improper. If we introduce enough delay to make an improper transfer function proper, then it can be used to process signals. Therefore, strictly speaking, transfer functions used in practice are all proper transfer functions.

If a system has a proper transfer function, then no output can appear before the application of an input and the output $y(k)$ depends on the input $u(mT)$, with

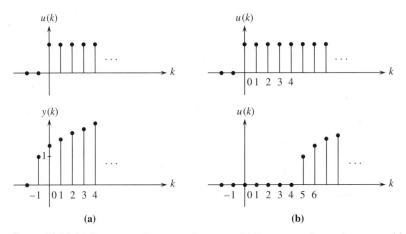

Figure 12.12 (a) Response of noncausal system. (b) Response of causal system with $r = 5$.

$m \leq k$. Such systems are called *causal systems*. We study in the remainder of this text only causal systems with proper digital transfer functions. Recall that in the continuous-time case, we also study only systems with proper transfer functions. However, the reasons are different. First, an improper analog transfer function cannot be easily built in practice. Second, it will amplify high-frequency noise, which often exists in analog systems. In the discrete-time case, we study proper digital transfer functions because of causality.

Consider a proper (biproper or strictly proper) transfer function $G(z) = N(z)/D(z)$. Let $r = \deg D(z) - \deg N(z)$. It is the difference between the degrees of the denominator and numerator. We call r the *pole-zero excess* of $G(z)$ because it equals the difference between the number of poles and the number of zeros. Let $y(k)$ be the step response of $G(z)$. If $r = 0$ or $G(z)$ is biproper, then $y(0) \neq 0$. If $r = 1$, then $y(0) = 0$ and $y(1) \neq 0$. In general, the step response of a digital transfer function with pole-zero excess r has the property

$$y(0) = 0 \qquad y(1) = 0 \cdots y(r - 1) = 0 \qquad y(r) \neq 0$$

as shown in Figure 12.12(b). In other words, a pole-zero excess of r will introduce a delay of r sampling instants. This phenomenon does not arise in the continuous-time case.

12.6 DISCRETE-TIME STATE EQUATIONS

Consider the discrete-time state-variable equation

$$\mathbf{x}(k + 1) = \mathbf{A}\mathbf{x}(k) + \mathbf{b}u(k) \tag{12.32a}$$

$$y(k) = \mathbf{c}\mathbf{x}(k) + du(k) \tag{12.32b}$$

This is a set of algebraic equations. Therefore the solution of the equation due to $\mathbf{x}(0)$ and $u(k)$, $k \geq 0$, can be obtained by direct substitution as

$$\mathbf{x}(1) = \mathbf{A}\mathbf{x}(0) + \mathbf{b}u(0)$$

$$\mathbf{x}(2) = \mathbf{A}\mathbf{x}(1) + \mathbf{b}u(1) = \mathbf{A}[\mathbf{A}\mathbf{x}(0) + \mathbf{b}u(0)] + \mathbf{b}u(1)$$
$$= \mathbf{A}^2\mathbf{x}(0) + [\mathbf{A}\mathbf{b}u(0) + \mathbf{b}u(1)]$$

$$\mathbf{x}(3) = \mathbf{A}\mathbf{x}(2) + \mathbf{b}u(2)$$
$$= \mathbf{A}^3\mathbf{x}(0) + [\mathbf{A}^2\mathbf{b}\mathbf{x}(0) + \mathbf{A}\mathbf{b}u(1) + \mathbf{b}u(2)]$$

and, in general,

$$\mathbf{x}(k) = \underbrace{\mathbf{A}^k\mathbf{x}(0)}_{\text{Zero-Input Response}} + \underbrace{\sum_{m=0}^{k-1} \mathbf{A}^{k-1-m}\mathbf{b}u(m)}_{\text{Zero-State Response}} \tag{12.33}$$

The application of the z-transform to (12.32a) yields

$$z[\mathbf{X}(z) - \mathbf{x}(0)] = \mathbf{A}\mathbf{X}(z) + \mathbf{b}U(z)$$

which implies

$$(z\mathbf{I} - \mathbf{A})\mathbf{X}(z) = z\mathbf{x}(0) + \mathbf{b}U(z)$$

Thus we have

$$\mathbf{X}(z) = (z\mathbf{I} - \mathbf{A})^{-1}z\mathbf{x}(0) + (z\mathbf{I} - \mathbf{A})^{-1}\mathbf{b}U(z) \tag{12.34a}$$

The substitution of (12.34a) into the z-transform of (12.32b) yields

$$Y(z) = \mathbf{c}(z\mathbf{I} - \mathbf{A})^{-1}z\mathbf{x}(0) + [\mathbf{c}(z\mathbf{I} - \mathbf{A})^{-1}\mathbf{b} + d]U(z) \tag{12.34b}$$

If $\mathbf{x}(0) = 0$, then (12.34b) reduces to

$$Y(z) = [\mathbf{c}(z\mathbf{I} - \mathbf{A})^{-1}\mathbf{b} + d]U(z) \tag{12.35}$$

Thus the transfer function of (12.32) is

$$G(z) = \frac{Y(z)}{U(z)} = \mathbf{c}(z\mathbf{I} - \mathbf{A})^{-1}\mathbf{b} + d \tag{12.36}$$

This is identical to the continuous-time case in (2.75) if z is replaced by s. The characteristic polynomial of $\mathbf{A}$ is defined, as in (2.76), as

$$\Delta(z) = \det (z\mathbf{I} - \mathbf{A})$$

Its roots are called the eigenvalues of $\mathbf{A}$.

12.6.1 Controllability and Observability

Consider the n-dimensional equation

$$\mathbf{x}(k + 1) = \mathbf{A}\mathbf{x}(k) + \mathbf{b}u(k) \tag{12.37a}$$

$$y(k) = \mathbf{c}\mathbf{x}(k) + du(k) \tag{12.37b}$$

The equation is *controllable* if we can transfer any state to any other state in a finite number of sampling instants by applying an input. The equation is *observable* if we can determine the initial state from the knowledge of the input and output over a finite number of sampling instants. The discrete-time equation is controllable if and only if the controllability matrix

$$\mathbf{U} = [\mathbf{b} \quad \mathbf{A}\mathbf{b} \quad \mathbf{A}^2\mathbf{b} \cdots \mathbf{A}^{n-1}\mathbf{b}] \tag{12.38}$$

has rank n. The equation is observable if and only if the observability matrix

$$\mathbf{V} = \begin{bmatrix} \mathbf{c} \\ \mathbf{c}\mathbf{A} \\ \vdots \\ \mathbf{c}\mathbf{A}^{n-1} \end{bmatrix} \tag{12.39}$$

has rank n. These conditions are identical to the continuous-time case. We prove the controllability part in the following. We write (12.33) explicitly for $k = n$ as

$$\mathbf{x}(n) - \mathbf{A}^n\mathbf{x}(0) = \sum_{m=0}^{n-1} \mathbf{A}^{n-1-m}\mathbf{b}u(m)$$

$$= \mathbf{b}u(n-1) + \mathbf{A}\mathbf{b}u(n-2) + \mathbf{A}^2\mathbf{b}u(n-3) \quad \text{(12.40)}$$
$$+ \cdots + \mathbf{A}^{n-1}\mathbf{b}u(0)$$

$$= [\mathbf{b} \quad \mathbf{A}\mathbf{b} \quad \mathbf{A}^2\mathbf{b} \cdots \mathbf{A}^{n-1}\mathbf{b}] \begin{bmatrix} u(n-1) \\ u(n-2) \\ u(n-3) \\ \vdots \\ u(0) \end{bmatrix}$$

For any $\mathbf{x}(0)$ and $\mathbf{x}(n)$, a solution $u(k)$, $k = 0, 1, \ldots, n-1$, exists in (12.40) if and only if the matrix $\mathbf{U}$ has rank n (Theorem B.1). This completes the proof. If an equation is controllable, then the transfer of a state to any other state can be achieved in n sampling periods and the input sequence can be computed from (12.40). Thus, the discrete-time case is considerably simpler than the continuous-time case. The observability part can be similarly established. See Problem 12.13.

If a state-variable equation is controllable and observable, then the equation is said to be a *minimal equation*. In this case, if we write

$$\mathbf{c}(z\mathbf{I} - \mathbf{A})^{-1}\mathbf{b} + d =: \frac{N(z)}{D(z)}$$

with

$$D(z) = \Delta(z) = \det (z\mathbf{I} - \mathbf{A})$$

then there are no common factors between $N(z)$ and $D(z)$, and the set of the eigenvalues of $\mathbf{A}$ [the roots of the characteristic polynomial $\Delta(z)$] equals the set of the poles of the transfer function. This situation is identical to the continuous-time case.

12.7   BASIC BLOCK DIAGRAMS AND REALIZATIONS

Every discrete-time state-variable equation can be easily built using the three elements shown in Figure 12.13. They are called *multipliers, summers* or *adders*, and *unit-delay elements*. The gain a of a multiplier can be positive or negative, larger or

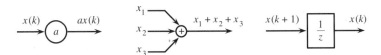

Figure 12.13 Three discrete basic elements.

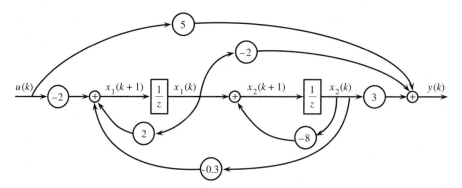

Figure 12.14 Basic block diagram of (12.41).

smaller than 1. An adder has two or more inputs and one and only one output. The output is simply the sum of all inputs. If the output of the unit-delay element is $x(k)$, then the input is $x(k + 1)$. A unit-delay element will be denoted by z^{-1}. These elements are quite similar to those in Figure 5.3. A block diagram which consists of only these three types of elements is called a *basic block diagram*.

We use an example to illustrate how to draw a basic block diagram for a discrete-time state-variable equation. Consider

$$\begin{bmatrix} x_1(k + 1) \\ x_2(k + 1) \end{bmatrix} = \begin{bmatrix} 2 & -0.3 \\ 0 & -8 \end{bmatrix} \begin{bmatrix} x_1(k) \\ x_2(k) \end{bmatrix} + \begin{bmatrix} -2 \\ 0 \end{bmatrix} u(k) \qquad (12.41a)$$

$$y(k) = \begin{bmatrix} -2 & 3 \end{bmatrix} \begin{bmatrix} x_1(k) \\ x_2(k) \end{bmatrix} + 5u(k) \qquad (12.41b)$$

It has dimension 2 and therefore needs two unit-delay elements. The outputs of the two elements are assigned as $x_1(k)$ and $x_2(k)$, as shown in Figure 12.14. Their inputs are $x_1(k + 1)$ and $x_2(k + 1)$. A basic block diagram of (12.41) can be obtained as shown in Figure 12.14. The procedure of developing the diagram is the same as the one in developing Figure 5.5. These two diagrams are in fact identical if integrators are replaced by delay elements or every $1/s$ is replaced by $1/z$.

Consider a transfer function $G(z)$. If we can find a state-variable equation

$$\mathbf{x}(k + 1) = \mathbf{A}\mathbf{x}(k) + \mathbf{b}u(k) \qquad (12.42a)$$

$$y(k) = \mathbf{c}\mathbf{x}(k) + du(k) \qquad (12.42b)$$

such that its transfer function from u to y equals $G(z)$, or

$$G(z) = \mathbf{c}(z\mathbf{I} - \mathbf{A})^{-1}\mathbf{b} + d \qquad (12.43)$$

then $G(z)$ is said to be *realizable* and (12.42) is a *realization* of $G(z)$. As in the continuous-time case, $G(z)$ is realizable if and only if $G(z)$ is a proper rational function. If $G(z)$ is improper, then any realization of $G(z)$ must be of the form

$$\mathbf{x}(k + 1) = \mathbf{A}\mathbf{x}(k) + \mathbf{b}u(k) \tag{12.44a}$$

$$y(k) = \mathbf{c}\mathbf{x}(k) + du(k) + d_1 u(k + 1) + d_2 u(k + 2) + \cdots \tag{12.44b}$$

This means that $y(k)$ will depend on the future input $u(k + l)$ with $l \geq 1$ and the system is not causal. Thus, we study only proper rational $G(z)$.

12.7.1 Realizations of $N(z)/D(z)$

Instead of discussing the general case, we use a transfer function of degree 4 to illustrate the realization procedure. The realization procedure is identical to the continuous-time case; therefore, we state only the result. Consider

$$G(z) = \frac{Y(z)}{U(z)} = \frac{b_4 z^4 + b_3 z^3 + b_2 z^2 + b_1 z + b_0}{a_4 z^4 + a_3 z^3 + a_2 z^2 + a_1 z + a_0} =: \frac{N(z)}{D(z)} \tag{12.45}$$

where a_i and b_i are real constants and $a_4 \neq 0$. The transfer function is biproper if $b_4 \neq 0$, and strictly proper if $b_4 = 0$. First, we express (12.45) as

$$G(z) = \frac{Y(z)}{U(z)} = G(\infty) + \frac{\bar{b}_3 z^3 + \bar{b}_2 z^2 + \bar{b}_1 z + \bar{b}_0}{z^4 + \bar{a}_3 z^3 + \bar{a}_2 z^2 + \bar{a}_1 z + \bar{a}_0} =: d + \frac{\bar{N}(z)}{\bar{D}(z)} \tag{12.46}$$

Then the following state-variable equation, similar to (5.17),

$$\mathbf{x}(k + 1) = \begin{bmatrix} -\bar{a}_3 & -\bar{a}_2 & -\bar{a}_1 & -\bar{a}_0 \\ 1 & 0 & 0 & 0 \\ 0 & 1 & 0 & 0 \\ 0 & 0 & 1 & 0 \end{bmatrix} \mathbf{x}(k) + \begin{bmatrix} 1 \\ 0 \\ 0 \\ 0 \end{bmatrix} u(k) \tag{12.47a}$$

$$y(k) = [\bar{b}_3 \quad \bar{b}_2 \quad \bar{b}_1 \quad \bar{b}_0]\mathbf{x}(k) + du(k) \tag{12.47b}$$

with $d = G(\infty)$, is a realization of (5.12). The value of $G(\infty)$ yields the direct transmission part. If $G(z)$ is strictly proper, then $d = 0$. Equation (12.47) is always controllable and is therefore called the *controllable-form realization*. If $N(z)$ and $D(z)$ in (12.45) have no common factors, then (12.46) is observable as well and the equation is called a *minimal equation*. Otherwise, the equation is not observable.

The following equation, which is similar to (5.18),

$$\mathbf{x}(k + 1) = \begin{bmatrix} -\bar{a}_3 & 1 & 0 & 0 \\ -\bar{a}_2 & 0 & 1 & 0 \\ -\bar{a}_1 & 0 & 0 & 1 \\ -\bar{a}_0 & 0 & 0 & 0 \end{bmatrix} \mathbf{x}(k) + \begin{bmatrix} \bar{b}_3 \\ \bar{b}_2 \\ \bar{b}_1 \\ \bar{b}_0 \end{bmatrix} u(k) \tag{12.48a}$$

$$y(k) = [1 \quad 0 \quad 0 \quad 0]\mathbf{x}(k) + du(k) \tag{12.48b}$$

is a different realization of (12.45). The equation is observable whether or not $N(z)$ and $D(z)$ have common factors. If $N(z)$ and $D(z)$ have no common factors, the equation is controllable as well and is a minimal equation. Equation (12.48) is called the *observable-form realization*.

Exercise 12.7.1

Find controllable- and observable-form realizations of

a. $\dfrac{2z + 10}{z + 2}$

b. $\dfrac{3z^2 - z + 2}{z^3 + 2z^2 + 1}$

c. $\dfrac{4z^3 + 2z + 1}{2z^3 + 3z^2 + 2}$

d. $\dfrac{5}{2z^2 + 4z + 3}$

The tandem and parallel realizations discussed in Section 5.5.2 can again be applied directly to discrete transfer functions, and the discussion will not be repeated.

To conclude this section, we mention that the same command tf2ss in MATLAB can be used to realize analog transfer functions and digital transfer functions. For example, if

$$G(z) = \frac{3z^2 - z + 2}{z^3 + 2z^2 + 1} = \frac{0 + 3z^{-1} - z^{-2} + 2z^{-3}}{1 + 2z^{-1} + 0 \cdot z^{-2} + z^{-3}} \tag{12.49}$$

then the following

```
num=[3 -1 2];den=[1 2 0 1];
[a,b,c,d]=tf2ss(num,den)
```

will generate its controllable-form realization. The command to compute the response of $G(s)$ due to a unit-step function is "step"; and $G(s)$ is expressed in descending powers of s. The command to compute the response of $G(z)$ due to a unit-step sequence is "dstep". Furthermore, $G(z)$ must be expressed in ascending powers of z^{-1}. Therefore,

```
num=[0 3 -1 2];den=[1 2 0 1];
y=dstep(num,den,20);
plot(y)
```

will generate 20 points of the unit-step response of (12.49).

12.8 STABILITY

A discrete-time system is said to be *bounded-input, bounded-output stable*, or simply *stable*, if every bounded input sequence excites a bounded output sequence. The condition for a system with digital transfer function $G(z)$ to be stable is that *every*

Table 12.2 The Jury Test

a_0	a_1	a_2	a_3	a_4	
a_4	a_3	a_2	a_1	a_0	$k_1 = \dfrac{a_4}{a_0}$
b_0	b_1	b_2	b_3	0	(1st a row) $- k_1$(2nd a row)
b_3	b_2	b_1	b_0		$k_2 = \dfrac{b_3}{b_0}$
c_0	c_1	c_2	0		(1st b row) $- k_2$(2nd b row)
c_2	c_1	c_0			$k_3 = \dfrac{c_2}{c_0}$
d_0	d_1	0			(1st c row) $- k_3$(2nd c row)
d_1	d_0				$k_4 = \dfrac{d_1}{d_0}$
e_0	0				(1st d row) $- k_4$(2nd d row)

pole of $G(z)$ must lie inside the unit circle of the z-plane or have a magnitude less than 1. This condition can be deduced from the continuous-time case, where stability requires every pole to lie inside the open left half s-plane. Because $z = e^{sT}$ maps the open left half s-plane into the interior of the unit circle in the z-plane, discrete stability requires every pole to lie inside the unit circle of the z-plane.

In the continuous-time case, we can use the Routh test to check whether all roots of a polynomial lie inside the open left half s-plane. In the discrete-time case, we have a similar test, called the Jury test. Consider the polynomial

$$D(z) = a_0 z^4 + a_1 z^3 + a_2 z^2 + a_3 z + a_4 \qquad a_0 > 0 \qquad (12.50)$$

It is a polynomial of degree 4 with a positive leading coefficient. We form the table in Table 12.2. The first row is simply the coefficients of $D(z)$ arranged in the descending power of z. The second row is the reversal of the order of the coefficients in the first row. We then take the ratio k_1 of the last elements of the first two rows as shown in the table. The subtraction of the product of k_1 and the second row from the first row yields the first b row. The last element of the b row will be zero and will be disregarded in the subsequent development. We reverse the order of the coefficients of the b row and repeat the process as shown in Table 12.2. If $D(z)$ is of degree n, then the table consists of $2n + 1$ rows.

The Jury Test

All roots of the polynomial of degree 4 and with a positive leading coefficient in (12.50) lie inside the unit circle if and only if the four leading coefficients (b_0, c_0, d_0, e_0) in Table 12.2 are all positive. ∎

Although this test is stated for a polynomial of degree 4, it can be easily extended to the general case. We use an example to illustrate its application.

Example 12.8.1

Consider a system with transfer function

$$G(z) = \frac{(z - 2)(z + 10)}{z^3 - 0.1z^2 - 0.12z - 0.4} \tag{12.51}$$

To check its stability, we use the denominator to form the table

1	-0.1	-0.12	-0.4	
-0.4	-0.12	-0.1	1	$k_1 = -0.4/1 = -0.4$
0.84	-0.148	-0.16	0	(1st row) + 0.4(2nd row)
-0.16	-0.148	0.84		$k_2 = -0.16/0.84 = -0.19$
0.8096	-0.176	0		(3rd row) + 0.19(4th row)
-0.176	0.8096			$k_3 = -0.176/0.8096 = -0.217$
0.771				(5th row) + 0.217(6th row)

The three leading coefficients 0.84, 0.8096, and 0.771 are all positive; thus, all roots of the denominator of $G(z)$ lie inside the unit circle. Thus the system is stable.

12.8.1 The Final-Value and Initial-Value Theorems

Let $F(z)$ be the z-transform of $f(k)$ and be a proper rational function. If $f(k)$ approaches a constant, zero or nonzero, then the constant can be obtained as

$$\lim_{k \to \infty} f(k) = \lim_{z \to 1} (z - 1)F(z) \tag{12.52}$$

This is called the *final-value theorem*. The theorem holds only if $f(k)$ approaches a constant. For example, if $f(k) = 2^k$, then $F(z) = z/(z - 2)$. For this z-transform pair, we have $f(\infty) = \infty$, but

$$\lim_{z \to 1} (z - 1) \cdot \frac{z}{z - 2} = 0 \cdot (-1) = 0$$

Thus (12.52) does not hold. The condition for $f(k)$ to approach a constant is that $(z - 1)F(z)$ is stable or, equivalently, all poles of $(z - 1)F(z)$ lie inside the unit circle. This implies that all poles of $F(z)$, except for a possible simple pole at $z = 1$, must lie inside the unit circle. As discussed in Figure 12.11, if all poles of $F(z)$ lie inside the unit circle, then the time response will approach zero. In this case, $F(z)$ has no pole $(z - 1)$ to cancel the factor $(z - 1)$, and the right-hand side of (12.52) is zero. If $F(z)$ has one pole at $z = 1$ and remaining poles inside the unit circle such as

$$F(z) = \frac{N(z)}{(z - 1)(z - a)(z - b)}$$

then it can be expanded, using partial fraction expansion, as

$$F(z) = k_1 \frac{z}{z - 1} + k_2 \frac{z}{z - a} + k_3 \frac{z}{z - b}$$

with $k_1 = \lim\limits_{z \to 1} \dfrac{z - 1}{z} F(z) = \lim\limits_{z \to 1} (z - 1) F(z)$. The inverse z-transform of $F(z)$ is

$$f(k) = k_1(1)^k + k_2 a^k + k_3 b^k$$

which approaches k_1 as $k \to \infty$ because $|a| < 1$ and $|b| < 1$. This establishes intuitively (12.52). For a different proof, see Reference [18].

Let $F(z)$ be the z-transform of $f(k)$ and be a proper rational function. Then we have

$$F(z) = f(0) + f(1)z^{-1} + f(2)z^{-2} + f(3)z^{-3} + \cdots$$

which implies

$$f(0) = \lim\limits_{z \to \infty} F(z) \qquad (12.53)$$

This is called the *initial-value theorem*.

12.9 STEADY-STATE RESPONSES OF STABLE SYSTEMS

Consider a discrete-time system with transfer function $G(z)$. The response of $G(z)$ as $k \to \infty$ is called the *steady-state response* of the system. If the system is stable, then the steady-state response of a step sequence will be a step sequence, not necessarily of the same magnitude. The steady-state response of a ramp sequence will be a ramp sequence; the steady-state response of a sinusoidal sequence will be a sinusoidal sequence with the same frequency. We establish these in the following.

Consider a system with discrete transfer function $G(z)$. Let the input be a step sequence with magnitude a—that is, $u(k) = a$, for $k = 0, 1, 2, \ldots$. Then $U(z) = az/(z - 1)$ and the output $y(k)$ is given by

$$Y(z) = G(z)U(z) = G(z) \cdot \frac{az}{z - 1}$$

To find the time response of $Y(z)$, we expand, using partial fraction expansion,

$$\frac{Y(z)}{z} = \frac{aG(z)}{z - 1} = \frac{aG(1)}{z - 1} + (\text{Terms due to poles of } G(z))$$

which implies

$$Y(z) = aG(1) \frac{z}{z - 1} + (\text{Terms due to poles of } G(z))$$

If $G(z)$ is stable, then every pole of $G(z)$ lies inside the unit circle of the z-plane and its time response approaches zero as $k \to \infty$. Thus we have

$$y_s(k) := \lim_{k \to \infty} y(k) = aG(1)(1)^k = aG(1) \qquad (12.54)$$

Thus, the steady-state response of a stable system with transfer function $G(z)$ due to a unit-step sequence equals $G(1)$. This is similar to (4.25) in the continuous-time case. Equation (12.54) can also be obtained by applying the final-value theorem. In order for the final-value theorem to be applicable, the poles of

$$(z - 1)Y(z) = (z - 1) \frac{aG(z)z}{z - 1} = azG(z)$$

must all lie inside the unit circle. This is the case because $G(z)$ is stable by assumption. Thus we have

$$y_s(k) := \lim_{k \to \infty} y(k) = \lim_{z \to 1} (z - 1)Y(z) = \lim_{z \to 1} azG(z) = aG(1)$$

This once again establishes (12.54).

Example 12.9.1

Consider the transfer function

$$G(z) = \frac{(z - 2)(z + 10)}{z^3 - 0.1z^2 - 0.12z - 0.4}$$

It is stable as shown in Example 12.8.1. If $u(k) = 1$, then the steady-state output is

$$y_s(k) = G(1) = \frac{(1 - 2)(1 + 10)}{1 - 0.1 - 0.12 - 0.4} = \frac{-11}{0.38} = -28.95$$

If $u(k) = akT$, for $k = 0, 1, 2, \ldots$, a ramp sequence, then

$$U(z) = \frac{aTz}{(z - 1)^2}$$

and

$$Y(z) = G(z)U(z) = G(z) \frac{aTz}{(z-1)^2}$$

We expand, using (A.8) in Appendix A,

$$\frac{Y(z)}{z} = \frac{aTG(z)}{(z-1)^2} = \frac{aTG(1)}{(z-1)^2} + \frac{aTG'(1)}{z-1} + \text{(Terms due to poles of } G(z)\text{)}$$

which can be written as

$$Y(z) = aG(1) \frac{Tz}{(z-1)^2} + aTG'(1) \frac{z}{z-1}$$

$$+ \text{(Terms due to poles of } G(z)\text{)} \tag{12.55}$$

where

$$G'(1) = \frac{dG(z)}{dz}\bigg|_{z=1}$$

Thus we conclude from (12.55) that if $G(z)$ is stable, then

$$y_s(k) = aG(1)kT + aTG'(1)$$

This equation is similar to (4.26a) in the continuous-time case. In the continuous-time case, the equation can also be expressed succinctly in terms of the coefficients of $G(s)$ as in (4.26b). This is, however, not possible in the discrete-time case.

If $G(z)$ is stable and if $u(k) = a \sin k\omega_o T$, then we have

$$y_s(k) = aA(\omega_o) \sin (k\omega_o T + \theta(\omega_o)) \tag{12.56}$$

with

$$A(\omega_o) = |G(e^{j\omega_o T})| \quad \text{and} \quad \theta(\omega_o) = \sphericalangle G(e^{j\omega_o T}) \tag{12.57}$$

In other words, if $G(z)$ is stable, its steady-state response due to a sinusoidal sequence approaches a sinusoidal sequence with the same frequency; its amplitude is modified by $A(\omega_o)$ and its phase by $\theta(\omega_o)$. The derivation of (12.56) is similar to (4.32) and will not be repeated.

The plot of $G(e^{j\omega T})$ with respect to ω is called the *frequency response* of the discrete-time system. The plot of its amplitude $A(\omega)$ is called the *amplitude characteristic* and the plot of $\theta(\omega)$, the *phase characteristic*. Because

$$e^{j\left(\omega + \frac{2\pi}{T}\right)T} = e^{j\omega T}e^{j2\pi} = e^{j\omega T}$$

$e^{j\omega T}$ is periodic with period $2\pi/T$. Consequently, so are $G(e^{j\omega T})$, $A(\omega)$, and $\theta(\omega)$. Therefore, we plot $A(\omega)$ and $\theta(\omega)$ only for ω from $-\pi/T$ to π/T. If all coefficients of $G(z)$ are real, as is always the case in practice, then $A(\omega)$ is symmetric and $\theta(\omega)$ is antisymmetric with respect to ω as shown in Figure 12.15. Therefore, we usually plot $A(\omega)$ and $\theta(\omega)$ only for ω from 0 to π/T or, equivalently, we plot $G(z)$ only along the upper circumference of the unit circle on the z-plane.

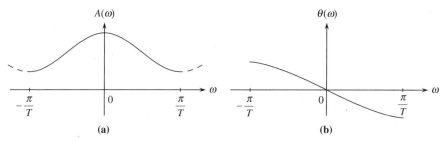

Figure 12.15 (a) Symmetric of $A(\omega)$. (b) Antisymmetric of $\theta(\omega)$.

12.9.1   Frequency Responses of Analog and Digital Systems

Consider a stable analog system with transfer function $\overline{G}(s)$. Let $g(t)$ be the inverse Laplace transform of $\overline{G}(s)$. The time function $g(t)$ is called the *impulse response* of the system. Let $G(z)$ denote the z-transform of $g(kT)$, the sample of $g(t)$ with sampling period T—that is,

$$G(z) = \mathcal{Z}[g(kT)] = \mathcal{Z}[[\mathcal{L}^{-1}\,\overline{G}(s)]_{t=kT}] \qquad (12.58\text{a})$$

For convenience, we write this simply as

$$G(z) = \mathcal{Z}[\overline{G}(s)] \qquad (12.58\text{b})$$

This is the transfer function of the discrete system whose impulse response equals the sample of the impulse response of the analog system. Now we discuss the relationship between the frequency response $\overline{G}(j\omega)$ of the analog system and the frequency response $G(e^{j\omega T})$ of the corresponding discrete system. It turns out that they are related by

$$G(e^{j\omega T}) = \frac{1}{T}\sum_{k=-\infty}^{\infty} \overline{G}\left(j\left(\omega - k\frac{2\pi}{T}\right)\right) = \frac{1}{T}\sum_{k=-\infty}^{\infty}\overline{G}(j(\omega - k\omega_s)) \quad (12.59)$$

where $\omega_s = 2\pi/T$ is called the *sampling frequency*. See Reference [18, p. 371; 13, p. 71.]. $\overline{G}(j(\omega - \omega_s))$ is the shifting of $\overline{G}(j\omega)$ to ω_s and $\overline{G}(j(\omega + \omega_s))$ is the shifting of $\overline{G}(j\omega)$ to $-\omega_s$. Thus, except for the factor $1/T$, $G(e^{j\omega T})$ is the sum of repetitions of $\overline{G}(j\omega)$ at $k\omega_s$ for all integers k. For example, if $\overline{G}(j\omega)$ is as shown in Figure 12.16(a), then the sum will be as shown in Figure 12.16(b). Note that the factor $1/T$ is not included in the sum, thus the vertical coordinate of Figure 12.16(b) is $TG(e^{j\omega T})$. The plot $\overline{G}(j\omega)$ in Figure 12.16(a) is zero for $|\omega| \geq \pi/T$, and its repetitions do not overlap with each other. In this case, sampling does not introduce *aliasing* and we have

$$\overline{G}(j\omega) = TG(e^{j\omega T}) \qquad \text{for } |\omega| \leq \pi/T \qquad (12.60)$$

The plot $\overline{G}(j\omega)$ in Figure 12.16(c) is not zero for $|\omega| \geq \pi/T$, and its repetitions do overlap with each other as shown in Figure 12.16(d). In this case, the sampling is said to cause *aliasing* and (12.60) does not hold. However, if the sampling period T is chosen to be sufficiently small, we have

$$\overline{G}(j\omega) \approx TG(e^{j\omega T}) \qquad \text{for } |\omega| \leq \pi/T \qquad (12.61)$$

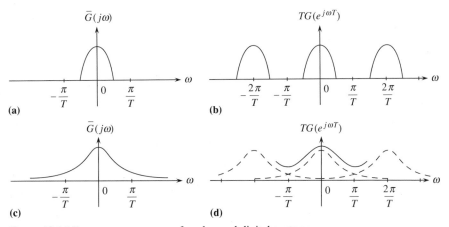

Figure 12.16 Frequency responses of analog and digital systems.

In conclusion, it is possible to obtain a discrete-time system with frequency response as close as desired to the frequency response of an analog system by choosing a sufficiently small sampling period or a sufficiently large sampling frequency.

12.10 LYAPUNOV STABILITY THEOREM

In this section, we study the stability of

$$\mathbf{x}(k + 1) = \mathbf{A}\mathbf{x}(k) \tag{12.62}$$

If all eigenvalues of $\mathbf{A}$ lie inside the unit circle, then $\mathbf{A}$ is said to be *stable*. If we compute its characteristic polynomial

$$\Delta(z) = \det(z\mathbf{I} - \mathbf{A}) \tag{12.63}$$

then the stability of $\mathbf{A}$ can be determined by applying the Jury test. Another way of checking the stability of $\mathbf{A}$ is applying the following theorem.

THEOREM 12.1 (Lyapunov Theorem)

All eigenvalues of $\mathbf{A}$ have magnitude less than 1 if and only if for any given positive definite symmetric matrix $\mathbf{N}$ or any given positive semidefinite symmetric matrix $\mathbf{N}$ with the property that $(\mathbf{A}, \mathbf{N})$ is observable, the matrix equation

$$\mathbf{A'MA} - \mathbf{M} = -\mathbf{N} \tag{12.64}$$

has a unique symmetric solution $\mathbf{M}$ and $\mathbf{M}$ is positive definite. ∎

To prove this theorem, we define the Lyapunov function

$$V(\mathbf{x}(k)) = \mathbf{x'}(k)\mathbf{M}\mathbf{x}(k) \tag{12.65}$$

We then use (12.62) to compute

$$\nabla V(\mathbf{x}(k)) := V(\mathbf{x}(k+1)) - V(\mathbf{x}(k)) = \mathbf{x}'(k+1)\mathbf{M}\mathbf{x}(k+1) - \mathbf{x}'(k)\mathbf{M}\mathbf{x}(k)$$
$$= \mathbf{x}'(k)\mathbf{A}'\mathbf{M}\mathbf{A}\mathbf{x}(k) - \mathbf{x}'(k)\mathbf{M}\mathbf{x}(k) = \mathbf{x}'(k)[\mathbf{A}'\mathbf{M}\mathbf{A} - \mathbf{M}]\mathbf{x}(k)$$
$$= -\mathbf{x}'(k)\mathbf{N}\mathbf{x}(k)$$

where we have substituted (12.64). The rest of the proof is similar to the continuous-time case in Section 11.8 and will not be repeated. We call (12.64) the *discrete Lyapunov equation*. The theorem can be used to establish the Jury test just as the continuous-time Lyapunov theorem can be used to establish the Routh test. The proof, however, is less transparent than in the continuous-time case. See Reference [15, p. 421].

PROBLEMS

12.1. Find the z-transforms of the following sequences, for $k = 0, 1, 2, \ldots$,

  a. $3\delta(k-3) + (-2)^k$

  b. $\sin 2k + e^{-0.2k}$

  c. $k(0.2)^k + (0.2)^k$

12.2. Find the z-transforms of the sequences obtained from sampling the following continuous-time signals with sampling period $T = 0.1$:

  a. $e^{-0.2t}\sin 3t + \cos 3t$

  b. $te^{0.1t}$

12.3. Use the direct division method and partial fraction method to find the inverse z-transforms of

  a. $\dfrac{z - 10}{(z + 1)(z - 0.1)}$

  b. $\dfrac{z}{(z + 0.2)(z - 0.3)}$

  c. $\dfrac{1}{z^3(z - 0.5)}$

12.4. Find the solution of the difference equation

$$y(k) + y(k-1) - 2y(k-2) = u(k-1) + 3u(k-2)$$

due to the initial conditions $y(-1) = 2$, $y(-2) = 1$, and the unit-step input sequence.

12.5. Repeat Problem 12.4 for the difference equation

$$y(k+2) + y(k+1) - 2y(k) = u(k+1) + 3u(k)$$

Is the result the same as the one in Problem 12.4?

12.6. Find the solution of the difference equation

$$y(k) + y(k - 1) - 2y(k - 2) = u(k - 1) + 3u(k - 2)$$

due to zero initial conditions (that is, $y(-1) = 0$, $y(-2) = 0$) and the unit-step input sequence. This is called the *unit-step response*.

12.7. Repeat Problem 12.6 for the difference equation

$$y(k + 2) + y(k + 1) - 2y(k) = u(k + 1) + 3u(k)$$

Is the result the same as the one in Problem 12.6?

12.8. Find the unit-step response of a system with transfer function

$$G(z) = \frac{z^2 - z + 1}{z + 0.9}$$

Will the response appear before the application of the input? A system with improper transfer function is a noncausal system. The output $y(k)$ of such a system depends on $u(l)$ with $l \geq k$—that is, present output depends on future input.

12.9. Consider

$$\begin{bmatrix} x_1(k + 1) \\ x_2(k + 1) \end{bmatrix} = \begin{bmatrix} 1 & 1 \\ 1 & 1 \end{bmatrix} \mathbf{x}(k) + \begin{bmatrix} 1 \\ 0 \end{bmatrix} u(k)$$

$$y(k) = [2 \quad 1]\mathbf{x}(k)$$

Compute its transfer function.

12.10. Consider

$$\begin{bmatrix} x_1(k + 1) \\ x_2(k + 1) \\ x_3(k + 1) \end{bmatrix} = \begin{bmatrix} 0 & 1 & 0 \\ 0 & 0 & 1 \\ 1 & 1 & 0 \end{bmatrix} \mathbf{x}(k) + \begin{bmatrix} 0 \\ 0 \\ 1 \end{bmatrix} u(k)$$

$$y(k) = [2 \quad 1 \quad 1]\mathbf{x}(k)$$

Compute its transfer function.

12.11. Is the equation in Problem 12.9 controllable? observable?

12.12. Is the equation in Problem 12.10 controllable? observable?

12.13. Consider (12.32) with $d = 0$. Show

$$\begin{bmatrix} y(0) \\ y(1) - \mathbf{c}bu(0) \\ \vdots \\ y(n - 1) - \mathbf{c}A^{n-2}\mathbf{b}u(0) - \cdots - \mathbf{c}bu(n - 2) \end{bmatrix} = \begin{bmatrix} \mathbf{c} \\ \mathbf{c}A \\ \vdots \\ \mathbf{c}A^{n-1} \end{bmatrix} \mathbf{x}(0)$$

Use this equation to establish the observability condition of (12.32).

12.14. Draw basic block diagrams for the equations in Problems 12.9 and 12.10.

12.15. Find realizations for the following transfer functions

a. $\dfrac{z^2 + 2}{4z^3}$

b. $\dfrac{3z^4 + 1}{2z^4 + 3z^3 + 4z^2 + z + 1}$

c. $\dfrac{(z + 3)^2}{(z + 1)^2(z + 2)}$

12.16. Are the transfer functions in Problems 12.15 stable?

12.17. Plot the frequency responses of

$$\overline{G}(s) = \frac{1}{s + 1} \qquad G(z) = \frac{z}{z - e^{-T}}$$

for $T = 1$ and $T = 0.1$.

12.18. Check the stability of

$$\mathbf{A} = \begin{bmatrix} 0.2 & 0.5 \\ 1 & 0 \end{bmatrix}$$

using the Jury test and Lyapunov Theorem.

13 *Discrete-Time System Design*

13.1 INTRODUCTION

Plants of control systems are mostly analog systems. However, because digital compensators have many advantages over analog ones, we may be asked to design digital compensators to control analog plants. In this chapter, we study the design of such compensators. There are two approaches to carrying out the design. The first approach uses the design methods discussed in the preceding chapters to design an analog compensator and then transform it into a digital one. The second approach first transforms analog plants into digital plants and then carries out design using digital techniques. The first approach performs discretization after design; the second approach performs discretization before design. We discuss the two approaches in order.

In this chapter, we encounter both analog and digital systems. To differentiate them, we use variables with an overbar to denote analog systems or signals and variables without an overbar to denote digital systems or signals. For example, $\bar{G}(s)$ is an analog transfer function and $G(z)$ is a digital transfer function; $\bar{y}(t)$ is an analog output and $y(kT)$ is a digital output. However, if $y(kT)$ is a sample of $\bar{y}(t)$, then $\bar{y}(kT) = y(kT)$ and the overbar will be dropped. If the same input is applied to an analog and a digital system, then we use $u(t)$ and $u(kT)$ to denote the inputs; no overbar will be used.

511

13.2   DIGITAL IMPLEMENTATIONS OF ANALOG COMPENSATORS—TIME-DOMAIN INVARIANCE

Consider the analog compensator with proper transfer function $\overline{C}(s)$ shown in Figure 13.1(a). The arrangement in Figure 13.1(b) implements the analog compensator digitally. It consists of three parts: an A/D converter, a digital system or an algorithm, and a D/A converter. The problem is to find a digital system such that for any input $e(t)$, the output $\overline{u}(t)$ of the analog compensator and the output $u(t)$ of the digital compensator are roughly equal. From Figure 13.1(b), we see that the output of the A/D converter equals $e(kT)$, the sample of $e(t)$ with sampling period T. We then search for a digital system which operates on $e(kT)$ to yield a sequence $\hat{u}(kT)$. The D/A converter then holds the value of $\hat{u}$ constant until the arrival of next data. Thus the output $u(t)$ of the digital compensator is stepwise as shown. The output of the analog compensator is generally not stepwise; therefore, the best we can achieve is that $u(t)$ approximately equals $\overline{u}(t)$.

In designing a digital system, ideally, for any input $e(t)$, $\hat{u}(kT)$ in Figure 13.1(b) should equal the sample of $\overline{u}(t)$. It is difficult, if not impossible, to design such a digital compensator that holds for all $e(t)$. It is, however, quite simple to design such a digital compensator for specific $e(t)$. In this section, we design such compensators for $e(t)$ to be an impulse and a step function.

Impulse-Invariance Method

Consider an analog compensator with a strictly proper transfer function $\overline{C}_s(s)$. If the input of $\overline{C}_s(s)$ is an impulse (its Laplace transform is 1), then the output is

$$U(s) = \overline{C}_s(s) \cdot 1 = \overline{C}_s(s)$$

Its inverse Laplace transform is actually the impulse response of the analog compensator. The z-transform of the sample of the impulse response yields a digital compensator with discrete transfer function

$$C(z) = \mathcal{Z}[\mathcal{L}^{-1}[\overline{C}_s(s)]|_{t=kT}]$$

or, using the notation in (12.7),

$$C(z) = \mathcal{Z}[\overline{C}_s(s)] \tag{13.1}$$

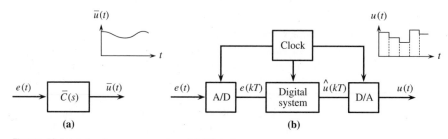

Figure 13.1 (a) Analog compensator. (b) Digital compensator.

As discussed in Section 12.9.1, if the sampling period is very small and the aliasing is negligible, then the frequency responses of $\overline{C}_s(s)$ and $C(z)$ will be of the same form but differ by the factor $1/T$ in the frequency range $|\omega| \leq \pi/T$. To take care of this factor, we introduce T into (13.1) to yield

$$C(z) = T\mathcal{Z}[\overline{C}_s(s)]$$

This yields an impulse-invariant digital compensator for a strictly proper $\overline{C}_s(s)$. If $\overline{C}(s)$ is biproper, then $\overline{C}(s)$ can be decomposed as

$$\overline{C}(s) = k + \overline{C}_s(s)$$

The inverse Laplace transform of k is $k\delta(t)$. The value of $\delta(t)$ is not defined at $t = 0$; therefore, its sample is meaningless. If we require the frequency response of $C(z)$ to equal the frequency response of k, then $C(z)$ is simply k. Thus the impulse-invariant digital compensator of $\overline{C}(s) = k + \overline{C}_s(s)$ is

$$C(z) = k + T\mathcal{Z}[\overline{C}_s(s)] \tag{13.2}$$

Note that the poles of $C(z)$ are obtained from the poles of $\overline{C}(s)$ or $\overline{C}_s(s)$ by the transformation $z = e^{sT}$ which maps the open left half s-plane into the interior of the unit circle of the z-plane; therefore, if all poles of $\overline{C}(s)$ lie inside the open left half s-plane, then all poles of $C(z)$ will lie inside the unit circle on the z-plane. Thus, if $\overline{C}(s)$ is stable, so is $C(z)$.

Example 13.2.1

Consider an analog compensator with transfer function

$$\overline{C}(s) = \frac{2(s - 2)}{(s + 1)(s + 3)} \tag{13.3}$$

To find its impulse response, we expand it, using partial fraction expansion, as

$$\overline{C}(s) = \frac{5}{s + 3} - \frac{3}{s + 1}$$

Thus, using (13.2) and Table 12.1, the impulse-invariant digital compensator is

$$C(z) = T\left[5 \cdot \frac{z}{z - e^{-3T}} - 3 \cdot \frac{z}{z - e^{-T}}\right] \tag{13.4}$$

The compensator depends on the sampling period. Different sampling periods yield different digital compensators. For example, if $T = 0.5$, then (13.4) becomes

$$C(z) = 0.5\left[\frac{5z}{z - 0.223} - \frac{3z}{z - 0.607}\right] = \frac{0.5z(2z - 2.366)}{(z - 0.223)(z - 0.607)} \tag{13.5}$$

It is a biproper digital transfer function.

Step-Invariance Method

Consider an analog compensator with transfer function $\overline{C}(s)$. We now develop a digital compensator $C(z)$ whose step response equals the samples of the step response of $\overline{C}(s)$. The Laplace transform of a unit-step function is $1/s$; the z-transform of a unit-step sequence is $z/(z-1)$. Thus, the step responses of both systems in the transform domains are

$$\overline{C}(s) \cdot \frac{1}{s} \quad \text{and} \quad C(z) \cdot \frac{z}{z-1}$$

If they are equal at sampling instants, then

$$\mathscr{Z}^{-1}\left[C(z) \cdot \frac{z}{z-1} \right] = \mathscr{L}^{-1}\left[\overline{C}(s) \cdot \frac{1}{s} \right]\Bigg|_{t=kT}$$

which, using the notation of (12.7), implies

$$C(z) = \frac{z-1}{z} \mathscr{Z}\left[\frac{\overline{C}(s)}{s} \right] = (1 - z^{-1})\mathscr{Z}\left[\frac{\overline{C}(s)}{s} \right] \tag{13.6}$$

This is called the *step-invariant digital compensator* of $\overline{C}(s)$.

Example 13.2.2

Find the step-invariant digital compensator of the analog compensator in (13.3). We first compute

$$\frac{\overline{C}(s)}{s} = \frac{2(s-2)}{s(s+1)(s+3)} = -\frac{4}{3s} + \frac{3}{s+1} - \frac{5}{3(s+3)}$$

Using Table 12.1, we have

$$\mathscr{Z}\left[\frac{\overline{C}(s)}{s} \right] = -\frac{4z}{3(z-1)} + \frac{3z}{z - e^{-T}} - \frac{5z}{3(z - e^{-3T})}$$

which becomes, after lengthy manipulation,

$$\mathscr{Z}\left[\frac{\overline{C}(s)}{s} \right] = \frac{z[(9e^{-T} - 5e^{-3T} - 4)z - (4e^{-4T} - 9e^{-3T} + 5e^{-T})]}{3(z-1)(z - e^{-T})(z - e^{-3T})}$$

Thus the step-invariant digital compensator of (13.1) is

$$C(z) = \frac{(9e^{-T} - 5e^{-3T} - 4)z - (4e^{-4T} - 9e^{-3T} + 5e^{-T})}{3(z - e^{-T})(z - e^{-3T})} \tag{13.7}$$

If the sampling period is 0.5, then the compensator is

$$C(z) = \frac{(9 \cdot 0.607 - 5 \cdot 0.223 - 4)z - (4 \cdot 0.135 - 9 \cdot 0.223 + 5 \cdot 0.607)}{3(z - 0.607)(z - 0.223)}$$

$$= \frac{0.116z - 0.523}{z^2 - 0.830z + 0.135} \tag{13.8}$$

This is a strictly proper transfer function. Although (13.5) and (13.8) have the same set of poles, their numerators are quite different. Thus the impulse-invariance and step-invariance methods implement a same analog compensator differently.

As can be seen from this example that the z-transform of $\overline{C}(s)/s$ will introduce an unstable pole at 1, which, however, will be cancelled by $(1 - z^{-1})$. Thus the poles of $C(z)$ in (13.6) consist of only the transformations of the poles of $\overline{C}(s)$ by $z = e^{sT}$. Thus if $\overline{C}(s)$ is stable, so is the step-invariant digital compensator.

The step-invariant digital compensator of $\overline{C}(s)$ can also be obtained using state-variable equations. Let

$$\dot{\mathbf{x}}(t) = \mathbf{A}\mathbf{x}(t) + \mathbf{b}e(t) \tag{13.9a}$$

$$\overline{u}(t) = \mathbf{c}\mathbf{x}(t) + de(t) \tag{13.9b}$$

be a realization of $\overline{C}(s)$. Note that the input of the analog compensator is $e(t)$ and the output is $\overline{u}(t)$. If the input is stepwise as shown in Figure 2.23(a), then the continuous-time state-variable equation in (13.9) can be described by, as derived in (2.89),

$$\mathbf{x}(k + 1) = \tilde{\mathbf{A}}\mathbf{x}(k) + \tilde{\mathbf{b}}e(k) \tag{13.10a}$$

$$\overline{u}(k) = \tilde{\mathbf{c}}\mathbf{x}(k) + \tilde{d}e(k) \tag{13.10b}$$

with

$$\tilde{\mathbf{A}} = e^{\mathbf{A}T} \qquad \tilde{\mathbf{b}} = \left(\int_0^T e^{\mathbf{A}\tau}\, d\tau \right)\mathbf{b} \qquad \tilde{\mathbf{c}} = \mathbf{c} \qquad \tilde{d} = d \tag{13.10c}$$

The output $\overline{u}(k)$ of (13.10) equals the sample of (13.9) if the input $e(t)$ is stepwise. Because a unit-step function is stepwise, the discrete-time state-variable equation in (13.10) describes the step-invariant digital compensator. The discrete transfer function of the compensator is

$$G(z) = \tilde{\mathbf{c}}(z\mathbf{I} - \tilde{\mathbf{A}})^{-1}\tilde{\mathbf{b}} + \tilde{d} \tag{13.11}$$

This is an alternative way of computing step-invariant digital compensators. This compensator can easily be obtained using MATLAB, as is illustrated by the following example.

Example 13.2.3

Find the step-invariant digital compensator for the analog compensator in (13.3). The controllable-form realization of $\overline{C}(s) = (2s - 4)/(s^2 + 4s + 3)$ is

$$\dot{\mathbf{x}}(t) = \begin{bmatrix} -4 & -3 \\ 1 & 0 \end{bmatrix} \mathbf{x}(t) + \begin{bmatrix} 1 \\ 0 \end{bmatrix} e(t) \tag{13.12a}$$

$$\overline{u}(t) = \begin{bmatrix} 2 & -4 \end{bmatrix}\mathbf{x}(t) \tag{13.12b}$$

This can also be obtained on MATLAB by typing

nu = [2 −4];de = [1 4 3];
[a,b,c,d] = tf2ss(nu,de)

Next we discretize (13.12a) with sampling period 0.5. The command

[da,db] = c2d(a,b,0.5)

will yield

$$\mathbf{da} = \begin{bmatrix} 0.0314 & -0.5751 \\ 0.1917 & 0.7982 \end{bmatrix} \quad \mathbf{db} = \begin{bmatrix} 0.1917 \\ 0.0673 \end{bmatrix}$$

where **da** and **db** denote discrete **a** and **b**. See Section 5.3. Thus, the step-invariant digital compensator is given by

$$\mathbf{x}(k + 1) = \begin{bmatrix} 0.0314 & -0.5751 \\ 0.1917 & 0.7982 \end{bmatrix} \mathbf{x}(k) + \begin{bmatrix} 0.1917 \\ 0.0673 \end{bmatrix} e(k) \qquad (13.13a)$$

$$\bar{u}(k) = [2 \quad -4]\mathbf{x}(k) \qquad (13.13b)$$

To compute its transfer function, we type

[num,den] = ss2tf(da,db,c,d,1)

Then MATLAB will yield

$$C(z) = \frac{0.1144z - 0.5219}{z^2 - 0.8297z + 0.1353} \qquad (13.14)$$

This transfer function is the same as (13.8), other than the discrepancy due to truncation errors. Therefore, step-invariant digital compensators can be obtained using either transfer functions or state-variable equations. In actual implementation, digital transfer functions must be realized as state-variable equations. Therefore, in using state-variable equations, we may stop after obtaining (13.13). There is no need to compute its transfer function.

13.2.1 Frequency-Domain Transformations

In addition to the time-domain invariance methods discussed in the preceding section, there are many other methods of implementing analog systems digitally. These methods will be obtained by transformations between s and z; therefore, they are grouped under the heading of frequency-domain transformations.

Consider an analog compensator with transfer function $\overline{C}(s)$. Let

$$\dot{\mathbf{x}}(t) = \mathbf{A}\mathbf{x}(t) + \mathbf{b}e(t) \qquad (13.15a)$$

$$\bar{u}(t) = \mathbf{c}\mathbf{x}(t) + de(t) \qquad (13.15b)$$

be its realization. In discretization of $\overline{C}(s)$ or (13.15), we are approximating an integration by a summation. Different approximations yield different discretizations and, consequently, different digital compensators. For convenience of discussion, we assume $e(t) = 0$ and $\mathbf{A}$ scalar.

Forward Approximation

The integration of (13.15a) from t_0 to $t_0 + T$ yields, with $e(t) = 0$,

$$\int_{t_0}^{t_0+T} \frac{d\mathbf{x}(t)}{dt} = \mathbf{x}(t_0 + T) - \mathbf{x}(t_0) = \int_{t_0}^{t_0+T} \mathbf{A}\mathbf{x}(t)dt \qquad (13.16)$$

Let $\mathbf{A}\mathbf{x}(t)$ be as shown in Figure 13.2. If the integration is approximated by the shaded area shown in Figure 13.2(a), then (13.16) becomes

$$\mathbf{x}(t_0 + T) - \mathbf{x}(t_0) = \mathbf{A}\mathbf{x}(t_0)T$$

or

$$\frac{\mathbf{x}(t_0 + T) - \mathbf{x}(t_0)}{T} = \mathbf{A}\mathbf{x}(t_0) \qquad (13.17)$$

This is the same as approximating the differentiation in (13.15a) by

$$\dot{\mathbf{x}}(t) \approx \frac{\mathbf{x}(t + T) - \mathbf{x}(t)}{T} \qquad (13.18)$$

Because $\mathcal{L}[\dot{\mathbf{x}}] = s\mathbf{X}(s)$ and

$$\mathcal{Z}\left[\frac{\mathbf{x}(t + T) - \mathbf{x}(t)}{T}\right] = \frac{z\mathbf{X}(z) - \mathbf{X}(z)}{T} = \frac{z - 1}{T}\mathbf{X}(\mathbf{z})$$

in the transform domains, Equation (13.18) is equivalent to

$$s = \frac{z - 1}{T} \qquad \text{(Forward difference)} \qquad (13.19)$$

Using this transformation, an analog compensator can easily be changed into a digital compensator. This is called the *forward-difference* or *Euler's method*. This transformation may not preserve the stability of $\overline{C}(s)$. For example, if $\overline{C}(s) = 1/(s + 2)$, then

$$C(z) = \frac{1}{\dfrac{z - 1}{T} + 2} = \frac{T}{z - 1 + 2T}$$

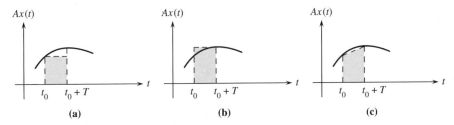

Figure 13.2 Various approximations of integration.

which is unstable for $T > 1$. Therefore, forward difference may not preserve the stability of $\overline{C}(s)$. In general, if the sampling period is sufficiently large, $C(z)$ may become unstable even if $\overline{C}(s)$ is stable.

The forward-difference transformation can easily be achieved using state-variable equations as in Section 5.2. Let

$$\dot{\mathbf{x}}(t) = \mathbf{A}\mathbf{x}(t) + \mathbf{b}e(t) \tag{13.20a}$$

$$\overline{u}(t) = \mathbf{c}\mathbf{x}(t) + de(t) \tag{13.20b}$$

be a realization of $\overline{C}(s)$, then

$$\mathbf{x}(k + 1) = (\mathbf{I} + T\mathbf{A})\mathbf{x}(k) + T\mathbf{b}e(k) \tag{13.21a}$$

$$u(k) = \mathbf{c}\mathbf{x}(k) + d e(k) \tag{13.21b}$$

is the digital compensator obtained by using the forward-difference method.

Example 13.2.4

Consider $\overline{C}(s) = 2(s - 2)/(s^2 + 4s + 3)$. Then

$$C(z) = \overline{C}(s)\big|_{s=(z-1)/T} = \cfrac{2\left(\cfrac{z-1}{T} - 2\right)}{\left(\cfrac{z-1}{T}\right)^2 + 4\,\cfrac{z-1}{T} + 3} \tag{13.22}$$

$$= \frac{2T(z - 1 - 2T)}{z^2 + (4T - 2)z + (3T^2 - 4T + 1)}$$

is a digital compensator obtained using the forward-difference method. If $T = 0.5$, then (13.22) becomes

$$C(z) = \frac{2 \cdot 0.5(z - 1 - 1)}{z^2 + (4 \cdot 0.5 - 2)z + (3 \cdot 0.25 - 2 + 1)} \tag{13.23}$$

$$= \frac{z - 2}{z^2 - 0.25} = \frac{z - 2}{(z + 0.5)(z - 0.5)}$$

This is the digital compensator. If we realize $\overline{C}(s)$ as

$$\dot{\mathbf{x}}(t) = \begin{bmatrix} -4 & -3 \\ 1 & 0 \end{bmatrix} \mathbf{x}(t) + \begin{bmatrix} 1 \\ 0 \end{bmatrix} e(t)$$

$$\overline{u}(t) = [2 \quad -4]\mathbf{x}(t)$$

Then

$$\mathbf{x}(k + 1) = \begin{bmatrix} 1 - 4T & -3T \\ T & 1 \end{bmatrix} \mathbf{x}(k) + \begin{bmatrix} T \\ 0 \end{bmatrix} e(k) \tag{13.24a}$$

$$u(k) = [2 \quad -4]\mathbf{x}(k) \tag{13.24b}$$

is the digital compensator. It can be shown that the transfer function of (13.24) equals (13.22). See Problem 13.5.

Backward Approximation

In the backward approximation, the integration in (13.16) is approximated by the shaded area shown in Figure 13.2(b). In this approximation, (13.16) becomes

$$\mathbf{x}(t_0 + T) - \mathbf{x}(t_0) = \mathbf{A}\mathbf{x}(t_0 + T)T$$

which can be written as

$$\frac{\mathbf{x}(t_0) - \mathbf{x}(t_0 - T)}{T} = \mathbf{A}\mathbf{x}(t_0)$$

Thus the differentiation in (13.15a) is approximated by

$$\dot{\mathbf{x}}(t) \approx \frac{\mathbf{x}(t_0) - \mathbf{x}(t_0 - T)}{T}$$

This is equivalent to, in the transform domains,

$$s = \frac{1 - z^{-1}}{T} = \frac{z - 1}{Tz} \quad \text{(Backward difference)} \tag{13.25}$$

This is called the *backward-difference method*. A pole $(s + \alpha + j\beta)$ in $\overline{C}(s)$ is transformed by (13.25) into

$$\frac{1 - z^{-1}}{T} + \alpha + j\beta = \frac{1}{T}[(1 + \alpha T) + j\beta T - z^{-1}]$$

$$= -\frac{(1 + \alpha T) + j\beta T}{zT}\left[z - \frac{1}{(1 + \alpha T) + j\beta T}\right] \tag{13.26}$$

If the pole is stable (that is, $\alpha > 0$), then the magnitude of $(1 + \alpha T) + j\beta T$ is always larger than 1; thus, the pole in (13.26) always lies inside the unit circle. Thus the transformation in (13.25) will transform a stable pole in $\overline{C}(s)$ into a stable pole in $C(z)$. Thus, if $\overline{C}(s)$ is stable, so is $C(z)$.

Trapezoid Approximation

In this method, the integration in (13.16) is approximated by the trapezoid shown in Figure 13.2(c) and (13.16) becomes

$$\mathbf{x}(t_0 + T) - \mathbf{x}(t_0) \approx \mathbf{A}\frac{\mathbf{x}(t_0 + T) + \mathbf{x}(t_0)}{2}T$$

Its z-transform domain is

$$z\mathbf{X}(z) - \mathbf{X}(z) = \frac{T}{2}\mathbf{A}[z\mathbf{X}(z) + \mathbf{X}(z)] = \frac{T(z + 1)}{2}\mathbf{A}\mathbf{X}(z)$$

which implies

$$\frac{2}{T} \cdot \frac{z - 1}{z + 1} \mathbf{X}(z) = \mathbf{A}\mathbf{X}(z)$$

The Laplace transform of (13.15a) with $e(t) = 0$ is $s\mathbf{X}(s) = \mathbf{A}\mathbf{X}(s)$. Thus, the approximation can be achieved by setting

$$s = \frac{2}{T} \frac{z - 1}{z + 1} \qquad \text{(Trapezoidal approximation)} \tag{13.27}$$

This is called the *trapezoidal approximation method*. Equation (13.27) implies

$$(z + 1)s\frac{T}{2} = (z - 1) \qquad \text{or} \qquad \left(1 - \frac{Ts}{2}\right) z = \left(1 + \frac{Ts}{2}\right)$$

Thus we have

$$z = \frac{1 + \dfrac{Ts}{2}}{1 - \dfrac{Ts}{2}} \tag{13.28}$$

For every z, we can compute a unique s from (13.27); for every s, we can compute a unique z from (13.28). Thus the mapping in (13.27) is a one-to-one mapping, called a *bilinear transformation*. Let $s = \sigma + j\overline{\omega}$. Then (13.28) becomes

$$z = \frac{1 + \dfrac{T\sigma}{2} + j\dfrac{T\overline{\omega}}{2}}{1 - \dfrac{T\sigma}{2} - j\dfrac{T\overline{\omega}}{2}}$$

and

$$|z| = \frac{\sqrt{\left(1 + \dfrac{T\sigma}{2}\right)^2 + \left(\dfrac{T\overline{\omega}}{2}\right)^2}}{\sqrt{\left(1 - \dfrac{T\sigma}{2}\right)^2 + \left(\dfrac{T\overline{\omega}}{2}\right)^2}} \tag{13.29}$$

This equation implies $|z| = 1$ if $\sigma = 0$, and $|z| < 1$ if $\sigma < 0$. Thus, the $j\omega$-axis on the s-plane is mapped onto the unit circle on the z-plane and the open left half s-plane is mapped onto the interior of the unit circle on the z-plane. To develop a precise relationship between the frequency in analog systems and the frequency in digital systems, we define $s = j\overline{\omega}$ and $z = e^{j\omega T}$. Then (13.27) becomes

$$j\overline{\omega} = \frac{2}{T} \frac{e^{j\omega T} - 1}{e^{j\omega T} + 1} = \frac{2}{T} \frac{e^{j0.5\omega T}(e^{j0.5\omega T} - e^{-j0.5\omega T})}{e^{j0.5\omega T}(e^{j0.5\omega T} + e^{-j0.5\omega T})}$$

$$= \frac{2}{T} \frac{2j \sin 0.5\omega T}{2 \cos 0.5\omega T} = \frac{2j}{T} \tan \frac{\omega T}{2}$$

which implies

$$\overline{\omega} = \frac{2}{T} \tan \frac{\omega T}{2}$$

(13.30)

This is plotted in Figure 13.3. We see that the analog frequency from $\overline{\omega} = 0$ to $\overline{\omega} = \infty$ is compressed into the digital frequency from $\omega = 0$ to $\omega = \pi/T$. This is called *frequency warping*. Because of this warping and the nonlinear relationship between $\overline{\omega}$ and ω, some simple manipulation, called *prewarping*, is needed in using bilinear transformations. This is a standard technique in digital filter design. See, for example, Reference [13].

Pole-Zero Mapping

Consider an analog compensator with pole p_i and zero q_i. In the pole-zero mapping, pole p_i is mapped into $e^{p_i T}$ and zero q_i is mapped into $e^{q_i T}$. For example, the compensator

$$\overline{C}(s) = \frac{2(s - 2)}{s^2 + 4s + 3} = \frac{2(s - 2)}{(s + 3)(s + 1)} = \frac{2(s - 2)}{(s - (-3))(s - (-1))}$$

is mapped into

$$C(z) = \frac{2(z - e^{2T})}{(z - e^{-3T})(z - e^{-T})}$$

Thus the transformation is very simple.

There is, however, one problem with this transformation. Consider

$$\overline{C}(s) = \frac{b}{D(s)} = \frac{b}{(s - a_1)(s - a_2)(s - a_3)}$$

(13.31)

Then its corresponding digital compensator is

$$C(z) = \frac{b}{(z - e^{a_1 T})(z - e^{a_2 T})(z - e^{a_3 T})}$$

(13.32)

Let $\overline{y}(t)$ be the step response of (13.31). Because $\overline{C}(s)$ has three more poles than

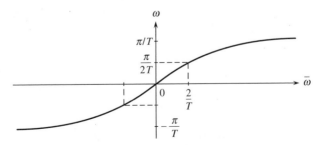

Figure 13.3 Analog and digital frequencies in (13.30).

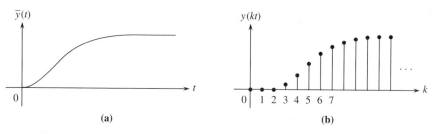

Figure 13.4 (a) Step response of (13.31). (b) Step response of (13.32).

zeros, it can be shown that $\bar{y}(0) = 0$, $\dot{\bar{y}}(0) = 0$, and $\ddot{\bar{y}}(0) = 0$ (Problem 13.7) and the unit-step response will be as shown in Figure 13.4(a). Let $y(kT)$ be the response of (13.32) due to a unit-step sequence. Then, because $C(z)$ has three more poles than zeros, the response will start from $k = 3$, as shown in Figure 13.4(b). In other words, there is a delay of 3 sampling instants. In general, if there is a pole-zero excess of r in $C(z)$, then there is a delay of r sampling instants and the response will start from $k = r$. In order to eliminate this delay, a polynomial of degree $r - 1$ is introduced into the numerator of $C(z)$ so that the response of $C(z)$ will start at $k = 1$. For example, we may modify (13.32) as

$$C(z) = \frac{bN(z)}{(z - e^{a_1 T})(z - e^{a_2 T})(z - e^{a_3 T})} \tag{13.33}$$

with deg $N(z) = 2$. If the zeros at $s = \infty$ in $\bar{C}(s)$ are considered to be mapped into $z = 0$, then we may choose $N(z) = z^2$. It is suggested to choose $N(z) = (z + 1)^2$ in Reference [52] and $N(z) = z^2 + 4z + 1$ in Reference [3]. Note that the steady-state response of $\bar{C}(s)$ in (13.31) due to a unit-step input is in general different from the steady-state response of $C(z)$ in (13.32) or (13.33). See Problem 13.8. If they are required to be equal, we may modify b in (13.33) to achieve this.

13.3   AN EXAMPLE

Consider the control system shown in Figure 13.5(a). The plant transfer function is

$$\bar{G}(s) = \frac{1}{s(s + 2)} \tag{13.34}$$

The overall transfer function is required to minimize a quadratic performance index and is computed in (9.38) as

$$\bar{G}_o(s) = \frac{3}{s^2 + 3.2s + 3} = \frac{3}{(s + 1.6 + j0.65)(s + 1.6 - j0.65)} \tag{13.35}$$

The compensator can be computed as

$$\bar{C}(s) = \frac{\bar{G}_o(s)}{\bar{G}(s)(1 - \bar{G}_o(s))} = \frac{3(s + 2)}{s + 3.2} = 3 - \frac{3.6}{s + 3.2} =: k + \bar{C}_s(s) \tag{13.36}$$

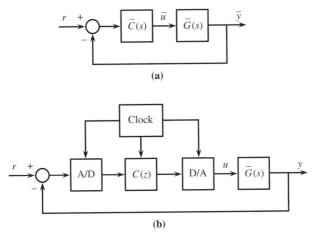

Figure 13.5 (a) Analog control system. (b) Digital control system.

Now we shall implement the analog compensator as a digital compensator as shown in Figure 13.5(b). The impulse-invariant digital compensator is, using (13.2),

$$C_a(z) = k + T\mathcal{Z}[\overline{C}_s(s)] = 3 - \frac{3.6Tz}{z - e^{-3.2T}} \qquad (13.37)$$

The step-invariant digital compensator is, using (13.6),

$$
\begin{aligned}
C_b(z) &= (1 - z^{-1})\mathcal{Z}\left[\frac{\overline{C}(s)}{s}\right] \\
&= (1 - z^{-1})\mathcal{Z}\left[\frac{1.875}{s} + \frac{1.125}{s + 3.2}\right] \\
&= (1 - z^{-1})\left[\frac{1.875z}{z - 1} + \frac{1.125z}{z - e^{-3.2T}}\right] \\
&= \frac{3z - 1.125 - 1.875e^{-3.2T}}{z - e^{-3.2T}}
\end{aligned}
\qquad (13.38)
$$

The forward-difference digital compensator is, by substituting $s = (z - 1)/T$,

$$C_c(z) = \frac{3(z - 1 + 2T)}{z - 1 + 3.2T} \qquad (13.39)$$

The backward-difference digital compensator is, by substituting $s = (z - 1)/Tz$,

$$C_d(z) = \frac{3(z - 1 + 2Tz)}{z - 1 + 3.2Tz} \qquad (13.40)$$

The digital compensator obtained by the bilinear transformation $s = 2(z - 1)/T(z + 1)$ is

$$C_e(z) = \frac{6((1 + T)z - 1 + T)}{(2 + 3.2T)z - 2 + 3.2T} \tag{13.41}$$

The digital compensator obtained by pole-zero mapping is

$$C_f(z) = \frac{3(z - e^{-2T})}{z - e^{-3.2T}} \tag{13.42}$$

Figure 13.6(a) through (f) shows the unit-step responses of the overall system in Figure 13.5(b) with $C_i(z)$ in (13.37) through (13.42) for $T = 0.1, 0.2, 0.4, 0.6,$ and 0.8. For comparison, the response of the analog system in Figure 13.5(a) is also plotted and is denoted by $T = 0$. We compare the compensators in the following

	$T = 0.1$	0.2	0.4	0.6	0.8
Impulse-invariant	A	A	C	C	F
Step-invariant	A	A	B	B	B
Forward difference	B	B	B	F	F
Backward difference	C	C	F	F	F
Bilinear	A−	A−	B	B	B+
Pole-zero	B+	B	C	C	C

where A denotes the best or closest to the analog system and F the worst or not acceptable. For $T = 0.1$ and 0.2, the responses in Figure 13.6(a) and (b) are very close to the analog response; therefore, they are given a grade of A. Although the responses in Figure 13.6(e) are quite good, they show some overshoot; therefore they are given a grade of A−. If T is large, the responses in Figure 13.6(a), (c), and (d) are unacceptable. Overall, the compensators obtained by using the step-invariant and bilinear methods yield the best results. The compensator obtained by pole-zero transformation is acceptable but not as good as the previous two. In conclusion, if the sampling period is sufficiently small, then any method can be used to digitize analog compensators. However, for a larger sampling period, it is better to use the step-invariant and bilinear transformation methods to digitize an analog compensator.

13.3.1 Selection of Sampling Periods

Although it is always possible to design a digital compensator to approximate an analog compensator by choosing a sufficiently small sampling period, it is not desirable, in practice, to choose an unnecessarily small one. The smaller the sampling period, the more computation it requires. Using a small sampling period may also

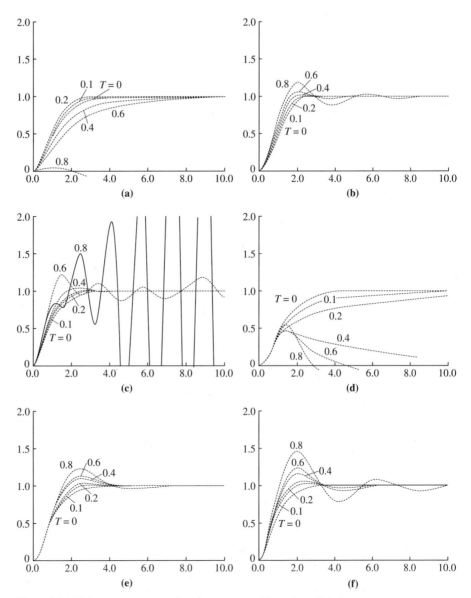

Figure 13.6 Unit-step responses of analog system with various digital compensators.

introduce computational problems, as is discussed in the next section. See also Reference [46]. How to choose an adequate sampling period has been widely discussed in the literature. References [46, 52] suggest that the sampling frequency or $1/T$ be chosen about ten times the bandwidth of the closed-loop transfer function. Reference [3] suggests the following rules: If the pole of an overall system is real—say,

$(s + a)$ with $a > 0$—then the sampling period may be chosen as

$$T = \frac{1}{(2 \sim 4) \times a} \tag{13.43}$$

Because $1/a$ is the time constant, (13.43) implies that two to four sampling points be chosen in one time constant. If the poles of an overall system are complex and have a damping ratio in the neighborhood of 0.7, then the sampling period can be chosen as

$$T = \frac{0.5 \sim 1}{\omega_n} \tag{13.44}$$

where ω_n is the natural frequency in radians per second. If an overall transfer function has more than one real pole and/or more than one pair of complex-conjugate poles, then we may use the real or complex-conjugate poles that are closest to the imaginary axis as a guide in choosing T.

The bandwidth of $G_o(s)$ in (13.35) is found, using MATLAB, as $B = 1.2$ radians. If the sampling frequency $1/T$ is chosen as ten times of $1.2/2\pi$, then $T = 2\pi/12 = 0.52$. The poles of the closed-loop system in (13.35) are $-1.6 \pm j0.65$. Its natural frequency is $\omega_n = \sqrt{3} = 1.73$, and its damping ratio is $\zeta = 3.2/2\omega_n = 0.46$. The damping ratio is not in the neighborhood of 0.7; therefore, strictly speaking, (13.44) cannot be used. However, as a comparison, we use it to compute the sampling period which ranges from 0.29 to 0.58. It is comparable to $T = 0.52$. If we use $T = 0.5$ for the system in the preceding section, then, as shown in Figure 3.6, the system with the digital compensator in (13.38) or (13.41) has a step response close to that of the original analog system but with a larger overshoot. If overshoot is not desirable, then the sampling period for the system should be chosen as 0.2, about 20 times the closed-loop bandwidth. If $T = 0.2$, the compensators in (13.37), (13.39), and (13.42) can also be used. Thus the selection of a sampling period depends on which digital compensator is used. In any case, the safest way of choosing a sampling period is to use computer simulations.

13.4   EQUIVALENT DIGITAL PLANTS

Digital compensators in the preceding sections are obtained by discretization of analog compensators. In the remainder of this chapter we discuss how to design digital compensators directly without first designing analog ones. In order to do this, we must first find an equivalent digital plant for a given analog plant. In digital control, a digital compensator generates a sequence of numbers, as shown in Figure 13.7(a). This digital signal is then transformed into an analog signal to drive the analog plant. There are a number of ways to change the digital signal into an analog one. If a number is held constant until the arrival of the next number, as shown in Figure 13.7(b), the conversion is called the *zero-order hold*. If a number is extrapolated by using the slope from the number and its previous value, as shown in Figure

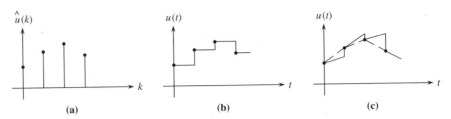

Figure 13.7 D/A conversions.

13.7(c), the conversion is called the *first-order hold*. Clearly, higher-order holds are also possible. However, the zero-order is the hold most widely used. The D/A converter discussed in Section 12.3 implements zero-order hold.

The Laplace transform of the pulse $p(t)$ with height 1 and width T shown in Figure A.3 is computed in (A.23) as

$$\frac{1}{s}(1 - e^{-sT})$$

If the input of a zero-order hold is 1, then the output is $p(t)$. Therefore, the zero-order hold can be considered to have transfer function

$$G_h(s) = \frac{1 - e^{-Ts}}{s} \tag{13.45}$$

In digital control, a plant is always connected to a zero-order hold (or a D/A converter) and the transfer function of the plant and hold becomes

$$\frac{(1 - e^{-Ts})}{s}\overline{G}(s) \tag{13.46}$$

as shown in Figure 13.8(a). Note that the input $u(kT)$ of Figure 13.8(a) must be modified as

$$u^*(t) = \sum_{k=0}^{\infty} u(kT)\overline{\delta}(t - kT)$$

as in (12.8) in order for the representation to be mathematically correct. The output $y(t)$ of the plant and hold is an analog signal. If we add an A/D converter after the plant as shown in Figure 13.8(a), then the output is $y(kT)$. If we consider $y(kT)$ as the output and $u(kT)$ as the input, then the analog plant and hold in Figure 13.8(a) can be modeled as a digital system with discrete transfer function

$$G(z) = \mathscr{Z}\left[(1 - e^{-Ts})\frac{\overline{G}(s)}{s}\right] = \mathscr{Z}\left[\frac{\overline{G}(s)}{s}\right] - \mathscr{Z}\left[e^{-Ts}\frac{\overline{G}(s)}{s}\right] \tag{13.47}$$

Because e^{-Ts} introduces only a time delay, we may move it outside the z-transform as

$$\mathscr{Z}\left[e^{-Ts}\frac{\overline{G}(s)}{s}\right] = e^{-Ts}\mathscr{Z}\left[\frac{\overline{G}(s)}{s}\right]$$

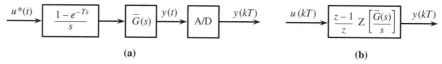

(a)   (b)

Figure 13.8 (a) Equivalent analog plant. (b) Equivalent digital plant.

Using $z = e^{Ts}$, (13.47) becomes

$$G(z) = (1 - z^{-1})\mathcal{Z}\left[\frac{\overline{G}(s)}{s}\right] = \frac{z-1}{z}\,\mathcal{Z}\left[\frac{\overline{G}(s)}{s}\right] \qquad (13.48)$$

This is a discrete transfer function. Its input is $u(kT)$ and its output is $y(kT)$, the sample of the analog plant output in Figure 13.8(a). The discrete system shown in Figure 13.8(b) is called the *equivalent discrete* or *digital plant*, and its discrete transfer function is given by (13.48).

By comparing (13.48) with (13.6), we see that the equivalent discrete plant transfer function $G(z)$ and the original analog plant transfer function $\overline{G}(s)$ are step invariant. As discussed in Section 13.2, the step-invariant digital plant $G(z)$ can also be obtained from analog plant $\overline{G}(s)$ by using state-variable equations. Let

$$\dot{\mathbf{x}}(t) = \mathbf{A}\mathbf{x}(t) + \mathbf{b}u(t) \qquad (13.49a)$$

$$y(t) = \mathbf{c}\mathbf{x}(t) + du(t) \qquad (13.49b)$$

be a realization of $\overline{G}(s)$. If the input is stepwise as in digital control, then the input $y(t)$ and output $u(t)$ at $t = kT$ can be described, as derived in (2.89), by

$$\mathbf{x}(k + 1) = \tilde{\mathbf{A}}\mathbf{x}(k) + \tilde{\mathbf{b}}u(k) \qquad (13.50a)$$

$$y(k) = \tilde{\mathbf{c}}\mathbf{x}(k) + \tilde{d}u(k) \qquad (13.50b)$$

with

$$\tilde{\mathbf{A}} = e^{\mathbf{A}T} \qquad \tilde{\mathbf{b}} = \left(\int_0^T e^{\mathbf{A}\tau}\,d\tau\right)\mathbf{b} \qquad \tilde{\mathbf{c}} = \mathbf{c} \qquad \tilde{d} = d \qquad (13.50c)$$

The transfer function of (13.50) equals (13.48). As shown in Example 13.2.3, this computation can easily be carried out by using the commands tf2ss, c2d, and ss2tf in MATLAB.

13.4.1   Hidden Dynamics and Non-Minimum-Phase Zeros

Once an equivalent digital plant is obtained, we can design a digital compensator to control the plant as shown in Figure 13.9(a) or, more generally, design an algorithm and use a digital computer to control the plant as shown in Figure 13.9(b). In the remainder of this chapter, we shall discuss how to carry out the design. Before proceeding, we use examples to illustrate the problems that may arise in using equivalent digital plants.

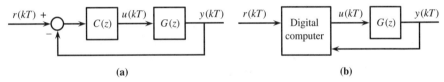

Figure 13.9 Direct design of digital control system.

Example 13.4.1

Consider an analog plant with transfer function

$$\overline{G}(s) = \frac{101}{(s + 1)^2 + 100} = \frac{101}{s^2 + 2s + 101} \tag{13.51}$$

with its poles plotted in Figure 13.10. Its step-invariant digital transfer function is given by

$$
\begin{aligned}
G(z) &= (1 - z^{-1})\mathcal{Z}\left[\frac{101}{s((s + 1)^2 + 100)}\right] \\
&= \frac{z - 1}{z}\mathcal{Z}\left[\frac{1}{s} - \frac{s + 1}{(s + 1)^2 + 100} - \frac{1}{(s + 1)^2 + 100}\right] \\
&= \frac{z - 1}{z}\left[\frac{z}{z - 1} - \frac{z^2 - ze^{-T}\cos 10T}{z^2 - 2ze^{-T}\cos 10T + e^{-2T}}\right. \\
&\qquad\qquad\left. - \frac{ze^{-T}\sin 10T}{10(z^2 - 2ze^{-T}\cos 10T + e^{-2T})}\right]
\end{aligned}
\tag{13.52}
$$

where we have used the z-transform pairs in Table 12.1. Now if the sampling period T is chosen as $10T = 2\pi$, then $\cos 10T = 1$, $\sin 10T = 0$ and $e^{-T} = e^{-0.2\pi} =$

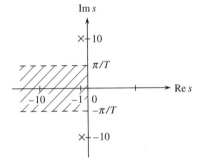

Figure 13.10 Poles of (13.51).

0.53. Thus (13.52) can be reduced as

$$
\begin{aligned}
G(z) &= \frac{z-1}{z}\left[\frac{z}{z-1} - \frac{z(z-e^{-T})}{z^2 - 2ze^{-T} + e^{-2T}}\right] \\
&= \frac{z-1}{z}\left[\frac{z}{z-1} - \frac{z(z-e^{-T})}{(z-e^{-T})^2}\right] \\
&= \frac{z-1}{z}\left[\frac{z}{z-1} - \frac{z}{z-e^{-T}}\right] = \frac{1-e^{-T}}{z-e^{-T}} = \frac{0.47}{z-0.53}
\end{aligned}
\tag{13.53}
$$

It is a transfer function with only one real pole, whereas the original analog plant transfer function in (13.51) has a pair of complex-conjugate poles. Figure 13.11 shows the unit-step responses of (13.51) and (13.53). We see that the oscillation in the analog plant does not appear in its step-invariant digital plant. Thus, some dynamics of an analog plant may disappear from or become hidden in its equivalent digital plant.

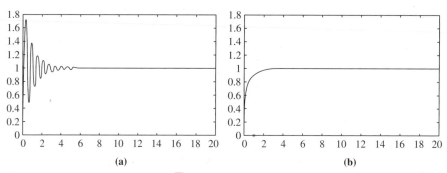

Figure 13.11 Unit-step responses of $\overline{G}(s)$ and $G(z)$.

The reason for the disappearance of the dynamics can easily be explained from the plot in Figure 13.10. Recall from Figure 12.9 that the mapping $z = e^{sT}$ is not a one-to-one mapping. If the sampling period T is chosen so that π/T equals half of the imaginary part of the complex poles as shown in Figure 13.10, then the complex poles will be mapped into real poles. Furthermore, the two poles are mapped into the same location. This is the reason for the disappearance of the dynamics. Knowing the reason, it becomes simple to avoid the problem. If the sampling period is chosen to be small enough that the primary strip (the region bounded between $-\pi/T$ and π/T as shown in Figure 12.9) covers all poles of $\overline{G}(s)$, then no dynamic will be lost in the sampling and its equivalent digital plant can be used in design.

Example 13.4.2

Loss of dynamics can also be explained using state-variable equations. Consider the transfer function in Example 13.4.1 or

$$\overline{G}(s) = \frac{101}{(s + 1)^2 + 100} = \frac{101}{(s + 1 + j10)(s + 1 - j10)} \quad (13.54)$$

$$= \frac{5.05j}{s + 1 + 10j} - \frac{5.05j}{s + 1 - 10j}$$

It is plotted in Figure 13.12. If we assign state variables as shown, then we can readily obtain the following state-variable equation

$$\dot{\mathbf{x}}(t) = \begin{bmatrix} -1 - 10j & 0 \\ 0 & -1 + 10j \end{bmatrix} \mathbf{x}(t) + \begin{bmatrix} 1 \\ 1 \end{bmatrix} u(t) \quad (13.55a)$$

$$y(t) = [5.05j \quad -5.05j]\mathbf{x}(t) \quad (13.55b)$$

This is a minimal realization of (13.54) and is controllable and observable.[1] Because matrix **A** is diagonal, the discretized equation of (13.55) with sampling period T can be easily obtained as

$$\mathbf{x}(k + 1) = \begin{bmatrix} e^{(-1 - 10j)\mathrm{T}} & 0 \\ 0 & e^{(-1 + 10j)T} \end{bmatrix} \mathbf{x}(k) + \begin{bmatrix} \dfrac{1}{1 + 10j}(1 - e^{(-1 - 10j)T}) \\ \dfrac{1}{1 - 10j}(1 - e^{(-1 + 10j)T}) \end{bmatrix} u(k)$$

$$y(k) = [5.05j \quad -5.05j]\mathbf{x}(k)$$

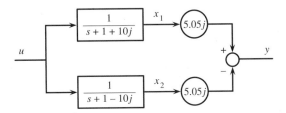

Figure 13.12 Block diagram of (13.54).

---

[1]Although we consider state-variable equations with only real coefficients in the preceding chapters, all concepts and results are equally applicable to equations with complex coefficients without any modification. See Reference [15].

If T is chosen as $10T = 2\pi$ or $T = 0.2\pi$, then $e^{-10jT} = 1$, and the equation reduces to

$$\mathbf{x}(k + 1) = \begin{bmatrix} e^{-T} & 0 \\ 0 & e^{-T} \end{bmatrix} \mathbf{x}(k) + \begin{bmatrix} \dfrac{1}{1 + 10j}(1 - e^{-T}) \\[2ex] \dfrac{1}{1 - 10j}(1 - e^{-T}) \end{bmatrix} u(k) \qquad (13.56a)$$

$$y(k) = [5.05j \quad -5.05j]\mathbf{x}(k) \tag{13.56b}$$

Because $\tilde{\mathbf{A}}$ is diagonal and has the same eigenvalues, the equation is neither controllable nor observable. See Problem 11.2. This can also be verified by checking the ranks of the controllability and observability matrices. The transfer function of (13.56) can be computed as $(1 - e^{-T})/(z - e^{-T})$, which is the same as (13.53). Thus controllability and observability of a state-variable equation may be destroyed after sampling. For a more general discussion, see Reference [15, p. 559].

We now discuss a different problem due to discretization. Consider $\overline{G}(s) = \overline{N}(s)/\overline{D}(s)$. Let $G(z) = N(z)/D(z)$ be its equivalent digital plant transfer function. Pole-zero excess is defined as the difference between the number of poles and the number of zeros. It turns out that if the pole-zero excess of $\overline{G}(s)$ is zero, so is $G(z)$. That is, if $\overline{G}(s)$ is biproper, so is $G(z)$. However, if $\overline{G}(s)$ is strictly proper, then the pole-zero excess of $G(z)$ is always 1 no matter what the pole-zero excess of $\overline{G}(s)$ is. Thus sampling will introduce zeros. The number of additional zeros introduced in $G(z)$ equals $r - 1$ where r is the pole-zero excess of $\overline{G}(s)$. We first give an example and then establish the assertion.

Example 13.4.3

Consider

$$\overline{G}(s) = \frac{1}{s(s^2 + 2s + 2)} \tag{13.57}$$

Its pole-zero excess is 3; therefore, the step-invariant discretization of $\overline{G}(s)$ will introduce two zeros into $G(z)$. We use MATLAB and state-variable equations to carry out discretization. The commands

```
nu = 1;de = [1 2 2 0];     (Express Ḡ(s) in numerator and denominator.)
[a,b,c,d] = tf2ss(nu,de);      (Yield controllable-form realization of Ḡ(s).)
[da,db] = c2d(a,b,0.5);     (Discretize a and b with sampling period 0.5.)
[dn,dd] = ss2tf(da,db,c,d,1);   (Compute the discrete transfer function G(z).)
[z,p,k] = tf2zp(dn,dd)     (Express G(z) in zero and pole form.)
```

yield[2]

$$G(z) = \frac{0.0161(z + 2.8829)(z + 0.2099)}{(z - 1.0000)(z - 0.5323 + 0.2908j)(z - 0.5323 - 0.2908j)} \tag{13.58}$$

The discretization does introduce two zeros. If the sampling period is chosen as $T = 0.1$, then

$$G(z) = \frac{1.585 \cdot 10^{-4}(z + 3.549)(z + 0.255)}{(z - 1)(z - 0.9003 + 0.0903j)(z - 0.9003 - 0.0903j)} \tag{13.59}$$

It also introduces two zeros.

Now we discuss why sampling will introduce additional zeros. The Laplace transform of the unit-step response $\bar{y}(t)$ of $\bar{G}(s)$ is $\bar{G}(s)/s$. Using the initial value theorem in Appendix A, its value at $t = 0$ can be computed as

$$\bar{y}(0) = \lim_{t \to \infty} s \, \frac{\bar{G}(s)}{s} = \bar{G}(\infty)$$

Let $y(kT)$ be the unit-step response of $G(z)$, that is,

$$Y(z) = G(z)U(z) = G(z) \, \frac{z}{z - 1}$$

Using the initial value theorem in (12.53), the value of $y(kT)$ at $k = 0$ can be obtained from $Y(z)$ as

$$y(0) = \lim_{z \to \infty} Y(z) = \lim_{z \to \infty} G(z) \, \frac{z}{z - 1} = G(\infty)$$

Because $y(kT)$ is the sample of $\bar{y}(t)$, we have $y(0) = \bar{y}(0)$ and, consequently, $\bar{G}(\infty)$ $= G(\infty)$. Now if $\bar{G}(s)$ is biproper, that is, $\bar{G}(\infty) \neq 0$, then $G(\infty) \neq 0$ and $G(z)$ is biproper. Similarly, if $\bar{G}(s)$ is strictly proper, so is $G(z)$. Let $G(z)$ be a strictly proper transfer function of degree 4:

$$G(z) = \frac{b_1 z^3 + b_2 z^2 + b_3 z + b_4}{z^4 + a_1 z^3 + a_2 z^2 + a_3 z + a_4} \tag{13.60}$$

Then its unit-step response is

$$Y(z) = \frac{b_1 z^3 + b_2 z^2 + b_3 z + b_4}{z^4 + a_1 z^3 + a_2 z^2 + a_3 z + a_4} \cdot \frac{z}{z - 1}$$

$$= 0 + b_1 z^{-1} + [b_2 - b_1 (a_1 - 1)]z^{-2} + \cdots \tag{13.61}$$

---

[2]The two steps ss2tf and tf2zp can be combined as ss2zp. However, I was not able to obtain correct answers using ss2zp for this problem, even though ss2zp is successfully used in Example 13.5.1. Therefore, the difficulty may be caused by numerical problems.

Thus we have $y(0) = 0$ and $y(T) = b_1$. The unit-step response $\bar{y}(t)$ of the analog system $\bar{G}(s)$ is generally of the form shown in Figure 13.4(a). No matter what its pole-zero excess is, generally $\bar{y}(T)$ is different from zero. If $y(kT)$ is a sample of $\bar{y}(t)$, then $y(T) = b_1 = \bar{y}(T) \neq 0$. Therefore, the pole-zero excess of $G(z)$ in (13.60) is 1. This establishes the assertion that if the pole-zero excess of $\bar{G}(s)$ is $r \geq 1$, the pole-zero excess of $G(z)$ is always 1. Thus, sampling will generate $r - 1$ additional zeros into $G(z)$.

In the analog case, a zero is called a *minimum-phase zero* if it lies inside the open left half s-plane, a *non-minimum-phase* zero if it lies inside the closed right half s-plane. Following this terminology, we call a zero inside the interior of the unit circle on the z-plane a *minimum-phase zero*, a zero on or outside the unit circle a *non-minimum-phase zero*. For the analog plant transfer function in (13.57), sampling introduces a minimum- and a non-minimum-phase zero as shown in (13.58) and (13.59). Non-minimum-phase zeros will introduce constraints in design, as will be discussed in a later section.

To conclude this section, we mention that the poles of $G(z)$ are transformed from the poles of $\bar{G}(s)$ by $z = e^{sT}$. Once the poles of $G(z)$ and, consequently, the coefficients a_i in (13.60) are computed, then the coefficients b_i in (13.60) can be computed from a_i and the samples of the unit-step response $\bar{y}(t)$ as

$$\begin{bmatrix} b_1 \\ b_2 \\ b_3 \\ b_4 \end{bmatrix} = \begin{bmatrix} 1 & 0 & 0 & 0 \\ a_1 - 1 & 1 & 0 & 0 \\ a_2 - a_1 & a_1 - 1 & 1 & 0 \\ a_3 - a_2 & a_2 - a_1 & a_1 - 1 & 1 \end{bmatrix} \begin{bmatrix} \bar{y}(T) \\ \bar{y}(2T) \\ \bar{y}(3T) \\ \bar{y}(4T) \end{bmatrix} \quad (13.62)$$

See Problem 13.11. Thus, the numerator of $G(z)$ in (13.60) is determined by the first four samples of $\bar{y}(t)$. If T is small, then these four samples are hardly distinguishable, as can be seen from Figure 13.4(a). Therefore, the possibility of introducing errors in b_i is large. Thus, using an unnecessarily small sampling period will not only increase the amount of computer computation, it will also increase computational error. Therefore, selection of a sampling period is not simple.

13.5   ROOT-LOCUS METHOD

In this section we discuss how to use the root-locus method to design a digital compensator directly from a given equivalent digital plant. The root-locus design method actually consists of two parts: searching for a desired pole region, and plotting the roots of $p(s) + kq(s)$ as a function of real k. The plot of root loci discussed in Section 7.4 for the continuous-time case is directly applicable to the discrete-time case; therefore, we discuss only the desired pole region in the digital case.

The desired pole region in Figure 7.4 for analog systems is developed from the specifications on the settling time, overshoot, and rise time. Settling time requires closed-loop poles to lie on the left-hand side of the vertical line passing through $-a = -4.5/t_s$, where t_s denotes the settling time. The vertical line is transformed by $z = e^{sT}$ into a circle with radius e^{-aT} as shown in Figure 13.13(a). Note that the

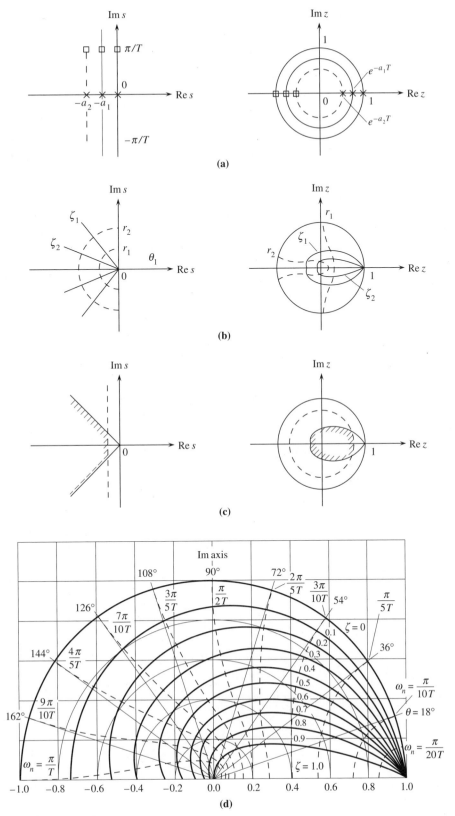

Figure 13.13 Desired pole region.

mapping $z = e^{sT}$ is not one-to-one, therefore we map only the primary strip (or $-\pi/T \leq \omega \leq \pi/T$) in the s-plane into the interior of the unit circle on the z-plane. The poles denoted by $\times$ on the s-plane are mapped into the positive real axis of the z-plane inside the unit circle. The poles with imaginary part π/T shown with small squares are mapped into the negative real axis inside the unit circle.

The overshoot is governed by the damping ratio ζ or the angle θ in the analog case. If we substitute $s = re^{j\theta}$ into $z = e^{sT}$ and plot z as a function of r (for a fixed θ) and as a function of θ (for a fixed r), then we will obtained the solid lines and dotted lines in Figure 13.13(b). Because the overshoot is governed by θ, the solid line in Figure 13.13(b) determines the overshoot. The distance from the origin or r is inversely proportional to the rise time; therefore, the dotted line in Figure 13.13(b) determines the rise time. Consequently, the desired pole region in the analog case can be mapped into the one shown in Figure 13.13(c) for the digital case. For convenience of design, the detailed relationship in Figure 13.13(b) is plotted in Figure 13.13(d). With the preceding discussion, we are ready to discuss design of digital compensators using the root-locus method. We use an example to illustrate the design.

Example 13.5.1

Consider the problem in Section 7.2.2—that is, given a plant with transfer function

$$\overline{G}(s) = \frac{1}{s(s + 2)} \tag{13.63}$$

use the root-locus method to design an overall system to meet

1. Position error $= 0$.
2. Overshoot $\leq 5\%$.
3. Settling time ≤ 9 seconds.
4. Rise time as small as possible.

First we compute the equivalent digital plant transfer function of (13.63):

$$G(z) = (1 - z^{-1})\mathcal{Z}\left[\frac{\overline{G}(s)}{s}\right] = (1 - z^{-1})\mathcal{Z}\left[\frac{1}{s^2(s + 2)}\right]$$

$$= (1 - z^{-1})\mathcal{Z}\left[\frac{0.5}{s^2} - \frac{0.25}{s} + \frac{0.25}{s + 2}\right] \tag{13.64}$$

$$= \frac{z - 1}{z}\left[\frac{0.5Tz}{(z - 1)^2} - \frac{0.25z}{z - 1} + \frac{0.25z}{z - e^{-2T}}\right]$$

If the sampling period is chosen as $T = 1$, then $e^{-2T} = 0.1353$ and (13.64) can be simplified as

$$G(z) = \frac{0.2838(z + 0.5232)}{(z - 1)(z - 0.1353)} \tag{13.65}$$

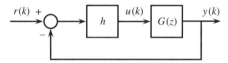

Figure 13.14 Unity-feedback system.

This can also be obtained using MATLAB by typing $\text{nu}=[1]$; $\text{de}=[1\ 2\ 0]$; $[\text{a,b,c,d}]=\text{tf2ss(nu,de)}$; $[\text{da,db}]=\text{c2d(a,b,1)}$; $[\text{z,p,k}]=\text{ss2zp(da,db,c,d)}$. The result is $z = -0.5232$, $p = 1, 0.1353$, and $k = 0.2838$, and is the same as (13.65). Next we choose the unity-feedback configuration in Figure 13.14 and find, if possible, a gain h to meet the design specifications. First we compute the overall transfer function:

$$G_o(z) = \frac{hG(z)}{1 + hG(z)} = \frac{0.2838h(z + 0.5232)}{(z - 1)(z - 0.1353) + 0.2838h(z + 0.5232)} \quad (13.66)$$

Because of the presence of factor $(z - 1)$ in the denominator, we have $G_o(1) = 1$ for all h. Thus if $G_o(z)$ is stable, and if $r(k) = a$, then

$$y_s(k) := \lim_{k \to \infty} y(k) = G_o(1)a = a$$

Thus the position error, as defined in (6.3), is

$$e_p = \lim_{k \to \infty} \left| \frac{r(k) - y_s(k)}{a} \right| = \left| \frac{a - a}{a} \right| = 0$$

and the overall system will automatically meet the specification on the position error so long as the system is stable. This situation is similar to the continuous-time case because the analog plant transfer function is of type 1. Thus if a digital plant transfer function has one pole at $z = 1$, and if the unity-feedback system in Figure 13.14 is stable, then the plant output will track asymptotically any step-reference input.

As discussed in Section 7.2.1, in order to have the overshoot less than 5%, the damping ratio ζ must be larger than 0.7. This can be translated into the curve denoted by 0.7 in Figure 13.15. In order to have the settling time less than 9 seconds, we require $\sigma \geq 4.5/9 = 0.5$. This can be translated into the circle with radius $e^{-0.5T} = 0.606$ as shown in Figure 13.15. We plot in the figure also the root loci of

$$\frac{-1}{h} = \frac{0.2838(z + 0.5232)}{(z - 1)(z - 0.1353)} \quad (13.67)$$

The one in Figure 13.15(a) is the complete root loci; the one in Figure 13.15(b) shows only the critical part. The root loci have two breakaway points at 0.48 and -1.52 and consist of a circle centered at -0.5232 and with radius 1. From the root loci, we see that if

$$h_1 \leq h \leq h_2$$

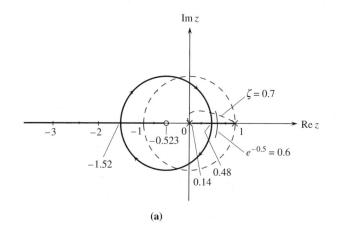

(a)

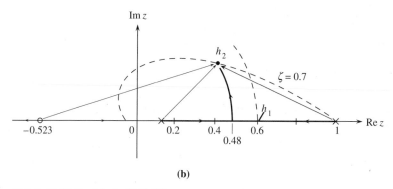

(b)

Figure 13.15 Root loci of (13.65).

where h_1 and h_2 are indicated on the plot, the system will meet the specifications on overshoot and settling time. Now the system is required to have a rise time as small as possible, therefore the closest pole should be as far away as possible from the origin of the s-plane or on the dotted line in Figure 13.13(d) with ω_n as large as possible. Thus we choose $h = h_2$. By drawing vectors from the two poles and one zero to h_2 as shown in Figure 13.15(b) and measuring their magnitudes, we obtain, from (13.67),

$$h_2 = \frac{0.67 \times 0.42}{0.2838 \times 0.96} = 1.03$$

Thus by choosing $h = 1.03$, the overall system will meet all design specifications.

This example shows that the root-locus method discussed in Chapter 7 can be directly applied to design digital control systems. Note that the result $h = 1.03$ in digital design is quite different from the result $h = 2$ obtained in analog design in Section 7.2.2. This discrepancy may be caused by the selection of the sampling period $T = 1$. To see the effect of the sampling period, we repeat the design by choosing $T = 0.2$. Then the equivalent digital transfer function is

$$G(z) = \frac{0.0176(z + 0.8753)}{(z - 1)(z - 0.6703)} = \frac{0.0176z + 0.0154}{z^2 - 1.6703z + 0.6703} \qquad (13.68)$$

From its root loci and using the same argument, the gain h that meets all the specifications can be found as $h = 1.71$. This is closer to the result of the analog design. To compare the analog and digital designs, we plot in Figure 13.16 the plant outputs and actuating signals of Figure 13.14 due to a unit-step reference input for $T = 1$ with $h = 1.03$ and $T = 0.2$ with $h = 1.71$. We also plot in Figure 13.16 the unit-step response and actuating signal for the analog design, denoted by $T = 0$ and $h = 2$. We see that the digital design with $T = 0.2$ and $h = 1.71$ is almost identical to the analog design with $T = 0$ and $h = 2$. The maximum value of the actuating signal in the digital design, however, is smaller.

In conclusion, the root-locus method can be applied to design digital control systems. The result, however, depends on the sampling period. If the sampling period is sufficiently small, then the result will be close to the one obtained by analog design. There is one problem in digital design, however. If the sampling period is small, then the possibility of introducing numerical error will be larger. For example, for the problem in the preceding example, as T decreases, the design will be carried out in a region closer to $z = 1$, where the solid lines in Figure 13.13(d) are more clustered. Therefore, the design will be more sensitive to numerical errors.

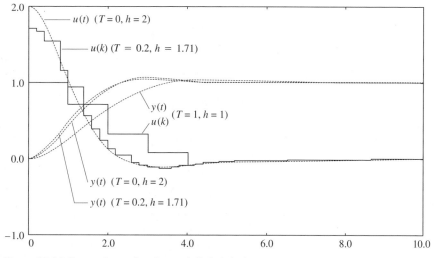

Figure 13.16 Comparison of analog and digital designs.

13.6   FREQUENCY-DOMAIN DESIGN

Frequency-domain design methods, specially, the Bode plot method, are useful in designing analog control systems. Because the frequency response $\overline{G}(j\omega)$ of analog transfer functions is a rational function of $j\omega$ and because the Bode plot of linear factors can be approximated by two asymptotes—which are obtained by considering two extreme cases: $\omega \to 0$ and $\omega \to \infty$—the plot of Bode plots is relatively simple. In the digital case, the frequency response $G(e^{j\omega T})$ is an irrational function of ω. Furthermore, the frequency of interest is limited to the range from 0 to π/T. Therefore, the procedure of plotting Bode plots in Chapter 8 cannot be directly applied to the discrete-time case.

One way to overcome this difficulty is to carry out a transformation. Consider

$$w = \frac{2}{T} \frac{z-1}{z+1} \tag{13.69a}$$

or

$$z = \frac{1 + \dfrac{wT}{2}}{1 - \dfrac{wT}{2}} \tag{13.69b}$$

This is the bilinear transformation studied in (13.27). Let $z = e^{j\omega T}$ and $w = jv$. Then, we have, similar to (13.30),

$$v = \frac{2}{T} \tan \frac{\omega T}{T} \tag{13.70}$$

Thus the transformation transforms ω from 0 to π/T to v from 0 to ∞ as shown in Figure 13.3. Define

$$\hat{G}(w) = G(z)|_{z=(1+wT/2)/(1-wT/2)} \tag{13.71}$$

Then $\hat{G}(jv)$ will be a rational function of v and v ranges from 0 to ∞. Thus the Bode design method can be applied. In conclusion, design of digital compensators using the Bode plot method consists of the following steps: (1) Sample the analog plant to obtain an equivalent digital plant transfer function $G(z)$. (2) Transform $G(z)$ to $\hat{G}(w)$ using (13.69). (3) Use the Bode method on $\hat{G}(jv)$ to find a compensator $\hat{C}(w)$. (4) Transform $\hat{C}(w)$ to $C(z)$ using (13.69). This completes the design.

The key criteria in the Bode design method are the phase margin, gain margin, and gain-crossover frequency. Because the transformations from $\overline{G}(s)$ to $G(z)$ by $z = e^{sT}$ and from $G(z)$ to $\hat{G}(w)$ by (13.69) are not linear, considerable distortions occur. Furthermore, the sampling of $\overline{G}(s)$ may introduce non-minimum-phase zeros into $G(z)$; this may also cause difficulty in using the Bode method. Thus, even though the Bode method can be used to design digital compensators, care must be exercised and the result must be simulated before actual implementation of digital compensators.

13.7 STATE FEEDBACK, STATE ESTIMATOR, AND DEAD-BEAT DESIGN

The design methods in Sections 11.4 and 11.6 for analog systems can be applied to digital systems without any modification. Consider the discrete-time state-variable equation

$$\mathbf{x}(k + 1) = \mathbf{A}\mathbf{x}(k) + \mathbf{b}u(k) \tag{13.72a}$$

$$y(k) = \mathbf{c}\mathbf{x}(k) \tag{13.72b}$$

If we introduce the state feedback

$$u(k) = r(k) - \mathbf{k}\mathbf{x}(k)$$

then (13.72) becomes

$$\mathbf{x}(k + 1) = (\mathbf{A} - \mathbf{b}\mathbf{k})\mathbf{x}(k) + \mathbf{b}r(k) \tag{13.73a}$$

$$y(k) = \mathbf{c}\mathbf{x}(k) \tag{13.73b}$$

with transfer function

$$G_o(z) = \mathbf{c}(z\mathbf{I} - \mathbf{A} + \mathbf{b}\mathbf{k})^{-1}\mathbf{b} \tag{13.74}$$

If $(\mathbf{A}, \mathbf{b})$ is controllable, by choosing a real feedback gain $\mathbf{k}$, all eigenvalues of $(\mathbf{A} - \mathbf{b}\mathbf{k})$ or, equivalently, all poles of $G_o(z)$ can be arbitrarily assigned, provided complex-conjugate poles are assigned in pairs. The design procedure in Section 11.4 can be used to find $\mathbf{k}$ without any modification. If $(\mathbf{A}, \mathbf{c})$ is observable, a full-dimensional or a reduced-dimensional state estimator can be designed to generate an estimated state; the rate for the estimated state to approach the original state can be arbitrarily chosen by selecting the eigenvalues of the estimator. The procedure for designing a full-dimensional estimator in Section 11.6 and the procedure for designing a reduced-dimensional estimator in Section 11.6.1 can be used without any modification. The only difference is that poles are chosen to lie inside the open left half plane in the analog case and to lie inside the unit circle in the digital case.

In the digital case, if all poles of $G_o(z)$ are chosen to be located at the origin, it is called the *dead-beat* design. In this case, the overall transfer function is of the form

$$G_o(z) = \frac{N_o(z)}{z^n} \tag{13.75}$$

with deg $N_o(z) \leq n$. The unit-step response of (13.75) is

$$Y(z) = \frac{N_o(z)}{z^n} \frac{z}{z - 1}$$

$$= c_0 + c_1 z^{-1} + c_2 z^{-2} + \cdots + c_{n-1} z^{-n+1} + c_n z^{-n} + \frac{N_o(1)z}{z - 1} \tag{13.76}$$

The transient response consists of a finite number of terms; it becomes identically zero after the nth sampling instant. Thus, if the input is a step sequence, the output of (13.75) will become a step sequence after the nth sampling instant. This cannot happen in analog systems; the step response of any stable analog system will become a step function only at $t \rightarrow \infty$. Note that a pole at $z = 0$ in digital systems corresponds to a pole at $s = -\infty$ in analog systems. It is not possible to design an analog system with a *proper* transfer function that has all poles at $s = -\infty$. Thus dead-beat design is not possible in analog systems.

Example 13.7.1

Consider the discrete-time state-variable equation

$$\mathbf{x}(k + 1) = \begin{bmatrix} 0 & 1 \\ 0 & -1 \end{bmatrix} \mathbf{x}(k) + \begin{bmatrix} 0 \\ 10 \end{bmatrix} u(k) \qquad (13.77\text{a})$$

$$y(k) = [1 \quad 0]\mathbf{x} \qquad (13.77\text{b})$$

with transfer function $G(z) = 10/z(z + 1)$. This equation is identical to (11.41) except that it is in the discrete-time domain. Find a feedback gain $\mathbf{k}$ in $u = r - \mathbf{kx}$ so that the resulting system has all eigenvalues at $z = 0$. We use the procedure in Section 11.4 to carry out the design. We compute

$$\Delta(z) = \det(z\mathbf{I} - \mathbf{A}) = \det \begin{bmatrix} z & -1 \\ 0 & z+1 \end{bmatrix} = z(z + 1) = z^2 + z + 0$$

and

$$\overline{\Delta}(z) = z^2 + 0 \cdot z + 0$$

Thus we have

$$\overline{\mathbf{k}} = [0 - 1 \quad 0 - 0] = [-1 \quad 0]$$

The similarity transformation is identical to the one in Example 11.4.2 and equals

$$\mathbf{P}^{-1} = \begin{bmatrix} 0 & 10 \\ 10 & 0 \end{bmatrix}$$

Thus the feedback gain is

$$\mathbf{k} = \overline{\mathbf{k}}\mathbf{P} = [-1 \quad 0] \begin{bmatrix} 0 & 0.1 \\ 0.1 & 0 \end{bmatrix} = [0 \quad -0.1] \qquad (13.78)$$

This completes the design of state feedback. Next we use the procedure in Section 11.6.1 to design a reduced-dimensional state estimator. The procedure is identical to the one in Example 11.6.1. Because (13.77) has dimension $n = 2$, its reduced-dimensional estimator has dimension 1. Because the eigenvalue of $\mathbf{F}$ is required to be different from those of $\mathbf{A}$, we cannot choose $\mathbf{F} = 0$. If we choose $\mathbf{F} = 0$, then $\mathbf{TA} - \mathbf{FT} = \mathbf{gc}$ has no solution. See Problem 13.14. We choose, rather arbitrarily,

$\mathbf{F} = 0.1$ and choose $\mathbf{g} = 1$. Clearly $(\mathbf{F}, \mathbf{g})$ is controllable. The matrix $\mathbf{T}$ is 1×2. Let $\mathbf{T} = [t_1 \quad t_2]$. Then it can be solved from

$$[t_1 \quad t_2] \begin{bmatrix} 0 & 1 \\ 0 & -1 \end{bmatrix} - 0.1 \times [t_1 \quad t_2] = 1 \times [1 \quad 0]$$

or

$$[0 \quad t_1 - t_2] - [0.1t_1 \quad 0.1t_2] = [1 \quad 0]$$

Thus we have $-0.1t_1 = 1$ and $t_1 - 1.1t_2 = 0$ which imply $t_1 = -10$ and $t_2 = t_1/1.1 = -9.09$. The matrix

$$\mathbf{P} := \begin{bmatrix} \mathbf{c} \\ \mathbf{T} \end{bmatrix} = \begin{bmatrix} 1 & 0 \\ -10 & -9.09 \end{bmatrix}$$

is clearly nonsingular and its inverse is

$$\begin{bmatrix} 1 & 0 \\ -10 & -9.09 \end{bmatrix}^{-1} = \begin{bmatrix} 1 & 0 \\ -1.1 & -0.11 \end{bmatrix}$$

We compute

$$\mathbf{h} = \mathbf{Tb} = [-10 \quad -9.09] \begin{bmatrix} 0 \\ 10 \end{bmatrix} = -90.9$$

Thus the 1-dimensional state estimator is

$$\mathbf{z}(k + 1) = 0.1\mathbf{z}(k) + y(k) - 90.9u(k) \tag{13.79a}$$

$$\hat{\mathbf{x}}(k) = \begin{bmatrix} 1 & 0 \\ -1.1 & -0.11 \end{bmatrix} \begin{bmatrix} y(k) \\ \mathbf{z}(k) \end{bmatrix} = \begin{bmatrix} y(k) \\ -1.1y(k) - 0.11\mathbf{z}(k) \end{bmatrix} \tag{13.79b}$$

This completes the design. We see that the design procedure is identical to the analog case. We plot in Figure 13.17 the state estimator in (13.79) and the state feedback

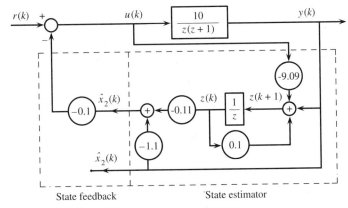

Figure 13.17 State feedback and state estimator.

in (13.78) applying at the output of the estimator. To verify the result, we use Mason's formula to compute the transfer function from r to y. It is computed as

$$G_o(z) = \frac{10}{z^2}$$

Thus the result is correct.

To conclude this section, we mention that in dead-beat design, the actuating signal usually will be large and the plant may saturate. This is especially true if the sampling period is chosen to be small. If the sampling period is large, the possibility of saturation will be smaller, but the design may be poorer. Therefore, computer simulation of dead-beat design is recommended before it is implemented in practice.

13.8   MODEL MATCHING

Pole placement and model matching, discussed in Chapter 10 for analog systems, can be applied directly to digital systems. To discuss model matching, we must discuss physical constraints in implementation. As in the analog case, we require

1. All compensators used have proper rational transfer functions.
2. The resulting system is well posed.
3. The resulting system is totally stable.
4. There is no plant leakage in the sense that all forward paths from r to y pass through the plant.

As was discussed in Section 12.5.2, a compensator with an improper transfer function is not a causal system and cannot be built to operate on real time. If a resulting system is not well posed, then the system has at least one improper closed-loop transfer function and the system can predict what input will be applied in the future. This is not possible in the real world. If a resulting system is not totally stable, then the response of the system may grow without bound if noise or disturbance enters the system. The fourth constraint implies that all power must pass through the plant and that no compensator may be introduced in parallel with the plant. For a more detailed discussion of these constraints, see Chapter 6.

Given a digital plant transfer function

$$G(z) = \frac{N(z)}{D(z)} \tag{13.80}$$

an overall transfer function

$$G_o(z) = \frac{N_o(z)}{D_o(z)} \tag{13.81}$$

is said to be implementable if there exists a control configuration such that $G_o(z)$ can be implemented without violating any of the preceding four constraints. This definition is similar to the analog case. The necessary and sufficient conditions for $G_o(z)$ = $N_o(z)/D_o(z)$ to be implementable are

1. $\deg D_o(z) - \deg N_o(z) \geq \deg D(z) - \deg N(z)$ (pole-zero excess inequality).
2. All zeros of $N(z)$ on and outside the unit circle must be retained in $N_o(z)$ (retainment of non-minimum-phase zeros).
3. All roots of $D_o(z)$ must lie inside the unit circle.

Condition (1) implies that if the plant has a delay of $r := [\deg D(z) - \deg N(z)]$ sampling instants, then $G_o(z)$ must have a delay of at least r sampling instants. If a control configuration has no plant leakage, then the roots of $N(z)$ will not be affected by feedback. Therefore, the only way to eliminate zeros of $N(z)$ is by direct pole-zero cancellations. Consequently, zeros on or outside the unit circle of the z-plane should be retained in $N_o(z)$. In fact, zeros that do not lie inside the desired pole region discussed in Figure 13.13(c) should be retained even if they lie inside the unit circle. Thus the definition and conditions of implementable transfer functions in the digital case are identical to the analog case discussed in Section 9.2.

As in the analog case, the unity-feedback configuration in Figure 10.1 can be used to implement every pole placement but not every model matching. The two-parameter configuration in Figure 10.6 and the plant input/output feedback configuration in Figure 10.15 can be used to implement any model matching. The design procedures in Chapter 10 for the analog case are directly applicable to the digital case. We use an example to illustrate the procedures.

Example 13.8.1

Consider

$$G(z) = \frac{z + 0.6}{(z - 1)(z + 0.5)} = \frac{N(z)}{D(z)} \tag{13.82}$$

Implement the overall transfer function

$$G_o(z) = \frac{0.25(z + 0.6)}{z^2 - 0.8z + 0.2} = \frac{N_o(z)}{D_o(z)} \tag{13.83}$$

Because $G_o(1) = 0.25(1 + 0.6)/(1 - 0.8 + 0.2) = 1$, the output of $G_o(z)$ will track any step-reference input without an error. Note that even though zero $(z + 0.6)$ lies inside the unit circle, it lies outside the desired pole region shown in Figure 13.13(c); therefore, it is retained in $G_o(z)$. We first use the unity-feedback configuration shown in Figure 13.18(a) to implement (13.83). Using (10.2) with s

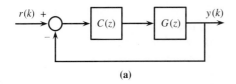

(a)

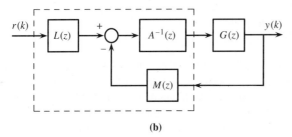

(b)

Figure 13.18 (a) Unity-feedback system. (b) Two-parameter feedback system.

replaced by z, we have

$$C(z) = \frac{G_o(z)}{G(z)(1 - G_o(z))}$$

$$= \frac{\dfrac{0.25(z + 0.6)}{z^2 - 0.8z + 0.2}}{\dfrac{(z + 0.6)}{(z - 1)(z + 0.5)} \cdot \left(1 - \dfrac{0.25(z + 0.6)}{z^2 - 0.8z + 0.2}\right)}$$ (13.84)

$$= \frac{0.25(z - 1)(z + 0.5)}{z^2 - 0.8z + 0.2 - 0.25(z + 0.6)}$$

$$= \frac{0.25(z - 1)(z + 0.5)}{(z - 1)(z - 0.05)} = \frac{0.25(z + 0.5)}{z - 0.05}$$

This is a proper compensator. We mention that the design has a pole-zero cancellation of $(z + 0.5)$. The pole is dictated by the plant transfer function. Because the pole is stable, the system is totally stable.

Next we implement $G_o(z)$ in the two-parameter configuration shown in Figure 13.18(b). We use the procedure in Section 10.4.1 with s replaced by z. We compute

$$\frac{G_o(z)}{N(z)} = \frac{0.25(z + 0.6)}{(z^2 - 0.8z + 0.2)(z + 0.6)} = \frac{0.25}{(z^2 - 0.8z + 0.2)} =: \frac{N_p(z)}{D_p(z)}$$

Next we find a $\overline{D}_p(z)$ such that the degree of $D_p(z)\overline{D}_p(z)$ is at least $2n - 1 = 3$. Because the degree of $D_p(z)$ is 2, the degree of $\overline{D}_p(z)$ can be chosen as 1. We choose $\overline{D}_p(s) = z$. Then we have

$$L(z) = N_p(z)\overline{D}_p(z) = 0.25 \times z = 0.25z$$

The polynomials $A(z) = A_0 + A_1 z$ and $M(z) = M_0 + M_1 z$ can be solved from

$$A(z)D(z) + M(z)N(z) = D_p(z)\overline{D}_p(z)$$

with

$$F(z) := D_p(z)\overline{D}_p(z) = (z^2 - 0.8z + 0.2)z = z^3 - 0.8z^2 + 0.2z$$

or from the following linear algebraic equation

$$
\begin{bmatrix}
D_0 & N_0 & 0 & 0 \\
D_1 & N_1 & D_0 & N_0 \\
D_2 & N_2 & D_1 & N_1 \\
0 & 0 & D_2 & N_2
\end{bmatrix}
\begin{bmatrix}
A_0 \\
M_0 \\
A_1 \\
M_1
\end{bmatrix}
=
\begin{bmatrix}
-0.5 & 0.6 & 0 & 0 \\
-0.5 & 1 & -0.5 & 0.6 \\
1 & 0 & -0.5 & 1 \\
0 & 0 & 1 & 0
\end{bmatrix}
\begin{bmatrix}
A_0 \\
M_0 \\
A_1 \\
M_1
\end{bmatrix}
=
\begin{bmatrix}
0 \\
0.2 \\
-0.8 \\
1
\end{bmatrix}
$$

The solution is $A(z) = A_0 + A_1 z = -3.3 + z$ and $M(z) = M_0 + M_1 z = -2.75 + 3z$. Thus the compensator is

$$[C_1(z) \quad -C_2(z)] = \left[\frac{0.25z}{z - 3.3} \quad -\frac{3z - 2.75}{z - 3.3} \right] \tag{13.85}$$

This completes the design. As for the unity-feedback configuration, this design also involves one pole-zero cancellation. However, this pole is chosen by the designer. We have chosen it as $z = 0$ for this problem.

From this example, we see that the design procedures in Chapter 10 for analog systems can be directly applied to design digital systems.

13.9 CONCLUDING REMARKS

This chapter introduced two approaches to design digital compensators. The first approach is to design an analog compensator and then transform it into a digital one. We discussed six methods in Section 13.2 to carry out the transformation. Among these, the step-invariant and bilinear transformation methods appear to yield the best results. The second approach is to transform an analog plant into an equivalent step-invariant digital plant. We then design digital compensators directly by using the root-locus method, state-space method, or linear algebraic method. The Bode design method cannot be directly applied because $G(e^{j\omega T})$ is an irrational function of ω and because ω is limited to $(-\pi/T, \pi/T)$. However, it can be so used after a bilinear transformation.

A question may be raised at this point: Which approach is simpler and yields a better result? In the second approach, we first discretize the plant and then carry out design. If the result is found to be unsatisfactory, we must select a different sampling period, again discretize the analog plant, and then repeat the design. In the first approach, we carry out analog design and then discretize the analog compensator.

If the discretized compensator does not yield a satisfactory result, we select a different sampling period and again discretize the analog compensator. There is no need to repeat the design; therefore, the first approach appears to be simpler. There is another problem with the second approach. In discretization of an analog plant, if the sampling period is small, the resulting digital plant will be more prone to numerical error. In analog design, the desired pole region consists of a good portion of the left half s-plane. In digital design, the desired pole region consists of only a portion of the unit circle of the z-plane. Therefore, digital design is clustered in a much smaller region and, consequently, the possibility of introducing numerical error is larger. In conclusion, the first approach (discretization after design) may be simpler than the second approach (discretization before design).

PROBLEMS

13.1. Find the impulse-invariant and step-invariant digital compensators of the following two analog compensators with sampling period $T = 1$ and $T = 0.5$:

    a. $\dfrac{2}{(s + 2)(s + 4)}$

    b. $\dfrac{2s - 3}{s + 4}$

13.2. Use state-variable equations to find step-invariant digital compensators for Problem 13.1.

13.3. Use forward-difference, backward-difference, and bilinear transformation methods to find digital compensators for Problem 13.1.

13.4. Use state-variable equations to find forward-difference digital compensators for Problem 13.1.

13.5. **a.** Find the controllable-form realization for the transfer function in (13.22).

      b. Show that the realization and the state-variable equation in (13.24) are equivalent. Can you conclude that the transfer function of (13.24) equals (13.22)? [*Hint:* The similarity transformation is
$$\begin{bmatrix} T & -T \\ 0 & T^2 \end{bmatrix}$$

13.6. Use pole-zero mapping to find digital compensators for Problem 13.1.

13.7. Use the analog initial value theorem to show that the step response of (13.31) has the property $\bar{y}(0) = 0$, $\dot{\bar{y}}(0) = 0$, and $\ddot{\bar{y}}(0) = 0$. Thus the value of $\bar{y}(T)$ is small if T is small.

13.8. **a.** Compute the steady-state response of the analog compensator in (13.31) due to a unit-step input.

b. Compute the steady-state response of the digital compensator in (13.33) with $N(z) = (z + 1)^2$ due to a unit-step sequence.

c. Are the results in (a) and (b) the same? If not, modify b in (13.33) so that they have the same steady-state responses.

13.9. Find equivalent digital plant transfer functions for the following analog plant transfer functions:

a. $\dfrac{1}{s^2}$

b. $\dfrac{10}{s(s + 2)}$

c. $\dfrac{8}{s(s + 4)(s - 2)}$

What are their pole-zero excesses?

13.10. Consider an analog plant with transfer function

$$\overline{G}(s) = \frac{1}{s(s + 1 + j10)(s + 1 - j10)}$$

Find its equivalent digital plant transfer functions for $T = 0.2\pi$ and $T = 0.1$. Can they both be used in design?

13.11. Use (13.61) to establish (13.62).

13.12. Given a digital plant

$$G(z) = \frac{z + 1}{(z - 0.3)(z - 1)}$$

Use the root-locus method to design a system to meet (1) position error $= 0$, (2) overshoot $\leq 10\%$, (3) settling time ≤ 5 seconds, and (4) rise time as small as possible.

13.13. Design a state feedback and a full-dimensional state estimator for the plant in Problem 13.12 such that all poles of the state feedback and state estimator are at $z = 0$.

13.14. In the design of a reduced-dimensional state estimator for the digital plant in Example 13.7.1, show that if $\mathbf{F} = 0$ and $\mathbf{g} = 1$, then the equation $\mathbf{TA} - \mathbf{FT} = \mathbf{gc}$ has no solution $\mathbf{T}$.

13.15. Design a state feedback and a reduced-dimensional state estimator for the plant in Problem 13.12 such that all poles of the state feedback are at $z = 0$. Can you choose the eigenvalue of the estimator as $z = 0$? If not, choose it as $z = 0.1$.

13.16. Repeat Problem 13.15 using the linear algebraic method. Are the results the same? Which method is simpler?

13.17. Consider the plant in Problem 13.12. Use the unity-feedback configuration to find a compensator such that all poles are located at $z = 0$. What zero, if any, is introduced in $G_o(z)$? Will $G_o(z)$ track any step-reference input without an error?

13.18. Consider the plant in Problem 13.12. Implement the following overall transfer function

$$G_o(z) = \frac{0.5(z + 1)}{z^2}$$

Will this system track any step-reference input?

13.19. Consider

$$G(z) = \frac{z + 2}{(z - 1)(z - 3)} \qquad G_o(z) = \frac{h(z + 2)}{z(z + 0.2)}.$$

Find h so that $G_o(1) = 1$. Implement $G_o(z)$ in the unity-feedback configuration and the two-parameter feedback configuration. Are they both acceptable in practice?

14 PID Controllers

14.1 INTRODUCTION

In this text, we have introduced a number of methods to design control systems. If a plant is single-variable and can be modeled as linear time-invariant and lumped, then the methods can be used to design a good control system. If a plant cannot be so modeled (for example, if it is nonlinear), then the design methods cannot be used. What should we do in this situation? Basically, there are three approaches:

1. Use nonlinear techniques. Because of the complexity of nonlinear problems, no simple and general design method is available to design all nonlinear control systems. However, a number of special techniques such as the phase-plane method, the describing function method, and the Lyapunov stability method are available for analysis and design. This is outside the scope of this text. The reader is referred, for example, to References [6, 56].

2. Find a linearized model, even if it is very crude, and carry out linear design, then apply the computed compensator to the actual plant. Clearly, there is no guarantee that the resulting system will be satisfactory. Then adjust the parameters of the compensator and hope that a satisfactory system can be obtained. If the nonlinear plant can be simulated on a computer, carry out the adjustment on the computer. This method may become unmanageable if the number of parameters to be adjusted is large.

3. Use the proportional-integral-derivative (PID) controller, adjust the parameters, and hope that a satisfactory overall system can be obtained. This approach is widely used in industry, and a number of PID controllers are available on the market.

551

In this chapter, we discuss first analog PID controllers and the adjustment of their parameters. We then discuss digital implementations of these controllers. Before proceeding, we mention that what will be discussed for nonlinear control systems is mostly extrapolated from linear control systems. Therefore, basic knowledge of linear control systems is essential in studying nonlinear systems. We also mention that, unlike linear systems, a result in a nonlinear system in a particular situation is not necessarily extendible to other situations. For example, the response of the nonlinear system in Figure 6.14 due to $r(t) = 0.3$ approaches 0.3, but the response due to $r(t) = 4 \times 0.3$ does not approach 4×0.3. Instead, it approaches infinity. Thus the response of nonlinear systems depends highly on the magnitude of the input. In this chapter, all initial conditions are assumed to be zero and the reference input is a unit-step function.

14.2   PID CONTROLLERS IN INDUSTRIAL PROCESSES

Consider the system shown in Figure 14.1. It is assumed that the plant cannot be adequately modeled as a linear time-invariant lumped system. This is often the case if the plant is a chemical process. Most chemical processes have a long time delay in responses and measurements; therefore, they cannot be modeled as lumped systems. Chemical reactions and liquid flow in pipes may not be describable by linear equations. Control valves are nonlinear elements; they saturate when fully opened. Therefore, many chemical processes cannot be adequately modeled as linear time-invariant and lumped systems.

Many industrial processes, including chemical processes, can be controlled in the open-loop configuration shown in Figure 14.1(a). In this case, the plant and the controller must be stable. The parameters of the controller can be adjusted manually or by a computer. In the open-loop system, the parameters generally must be readjusted whenever there are changes in the set point and load. If we introduce feedback around the nonlinear plant as shown in Figure 14.1(b), then the readjustment may not be needed. Feedback may also improve performance, as in the linear case, and make the resulting systems less sensitive to plant perturbations and external disturbances. Therefore, it is desirable to introduce feedback for nonlinear plants.

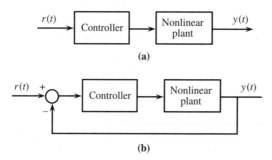

Figure 14.1 (a) Open-loop nonlinear system. (b) Unity-feedback nonlinear system.

The design of nonlinear feedback systems is usually carried out by trial and error. Whenever we use a trial-and-error method, we begin with simple control configurations such as the unity-feedback configuration shown in Figure 14.1(b) and use simple compensators or controllers. If not successful, we then try more complex configurations and controllers. We discuss in the following some simple controllers.

Proportional Controller

The transfer function of a proportional controller is simply a gain, say k_p. If the input of the controller is $e(t)$, then the output is $u(t) = k_p e(t)$ or, in the Laplace transform domain, $U(s) = k_p E(s)$. To abuse the terminology, we call a nonlinear system *stable*[1] if its output excited by a unit-step input is bounded and approaches a constant as $t \to \infty$. Now if a plant is stable, as is often the case for chemical processes, the feedback system in Figure 14.1 will remain stable for a range of k_p. As k_p increases, the unit-step response may become faster and eventually the feedback system may become unstable. This is illustrated by an example.

Example 14.2.1

Consider the system shown in Figure 14.2. The plant consists of a saturation non-linearity, such as a valve, followed by a linear system with transfer function

$$G(s) = \frac{-s + 1}{(5s + 1)(s + 1)}$$

The controller is a proportional controller with gain k_p. We use Protoblock[2] to simulate the feedback system. Its unit-step responses with $k_p = 0.1, 1, 4.5, 9$, are shown

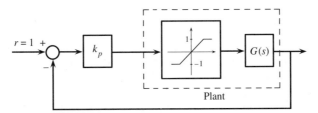

Figure 14.2 Unity-feedback system with nonlinear plant.

---

[1]The stability of nonlinear systems is much more complex than that of linear time-invariant lumped systems. The response of a nonlinear system depends on whether or not initial conditions are zero and on the magnitude of the input. Therefore, the concept of bounded-input, bounded-output stability defined for linear systems is generally not used for nonlinear systems. Instead, we have asymptotic stability, Lyapunov stability, and stability of limit cycles. This is outside the scope of this text and will not be discussed.

[2]A trademark of the Grumman Corporation. This author is grateful to Dr. Chien Y. Huang for carrying out the simulation.

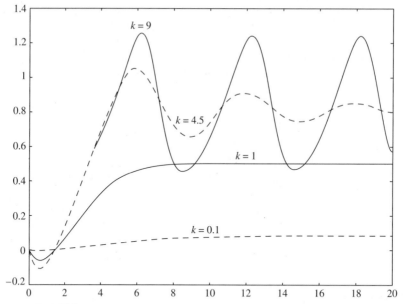

Figure 14.3 Unit-step responses.

in Figure 14.3. We see that as k_p increases, the response becomes faster but more oscillatory. If k_p is less than 9, the response will eventually approach a constant, and the system is stable. If k_p is larger than 9, the response will approach a sustained oscillation and the system is not stable. Recall that we have defined a nonlinear system to be unstable if its unit-step response does not approach a constant. The smallest k_p at which the system becomes unstable is called the *ultimate gain*.

For some nonlinear plants, it may be possible to obtain good control systems by employing only proportional controllers. There is, however, one problem in using such controllers. We see from Figure 14.3 that, for the same unit-step reference input, the steady-state plant outputs are different for different k_p. Therefore, if the steady-state plant output is required to be, for example, 1, then for different k_p, the magnitude of the step-reference input must be different. Consequently, we must adjust the magnitude of the reference input or *reset* the set point for each k_p. Therefore, manual resetting of the set point is often required in using proportional controllers.

The preceding discussion is in fact extrapolated from the linear time-invariant lumped systems discussed in Section 6.3.2. In the unity-feedback configuration in Figure 6.4(a), if the plant is stable and the compensator is a proportional controller, then the loop transfer function is of type 0. In this case, the position error is different from zero and calibration or readjustment of the magnitude of the step-reference input is needed to have the plant output reach a desired value. If the loop transfer function is of type 1, then the position error is zero and the readjustment of the reference input is not necessary. This is the case if the controller consists of an integrator, as will be discussed in the next paragraph.

Integral Controller

If the input of an integral controller with gain k_i is $e(t)$, then the output is

$$u(t) = k_i \int_0^t e(\tau)d\tau \qquad (14.1)$$

or $U(s) = k_i E(s)/s$. Thus the transfer function of the integral controller is k_i/s. For linear unity-feedback systems, if the forward path has a pole at $s = 0$ or of type 1, and if the feedback system is stable, then the steady-state error due to any step-reference input is zero for any k_i. We may have the same situation for nonlinear unity-feedback systems. In other words, if we introduce an integral controller in Figure 14.1(b) and if the system is stable, then for every k_i, the steady-state error may be zero, and there is no need to reset the set point. For this reason, the integral control is also called the *reset* control and k_i is called the *reset rate*.

Although the integral controller will eliminate the need of resetting the set point, its presence will make the stabilization of feedback systems more difficult. Even if it is stable, the speed of response may decrease and the response may be more oscillatory, as is generally the case for linear time-invariant lumped systems. In addition, it may generate the phenomenon of *integral windup* or *reset windup*. Consider the system shown in Figure 14.4(a) with an integral controller and a valve saturation nonlinearity with saturation level $\bar{u}_m$. Suppose the error signal $e(t)$ is as shown in Figure 14.4(a). Then the corresponding controller output $u(t)$ and valve output $\bar{u}(t)$ are as shown. Ideally, if the error signal changes from positive to negative at t_0, its effect on $\bar{u}(t)$ will appear at the same instant t_0. However, because of the integration, the value of $u(t)$ at $t = t_0$ is quite large as shown, and it will take awhile, say until time t_2, for the signal to unwind to $\bar{u}_m$. Therefore, the effect of $e(t)$ on $\bar{u}(t)$ will appear only after t_2 as shown in Figure 14.4(a) and there is a delay of $t_2 - t_0$. In conclusion, because of the integration, the signal $u(t)$ winds up over the saturation level and must be unwound before the error signal can affect on $\bar{u}(t)$. This is called *integral windup* or *reset windup*.

If the error signal $e(t)$ in Figure 14.4 is small and approaches zero rapidly, integral windup may not occur. If integral windup does occur and makes the feedback system unsatisfactory, it is possible to eliminate the problem. The basic idea is to disable the integrator when the signal $u(t)$ reaches the saturation level. For example, consider the arrangement shown Figure 14.4(b). The input $\bar{e}(t)$ of the integral controller is $\bar{e}(t) = e(t) \times w(t)$, where $w(t) = 1$ if $|u(t)| \leq \bar{u}_m$ and $w(t) = 0$ if $|u(t)| > \bar{u}_m$. Such $w(t)$ can be generated by a computer program or by the arrangement shown in Figure 14.4(b). The arrangement disables the integrator when its output reaches the saturation level, thus $u(t)$ will not wind up over the saturation level, and when $e(t)$ changes sign at t_0, its effect appears immediately at $\bar{u}(t)$. Hence, the performance of the feedback system may be improved. This is called *antiwindup* or *integral windup prevention*.

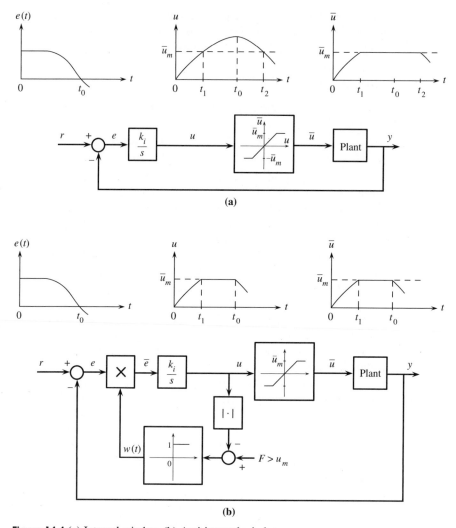

Figure 14.4 (a) Integral windup. (b) Anti-integral windup.

Derivative Controller

If the input of a derivative controller with derivative constant k_d is $e(t)$, then its output is $k_d \, de(t)/dt$ or, in the Laplace transform domain, $k_d s E(s)$. Therefore, the transfer function of the derivative controller is $k_d s$. This is an improper transfer function and is difficult to implement. In practice, it is built as

$$\frac{k_d s}{1 + \dfrac{k_d s}{N}} \tag{14.2}$$

where N, ranging from 3 to 10, is determined by the manufacturer and is called the *taming factor*. This taming factor makes the controller easier to build. It also limits high-frequency gain; therefore, high-frequency noise will not be unduly amplified. For control signals, which are generally of low frequency, the transfer function in (14.3) can be approximated as $k_d s$.

The derivative controller is rarely used by itself in feedback control systems. Suppose the error signal $e(t)$ is very large and changes slowly or, in the extreme case, is a constant. In this case, a good controller should generate a large actuating signal to force the plant output to catch up with the reference signal so that the error signal will be reduced. However, if we use the derivative controller, the actuating signal will be zero, and the error signal will remain large. For this reason, the derivative controller is not used by itself in practice. If we write the derivative of $e(t)$ at $t = t_0$ as

$$\frac{de(t)}{dt}\bigg|_{t=t_0} = \lim_{a \to 0} \frac{e(t_0 + a) - e(t_0)}{a}$$

with $a > 0$, then its value depends on the future $e(t)$. Thus the derivative or rate controller is also called the *anticipatory controller*.

The combination of the proportional, integral, and derivative controllers is called a *PID controller*. Its transfer function is given by

$$k_p + \frac{k_i}{s} + k_d s = k_p \left(1 + \frac{1}{T_i s} + T_d s \right) \tag{14.3}$$

where T_i is called the *integral time constant* and T_d the *derivative time constant*. The PID controller can be arranged as shown in Figure 14.5(a). This arrangement is discussed in most control texts and is called the "textbook PID controller" in Reference [3]. The arrangement is not desirable if the reference input r contains discontinuities, such as in a step function. In this case, $e(t)$ will be discontinuous and its differentiation will generate an impulse or a very large actuating signal. An alternative arrangement, called the *derivative-of-output controller*, is shown in Figure 14.5(b) where only the plant output $y(t)$ is differentiated. In this arrangement, the discontinuity of r will appear at u through the proportional gain but will not be

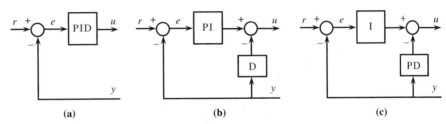

Figure 14.5 (a) Textbook PID controller. (b) Derivative-of-output controller. (c) Set-point-on-I controller.

amplified by differentiation. Yet another arrangement is shown in Figure 14.5(c), where the error e is integrated. It is called the *set-point-on-I-only controller*. In this case, the discontinuity of r will be smoothed by the integration.

There are three parameters to be adjusted or tuned in a PID controller. It may be tuned by trial and error. We may first set $k_i = 0$ and $k_d = 0$ and vary k_p to see whether a satisfactory feedback system can be obtained. If not, we may then also vary k_i. If we still cannot obtain a satisfactory system, we then vary all three parameters. This is a trial-and-error method.

14.2.1 Rules of Ziegler and Nichols

The PID controller has been widely used in industry, especially in chemical processes. Therefore, various rules for adjusting the three parameters have been developed. We discuss in this subsection two sets of rules developed by Ziegler and Nichols [69]. They found that if the unit-step response of a system is of the form shown in Figure 14.6(a)—that is, the second overshoot is roughly 25% of the first overshoot—then the integral of the absolute error (IAE) or

$$ J = \int_0^\infty |e(t)| \, dt = \int_0^\infty |r(t) - y(t)| \, dt \qquad (14.4) $$

with $r(t) = 1$, is minimized. This is called the *quarter-decay criterion*. Ziegler and Nichols used this criterion to develop their rules. These rules were developed mainly from experiment.

Closed-Loop Method   Consider the system shown in Figure 14.1(b). The controller consists of only a proportional controller with gain k_p. It is assumed that the system is stable for $0 \le k_p < k_u$ and the unit-step response of the system with $k_p = k_u$ is of the form shown in Figure 14.6(b). It has a sustained oscillation with period T_u. We call k_u the *ultimate gain* and T_u the *ultimate period*. Then the rules of Ziegler and Nichols for tuning the PID controller are as shown in Table 14.1.

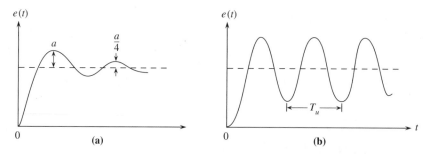

Figure 14.6 (a) Quarter-decay response. (b) Ultimate gain and period.

Table 14.1 Closed-Loop Method

Controller	k_p	T_i	T_d
P	$0.5K_u$		
PI	$0.4K_u$	$0.8T_u$	
PID	$0.6K_u$	$0.5T_u$	$0.12T_u$

Open-Loop Method   In this method, we measure the unit-step response of the plant without closing the loop. It is assumed that the response is of the form shown in Figure 14.7. If the response is not of that form, the method is not applicable. The response is approximated by straight lines, with K, L, and T indicated as shown. Then the response can be approximated by the unit-step response of a system with transfer function

$$G(s) = \frac{Ke^{-Ls}}{Ts + 1} \tag{14.5}$$

This transfer function contains an irrational function e^{-Ls}, which is due to the time delay of L seconds. The rest is a linear time-invariant lumped system with gain K and time constant T. Then the rules of Ziegler and Nichols for tuning the PID controller are as shown in Table 14.2. These two sets of rules are quite simple and can be easily applied. Certainly, there is no guarantee that they will yield good control systems, but they can always be used as an initial set of parameters for subsequent adjustment or tuning.

14.2.2   Rational Approximations of Time Delays

Industrial processes often contain time delays, which are also called *dead times* or *transport lags*. The transfer function of the plants will then contain the factor e^{-Ls} as shown in (14.5). The function e^a can be expressed as

$$e^a = 1 + \frac{a}{1!} + \frac{a^2}{2!} + \frac{a^3}{3!} + \cdots + \frac{a^n}{n!} + \cdots$$

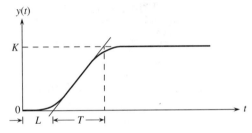

Figure 14.7 Unit-step response.

Table 14.2 Open-Loop Method

Controller	k_p	T_i	T_d
P	TL		
PI	$0.9TL$	$0.3L$	
PID	$1.2TL$	$2L$	$0.5L$

where $n! = n(n - 1)(n - 2) \cdots 2 \cdot 1$. Taking the first two or three terms, we obtain the following rational function approximations:

$$e^{-Ls} = \frac{1}{e^{Ls}} \approx \frac{1}{1 + Ls} \tag{14.6a}$$

$$e^{-Ls} = \frac{e^{-Ls/2}}{e^{Ls/2}} \approx \frac{1 - \dfrac{Ls}{2}}{1 + \dfrac{Ls}{2}} = \frac{2 - Ls}{2 + Ls} \tag{14.6b}$$

and

$$e^{-Ls} = \frac{e^{-Ls/2}}{e^{Ls/2}} \approx \frac{1 - \dfrac{Ls}{2} + \dfrac{(Ls)^2}{8}}{1 + \dfrac{Ls}{2} + \dfrac{(Ls)^2}{8}} = \frac{8 - 4Ls + L^2 s^2}{8 + 4Ls + L^2 s^2} \tag{14.6c}$$

Note that these approximations are good for s small or s approaching zero. Because s small governs the time response as t approaches infinity, as can be seen from the final-value theorem, these equations give good approximations for the steady-state response but, generally, poor approximations for the transient response. For example, Figure 14.8 shows the unit-step responses of

$$G(s) = \frac{e^{-3s}}{2s + 1}$$

and its rational function approximations

$$G_1(s) = \frac{1}{(2s + 1)(1 + 3s)} \qquad G_2(s) = \frac{2 - 3s}{(2s + 1)(2 + 3s)}$$

$$G_3(s) = \frac{8 - 12s + 9s^2}{(2s + 1)(8 + 12s + 9s^2)}$$

All of them approach the same steady-state value, but their transient responses are quite different. The unit-step response of $G_3(s)$ is closest to the one of $G(s)$.

Once a plant transfer function with time delay is approximated by a rational transfer function, then all methods introduced in this text can be used to carry out the design. This is the second approach mentioned in Section 14.1.

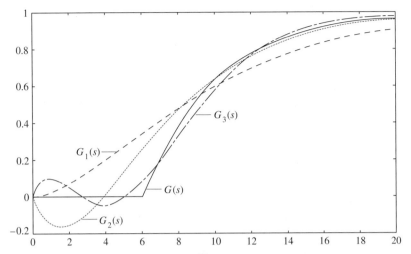

Figure 14.8 Unit-step response of rational function approximations.

14.3 PID Controllers for Linear Time-Invariant Lumped Systems

The PID controller certainly can also be used in the design of linear time-invariant lumped systems. In fact, the proportional controller is always the first controller to be tried in using the root-locus and Bode plot design methods. If we use the PID controller for linear time-invariant lumped systems, then the tuning formula in Table 14.1 can also be used. In this case, the ultimate gain K_u and ultimate period T_u can be obtained by measurement or from the Bode plot or root loci. If the Bode plot of the plant is as shown in Figure 14.9(a) with phase-crossover frequency ω_p and gain margin a dB, then the ultimate gain and ultimate period are given by

$$K_u = 10^{a/20} \qquad T_u = \frac{2\pi}{\omega_p}$$

If the root loci of the plant are as shown in Figure 14.9(b), then from the intersection of the root loci with the imaginary axis, we can readily obtain the ultimate gain K_u and the ultimate period as $T_u = 2\pi/\omega_u$. Once K_u and T_u are obtained, then the tuning rule in Table 14.1 can be employed.

Even though PID controllers can be directly applied to linear time-invariant lumped systems, there seems no reason to restrict compensators to them. The transfer function of PI controllers is

$$k_p + \frac{k_i}{s} = \frac{k_p s + k_i}{s}$$

This is a special case of the phase-lag compensator with transfer function $k(s + b)/(s + a)$ and $b > a \geq 0$. Therefore, phase-lag compensators are more general than PI compensators and it should be possible to design better systems without restricting $a = 0$. This is indeed the case for the system in Example 8.10.1. See the responses

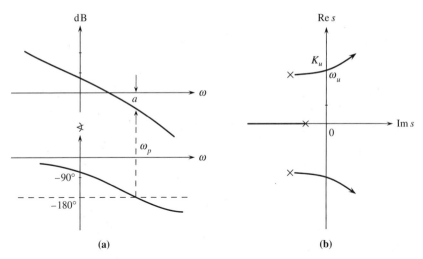

Figure 14.9 (a) Bode plot. (b) Root loci.

in Figure 8.39. The phase-lag controller yields a better system than the PI controller does.

If we use the more realistic derivative controller in (14.2), then the transfer function of PD controllers is

$$k_p + \frac{k_d s}{1 + \dfrac{k_d s}{N}} = \frac{(N + k_p)\left(s + \dfrac{Nk_p}{k_d(N + k_p)}\right)}{s + \dfrac{N}{k_d}} =: \frac{k(s + ac)}{s + a} \qquad (14.7)$$

where $k = N + k_p$, $a = N/k_d$, and $c = k_p/(N + k_p) < 1$. This is a special case of the phase-lead compensator with transfer function $k(s + b)/(s + a)$ and $0 \le b < a$. Similarly, the transfer function of realistic PID controllers is a special case of the following compensator of degree 2:

$$\frac{k(s^2 + b_1 s + b_0)}{s^2 + a_1 s + a_0} \qquad (14.8)$$

Thus if we use general controllers, then the resulting systems should be at least as good as those obtained by using PID controllers. Furthermore, systematic design methods are available to compute general controllers. Therefore, for linear time-invariant lumped systems, there seems no reason to restrict controllers to PID controllers.

There is, however, one situation where PID controllers are useful even if a plant can be adequately modeled as an LITL system. PID controllers can be built using hydraulic or pneumatic devices; general proper transfer functions, however, cannot be so built. If control systems are required to use hydraulic systems, such as in all existing Boeing commercial aircrafts, then we may have to use PID controllers. A new model of AirBus uses control by wire (by electrical wire) rather than hydraulic tube. In such a case there seems no reason to restrict controllers to PID controllers, because controllers with any proper transfer functions can be easily built using electrical circuits as discussed in Chapter 5. Using control by wire, the controller can be more complex and the performance of the system can be improved.

14.4 DIGITAL PID CONTROLLERS

In this section we discuss digital implementations of PID controllers. As in Chapter 13, we use $\overline{G}(s)$ to denote analog PID controllers and $G(z)$, digital PID controllers. Consider the analog PID controller discussed in (14.3)

$$\overline{G}(s) = k_p \left(1 + \frac{1}{T_i s} + T_d s \right) \tag{14.9}$$

We have discussed a number of discretization methods in Section 13.2, all of which can be used to discretize (14.9). If we use the forward difference in (13.19) for both the integrator and differentiator, then (14.9) becomes

$$G(z) = k_p \left[1 + \frac{T}{T_i(z - 1)} + \frac{T_d(z - 1)}{T} \right] \tag{14.10}$$

This is the simplest digital implementation of the PID controller. Another possibility is to use the trapezoidal approximation in (13.27) for the integrator and the backward difference in (13.25) for the differentiator; then (14.9) becomes

$$\begin{aligned}
G(z) &= k_p \left[1 + \frac{T}{2T_i} \frac{z + 1}{z - 1} + \frac{T_d(z - 1)}{Tz} \right] \\
&= k_p \left[1 + \frac{T}{2T_i} \frac{2 - (1 - z^{-1})}{1 - z^{-1}} + \frac{T_d(1 - z^{-1})}{T} \right] \\
&= k_p \left[1 - \frac{T}{2T_i} + \frac{T}{T_i} \frac{1}{1 - z^{-1}} + \frac{T_d(1 - z^{-1})}{T} \right]
\end{aligned} \tag{14.11}$$

If we define

$$\hat{k}_p = k_p \left(1 - \frac{T}{2T_i} \right) \qquad \text{(Proportional gain)} \tag{14.12a}$$

$$\hat{k}_i = \frac{k_p T}{T_i} \qquad \text{(Integral gain)} \tag{14.12b}$$

$$\hat{k}_d = \frac{k_p T_d}{T} \qquad \text{(Derivative gain)} \tag{14.12c}$$

then (14.11) becomes

$$G(z) = \hat{k}_p + \frac{\hat{k}_i}{1 - z^{-1}} + \hat{k}_d(1 - z^{-1}) \tag{14.13}$$

This is one commonly used digital PID controller. Note that digital $\hat{k}_p$ differs from analog k_p by the amount $k_p T/2T_i$ which is small if the sampling period T is small. However, if T is small, then the derivative gain $\hat{k}_d$ will be large. This problem will not arise if the analog differentiator is implemented as in (14.2). In this case, the transfer function of analog PID controllers becomes

$$\overline{G}(s) = k_p \left[1 + \frac{1}{T_i s} + \frac{T_d s}{1 + \frac{T_d s}{N}} \right] \tag{14.14a}$$

$$= k_p \left[1 + \frac{1}{T_i s} + \frac{N(s - 0)}{s - \left(-\frac{N}{T_d}\right)} \right] \tag{14.14b}$$

If we use the impulse-invariant method for the integrator and the pole-zero mapping for the differentiator, then we have

$$G(z) = k_p \left[1 + \frac{T}{T_i(z - 1)} + \frac{N(z - 1)}{z - \alpha} \right] \tag{14.15a}$$

with

$$\alpha = e^{-\frac{NT}{T_d}} \tag{14.15b}$$

If we use the forward difference for the integrator and the backward difference for the differentiator, then we have

$$G(z) = k_p \left[1 + \frac{T}{T_i(z - 1)} + \frac{NT_d}{T_d + NT} \frac{z - 1}{z - \beta} \right] \tag{14.16a}$$

with

$$\beta = \frac{T_d}{T_d + NT} \tag{14.16b}$$

This is a commonly used digital PID controller. We mention that if T is very small, then (14.15) and (14.16) yield roughly the same transfer function. In (14.16a), if we use forward difference for both the integrator and differentiator, then the resulting digital differentiator may become unstable. This is the reason we use forward difference for the integrator and backward difference for the differentiator.

The digital PID controllers in (14.10), (14.13), (14.15), and (14.16) are said to be in *position form*. Now we develop a different form, called *velocity form*. Let the input and output of the digital PID controller in (14.13) be $e(k) := e(kT)$ and $u(k) := u(kT)$. Then we have

$$U(z) = \left[\hat{k}_p + \frac{\hat{k}_i}{1 - z^{-1}} + \hat{k}_d(1 - z^{-1}) \right] E(z) \tag{14.17}$$

which can be written as

$$(1 - z^{-1})U(z) = \hat{k}_p(1 - z^{-1})E(z) + \hat{k}_i E(z) + \hat{k}_d(1 - 2z^{-1} + z^{-2})E(z)$$

In the time domain this becomes

$$\begin{aligned} u(k) - u(k - 1) &= \hat{k}_p[e(k) - e(k - 1)] + \hat{k}_i e(k) \\ &+ \hat{k}_d[e(k) - 2e(k - 1) + e(k - 2)] \end{aligned} \tag{14.18}$$

In the unity-feedback configuration, we have

$$e(k) = r(k) - y(k) \tag{14.19}$$

where $r(k)$ is the reference input sequence and $y(k)$ is the plant output. If the reference input is a step sequence, then $r(k) = r(k - 1) = r(k - 2)$ and

$$\begin{aligned} e(k) - e(k - 1) &= r(k) - y(k) - [r(k - 1) - y(k - 1)] \\ &= -[y(k) - y(k - 1)] \end{aligned} \tag{14.20}$$

and

$$\begin{aligned} e(k) - 2e(k - 1) + e(k - 2) &= r(k) - y(k) - 2[r(k - 1) - y(k - 1)] \\ &+ r(k - 2) - y(k - 2) \\ &= -[y(k) - 2y(k - 1) + y(k - 2)] \end{aligned} \tag{14.21}$$

The substitution of (14.19) through (14.21) into (14.18) yields

$$\begin{aligned} u(k) - u(k - 1) &= -\hat{k}_p[y(k) - y(k - 1)] + \hat{k}_i[r(k) \\ &- y(k)] - \hat{k}_d[y(k) - 2y(k - 1) + y(k - 2)] \end{aligned} \tag{14.22}$$

The z-transform of (14.22) is

$$(1 - z^{-1})U(z) = -\hat{k}_p(1 - z^{-1})Y(z) + \hat{k}_i[R(z) - Y(z)] - \hat{k}_d(1 - z^{-1})^2 Y(z)$$

which implies

$$U(z) = -\hat{k}_p Y(z) + \frac{\hat{k}_i}{1 - z^{-1}} E(z) - \hat{k}_d(1 - z^{-1})Y(z)$$

This is plotted in Figure 14.10. We see that only the integration acts on the error signal; the proportional and derivative actions act only on the plant output. This is called the *velocity-form PID controller*. This is the set-point-on-I-only controller shown in Figure 14.5(c).

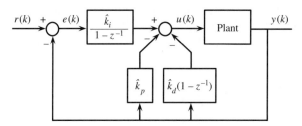

Figure 14.10 Velocity-form PID controller.

If the sampling period is small, then the effects of analog and digital PID controllers will be close; therefore, the tuning methods discussed for the analog case can be used to tune digital PID controllers. Because the dynamics of industrial processes are complex and not necessarily linear, no analytical methods are available to determine parameters of PID controllers; therefore, their determinations will involve trial and error. At present, active research has been going on to tune these parameters automatically. See, for example, References [3, 31].

The Laplace Transform

A.1 DEFINITION

In this appendix, we give a brief introduction of the Laplace transform and discuss its application in solving linear time-invariant differential equations. The introduction is not intended to be complete; it covers only the material used in this text.

Consider a function $f(t)$ defined for $t \geq 0$. The Laplace transform of $f(t)$, denoted by $F(s)$, is defined as

$$F(s) := \mathcal{L}[f(t)] := \int_{0^-}^{\infty} f(t)e^{-st}\, dt \tag{A.1}$$

where s is a complex variable and is often referred to as the *Laplace-transform variable*. The lower limit 0^- of the integral denotes that the limit approaches zero from a negative value. There are two reasons for using 0^- rather than 0 as the lower limit, as will be explained later.

Example A.1.1

Consider $f(t) = e^{-at}$, for $t \geq 0$. Its Laplace transform is

$$F(s) = \mathcal{L}[e^{-at}] = \int_{0^-}^{\infty} e^{-at}e^{-st}\, dt = \frac{-1}{s+a} e^{-(a+s)t} \bigg|_{t=0^-}^{\infty}$$

$$= \frac{-1}{s + a} [e^{-(a+s)t}|_{t=\infty} - e^{-(a+s)t}|_{t=0^-}] \qquad (A.2)$$

$$= \frac{-1}{s + a} [0 - 1] = \frac{1}{s + a}$$

where we have used $e^{-(s+a)t}|_{t=\infty} = 0$. This holds only if Re $s >$ Re $(-a)$, where Re stands for the real part. This condition, called the *region of convergence*, is often disregarded, however. See Reference [18] for a justification. Thus, the Laplace transform of e^{-at} is $1/(s + a)$. This transform holds whether a is real or complex.

Consider the two functions defined in Figure A.1(a) and (b). The one in Figure A.1(a) is a pulse with width ϵ and height $1/\epsilon$. Thus the pulse has area 1, for all $\epsilon > 0$. The function in Figure A.1(b) consists of two triangles with total area equal to 1, for all $\epsilon > 0$. The *impulse* or *delta function* is defined as

$$\delta(t) = \lim_{\epsilon \to 0} \delta_\epsilon(t)$$

where $\delta_\epsilon(t)$ can be either the function in Figure A.1(a) or that in (b). The impulse is customarily denoted by an arrow, as shown in Figure A.1(c). If the area of the function in Figure A.1(a) or in (b) is 1, then the impulse is said to have weight 1. Note that $\delta(t) = 0$, for $t \neq 0$. Because $\delta(0)$ may assume the value of ∞, if Figure A.1(a) is used, or 0, if Figure A.1(b) is used, $\delta(t)$ is not defined at $t = 0$.

The impulse has the following properties

$$\int_{-\infty}^{\infty} \delta(t)dt = \int_{-\epsilon}^{\epsilon} \delta(t)dt = 1 \qquad (A.3)$$

for every $\epsilon > 0$, and

$$\int_{-\infty}^{\infty} f(t)\delta(t)dt = \int_{-\infty}^{\infty} f(t)\delta(t - 0)dt = f(t)|_{t=0} = f(0) \qquad (A.4)$$

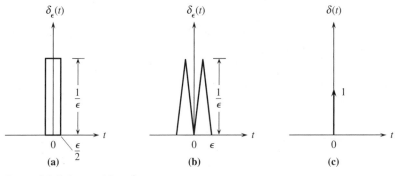

Figure A.1 Pulses and impulse

if $f(t)$ is continuous at $t = 0$. Strictly speaking, the impulse should be defined using (A.3) and (A.4).

Using (A.4), the Laplace transform of the impulse is

$$\Delta(s) := \mathscr{L}[\delta(t)] = \int_{0^-}^{\infty} \delta(t)e^{-st}\,dt = e^{-st}\big|_{t=0} = 1 \qquad \text{(A.5)}$$

Note that if the lower limit of (A.1) is 0 rather than 0^-, then the Laplace transform of $\delta(t)$ could be 0.5 or some other value. If we use 0^-, then the impulse will be included wholly in the Laplace transform and no ambiguity will arise in defining $\Delta(s)$. This is one of the reasons for using 0^- as the lower limit of (A.1).

The Laplace transform is defined as an integral. Because the integral is a linear operator, so is the Laplace transform—that is,

$$\mathscr{L}[a_1 f_1(t) + a_2 f_2(t)] = a_1 \mathscr{L}[f_1(t)] + a_2 \mathscr{L}[f_2(t)]$$

for any constants a_1 and a_2. We list in Table A.1 some of the often used Laplace-transform pairs.

Table A.1 Laplace-Transform Pairs

$f(t)$, $t \geq 0$	$F(s)$
$\delta(t)$ (impulse)	1
1 (unit-step function)	$\dfrac{1}{s}$
t^n (n = positive integer)	$\dfrac{n!}{s^{n+1}}$
e^{-at} (a = real or complex)	$\dfrac{1}{s + a}$
$t^n e^{-at}$	$\dfrac{n!}{(s + a)^{n+1}}$
$\sin \omega t$	$\dfrac{\omega}{s^2 + \omega^2}$
$\cos \omega t$	$\dfrac{s}{s^2 + \omega^2}$
$e^{-at} \sin \omega t$	$\dfrac{\omega}{(s + a)^2 + \omega^2}$
$e^{-at} \cos \omega t$	$\dfrac{s + a}{(s + a)^2 + \omega_2}$
$f(t)e^{at}$	$F(s - a)$

A.2   INVERSE LAPLACE TRANSFORM—PARTIAL FRACTION EXPANSION

The computation of $f(t)$ from its Laplace transform $F(s)$ is called the *inverse Laplace transform*. Although $f(t)$ can be computed from

$$f(t) = \frac{1}{2\pi j} \int_{c-j\infty}^{c+j\infty} F(s)e^{st}\, ds$$

the formula is rarely used in engineering. It is much simpler to find the inverse of $F(s)$ by looking it up in a table. However, before using a table, we must express $F(s)$ as a sum of terms available in the table.

Consider the Laplace transform

$$F(s) = \frac{N(s)}{D(s)} = \frac{N(s)}{(s-a)(s-b)^2\overline{D}(s)} \tag{A.6}$$

where $N(s)$ and $D(s)$ are two polynomials with deg $N(s) \le$ deg $D(s)$, where deg stands for the degree. We assume that $D(s)$ has a simple root at $s = a$ and a repeated root with multiplicity 2 at $s = b$, as shown in (A.6). Then $F(s)$ can be expanded as

$$F(s) = k_0 + \frac{k_a}{s-a} + \frac{k_{b1}}{s-b} + \frac{k_{b2}}{(s-b)^2}$$
$$+ \text{(Terms due to the roots of } \overline{D}(s)) \tag{A.7}$$

with

$$k_0 = F(\infty) \tag{A.8a}$$

$$k_a = F(s)(s-a)\big|_{s=a} \tag{A.8b}$$

$$k_{b2} = F(s)(s-b)^2\big|_{s=b} \tag{A.8c}$$

and

$$k_{b1} = \frac{d}{ds}[F(s)(s-b)^2]\big|_{s=b} \tag{A.8d}$$

This procedure is called *partial fraction expansion*. Using Table A.1, the inverse Laplace transform of $F(s)$ is

$$f(t) = k_0\delta(t) + k_ae^{at} + k_{b1}e^{bt} + k_{b2}te^{bt} + \text{(Terms due to the roots of } \overline{D}(s))$$

Note that (A.8b) is applicable for any simple root; (A.8c) and (A.8d) are applicable for any repeated root with multiplicity 2. Formulas for repeated roots with multiplicity 3 or higher and alternative formulas are available in Reference [18].

Example A.2.1

Find the inverse Laplace transform of

$$F(s) = \frac{s^3 - 2s + 3}{s^2(s+1)(s-2)}$$

We expand it as

$$F(s) = k_0 + \frac{k_1}{s + 1} + \frac{k_2}{s - 2} + \frac{k_{31}}{s} + \frac{k_{32}}{s^2}$$

with

$$k_0 = F(\infty) = 0$$

$$k_1 = F(s)(s + 1)\big|_{s=-1} = \frac{s^3 - 2s + 3}{s^2(s - 2)}\bigg|_{s=-1} = \frac{-1 + 2 + 3}{1 \cdot (-3)} = \frac{-4}{3}$$

$$k_2 = F(s)(s - 2)\big|_{s=2} = \frac{s^3 - 2s + 3}{s^2(s + 1)}\bigg|_{s=2} = \frac{8 - 4 + 3}{4 \cdot 3} = \frac{7}{12}$$

$$k_{32} = F(s)s^2\big|_{s=0} = \frac{s^3 - 2s + 3}{(s - 2)(s + 1)}\bigg|_{s=0} = \frac{3}{-2} = -1.5$$

$$k_{31} = \frac{d}{ds}[F(s)s^2]\big|_{s=0}$$

$$= \frac{(s + 1)(s - 2)(3s^2 - 2) - (s^3 - 2s + 3)(2s - 1)}{(s + 1)^2(s - 2)^2}\bigg|_{s=0} = \frac{7}{4}$$

Thus the inverse Laplace transform of $F(s)$ is

$$f(t) = \frac{-4}{3} e^{-t} + \frac{7}{12} e^{2t} - \frac{3}{2} t + \frac{7}{4}$$

for $t \geq 0$.

Exercise A.2.1

Find the inverse Laplace transforms of $2/s(s + 2)$ and $(s - 1)/s(s + 1)$.

[**Answers:** $1 - e^{-2t}, t \geq 0; -1 + 2e^{-t}, t \geq 0$.]

For a more detailed discussion of partial fraction expansion, see Reference [18].

A.3 SOME PROPERTIES OF THE LAPLACE TRANSFORM

In this section we discuss some properties of the Laplace transform.

Differentiation in Time

Let $F(s) = \mathcal{L}[f(t)]$. Then

$$\mathcal{L}\left[\frac{d}{dt} f(t)\right] = sF(s) - f(0^-) \tag{A.9a}$$

$$\mathcal{L}\left[\frac{d^2}{dt^2}f(t)\right] = s^2F(s) - sf(0^-) - f^{(1)}(0^-) \tag{A.9b}$$

and, in general

$$\mathcal{L}\left[\frac{d^n}{dt^n}f(t)\right] = s^nF(s) - s^{n-1}f(0^-)$$
$$- s^{n-2}f^{(1)}(0^-) - \cdots - f^{(n-1)}(0^-) \tag{A.9c}$$

where $f^{(i)}(0^-)$ denotes the ith derivative of $f(t)$ at $t = 0^-$. We see that if $f^{(i)}(0^-) = 0$ for $i = 0, 1, 2, \ldots$, then differentiation in the time domain is converted into multiplication by s in the Laplace-transform domain.

Integration in Time

The Laplace transform of the integral of $f(t)$ is given by

$$\mathcal{L}\left[\int_0^t f(t)dt\right] = \frac{1}{s}F(s)$$

Hence integration in the time domain is converted into division by s in the Laplace transform domain.

Final-Value Theorem

Let $f(t)$ be a function defined for $t \geq 0$. If $f(t)$ approaches a constant as $t \to \infty$, or if $sF(s)$ has no pole in the closed right half s-plane, then

$$\lim_{t \to \infty} f(t) = \lim_{s \to 0} sF(s) \tag{A.10}$$

This theorem is applicable only if $f(t)$ approaches a constant as $t \to \infty$. If $f(t)$ becomes infinite or remains oscillatory as $t \to \infty$, then the theorem cannot be used. For example, the Laplace transform of $f(t) = e^t$ is $F(s) = 1/(s - 1)$. Clearly we have

$$\lim_{s \to 0} sF(s) = 0$$

However, the function $f(t)$ approaches infinity as $t \to \infty$ and the equality in (A.10) does not hold.

Initial-Value Theorem

Let $f(t)$ be a function defined for $t \geq 0$, and let $F(s)$ be its Laplace transform. It is assumed that $F(s)$ is a rational function of s. If $F(s) = N(s)/D(s)$ is strictly proper, that is, deg $D(s) >$ deg $N(s)$, then

$$f(0^+) = \lim_{s \to \infty} sF(s)$$

Consider

$$F_1(s) = \frac{s + 3}{s^3 + 2s^2 + 4s + 2}$$

and

$$F_2(s) = \frac{2s^2 + 3s + 1}{s^3 + 2s^2 + 4s + 2}$$

They are both strictly proper. Therefore the initial-value theorem can be applied. The application of the theorem yields

$$f_1(0^+) = \lim_{s \to \infty} sF_1(s) = 0$$

and

$$f_2(0^+) = \lim_{s \to \infty} sF_2(s) = \frac{2s^3 + 3s^2 + s}{s^3 + 2s^2 + 4s + 2} = 2$$

The rational function $F_3(s) = (2s + 1)/(s + 1)$ is not strictly proper. The application of the initial-value theorem yields

$$f_3(0^+) = \lim_{s \to \infty} sF_3(s) = \lim_{s \to \infty} \frac{s(2s + 1)}{s + 1} = \infty \qquad \text{(A.11)}$$

The inverse Laplace transform of

$$F_3(s) = \frac{2s + 1}{s + 1} = 2 + \frac{-1}{s + 1}$$

is, using Table A.1,

$$f_3(t) = 2\delta(t) - e^{-t}$$

Because $\delta(t) = 0$ if $t \neq 0$, we have $f_3(0^+) = -1$, which is different from (A.11). Thus, if $F(s)$ is not strictly proper, the initial-value theorem cannot be directly employed.

A.4 SOLVING LTIL DIFFERENTIAL EQUATIONS

In this section we apply the Laplace transform to solve linear differential equations with constant coefficients. This is illustrated by examples.

Example A.4.1

Consider the first-order differential equation

$$\frac{d}{dt} y(t) + 2y(t) = 3 \frac{d}{dt} u(t) + 2u(t) \qquad \text{(A.12)}$$

The problem is to find $y(t)$ due to the initial condition $y(0^-) = 2$ and the input $u(t) = 1$, for $t \geq 0$. The application of the Laplace transform to (A.12) yields, using (A.9a),

$$sY(s) - y(0^-) + 2Y(s) = 3sU(s) - 3u(0^-) + 2U(s) \tag{A.13}$$

The Laplace transform of $u(t)$ is $1/s$. The substitution of $y(0^-) = 2$, $u(0^-) = 0$, and $U(s) = 1/s$ into (A.13) yields

$$(s + 2)Y(s) = y(0^-) + 3sU(s) - 3u(0^-) + 2U(s)$$

$$= 2 + 3 + \frac{2}{s} = 5 + \frac{2}{s}$$

which implies

$$Y(s) = \frac{5}{s + 2} + \frac{2}{s(s + 2)}$$

which can be simplified as, after partial fraction expansion of the second term,

$$Y(s) = \frac{5}{s + 2} + \frac{1}{s} - \frac{1}{s + 2} = \frac{4}{s + 2} + \frac{1}{s}$$

Thus we have, using Table A.1,

$$y(t) = 4e^{-2t} + 1 \tag{A.14}$$

for $t \geq 0$.

This example gives another reason for using 0^- rather than 0 as the lower limit in defining the Laplace transform. From (A.14), we have $y(0) = 5$, which is different from the initial condition $y(0^-) = 2$. If we had used $y(0) = 2$, then confusion would have arisen. In conclusion, the reason for using 0^- as the lower limit in (A.1) is twofold: First, to include impulses at $t = 0$ in the Laplace transform, and second, to avoid possible confusion in using initial conditions. If $f(t)$ does not contain impulses at $t = 0$ and is continuous at $t = 0$, then there is no difference in using either 0 or 0^- in (A.1).

Exercise A.4.1

Find the solution of (A.12) due to $y(0^-) = 2$, $u(0^-) = 0$, and $u(t) = \delta(t)$.

[**Answer:** $y(t) = 3\delta(t) - 2e^{-2t}, t \geq 0$.]

We give one more example to conclude this section.

Example A.4.2

Consider the second-order differential equation

$$\frac{d^2}{dt^2} y(t) + 2 \frac{d}{dt} y(t) + 5y(t) = \frac{d}{dt} u(t) \tag{A.15}$$

It is assumed that all initial conditions are zero. Find the response $y(t)$ due to $u(t) = e^{-t}, t \geq 0$. The application of the Laplace transform to (A.15) yields

$$s^2 Y(s) - sy(0^-) - y^{(1)}(0^-) + 2(sY(s) - y(0^-)) + 5Y(s) = sU(s) - u(0^-)$$

which, because all initial conditions are zero, reduces to

$$(s^2 + 2s + 5)Y(s) = sU(s)$$

or

$$Y(s) = \frac{s}{s^2 + 2s + 5} U(s) \tag{A.16}$$

The substitution of $U(s) = \mathscr{L}[u(t)] = 1/(s + 1)$ into (A.15) yields

$$Y(s) = \frac{s}{(s^2 + 2s + 5)(s + 1)} = \frac{s}{(s + 1)(s + 1 - j2)(s + 1 + j2)} \tag{A.17}$$

Thus, the remaining task is to compute the inverse Laplace transform of $Y(s)$. We expand it as

$$Y(s) = \frac{k_1}{s + 1} + \frac{k_2}{s + 1 - j2} + \frac{k_3}{s + 1 + j2}$$

with

$$k_1 = Y(s)(s + 1)\big|_{s=-1} = \frac{s}{(s + 1 - j2)(s + 1 + j2)}\bigg|_{s=-1} = \frac{-1}{4}$$

$$k_2 = Y(s)(s + 1 - j2)\big|_{s=-1+j2} = \frac{-1 + j2}{(j2)(j4)} = \frac{1 - j2}{8}$$

and

$$k_3 = Y(s)(s + 1 + j2)\big|_{s=-1-j2} = \frac{-1 - j2}{(-j2)(-j4)} = \frac{1 + j2}{8}$$

The computation of k_3 is in fact unnecessary once k_2 is computed, because k_3 equals the complex conjugate of k_2—that is, if $k_2 = a + jb$, then

$$k_3 = k_2^* := a - jb$$

In the subsequent development, it is simpler to use polar form $re^{j\theta}$. The polar form of $a + jb$ is

$$x = re^{j\theta}$$

where $r = \sqrt{a^2 + b^2}$ and $\theta = \tan^{-1}(b/a)$. For example, if $x = -2 + j1$, then

$$x = \sqrt{4 + 1}\, e^{j\tan^{-1}[1/(-2)]} = \sqrt{5}\, e^{j\tan^{-1}(-0.5)}$$

Because $\tan(-26.5°) = \tan 153.5° = -0.5$, one may incorrectly write x as $\sqrt{5}\, e^{-j26.5°}$. The correct x, however, should be

$$x = \sqrt{5}\, e^{j153.5°}$$

as can be seen from Figure A.2. In the complex plane, x is a vector with real part -2 and imaginary part 1 as shown. Thus its phase is $153.5°$ rather than $-26.5°$. In computing the polar form of a complex number, it is advisable to draw a rough graph to insure that we obtain the correct phase.

Now we shall express k_2 and k_3 in polar form as

$$k_2 = \frac{\sqrt{5}}{8}\, e^{-j63.5°} = \frac{\sqrt{5}}{8}\, e^{-j1.1 \text{ (radians)}} = k_3^*$$

Therefore the inverse Laplace transform of $Y(s)$ in (A.17) is

$$y(t) = -0.25e^{-t} + \frac{\sqrt{5}}{8}\, e^{-j1.1}\, e^{-(1-j2)t} + \frac{\sqrt{5}}{8}\, e^{j1.1}e^{-(1+j2)t} \qquad \text{(A.18)}$$

If all coefficients of $Y(s)$ are real, then $y(t)$ must also be real. The second and third terms in (A.18) are complex-valued. However, their sum must be real. Indeed, we have

$$y(t) = -0.25e^{-t} + \frac{\sqrt{5}}{8}\, e^{-t}[e^{j(2t-1.1)} + e^{-j(2t-1.1)}]$$

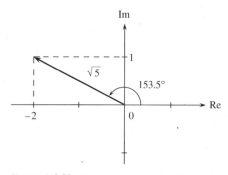

Figure A.2 Vector.

Using the identity

$$\cos \alpha = \frac{e^{j\alpha} + e^{-j\alpha}}{2}$$

we obtain

$$y(t) = -0.25e^{-t} + \frac{\sqrt{5}}{4} e^{-t} \cos(2t - 1.1) \qquad \text{(A.19)}$$

It is a real-valued function. Because the unit of ω in $\cos(\omega t + \theta)$ is in radians per second, the phase or angle θ should be expressed in radians rather than in degrees, as in (A.19). Otherwise ωt and θ cannot be directly added.

Exercise A.4.2

Find the solution of

$$\frac{d^2y(t)}{dt^2} + 2\frac{dy(t)}{dt} + 5y(t) = \frac{du(t)}{dt} - u(t)$$

due to $u(t) = e^{-t}$, $t \geq 0$. It is assumed that all initial conditions are zero.

[**Answer:** $y(t) = 0.5e^{-t} + 0.5e^{-t}\cos 2t + 0.5e^{-t}\sin 2t$.]

A.5   TIME DELAY

Consider a time function $f(t)$ which is zero for $t \leq 0$. Then $f(t - T)$ with $T \geq 0$ is a delay of $f(t)$ by T seconds as shown in Figure A.3. Let $F(s)$ be the Laplace transform of $f(t)$. Then we have

$$\mathcal{L}[f(t - T)] = e^{-Ts}F(s) \qquad \text{(A.20)}$$

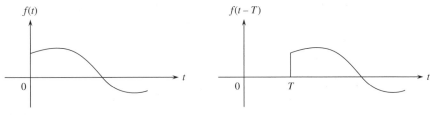

Figure A.3 Time function and its time delay.

Indeed, the Laplace transform of $f(t - T)$ is, by definition,

$$\mathcal{L}[f(t - T)] = \int_{0-}^{\infty} f(t - T)e^{-st}\, dt = e^{-Ts} \int_{0-}^{\infty} f(t - T)e^{-s(t-T)}\, dt$$

which becomes, by defining $v = t - T$ and using $f(t) = 0$ for $t < 0$,

$$\mathcal{L}[f(t - T)] = e^{-Ts} \int_{-T}^{\infty} f(v)e^{-sv}\, dv = e^{-Ts} \int_{0}^{\infty} f(v)e^{-vs}\, dv = e^{-Ts}F(s)$$

This shows (A.20). For example, because $\Delta(s) = \mathcal{L}[\delta(t)] = 1$, we have

$$\mathcal{L}[\delta(t - T)] = e^{-Ts} \tag{A.21}$$

Consider the pulse $p(t)$ shown in Figure A.4. Let $q(t)$ be a unit-step function, that is, $q(t) = 1$ for $t \geq 0$ and $q(t) = 0$, for $t < 0$. Then $p(t)$ can be expressed as

$$p(t) = q(t) - q(t - T) \tag{A.22}$$

Thus we have

$$P(s) := \mathcal{L}[p(t)] = \mathcal{L}[q(t)] - \mathcal{L}[q(t - T)] = \frac{1}{s} - \frac{e^{-Ts}}{s} = \frac{1 - e^{-Ts}}{s} \tag{A.23}$$

This formula will be used in Chapter 12.

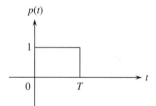

Figure A.4 Pulse.

B | Linear Algebraic Equations

In this appendix, we give a brief introduction to matrices and solutions of linear algebraic equations. It is introduced to the extent sufficient to solve the problems in this text. We also discuss briefly the problem of ill-conditioning and numerical stability of algorithms on computer computation. For a more detailed discussion, the reader is referred to References [15, 18].

B.1 MATRICES

A matrix is a rectangular array of elements such as

$$
\mathbf{A} = \begin{bmatrix}
a_{11} & a_{12} & \cdots & a_{1m} \\
a_{21} & a_{22} & \cdots & a_{2m} \\
\cdot & \cdot & \cdots & \cdot \\
\cdot & \cdot & \cdots & \cdot \\
\cdot & \cdot & \cdots & \cdot \\
a_{n1} & a_{n2} & \cdots & a_{nm}
\end{bmatrix} = [a_{ij}]
\tag{B.1}
$$

All a_{ij} are real numbers. The matrix has n rows and m columns and is called an $n \times m$ matrix. The element a_{ij} is located at the ith row and jth column and is called the (i, j)th element or entry. The matrix is called a square matrix of order n if $m = n$, a column vector or simply a column if $m = 1$, a row vector or a row if $n = 1$.

Two $n \times m$ matrices are equal if and only if all corresponding elements are the same. The addition and multiplication of matrices is defined as follows:

$$\underset{n \times m}{\mathbf{A}} + \underset{n \times m}{\mathbf{B}} = [a_{ij} + b_{ij}]_{n \times m}$$

$$\underset{1 \times 1}{c} \ \underset{n \times m}{\mathbf{A}} = \underset{n \times m}{\mathbf{A}} \ \underset{1 \times 1}{c} = [ca_{ij}]_{n \times m}$$

$$\underset{n \times m}{\mathbf{A}} \ \underset{m \times p}{\mathbf{B}} = \left[\sum_{k=1}^{m} a_{ik} b_{kj} \right]_{n \times p}$$

For a square matrix $\mathbf{A} = [a_{ij}]$, the entries a_{ii}, $i = 1, 2, 3, \ldots$, on the diagonal are called the *diagonal elements* of $\mathbf{A}$. A square matrix is called a *lower triangular matrix* if all entries above the diagonal elements are zero; an *upper triangular matrix* if all entries below the diagonal elements are zero. A square matrix is called a *diagonal matrix* if all entries, except the diagonal entries, are zero. A diagonal matrix is called a *unit matrix* if all diagonal entries equal 1. The transpose of $\mathbf{A}$, denoted by $\mathbf{A}'$, interchanges the rows and columns. For example, if

$$\mathbf{A} = \begin{bmatrix} a_{11} & a_{12} \\ a_{21} & a_{22} \\ a_{31} & a_{32} \end{bmatrix} \quad \text{then} \quad \mathbf{A}' = \begin{bmatrix} a_{11} & a_{21} & a_{31} \\ a_{12} & a_{22} & a_{32} \end{bmatrix}$$

Therefore, if $\mathbf{A}$ is $n \times m$, then $\mathbf{A}'$ is $m \times n$. To save space, the vector

$$\mathbf{x} = \begin{bmatrix} 1 \\ -4 \\ 3 \end{bmatrix}$$

is often written as $\mathbf{x}' = [1 \quad -4 \quad 3]$. In general, matrices do not commute, that is $\mathbf{AB} \neq \mathbf{BA}$. But we can move a scalar such as $c\mathbf{AB} = \mathbf{A}c\mathbf{B} = \mathbf{AB}c$.

B.2   DETERMINANT AND INVERSE

The determinant of a square matrix of order n is defined as

$$\det \mathbf{A} = \sum (-1)^j a_{1j_1} a_{2j_2} \cdots a_{nj_n}$$

where $j_1, j_2, \ldots, j_n$ are all possible orderings of the second subscripts $1, 2, \ldots, n$, and the integer j is the number of interchanges of two digits required to bring the ordering $j_1, j_2, \ldots, j_n$ into the natural ordering $1, 2, \ldots, n$. For example, we have

$$\det \begin{bmatrix} a_{11} & a_{12} \\ a_{21} & a_{22} \end{bmatrix} = (-1)^0 a_{11} a_{22} + (-1)^1 a_{12} a_{21} = a_{11} a_{22} - a_{12} a_{21}$$

and

$$\det \begin{bmatrix} a_{11} & a_{12} & a_{13} \\ a_{21} & a_{22} & a_{23} \\ a_{31} & a_{32} & a_{33} \end{bmatrix} = a_{11} a_{22} a_{33} + a_{12} a_{23} a_{31} + a_{13} a_{21} a_{32} - a_{13} a_{22} a_{31}$$

$$- a_{12} a_{21} a_{33} - a_{11} a_{23} a_{32}$$

The determinant of upper or lower triangular matrices equals the product of diagonal entries. For example, we have

$$\det \begin{bmatrix} a_{11} & 0 & 0 \\ a_{21} & a_{22} & 0 \\ a_{31} & a_{32} & a_{33} \end{bmatrix} = \det \begin{bmatrix} a_{11} & a_{12} & a_{13} \\ 0 & a_{22} & a_{23} \\ 0 & 0 & a_{33} \end{bmatrix} = a_{11}a_{22}a_{33} \tag{B.2}$$

Let $\mathbf{A}$ and $\mathbf{B}$ be square matrices. Then we have

$$\det (\mathbf{AB}) = \det \mathbf{A} \det \mathbf{B} \tag{B.3}$$

where $\mathbf{A}$ and $\mathbf{B}$ have the same order, and

$$\det \begin{bmatrix} \mathbf{A} & \mathbf{0} \\ \mathbf{C} & \mathbf{B} \end{bmatrix} = \det \begin{bmatrix} \mathbf{A} & \mathbf{D} \\ \mathbf{0} & \mathbf{B} \end{bmatrix} = \det \mathbf{A} \det \mathbf{B} \tag{B.4}$$

where $\mathbf{A}$ and $\mathbf{B}$ need not be of the same order. For example, if $\mathbf{A}$ is $n \times n$ and $\mathbf{B}$ is $m \times m$, then $\mathbf{C}$ is $m \times n$ and $\mathbf{D}$ is $n \times m$. The composite matrices in (B.4) may be called *block diagonal matrices*.

A square matrix is called *nonsingular* if its determinant is nonzero; *singular* if it is zero. For nonsingular matrices, we may define the inverse. The inverse of $\mathbf{A}$, denoted by $\mathbf{A}^{-1}$, has the property $\mathbf{A}^{-1}\mathbf{A} = \mathbf{AA}^{-1} = \mathbf{I}$. It can be computed by using the formula

$$\mathbf{A}^{-1} = \frac{1}{\det \mathbf{A}} \text{Adj } \mathbf{A} = \frac{1}{\det \mathbf{A}} [c_{ij}] \tag{B.5}$$

where

$$c_{ij} = (-1)^{i+j} \text{ (Determinant of the submatrix of } \mathbf{A}$$
$$\text{by deleting its } j\text{th row and } i\text{th column.)}$$

The matrix $[c_{ij}]$ is called the *adjoint* of $\mathbf{A}$. For example, we have

$$\begin{bmatrix} a_{11} & a_{12} \\ a_{21} & a_{22} \end{bmatrix}^{-1} = \frac{1}{a_{11}a_{22} - a_{12}a_{21}} \begin{bmatrix} a_{22} & -a_{12} \\ -a_{21} & a_{11} \end{bmatrix}$$

The inverse of 2×2 matrices is simple; we compute the determinant, interchange the diagonal elements, and change the signs of off-diagonal elements. The computation of the inverse of matrices of order 3 or higher is generally complicated.

The inverse of a triangular matrix is again a triangular matrix. For example, the inverse of

$$\mathbf{A} = \begin{bmatrix} a_{11} & 0 & 0 \\ a_{21} & a_{22} & 0 \\ a_{31} & a_{32} & a_{33} \end{bmatrix}$$

is of the form

$$\mathbf{B} := \mathbf{A}^{-1} = \begin{bmatrix} b_{11} & 0 & 0 \\ b_{21} & b_{22} & 0 \\ b_{31} & b_{32} & b_{33} \end{bmatrix}$$

By definition, we have

$$
\begin{bmatrix} b_{11} & 0 & 0 \\ b_{21} & b_{22} & 0 \\ b_{31} & b_{32} & b_{33} \end{bmatrix} \begin{bmatrix} a_{11} & 0 & 0 \\ a_{21} & a_{22} & 0 \\ a_{31} & a_{32} & a_{33} \end{bmatrix} = \begin{bmatrix} 1 & 0 & 0 \\ 0 & 1 & 0 \\ 0 & 0 & 1 \end{bmatrix}
$$

Equating the element (1, 1) yields $b_{11}a_{11} = 1$. Thus we have $b_{11} = a_{11}^{-1}$. Equating elements (2, 2) and (3, 3) yields $b_{22} = a_{22}^{-1}$ and $b_{33} = a_{33}^{-1}$. Equating element (2, 1) yields

$$
b_{21}a_{11} + b_{22}a_{21} = 0
$$

which implies

$$
b_{21} = -\frac{a_{21}}{a_{11}} \cdot b_{22} = \frac{-a_{21}}{a_{11}a_{22}}
$$

Proceeding forward, the inverse of triangular matrices can be easily computed.

The preceding procedure can be used to compute the inverse of block triangular matrices. For example, we have

$$
\begin{bmatrix} \mathbf{A} & \mathbf{D} \\ \mathbf{0} & \mathbf{B} \end{bmatrix}^{-1} = \begin{bmatrix} \mathbf{A}^{-1} & \alpha \\ \mathbf{0} & \mathbf{B}^{-1} \end{bmatrix}
\tag{B.6}
$$

where $\mathbf{A}\alpha + \mathbf{D}\mathbf{B}^{-1} = \mathbf{0}$. Thus we have

$$
\alpha = -\mathbf{A}^{-1}\mathbf{D}\mathbf{B}^{-1}
$$

B.3   THE RANK OF MATRICES

Consider the matrix in (B.1). Let $\mathbf{a}_{ir}$ denote its ith row, that is,

$$
\mathbf{a}_{ir} = [a_{i1} \quad a_{i2} \quad \cdots \quad a_{im}]
$$

The set of n row vectors in (B.1) is said to be *linearly dependent* if there exist n real numbers $\alpha_1, \alpha_2, \ldots, \alpha_n$, not all zero, such that

$$
\alpha_1\mathbf{a}_{1r} + \alpha_2\mathbf{a}_{2r} + \cdots + \alpha_n\mathbf{a}_{nr} = \mathbf{0}
\tag{B.7}
$$

where $\mathbf{0}$ is a $1 \times m$ vector with 0 as entries. If we cannot find n real numbers α_1, $\alpha_2, \ldots, \alpha_n$, not all zero, to meet (B.7), then the set is said to be *linearly independent*. Note that if $\alpha_i = 0$ for all i, then (B.7) always holds. Therefore the crucial point is whether we can find α_i, not all zero, to meet (B.7). For example, consider

$$
\begin{bmatrix} \mathbf{a}_{1r} \\ \mathbf{a}_{2r} \\ \mathbf{a}_{3r} \end{bmatrix} = \begin{bmatrix} 1 & 2 & 3 & 4 \\ 2 & -1 & 0 & 0 \\ 2 & 4 & 6 & 8 \end{bmatrix}
\tag{B.8}
$$

We have

$$
1 \times \mathbf{a}_{1r} + 0 \times \mathbf{a}_{2r} + (-0.5) \times \mathbf{a}_{3r} = [0 \quad 0 \quad 0 \quad 0]
$$

Therefore the three row vectors in (B.8) are linearly dependent. Consider

$$\begin{bmatrix} \mathbf{a}_{1r} \\ \mathbf{a}_{2r} \\ \mathbf{a}_{3r} \end{bmatrix} = \begin{bmatrix} 1 & 2 & 3 & 4 \\ 2 & -1 & 0 & 0 \\ 1 & 2 & 0 & 0 \end{bmatrix} \tag{B.9}$$

We have

$$\alpha_1 \mathbf{a}_{1r} + \alpha_2 \mathbf{a}_{2r} + \alpha_3 \mathbf{a}_{3r}$$
$$= [\alpha_1 + 2\alpha_2 + \alpha_3 \quad 2\alpha_1 - \alpha_2 + 2\alpha_3 \quad 3\alpha_1 \quad 4\alpha_1] \tag{B.10}$$
$$= [0 \quad 0 \quad 0 \quad 0]$$

The only α_i meeting (B.10) are $\alpha_i = 0$ for $i = 1, 2,$ and 3. Therefore, the three row vectors in (B.9) are linearly independent.

If a set of vectors is linearly dependent, then at least one of them can be expressed as a linear combination of the others. For example, the first row of (B.8) can be expressed as

$$\mathbf{a}_{1r} = 0 \times \mathbf{a}_{2r} + 0.5 \times \mathbf{a}_{3r}$$

This first row is a *dependent row*. If we delete all dependent rows in a matrix, the remainder will be linearly independent. The maximum number of linearly independent rows in a matrix is called the *rank* of the matrix. Thus, the matrix in (B.8) has rank 2 and the matrix in (B.9) has rank 3.

The rank of a matrix can also be defined as the maximum number of linearly independent *columns* in the matrix. Additionally, it can also be defined from determinants as follows: An $n \times m$ matrix has rank r if the matrix has an $r \times r$ submatrix with nonzero determinant and all square submatrices with higher order have determinants zero. Of course, these definitions all lead to the same rank. A consequence of these definitions is that for an $n \times m$ matrix, we have

$$\text{Rank } (\mathbf{A}) \leq \min (n, m) \tag{B.11}$$

An $n \times m$ matrix is said to have a *full row rank* if it has rank n or all its rows are linearly independent. A necessary condition for the matrix to have a full row rank is $m \geq n$. Thus, if a matrix has fewer rows than columns, then it cannot have a full row rank. If a square matrix has a full row rank, then it also has a full column rank and is called *nonsingular*.

To conclude this section, we discuss the use of MATLAB to compute the rank of matrices. Matrices are represented in MATLAB row by row, separated by semicolon. For example, the matrix in (B.8) is represented as

    a=[1 2 3 4;2 −1 0 0;2 4 6 8];

The command

    rank(a)

yields 2, the rank of the matrix in (B.8). The command

    rank([1 2 3 4;2 −1 0 0;1 2 0 0])

yields 3, the rank of the matrix in (B.9). Thus the use of the computer software is very simple.

The number of bits used in digital computers is finite, therefore numerical errors always occur in computer computation. As a result, two issues are important in computer computation. The first issue is whether the problem is ill conditioned or not. For example, we have

$$\text{Rank} \begin{bmatrix} 1/3 & 2 \\ 1 & 6 \end{bmatrix} = 1 \qquad \text{Rank} \begin{bmatrix} 0.33333 & 2 \\ 1 & 6 \end{bmatrix} = 2$$

We see that small changes in parameters yield an entirely different result. Such a problem is said to be *ill conditioned*. Thus, the computation of the rank is not a simple problem on a digital computer. The second issue is the computational method. A method is said to be *numerically stable* if the method will suppress numerical errors in the process of computation. It is *numerically unstable* if it will amplify numerical errors and yield erroneous results. The most reliable method of computing the rank is to use the singular value decomposition. See Reference [15].

B.4   LINEAR ALGEBRAIC EQUATIONS

Consider the set of linear algebraic equations

$$a_{11}x_1 + a_{12}x_2 + \cdots + a_{1m}x_m = y_1$$
$$a_{21}x_1 + a_{22}x_2 + \cdots + a_{2m}x_m = y_2$$
$$\vdots$$
$$a_{n1}x_1 + a_{n2}x_2 + \cdots + a_{nm}x_m = y_n$$

where a_{ij} and y_i are known and x_i are unknown. This set of equations can be written in matrix form as

$$\mathbf{Ax} = \mathbf{y} \tag{B.12}$$

where

$$\mathbf{A} = \begin{bmatrix} a_{11} & a_{12} & \cdots & a_{1m} \\ a_{21} & a_{22} & \cdots & a_{2m} \\ \vdots & \vdots & \cdots & \vdots \\ a_{n1} & a_{n2} & \cdots & a_{nm} \end{bmatrix} = [a_{ij}] \qquad \mathbf{x} = \begin{bmatrix} x_1 \\ x_2 \\ \vdots \\ x_m \end{bmatrix} \qquad \mathbf{y} = \begin{bmatrix} y_1 \\ y_2 \\ \vdots \\ y_n \end{bmatrix}$$

The set has n equations and m unknowns. $\mathbf{A}$ is an $n \times m$ matrix, $\mathbf{x}$ is an $m \times 1$ vector, and $\mathbf{y}$ is an $n \times 1$ vector.

THEOREM B.1

For *every* $\mathbf{y}$, a solution $\mathbf{x}$ exists in $\mathbf{Ax} = \mathbf{y}$ if and only if $\mathbf{A}$ has a full row rank. ■

For a proof of this theorem, see Reference [15]. We use an example to illustrate its implication. Consider

$$
\begin{bmatrix} 1 & 2 & 3 & 4 \\ 2 & -1 & 0 & 0 \\ 3 & 4 & 6 & 8 \end{bmatrix} \begin{bmatrix} x_1 \\ x_2 \\ x_3 \\ x_4 \end{bmatrix} = \begin{bmatrix} y_1 \\ y_2 \\ y_3 \end{bmatrix}
$$

Although this equation has a solution ($\mathbf{x}' = [1 \quad 1 \quad 0 \quad 1]$) for $\mathbf{y}' = [7 \quad 1 \quad 14]$, it does not have a solution for $\mathbf{y}' = [0 \quad 0 \quad 1]$. In other words, the equation has solutions for some $\mathbf{y}$, but not for every $\mathbf{y}$. This follows from Theorem B.1 because the 3×4 matrix does not have a full row rank.

Consider again (B.12). It is assumed that $\mathbf{A}$ has a full row rank. Then for any $\mathbf{y}$, there exists an $\mathbf{x}$ to meet the equation. Now if $n = m$, the solution is unique. If $n < m$ or, equivalently, (B.12) has more unknowns than equations, then solutions are not unique; $(m - n)$ number of the parameters of the solutions can be arbitrarily assigned. For example, consider

$$
\mathbf{A}\mathbf{x} = \begin{bmatrix} 2 & 1 & -4 \\ -1 & 0 & 2 \end{bmatrix} \begin{bmatrix} x_1 \\ x_2 \\ x_3 \end{bmatrix} = \begin{bmatrix} 3 \\ -1 \end{bmatrix} \tag{B.13}
$$

The matrix $\mathbf{A}$ in (B.13) has a full row rank. It has three unknowns and two equations, therefore one of x_1, x_2, and x_3 can be arbitrarily assigned. It is important to mention that not every one of x_1, x_2, or x_3 can be assigned. For example, if we assign $x_2 = 3$, then (B.13) becomes

$$
2x_1 + 3 - 4x_3 = 3 \qquad -x_1 + 2x_3 = -1
$$

or

$$
2x_1 - 4x_3 = 0 \qquad -x_1 + 2x_3 = -1
$$

These equations are inconsistent, therefore we cannot assign x_2 arbitrarily. It turns out that either x_1 or x_3 in (B.13) can be arbitrarily assigned. The reason is as follows: The first column of $\mathbf{A}$ in (B.13) is linearly dependent on the remaining two columns. If we delete the first column, the remaining matrix still has a full row rank. Therefore the coefficient corresponding to the first column—namely, x_1—can be arbitrarily assigned. If we assign it as $x_1 = 10$, then (B.13) becomes

$$
20 + x_2 - 4x_3 = 3 \qquad -10 + 2x_3 = -1
$$

or

$$
x_2 - 4x_3 = -17 \qquad 2x_3 = 9
$$

which imply $x_3 = 4.5$ and $x_2 = 1$. Thus $\mathbf{x}' = [10 \quad 1 \quad 4.5]$ is a solution. If we choose a different x_1, we will obtain a different solution. Similarly, we can assign x_3 arbitrarily, but we cannot assign both x_1 and x_3 arbitrarily.

Exercise B.1

In (B.13), if we assign $x_3 = 10$, what is the solution? If we assign $x_1 = 1$ and $x_3 = 10$, does (B.13) have a solution?

[**Answers:** $x_1 = 21$, $x_2 = 1$, $x_3 = 10$; no.]

Consider again (B.12) with $\mathbf{A}$ square. If $\mathbf{y} = \mathbf{0}$, (B.12) reduces to

$$\mathbf{A}\mathbf{x} = \mathbf{0} \tag{B.14}$$

This is called a *homogeneous equation*. It is clear that $\mathbf{x} = \mathbf{0}$ is always a solution of (B.14) whether or not $\mathbf{A}$ is nonsingular. This solution is called the *trivial* solution. A nonzero $\mathbf{x}$ meeting $\mathbf{A}\mathbf{x} = \mathbf{0}$ is called a *nontrivial solution*.

THEOREM B.2

A nontrivial solution exists in $\mathbf{A}\mathbf{x} = \mathbf{0}$ if and only if $\mathbf{A}$ is singular. Or, equivalently, $\mathbf{x} = \mathbf{0}$ is the *only* solution of $\mathbf{A}\mathbf{x} = \mathbf{0}$ if and only if $\mathbf{A}$ is nonsingular. ∎

B.5   ELIMINATION AND SUBSTITUTION

There are many ways to compute the solution of $\mathbf{A}\mathbf{x} = \mathbf{y}$. We discuss in this section the method of Gaussian elimination. This method is applicable no matter whether or not $\mathbf{A}$ is nonsingular. It can also be used to compute nontrivial solutions of $\mathbf{A}\mathbf{x} = \mathbf{0}$. This is illustrated by examples.

Example B.1

Find a solution of

$$x_1 + 2x_2 + x_3 = 10 \tag{B.15}$$

$$2x_1 + 5x_2 - 2x_3 = 3 \tag{B.16}$$

$$x_1 + 3x_2 = 0 \tag{B.17}$$

Subtraction of the product of 2 and (B.15) from (B.16), and subtraction of (B.15) from (B.17) yield

$$x_1 + 2x_2 + x_3 = 10 \tag{B.15'}$$

$$x_2 - 4x_3 = -17 \tag{B.16'}$$

$$x_2 - x_3 = -10 \tag{B.17'}$$

Subtraction of (B.16′) from (B.17′) yields

$$x_1 + 2x_2 + x_3 = 10 \tag{B.15″}$$

$$x_2 - 4x_3 = -17 \tag{B.16″}$$

$$3x_3 = 7 \tag{B.17″}$$

This process is called *Gaussian elimination*. Once this step is completed, the solution can easily be obtained as follows. From (B.17″), we have

$$x_3 = \frac{7}{3}$$

Substitution of x_3 into (B.16″) yields

$$x_2 = -17 + 4x_3 = -17 + \frac{28}{3} = -\frac{23}{3}$$

Substitution of x_3 and x_2 into (B.15″) yields

$$x_1 = 10 - 2x_2 - x_3 = 10 + \frac{46}{3} - \frac{7}{3} = 23$$

This process is called *back substitution*. Thus, the solution of linear algebraic equations can be obtained by Gaussian elimination and then back substitution.

Example B.2

Find a nontrivial solution, if it exists, of

$$x_1 + 2x_2 + 3x_3 = 0 \tag{B.18}$$

$$2x_1 + 5x_2 - 2x_3 = 0 \tag{B.19}$$

$$3x_1 + 7x_2 + x_3 = 0 \tag{B.20}$$

Subtraction of the product of 2 and (B.18) from (B.19) and subtraction of the product of 3 and (B.18) from (B.20) yield

$$x_1 + 2x_2 + 3x_3 = 0 \tag{B.18′}$$

$$x_2 - 8x_3 = 0 \tag{B.19′}$$

$$x_2 - 8x_3 = 0 \tag{B.20′}$$

We see that (B.19′) and (B.20′) are identical. In other words, the two unknowns x_2 and x_3 are governed by only one equation. Thus either one can be arbitrarily assigned. Let us choose $x_3 = 1$. Then $x_2 = 8$. The substitution of $x_3 = 1$ and $x_2 = 8$ into (B.18′) yields

$$x_1 = -2x_2 - 3x_3 = -16 - 3 = -19$$

Thus $x_1 = -19$, $x_2 = 8$, $x_3 = 1$ is a nontrivial solution.

Gaussian elimination is not a numerically stable method and should not be used on computer computation. The procedure, however, is useful in hand computation. In hand calculation, there is no need to eliminate x_i in the order of x_1, x_2, and x_3. They should be eliminated in the order which requires less computation. In Example B.1, for instance, x_3 does not appear in (B.17). Therefore we should use (B.15) and (B.16) to eliminate x_3 to yield

$$4x_1 + 9x_2 = 23$$

We use this equation and (B.17) to eliminate x_2 to yield $x_1 = 23$. The substitution of x_1 into (B.17) yields $x_2 = -23/3$. The substitution of x_1 and x_2 into (B.15) yields $x_3 = 7/3$. This modified procedure is simpler than Gaussian elimination and is suitable for hand calculation. For a more detailed discussion, see Reference [18].

B.6   GAUSSIAN ELIMINATION WITH PARTIAL PIVOTING

In this section, we modify Gaussian elimination to yield a numerically stable method. Consider

$$\begin{bmatrix} a_{11} & a_{12} & a_{13} & a_{14} \\ a_{21} & a_{22} & a_{23} & a_{24} \\ a_{31} & a_{32} & a_{33} & a_{34} \\ a_{41} & a_{42} & a_{43} & a_{44} \end{bmatrix} \begin{bmatrix} x_1 \\ x_2 \\ x_3 \\ x_4 \end{bmatrix} = \begin{bmatrix} y_1 \\ y_2 \\ y_3 \\ y_4 \end{bmatrix} \tag{B.21}$$

Before carrying out elimination in the first column (corresponding to the elimination of x_1 from the 2nd, 3rd, and 4th equations), we search for the element with the largest magnitude in the first column, say a_{31}, and then interchange the first and third equations. This step is called *partial pivoting*. The element a_{31}, which is now located at position $(1, 1)$, is called the *pivot*. We then divide the first equation by a_{31} to normalize the pivot to 1. After partial pivoting and normalization, we carry out elimination to yield

$$\begin{bmatrix} 1 & a_{12}^{(1)} & a_{13}^{(1)} & a_{14}^{(1)} \\ 0 & a_{22}^{(1)} & a_{23}^{(1)} & a_{24}^{(1)} \\ 0 & a_{32}^{(1)} & a_{33}^{(1)} & a_{34}^{(1)} \\ 0 & a_{42}^{(1)} & a_{43}^{(1)} & a_{44}^{(1)} \end{bmatrix} \begin{bmatrix} x_1 \\ x_2 \\ x_3 \\ x_4 \end{bmatrix} = \begin{bmatrix} y_1^{(1)} \\ y_2^{(1)} \\ y_3^{(1)} \\ y_4^{(1)} \end{bmatrix} \tag{B.22}$$

In the elimination, the same operations must be applied to y_i. Next we repeat the same procedure to the submatrix bounded by the dashed lines. If the element with the largest magnitude among $a_{22}^{(1)}$, $a_{32}^{(1)}$, and $a_{42}^{(1)}$ is nonzero, we bring it to position $(2, 2)$, normalize it to 1, and then carry out elimination to yield

$$
\begin{bmatrix}
1 & a_{12}^{(1)} & a_{13}^{(1)} & a_{14}^{(1)} \\
0 & 1 & a_{23}^{(2)} & a_{24}^{(2)} \\
0 & 0 & a_{33}^{(2)} & a_{34}^{(2)} \\
0 & 0 & a_{43}^{(2)} & a_{44}^{(2)}
\end{bmatrix}
\begin{bmatrix}
x_1 \\ x_2 \\ x_3 \\ x_4
\end{bmatrix}
=
\begin{bmatrix}
y_1^{(1)} \\ y_2^{(2)} \\ y_3^{(2)} \\ y_4^{(2)}
\end{bmatrix}
\tag{B.23a}
$$

If the three elements $a_{22}^{(1)}$, $a_{32}^{(1)}$, and $a_{42}^{(1)}$ in (B.22) are all zero, (B.22) becomes

$$
\begin{bmatrix}
1 & a_{12}^{(1)} & a_{13}^{(1)} & a_{14}^{(1)} \\
0 & 0 & a_{23}^{(1)} & a_{24}^{(1)} \\
0 & 0 & a_{33}^{(1)} & a_{34}^{(1)} \\
0 & 0 & a_{43}^{(1)} & a_{44}^{(1)}
\end{bmatrix}
\begin{bmatrix}
x_1 \\ x_2 \\ x_3 \\ x_4
\end{bmatrix}
=
\begin{bmatrix}
y_1^{(1)} \\ y_2^{(1)} \\ y_3^{(1)} \\ y_4^{(1)}
\end{bmatrix}
\tag{B.23b}
$$

In this case, no elimination is needed and the pivot is zero. We repeat the same procedure to the submatrix bounded by the solid lines in (B.23). Finally we can obtain

$$
\begin{bmatrix}
1 & p & p & p \\
0 & 0 & p & p \\
0 & 0 & 1 & p \\
0 & 0 & 0 & 1
\end{bmatrix}
\begin{bmatrix}
x_1 \\ x_2 \\ x_3 \\ x_4
\end{bmatrix}
=
\begin{bmatrix}
y_1^{(1)} \\ y_2^{(2)} \\ y_3^{(3)} \\ y_4^{(4)}
\end{bmatrix}
\tag{B.24}
$$

where p denotes possible nonzero elements. The transformation of (B.21) into the form in (B.24) is called *Gaussian elimination with partial pivoting*. It is a numerically stable method. It can easily be programmed on a digital computer and is widely used in practice. Once (B.21) is transformed into (B.24), the solution x_i can be easily obtained by back substitution.

Now we discuss the use of MATLAB to solve a set of linear algebraic equations. We rewrite (B.15), (B.16), and (B.17) in matrix form as

$$
\begin{bmatrix}
1 & 2 & 1 \\
2 & 5 & -2 \\
1 & 3 & 0
\end{bmatrix}
\begin{bmatrix}
x_1 \\ x_2 \\ x_3
\end{bmatrix}
=
\begin{bmatrix}
10 \\ 3 \\ 0
\end{bmatrix}
$$

The commands

```
a=[1 2 1;2 5 −2;1 3 0]; b=[10;3;0]; a\b
```

yield $x_1 = 23.000$, $x_2 = -7.6667$, $x_3 = 2.3333$. Thus the use of MATLAB is very simple.

References

[1]   Anderson, B. D. O. and J. B. Moore. *Linear Optimal Control*, Englewood Cliffs, NJ: Prentice-Hall, 1971.

[2]   Åström, K. J. "Robustness of a design based on assignment of poles and zeros," *IEEE Trans. Automatic Control*, Vol. AC-25, pp. 588–591, 1980.

[3]   Åström, K. J. and B. Wittenmark. *Computer Controlled Systems: Theory and Design*, Englewood Cliffs, NJ: Prentice-Hall, 1984.

[4]   ———. *Adaptive Control*, Reading, MA: Addison-Wesley, 1989.

[5]   Athans, M. and P. L. Falb. *Optimal Control*, New York: McGraw-Hill, 1966.

[6]   Atherton, D. P. *Nonlinear Control Engineering*, London: Van Nostrand Reinhold, 1982.

[7]   Bode, H. W. *Network Analysis and Feedback Amplifier Design*, Princeton, NJ: Van Nostrand, 1945.

[8]   Boyd, S. P. et al. "A new CAD method and associated architectures for linear controllers," *IEEE Trans. Automatic Control*, Vol. 33, pp. 268–283, 1988.

[9]   Callier, F. M. and C. A. Desoer. *Multivariable Feedback Systems*, New York: Springer-Verlag, 1982.

[10] Chang, S. S. L. *Synthesis of Optimal Control Systems*, New York: McGraw-Hill, 1961.

[11] Chen, C. T. "Design of feedback control systems," *Proc. Natl. Electronic Conf.*, Vol. 57, pp. 46–51, 1969.

[12] ———. *Analysis and Synthesis of Linear Control Systems*, New York: Holt, Rinehart and Winston, 1975.

[13] ———. *One-Dimensional Digital Signal Processing*, New York: Dekker, 1979.

[14] ———. "A contribution to the design of linear time-invariant multivariable systems," *Proc. Am. Automatic Control Conf.*, June, 1983.

[15] ———. *Linear System Theory and Design*, New York: Holt, Rinehart and Winston, 1984.

[16] ———. *Control System Design: Conventional, Algebraic and Optimal Methods*, Stony Brook, NY: Pond Woods Press, 1987.

[17] ———. "Introduction to the Linear Algebraic Method for Control System Design," *IEEE Control Systems Magazine*, Vol. 7, No. 5, pp. 36–42, 1987.

[18] ———. *System and Signal Analysis*, New York: Holt, Rinehart and Winston, 1989.

[19] Chen, C. T. and B. Seo. "Applications of the Linear Algebraic Method for Control System Design," *IEEE Control Systems Magazine*, Vol. 10, No. 1, pp. 43–47, 1989.

[20] ———. "The Inward Approach in the Design of Control Systems," *IEEE Trans. on Education*, Vol. 33, pp. 270–278, 1990.

[21] Chen, C. T. and S. Y. Zhang. "Various implementations of implementable transfer matrices," *IEEE Trans. Automatic Control*, Vol. AC-30, pp. 1115–1118, 1985.

[22] Coughanowr, D. R. *Process Systems Analysis and Control*, 2nd ed., New York: McGraw-Hill, 1991.

[23] Craig, J. J. *Introduction to Robotics*, 2nd ed., Reading MA: Addison-Wesley, 1989.

[24] Daryanani, G. *Principles of Active Network Synthesis and Design*, New York: Wiley, 1976.

[25] Doebelin, E. O. *Control System Principles and Design*, New York: John Wiley, 1985.

[26] Doyle, J. and G. Stein. "Multivariable feedback design: Concepts for classical/modern analysis," *IEEE Trans. on Automatic Control*, Vol. 26, No. 1, pp. 4–16, 1981.

[27] Evans, W. R. "Control system synthesis by the root locus method," *Trans. AIEE*, Vol. 69, pp. 67–69, 1950.

[28] Franklin, G. F., J. D. Powell, and M. L. Workman. *Digital Control of Dynamic Systems*, 2nd ed., Reading, MA: Addison-Wesley, 1990.

[29] Franklin, G. F., J. D. Powell, and A. Emami-Naeini. *Feedback Control of Dynamic Systems*, Reading, MA: Addison-Wesley, 1986.

[30] Frederick, D. K., C. J. Herget, R. Kool, and C. M. Rimvall. "The extended list of control software," *Electronics*, Vol. 61, No. 6, p. 77, 1988.

[31] Gawthrop, P. J. and P. E. Nomikos. "Automatic tuning of commercial PID controllers for single-loop and multiloop applications," *IEEE Control Systems Magazine*, Vol. 10, No. 1, pp. 34–42, 1990.

[32] Gayakwad, R. and L. Sokoloff. *Analog and Digital Control Systems*, Englewood Cliffs, NJ: Prentice-Hall, 1988.

[33] Graham, D. and R. C. Lathrop. "The synthesis of optimum response: Criteria and standard forms," *AIEE*, Vol. 72, Pt. II, pp. 273–288, 1953.

[34] Hang, C. C. "The choice of controller zeros," *IEEE Control Systems Magazine*, Vol. 9, No. 1, pp. 72–75, 1989.

[35] Hang, C. C. and K. K. Sin. "A comparative performance study of PID auto-tuners," *IEEE Control Systems Magazine*, Vol. 11, No. 5, pp. 41–47, 1991.

[36] Horowitz, I. M. *Synthesis of Feedback Systems*, New York: Academic Press, 1963.

[37] Jacob, J. M. *Industrial Control Electronics: Application and Design*, Englewood Cliffs, NJ: Prentice-Hall, 1989.

[38] Jury, E. I. *Sampled-Data Control Systems*, New York: Wiley, 1958.

[39] Killian, M. J. "Analysis and Synthesis of Phase-locked Loop Motor Speed Control Systems," Master Thesis, State University of New York, 1979.

[40] Ku, Y. H. *Analysis and Control of Linear Systems*, Scranton, PA: International Textbook, 1962.

[41] Kučera, V. *Discrete Linear Control—The Polynomial Equation Approach*, Chichester, UK: Wiley, 1979.

[42] Kuo, B. C. *Automatic Control Systems*, 3rd ed. Englewood Cliffs, NJ: Prentice-Hall, 1975.

[43] Kuo, B. C. *Digital Control Systems*, New York: Holt, Rinehart and Winston, 1980.

[44] Lin, J. L., J. J. Sheen, and C. T. Chen. "Comparisons of various implementations of quadratic optimal systems," Technical Report, National Cheng-Kung University, Taiwan, 1986.

[45] Luenberger, D. G. "Observing the state of a linear system," *IEEE Trans. Military Electronics*, Vol. MIL-8, pp. 74–80, 1964.

[46] Middleton, R. H. and G. C. Goodwin. *Digital Control and Estimation, A Unified Approach*, Englewood Cliffs, NJ: Prentice-Hall, 1990.

[47] Moler, C. and C. F. Van Loan. "Nineteen dubious ways to compute the exponential of a matrix," *SIAM Review*, Vol. 20, pp. 801–836, 1978.

[48] Newton, G. C. Jr., L. A. Gould, and J. F. Kaiser. *Analytical Design of Linear Feedback Controls*, New York: Wiley, 1959.

[49] Norimatsu, T. and M. Ito. "On the zero non-regular control system," *J. Inst. Elec. Eng. Japan*, Vol. 81, pp. 566–575, 1961. See also B. Porter, "Comments on 'On undershoot and nonminimum phase zeros'," *IEEE Trans. on Automatic Control*, Vol. AC-32, pp. 271–272, 1987.

[50] Northrop, R. B. *Analog Electronic Circuits, Analysis and Application*, Reading, MA: Addison-Wesley, 1990.

[51] Nyquist, H. "Regeneration theory," *Bell Syst. Tech. J.*, Vol. 11, pp. 126–147, 1932.

[52] Ogata, K. *Discrete-Time Control Systems*, Englewood Cliffs, NJ: Prentice-Hall, 1987.

[53] Rosenbrock, H. H. *State-Space and Multivariable Theory*, New York: Wiley-Interscience, 1970.

[54] Seo, B. "Computer-Aided Design of Control Systems" Ph.D. dissertation, SUNY, Stony Brook, NY, 1989.

[55] Shipley, D. P. "A unified approach to synthesis of linear systems," *IEEE Trans. on Automatic Control*, Vol. AC-8, pp. 114–125, April, 1963.

[56] Slotine, J-J. E. and W. Li. *Applied Nonlinear Control*, Englewood Cliffs, NJ: Prentice-Hall, 1990.

[57] Truxal, J. G. *Automatic Feedback Control System Synthesis*, New York: McGraw-Hill, 1955.

[58] ———. *Introductory System Engineering*, New York: McGraw-Hill, 1972.

[59] Tsui, C. C. "An analytical solution to the equation TA-FT=LC and its applications," *IEEE Trans. on Automatic Control*, Vol. 32, No. 8, pp. 742–744, 1987.

[60] ———. "On preserving the robustness of optimal control system with observers," *IEEE Trans. on Automatic Control*, Vol. 32, No. 9, pp. 823–826, 1987.

[61] Van de Vegte, J. *Feedback Control Systems*, 2nd ed., Englewood Cliffs, NJ: Prentice-Hall, 1990.

[62] Vidyasagar, M. "On the well-posedness of large-scale interconnected systems," *IEEE Trans. on Automatic Control*, Vol. AC-25, pp. 413–421, 1980.

[63] ———. *Control System Synthesis: A Factorization Approach*, Cambridge, MA: MIT Press, 1985.

[64] Willems, J. C. *The Analysis of Feedback Systems*, Cambridge, MA: MIT Press, 1971.

[65] Wolovich, W. A. *Linear Multivariable Systems*, New York: Springer-Verlag, 1974.

[66] Wonham, W. M. "On pole assignment in multi-input controllable linear systems," *IEEE Trans. on Automatic Control*, Vol. AC-12, pp. 660–665, 1967.

[67] ———. *Linear Multivariable Control*, New York: Springer-Verlag, 1985.

[68] Youla, D. C., J. J. Bongiorno, Jr., and C. N. Lu. "Single-loop feedback-stabilization of linear multivariable dynamical plants," *Automatica*, Vol. 10, pp. 159–173, 1974.

[69] Ziegler, J. G. and N. B. Nichols. "Optimum settings for automatic controllers," *Trans. ASME*, Vol. 64, pp. 759–768, 1942.

[70] ———. "Process lags in automatic control circuits," *Trans. ASME,* Vol. 65, pp. 433–442, 1943.

Index